Applied Science

Applied Science

Volume 2

Editor

Donald R. Franceschetti, Ph.D.

The University of Memphis

SALEM PRESS
A Division of EBSCO Publishing

Ipswich, Massachusetts   Hackensack, New Jersey

Cover Photo: Aerobraking Simulation on a Supercomputer (© Roger Ressmeyer/CORBIS)

Library of Congress Cataloging-in-Publication Data

Applied science / editor, Donald R. Franceschetti.
    p. cm.
 ISBN 978-1-58765-781-8 (set) — ISBN 978-1-58765-782-5 (v. 1) — ISBN 978-1-58765-783-2 (v. 2) — ISBN 978-1-58765-784-9 (v. 3) — ISBN 978-1-58765-785-6 (v. 4) — ISBN 978-1-58765-786-3 (v. 5)
 1. Engineering. 2. Technology. I. Franceschetti, Donald R., 1947-
  600--dc23

2012002375

FIRST PRINTING
PRINTED IN THE UNITED STATES OF AMERICA

CONTENTS

COMMON UNITS OF MEASURE

Common prefixes for metric units—which may apply in more cases than shown below—include *giga-* (1 billion times the unit), *mega-* (one million times), *kilo-* (1,000 times), *hecto-* (100 times), *deka-* (10 times), *deci-* (0.1 times, or one tenth), *centi-* (0.01, or one hundredth), *milli-* (0.001, or one thousandth), and *micro-* (0.0001, or one millionth).

Unit	Quantity	Symbol	Equivalents
Acre	Area	ac	43,560 square feet 4,840 square yards 0.405 hectare
Ampere	Electric current	A *or* amp	1.00016502722949 international ampere 0.1 biot *or* abampere
Angstrom	Length	Å	0.1 nanometer 0.0000001 millimeter 0.000000004 inch
Astronomical unit	Length	AU	92,955,807 miles 149,597,871 kilometers (mean Earth-Sun distance)
Barn	Area	b	10^{-28} meters squared (approx. cross-sectional area of 1 uranium nucleus)
Barrel (dry, for most produce)	Volume/capacity	bbl	7,056 cubic inches; 105 dry quarts; 3.281 bushels, struck measure
Barrel (liquid)	Volume/capacity	bbl	31 to 42 gallons
British thermal unit	Energy	Btu	1055.05585262 joule
Bushel (U.S., heaped)	Volume/capacity	bsh *or* bu	2,747.715 cubic inches 1.278 bushels, struck measure
Bushel (U.S., struck measure)	Volume/capacity	bsh *or* bu	2,150.42 cubic inches 35.238 liters
Candela	Luminous intensity	cd	1.09 hefner candle
Celsius	Temperature	C	1° centigrade
Centigram	Mass/weight	cg	0.15 grain
Centimeter	Length	cm	0.3937 inch
Centimeter, cubic	Volume/capacity	cm³	0.061 cubic inch
Centimeter, square	Area	cm²	0.155 square inch
Coulomb	Electric charge	C	1 ampere second
Cup	Volume/capacity	C	250 milliliters 8 fluid ounces 0.5 liquid pint

Unit	Quantity	Symbol	Equivalents
Deciliter	Volume/capacity	dl	0.21 pint
Decimeter	Length	dm	3.937 inches
Decimeter, cubic	Volume/capacity	dm³	61.024 cubic inches
Decimeter, square	Area	dm²	15.5 square inches
Dekaliter	Volume/capacity	dal	2.642 gallons 1.135 pecks
Dekameter	Length	dam	32.808 feet
Dram	Mass/weight	dr *or* dr avdp	0.0625 ounce 27.344 grains 1.772 grams
Electron volt	Energy	eV	$1.5185847232839 \times 10^{-22}$ Btus $1.6021917 \times 10^{-19}$ joules
Fermi	Length	fm	1 femtometer 1.0×10^{-15} meters
Foot	Length	ft *or* '	12 inches 0.3048 meter 30.48 centimeters
Foot, square	Area	ft²	929.030 square centimeters
Foot, cubic	Volume/capacity	ft³	0.028 cubic meter 0.0370 cubic yard 1,728 cubic inches
Gallon (British Imperial)	Volume/capacity	gal	277.42 cubic inches 1.201 U.S. gallons 4.546 liters 160 British fluid ounces
Gallon (U.S.)	Volume/capacity	gal	231 cubic inches 3.785 liters 0.833 British gallon 128 U.S. fluid ounces
Giga-electron volt	Energy	GeV	$1.6021917 \times 10^{-10}$ joule
Gigahertz	Frequency	GHz	—
Gill	Volume/capacity	gi	7.219 cubic inches 4 fluid ounces 0.118 liter
Grain	Mass/weight	gr	0.037 dram 0.002083 ounce 0.0648 gram
Gram	Mass/weight	g	15.432 grains 0.035 avoirdupois ounce

Unit	Quantity	Symbol	Equivalents
Hectare	Area	ha	2.471 acres
Hectoliter	Volume/capacity	hl	26.418 gallons 2.838 bushels
Hertz	Frequency	Hz	$1.08782775707767 \times 10^{-10}$ cesium atom frequency
Hour	Time	h	60 minutes 3,600 seconds
Inch	Length	in *or* "	2.54 centimeters
Inch, cubic	Volume/capacity	in^3	0.554 fluid ounce 4.433 fluid drams 16.387 cubic centimeters
Inch, square	Area	in^2	6.4516 square centimeters
Joule	Energy	J	$6.2414503832469 \times 10^{18}$ electron volt
Joule per kelvin	Heat capacity	J/K	$7.24311216248908 \times 10^{22}$ Boltzmann constant
Joule per second	Power	J/s	1 watt
Kelvin	Temperature	K	-272.15° Celsius
Kilo-electron volt	Energy	keV	$1.5185847232839 \times 10^{-19}$ joule
Kilogram	Mass/weight	kg	2.205 pounds
Kilogram per cubic meter	Mass/weight density	kg/m^3	$5.78036672001339 \times 10^{-4}$ ounces per cubic inch
Kilohertz	Frequency	kHz	—
Kiloliter	Volume/capacity	kl	—
Kilometer	Length	km	0.621 mile
Kilometer, square	Area	km^2	0.386 square mile 247.105 acres
Light-year (distance traveled by light in one Earth year)	Length/distance	lt-yr	5,878,499,814,275.88 miles 9.46×10^{12} kilometers
Liter	Volume/capacity	L	1.057 liquid quarts 0.908 dry quart 61.024 cubic inches
Mega-electron volt	Energy	MeV	—
Megahertz	Frequency	MHz	—
Meter	Length	m	39.37 inches
Meter, cubic	Volume/capacity	m^3	1.308 cubic yards

Unit	Quantity	Symbol	Equivalents
Meter per second	Velocity	m/s	2.24 miles per hour 3.60 kilometers per hour
Meter per second per second	Acceleration	m/s^2	12,960.00 kilometers per hour per hour 8,052.97 miles per hour per hour
Meter, square	Area	m^2	1.196 square yards 10.764 square feet
Metric. *See* unit name			
Microgram	Mass/weight	mcg *or* μg	0.000001 gram
Microliter	Volume/capacity	μl	0.00027 fluid ounce
Micrometer	Length	μm	0.001 millimeter 0.00003937 inch
Mile (nautical international)	Length	mi	1.852 kilometers 1.151 statute miles 0.999 U.S. nautical miles
Mile (statute or land)	Length	mi	5,280 feet 1.609 kilometers
Mile, square	Area	mi^2	258.999 hectares
Milligram	Mass/weight	mg	0.015 grain
Milliliter	Volume/capacity	ml	0.271 fluid dram 16.231 minims 0.061 cubic inch
Millimeter	Length	mm	0.03937 inch
Millimeter, square	Area	mm^2	0.002 square inch
Minute	Time	m	60 seconds
Mole	Amount of substance	mol	6.02×10^{23} atoms or molecules of a given substance
Nanometer	Length	nm	1,000,000 fermis 10 angstroms 0.001 micrometer 0.00000003937 inch
Newton	Force	N	0.224808943099711 pound force 0.101971621297793 kilogram force 100,000 dynes
Newton meter	Torque	N·m	0.7375621 foot-pound
Ounce (avoirdupois)	Mass/weight	oz	28.350 grams 437.5 grains 0.911 troy or apothecaries' ounce

Unit	Quantity	Symbol	Equivalents
Ounce (troy)	Mass/weight	oz	31.103 grams 480 grains 1.097 avoirdupois ounces
Ounce (U.S., fluid or liquid)	Mass/weight	oz	1.805 cubic inch 29.574 milliliters 1.041 British fluid ounces
Parsec	Length	pc	30,856,775,876,793 kilometers 19,173,511,615,163 miles
Peck	Volume/capacity	pk	8.810 liters
Pint (dry)	Volume/capacity	pt	33.600 cubic inches 0.551 liter
Pint (liquid)	Volume/capacity	pt	28.875 cubic inches 0.473 liter
Pound (avoirdupois)	Mass/weight	lb	7,000 grains 1.215 troy or apothecaries' pounds 453.59237 grams
Pound (troy)	Mass/weight	lb	5,760 grains 0.823 avoirdupois pound 373.242 grams
Quart (British)	Volume/capacity	qt	69.354 cubic inches 1.032 U.S. dry quarts 1.201 U.S. liquid quarts
Quart (U.S., dry)	Volume/capacity	qt	67.201 cubic inches 1.101 liters 0.969 British quart
Quart (U.S., liquid)	Volume/capacity	qt	57.75 cubic inches 0.946 liter 0.833 British quart
Rod	Length	rd	5.029 meters 5.50 yards
Rod, square	Area	rd^2	25.293 square meters 30.25 square yards 0.00625 acre
Second	Time	s or sec	$1/60$ minute $1/3600$ hour
Tablespoon	Volume/capacity	T or tb	3 teaspoons 4 fluid drams
Teaspoon	Volume/capacity	t or tsp	0.33 tablespoon 1.33 fluid drams

Unit	Quantity	Symbol	Equivalents
Ton (gross or long)	Mass/weight	t	2,240 pounds 1.12 net tons 1.016 metric tons
Ton (metric)	Mass/weight	t	1,000 kilograms 2,204.62 pounds 0.984 gross ton 1.102 net tons
Ton (net or short)	Mass/weight	t	2,000 pounds 0.893 gross ton 0.907 metric ton
Volt	Electric potential	V	1 joule per coulomb
Watt	Power	W	1 joule per second 0.001 kilowatt $2.84345136093995 \times 10^{-4}$ ton of refrigeration
Yard	Length	yd	0.9144 meter
Yard, cubic	Volume/capacity	yd^3	0.765 cubic meter
Yard, square	Area	yd^2	0.836 square meter

COMPLETE LIST OF CONTENTS

Volume 1

Volume 2

Volume 3

Volume 4

Volume 5

Contents v

Applied Science

COMPUTER ENGINEERING

FIELDS OF STUDY

Computer engineering; computer science; computer programming; computer information systems; electrical engineering; information systems; computer information technology; software engineering.

SUMMARY

Computer engineering refers to the field of designing hardware and software components that interact to maximize the speed and processing capabilities of the central processing unit (CPU), memory, and the peripheral devices, which include the keyboard, monitor, disk drives, mouse, and printer. Because the first computers were based on the use of on-and-off mechanical switches to control electrical circuits, computer hardware is still based on the binary number system. Computer engineering involves the development of operating systems that are able to interact with compilers that translate the software programs written by humans into the machine instructions that depend on the binary number system to control electrical logic circuits and communication ports to access the Internet.

KEY TERMS AND CONCEPTS

- **Basic Input-Output System (BIOS):** Computer program that allows the central processing unit (CPU) of a computer to communicate with other computer hardware.
- **Browser:** Software program that is used to view Web pages on the Internet.
- **Icon:** Small picture that represents a file, program, disk, menu, or option.
- **Internet Service Provider (ISP):** Organization that provides paid access to the Internet.
- **Logical Topology:** Pathway within the physical network devices that directs the flow of data. The bus and ring are the only two types of logical topology.
- **Mainframe:** Large, stand-alone computer that completes batch processing (groups of computer instructions completed at once).
- **Protocol:** Set of rules to be followed that allows for communication.
- **Server:** Computer that is dedicated to managing resources shared by users (clients).

DEFINITION AND BASIC PRINCIPLES

Much of the work within the field of computer engineering focuses on the optimization of computer hardware, which is the general term that describes the electronic and mechanical devices that make it possible for a computer user (client) to utilize the power of a computer. These physical devices are based on binary logic. Humans use the decimal system for numbers, instead of the base two-number system of binary logic, and naturally humans communicate with words. A great deal of interface activity is necessary to bridge this communication gap, and computer engineering involves additional types of software (programs) that function as intermediate interfaces to translate human instructions into hardware activity. Examples of these types of software include operating systems, drivers, browsers, compilers, and linkers.

Computer hardware and software generally can be arranged in a series of hierarchical levels, with the lowest level of software being the machine language, consisting of numbers and operands that the processor executes. Assembly language is the next level, and it uses instruction mnemonics, which are machine-specific instructions used to communicate with the operating system and hardware. Each instruction written in assembly language corresponds to one instruction written in machine code, and these instructions are used directly by the processor. Assembly language is also used to optimize the runtime execution of application programs. At the next level is the operating system, which is a computer program written so that it can manage resources, such as disk drives and printers, and can also function as an interface between a computer user and the various pieces of hardware. The highest level includes applications that humans use on a daily basis. These are considered the highest level because they consist of statements written in English and are very close to human language.

BACKGROUND AND HISTORY

The first computers used vacuum tubes and mechanical relays to indicate the switch positions of *on* or *off* as the logic units corresponding to the binary digits of 0 or 1, and it was necessary to reconfigure them each time a new task was approached. They were large enough to occupy entire rooms and required huge amounts of electricity and cooling. In the 1930's the Atanasoff-Berry Computer (ABC) was created at Iowa State University to solve simultaneous numerical equations, and it was followed by the electronic numerical integrator and computer (ENIAC), developed by the military for mathematical operations.

The transistor was invented in 1947 by John Bardeen, Walter Brattain, and William Shockley, which led to the use of large transistors as the logic units in the 1950's. The integrated circuit chip was invented by Jack St. Clair Kilby and Robert Norton Noyce in 1958 and caused integrated circuits to come into usage in the 1960's. These early integrated circuits were still quite large and included transistors, diodes, capacitors, and transistors. Modern silicon chips can hold these components and as many as 55 million transistors. Silicon chips are called microprocessors, because each microprocessor can hold these logic units within just over a square inch of space.

HOW IT WORKS

Hardware. The hardware, or physical components, of a computer can be classified according to their general uses of input, output, processing, and storage. Typical input devices include the mouse, keyboard, scanner, and microphone that facilitate communication of information between the human user and the computer. The operation of each of these peripheral devices requires special software called a driver, which is a type of controller, that is able to translate the input data into a form that can be communicated to the operating system and controls input and output peripheral devices.

A read-only memory (ROM) chip contains instructions for the basic input-output system (BIOS) that all the peripheral devices use to interact with the CPU. This process is especially important when a user first turns on a computer for the boot process.

When a computer is turned on, it first activates the BIOS, which is software that facilitates the interactions between the operating system, hardware, and peripherals. The BIOS accomplishes this interaction by first running the power-on self test (POST), which is a set of routines that are always available at a specific

Computer circuit board. (Pasieka/Photo Researchers, Inc.)

memory address in the read-only memory. These routines communicate with the keyboard, monitor, disk drives, printer, and communication ports to access the Internet. The BIOS also controls the time-of-day clock. These tasks completed by the BIOS are sometimes referred to as booting up (from the old expression "lift itself up by its own bootstraps"). The last instruction within the BIOS is to start reading the operating system from either a boot disk in a diskette drive or the hard drive. When shutting down a computer there are also steps that are followed to allow for settings to be stored and network connections to be terminated.

The CPU allows instructions and data to be stored in memory locations, called registers, which facilitate the processing of information as it is exchanged between the control unit, arithmetic-logic unit, and any peripheral devices. The processor interacts continuously with storage locations, which can be classified as either volatile or nonvolatile types of memory. Volatile memory is erased when the computer is turned off and consists of main memory, called random-access memory (RAM) and cache memory. The fundamental unit of volatile memory is the flip-flop, which can store a value of 0 or 1 when the computer is on. This value can be flipped when the computer needs to change it. If a series of 8 flip-flops is hooked together, an 8-bit number can be stored in a register. Registers can store only a small amount of data on a temporary basis while the computer is actually on. Therefore, the RAM is needed for larger amounts of information. However, it takes longer to access data stored in the RAM because it is outside the processor and needs to be retrieved, causing a lag time. Another type of memory, called cache, is located in the processor and can be considered an intermediate type of memory between registers and main memory.

Nonvolatile memory is not erased when a computer is turned off. It consists of hard disks that make up the hard drive or flash memory. Although additional nonvolatile memory can be purchased for less money than volatile memory, it is slower. The hard drive consists of several circular discs called platters that are made from aluminum, glass, or ceramics and covered by a magnetic material so that they can develop a magnetic charge. There are read and write heads made of copper so that a magnetic field develops that is able to read or write data when interacting with the platters. A spindle motor causes the platters to spin at a constant rate, and either a stepper motor or voice coil is used as the head actuator to initiate interaction with the platters.

The control unit of the CPU manages the circuits for completing operations of the arithmetic-logic unit. The arithmetic-logic unit of the CPU also contains circuits for completing the logical operations, in addition to data operations, causing the CPU essentially to function as the brain of the computer. The CPU is located physically on the motherboard. The motherboard is a flat board that contains all the chips needed to run a computer, including the CPU, BIOS, and RAM, as well as expansion slots and power- supply connectors. A set of wires, called a bus, etched into the motherboard connects these components. Expansion slots are empty places on the motherboard that allow upgrades or the insertion of expansion cards for various video and voice controllers, memory expansion cards, fax boards, and modems without having to reconfigure the entire computer. The motherboard is the main circuit board for the entire computer.

Circuit Design and Connectivity. Most makers of processor chips use the transistor-transistor logic (TTL) because this type of logic gate allows for the output from one gate to be used directly as the input for another gate without additional electronic input, which maximizes possible data transmission while minimizing electronic complications. The TTL makes this possible because any value less than 0.5 volt is recognized as the logic value of 0, while any value that exceeds 2.7 volts indicates the logic value

---

Fascinating Facts About Computer Engineering

- It is possible to warm up the grill before arriving home by connecting to the Internet and attaching an iGrill to a USB port. Recipes can also be downloaded directly from the Internet for the iGrill.
- The average laptop computer has more processing power than the ENIAC, which had more than 20,000 vacuum tubes and was large enough to fill entire rooms.
- By about 2015, batteries in digital devices will consume oxygen from the air in order to maintain their power.
- By 2015, it is expected that some mobile devices will be powered by kinetic energy and will no longer depend on batteries.
- A job growth rate of 31 percent is projected for computer engineers by 2018.

---

of 1. The processor chips interact with external computer devices via connectivity locations called ports. One of the most important of these ports is called the Universal Serial Bus (USB) port, which is a high-speed, serial, daisy-chainable port in newer computers used to connect keyboards, printers, mice, external disk drives, and additional input and output devices.

Software. The operating system consists of software programs that function as an interface between the user and the hardware components. The operating system also assists the output devices of printers and monitors. Most of the operating systems being used also have an application programming interface (API), which includes graphics and facilitates use. APIs are written in high-level languages (using statements approximating human language), such as C++, Java, and Visual Basic.

APPLICATIONS AND PRODUCTS

Stand-Alone Computers. Most computer users rely on relatively small computers such as laptops and personal computers (microcomputers). Companies that manufacture these relatively inexpensive computers have come into existence only since the early 1980's and have transformed the lives of average Americans by making computer usage a part of everyday life. Before microcomputers came into such wide usage, the workstation was the most accessible smaller-size computer. It is still used primarily by small and medium-size businesses that need the additional memory and speed capabilities. Larger organizations such as universities use mainframe computers to handle their larger power requirements. Mainframes generally occupy an entire room. The most powerful computers are referred to as supercomputers, and they are so expensive that often several universities will share them for scientific and computational activities. The military uses them as well. They often require the space of several rooms.

Inter-Network Service Architecture, Interfaces, and Inter-Network Interfaces. A network consists of two or more computers connected together in order to share resources. The first networks used coaxial cable, but now wireless technologies allow computer devices to communicate without the need to be physically connected by a coaxial cable. The Internet has been a computer-engineering application that has transformed the way people live. Connecting to the Internet first involved the same analogue transmission used by the plain old telephone service (POTS), but connections have evolved to the use of fiber-optic technology and wireless connections. Laptop computers, personal digital assistants (PDAs), cell phones, smart phones, RFID (radio frequency identification), iPods, iPads, and Global Positioning Systems (GPS) are able to communicate, and their developments have been made possible by the implementation of the fundamental architectural model for inter-network service connections called the Open Systems Interconnection (OSI) model. OSI is the layered architectural model for connecting networks. It was developed in 1977 and is used to make troubleshooting easier so that if a component fails on one computer, a new, similar component can be used to fix the problem, even if the component was manufactured by a different company.

OSI's seven layers are the application, presentation, session, transport, network, data link, and physical layers.

The application and presentation layers work together. The application layer synchronizes applications and services in use by a person on an individual computer with the applications and services shared with others via a server. The services include e-mail, the World Wide Web, and financial transactions. One of the primary functions of the presentation layer is the conversion of data in a native format such as extended binary coded decimal interchange code (EBCDIC) into a standard format such as American Standard Code for Information Interchange (ASCII).

As its name implies, the primary purpose of the session layer is to control the dialog sessions between two devices. The network file system (NFS) and structured query language (SQL) are examples of tools used in this layer.

The transport layer controls the connection-oriented flow of data by sending acknowledgments to data senders once the recipient has received data and also makes sure that segments of data are retransmitted if they do not reach the intended recipient. A router is one of the fundamental devices that works in this layer, and it is used to connect two or more networks together physically by providing a high degree of security and traffic control.

The data link layer translates the data transmitted by the components of the physical layer, and it is within the physical layer that the most dramatic technological advances made possible by computer engineering have had the greatest impact. The coaxial cable originally used has been replaced by fiber-optic technology and wireless connections.

Fiber Distributed Data Interface (FDDI) is the fiber-optic technology that is a high-speed method of networking, composed of two concentric rings of fiber-optic cable, allowing it to transmit over a longer distance but at greater expense. Fiber-optic cable uses glass threads or plastic fibers to transmit data. Each cable contains a bundle of threads that work by reflecting light waves. They have much greater bandwidth than the traditional coaxial cables and can carry data at a much faster rate of 100 gigabits per second because light travels faster than electrical signals. Fiber-optic cables are also not susceptible to electromagnetic interference and weigh less than coaxial cables.

Wireless networks use radio or infrared signals for cell phones and a rapidly growing variety of handheld devices, including iPhones, iPods, and tablets. New technologies using Bluetooth and Wi-Fi for mobile connections are leading to the next phase of inter-network communications with the Internet called cloud computing, which is the direction for the most economic growth. Cloud computing is basically a wireless Internet application where servers supply resources, software, and other information to users on demand and for a fee.

Smart phones that use Google's new Android operating system for mobile phones have a PlayStation emulator called PSX4Droid that allows PlayStation games to be played on these phones. Besides making games easily accessible, cloud computing is also making it easier and cheaper to do business all around the world with applications such as Go To Meeting, which is one example of video conferencing technology used by businesses.

IMPACT ON INDUSTRY

The first developments in technology that made personal computers, the Internet, and cell phones more easily accessible to the average consumer have primarily been made by American companies. Computer manufacturers include IBM and Microsoft, which were followed by Dell, Compaq, Gateway, Intel, Hewlett-Packard, Sun Microsystems, and Cisco. Cisco continues to be the primary supplier of the various hardware devices necessary for connecting networks, such as routers, switches, hubs, and bridges. Mobile computing is being led by American companies Apple, with its iPhone, and Research in Motion (RIM), with its Blackberry. These mobile devices require collaboration with phone companies such as AT&T and Verizon, causing them to grow in worldwide dominance as well. Most U.S. technology companies are international in scope, with expansion still expected in the less-developed countries of Southeast Asia and Eastern Europe.

Microsoft was founded in 1975 by Bill Gates and Paul Allen. All the personal computers (PCs) that became more accessible to consumers depended on Microsoft's operating system, Windows. Over the years, Microsoft has made numerous improvements with each new version of its operating system, and it has continued to dominate the PC market to such an extent that the U.S. government sued Microsoft, charging it with violating antitrust laws. As of 2011, Microsoft remains the largest software company in the world, although it has been losing some of its dominance as an innovative technology company because of the growth of Google, with its Internet search engine software, and Apple, with its new, cutting-edge consumer electronics, including the

iPod, iPad, and iPhone. In 2010, Microsoft started selling the Kinect for Xbox 360, which is a motion-capture device used for games that easily connects to a television and the Internet. More than 2.5 million of the Kinect for Xbox 360 units have been sold in the United States since November, 2010. To try to compete with Apple's dominance in the mobile-technology market, Microsoft announced in January, 2011, that it would collaborate with chip manufacturer Micron, to incorporate Micron's flash memory into mobile devices. A new version of Microsoft's Windows operating system was to use processor chips manufactured by ARM Holdings rather than those by Intel, because the ARM chips contain lower power, which facilitates mobile communication. Microsoft continues to dominate the PC operating-system market worldwide.

Apple is one of the most successful U.S. companies because of its Macintosh computer and iPod. Introduced in 2001, the iPod uses various audio file formats, such as MP3 and WAV, to function as a portable media player. To further expand on the popularity of the iPod, Apple opened its own online media and applications stores to allow consumers to access music and video applications via the Internet. In 2007, Apple expanded its market share by introducing the iPhone, and adding visual text messaging, emailing of digital photos using its own camera capabilities, and a multi-touch screen for enhanced access to the Internet. The Apple iPhone is essentially a miniature handheld computer that also functions as a phone. The apps that are be downloaded from the Apple Store online are computer programs that can be designed to perform just about any specific task. There are more than 70,000 games available for purchase for the iPhone, and more than 7 million apps have been sold as of January, 2011.

Google was started by Larry Page and Sergey Brin, who met as graduate students in Stanford University's computer science department, in 1996. Ten years later, the Oxford dictionary added the term "google," a verb meaning to do an Internet search on a topic using the Google search engine. Despite attempts by several competitors, including Microsoft, to develop search engine software, Google has dominated the market. In addition to its search engine, Google has developed many new innovative software applications, including Gmail, Google Maps, and its own operating system called Android, designed specifically to assist Google with its move into the highly lucrative mobile-technology market.

CAREERS AND COURSE WORK

Knowledge of the design of electrical devices and logic circuits is important to computer engineers, so a strong background in mathematics and physics is helpful. Since there is a great deal of overlap between computer engineering and electrical engineering, any graduate with a degree in electrical engineering may also find work within the computer field.

One of the most rapidly expanding occupations within the computer engineering field is that of network systems and data communication analyst. According to the Bureau of Labor Statistics, this occupation is projected to grow by 53.4 percent to almost 448,000 workers between by 2018, making it the second-fastest growing occupation in the United States. The related occupation of computer systems analyst is also expected to grow faster than the average for all jobs through 2014. Related jobs include network security specialists and software engineers. All of these occupations require at least a bachelor's degree, typically in computer engineering, electrical engineering, software engineering, computer science, information systems, or network security.

SOCIAL CONTEXT AND FUTURE PROSPECTS

The use of cookies, a file stored on an Internet user's computer that contains information about that user, an identification code or customized preferences, that is recognized by the servers of the Web sites the user visits, as a tool to increase online sales by allowing e-businesses to monitor the preferences of customers as they access different Web pages is becoming more prevalent. However, this constant online monitoring also raises privacy issues. In the future there is the chance that one's privacy could be invaded with regard to medical or criminal records and all manner of financial information. In addition, wireless technologies are projected to increase the usage of smart phones, the iPad, the iPod, and many new consumer gadgets, which can access countless e-business and social networking Web sites. Thus, the rapid growth of Internet applications that facilitate communication and financial transactions will continue to be accompanied by increasing rates of identity theft and other cyber crimes, ensuring that network security will continue to be an important application of computer engineering.

Jeanne L. Kuhler, B.S., M.S., Ph.D.

FURTHER READING

Cassel, Lillian N., and Richard H. Austing. *Computer Networks and Open Systems: An Application Development Perspective.* Sudbury, Mass.: Jones & Bartlett Learning, 2000. This textbook describes the OSI architecture model.

Das, Sumitabha. *Your UNIX: The Ultimate Guide.* New York: McGraw-Hill, 2005. This introductory textbook provides background and instructions for using the UNIX operating system.

Dhillon, Gurphreet. "Dimensions of Power and IS Implementation." *Information and Management* 41, no. 5 (2004): 635-644. Describes some of the work done by managers when choosing which computer languages and tools to implement.

Irvine, Kip R. *Assembly Language for Intel-Based Computers.* 5th ed. Upper Saddle River, N.J.: Prentice Hall, 2006. This introductory textbook provides instruction for using assembly language for Intel processors.

Kerns, David V., Jr., and J. David Irwin. *Essentials of Electrical and Computer Engineering* 2d ed. Upper Saddle River, N.J.: Prentice Hall, 2004. A solid introductory text that integrates conceptual discussions with modern, relevant technological applications.

Magee, Jeff, and Jeff Kramer. *Concurrency: State Models and Java Programming.* Hoboken, N.J.: John Wiley & Sons, 2006. This intermediate-level textbook describes the software engineering techniques involving control of the timing of different processes.

Silvester, P. P., and D. A. Lowther. *Computer Engineering Circuits, Programs, and Data.* New York: Oxford University Press, 1989. This is an introductory text for engineering students.

Sommerville, Ian. *Software Engineering.* 9th ed. Boston: Addison-Wesley, 2010. Text presents a broad, up-to-date perspective of software engineering that focuses on fundamental processes and techniques for creating software systems. Includes case studies and extensive Web resources.

WEB SITES

Computer Society
http://www.computer.org

Institute of Electrical and Electronics Engineers
http://www.ieee.org

Software Engineering Institute
http://www.sei.cmu.edu

See also: Computer Graphics; Computer Languages, Compilers, and Tools; Computer Networks; Computer Science; Electrical Engineering; Engineering; Human-Computer Interaction; Internet and Web Engineering; Pattern Recognition; Software Engineering.

COMPUTER GRAPHICS

FIELDS OF STUDY

Three-dimensional (3-D) design; calculus; computer programming; computer animation; digital modeling; graphic design; multimedia applications; software engineering; Web development; vector graphics and design; drawing; animation.

SUMMARY

Computer graphics involves the creation, display, and storage of images on a computer with the use of specialized software. Computer graphics fills an essential role in many everyday applications. 3-D animation has revolutionized video games and has resulted in box-office hits in theaters. Virtual images of people's bodies and tissues are used in medicine for teaching, surgery simulations, and diagnoses. Educators and scientists are able to develop 3-D models that illustrate principles in a more comprehensible manner than a two-dimensional (2-D) image can. Through the use of such imagery, architects and engineers can prepare virtual buildings and models to test options prior to construction. Finally, businesses use computer graphics to prepare charts and graphs for better comprehension during presentations.

KEY TERMS AND CONCEPTS

- **Application Programming Interface (API):** Set of functions built into a software program that allows communication with another software program.
- **Computer-Aided Design (CAD):** Software used by architects, engineers, and artists to create drawings and plans.
- **Computer Animation:** Art of creating moving images by means of computer technology.
- **Graphical User Interface (GUI):** Program by which a user interacts with a computer by controlling visual symbols on the screen.
- **Graphics Processing Unit (GPU):** Specialized microprocessor typically found on a video card that accelerates graphics processing.
- **Pixels:** Abbreviation for "picture elements," the smallest discrete components of a graphic image appearing on a computer screen or other graphics output device.
- **Raster Graphics:** Digital images that use pixels arranged in a grid formation to represent an image.
- **Rendering:** Process of generating an image from a model by means of computer programs.
- **Three-Dimensional (3-D) model:** Representation of any 3-D surface of an object using graphics software.
- **Vector Graphics:** Field of computer graphics that uses mathematical relationships between points and the paths connecting them to describe an image.

DEFINITION AND BASIC PRINCIPLES

The field of computer graphics uses computers to create digital images or to modify and use images obtained from the real world. The images are created from internal models by means of computer programs. Two types of graphics data can be stored in a computer. Vector graphics are based on mathematical formulas that generate geometric images by joining straight lines. Raster graphics are based on a grid of dots known as pixels, or picture elements. Computer graphics can be expressed as 2-D, 3-D, or animated images.

The graphic data must be processed in order to render the image and display on a computer movie screen. The work of computer-graphics programmers has been facilitated by the development of application programming interfaces (APIs), notably the Open Graphics Library (OpenGL). OpenGL provides a set of standard operations to render images across a wide variety of platforms (operating systems). The graphics processing unit (GPU) found in a video card facilitates presentation of the image.

There are subtle differences between the responsibilities of the computer-graphics specialist and graphic or Web designers. Computer-graphics specialists develop programs to display visual images or models, while designers are creative artists who use programs to communicate a specific message effectively. The end products of graphic designers are seen in various print media, while Web designers produce digital media.

BACKGROUND AND HISTORY

The beginning of computer graphics has been largely attributed to Ivan Sutherland, who developed a computer drawing program in 1961 for his dissertation work at the Massachusetts Institute of Technology. This program, Sketchpad, was a seminal event in the area of human-computer interaction, as it was one of the first to use graphical user interfaces (GUIs). Sutherland used a light pen containing a photoelectric cell that interacted with elements on the monitor. His method was based on vector graphics. Sketchpad provided vastly greater possibilities for the designer or engineer over previous methods based on pen and paper.

Other researchers further developed vector-graphics capabilities. Raster-based graphics using pixels was later developed and is the primary technology being used. The mouse was invented and proved more convenient than a light pen for selecting icons and other elements on a computer screen. By the early 1980's, Microsoft's personal computer (PC) and Apple's Macintosh were marketed using operating systems that incorporated GUIs and input devices that included the mouse as well as the standard keyboard.

Major corporations developed an early interest in computer graphics. Engineers at Bell Telephone Laboratories, Lawrence Berkeley National Laboratory, and Boeing developed films to illustrate satellite orbits, aircraft vibrations, and other physics principles. Flight simulators were developed by Evans & Sutherland and General Electric.

The invention of video graphics cards in the late 1980's followed by continual improvements gave rise to advances in animation. Video games and full-length animated motion pictures have become large components of popular culture.

HOW IT WORKS

Types of Images. Vector graphics uses mathematical formulas to generate lines or paths that are connected at points called vertices to form geometric shapes (usually triangles). These shapes are joined in a meshwork on the surfaces of figures. Surfaces on one plane are two-dimensional, while connecting vertices in three dimensions will produce 3-D images. Two-dimensional images are more useful for applications such as advertising, technical drawing, and cartography. Raster images, on the other hand, develop images based on pixels, or picture elements. Pixels can be thought of as tiny dots or cells that contain minute portions of the image and together compose the image on the computer screen. The bits of information that the pixels are able to process determine the resolution or sharpness of the image. Raster images are much more commonly used in computer graphics and are essential for 3-D and animation work.

Graphics Pipeline. The process of creating an image from data is known as the graphics pipeline. The pipeline consists of three main stages: modeling, rendering, and display.

Modeling begins with a specification of the objects, or components of a scene, in terms of shape, size, color, texture, and other parameters. These objects then undergo a transformation involving their correct placement in a scene.

Rendering is the process of creating the actual image or animation from the scene. Rendering is analogous to a photograph or an artist's drawing of a scene. Aspects such as proper illumination and the visibility of objects are important at this stage.

The final image is displayed on a computer monitor with the use of advanced software, as well as computer hardware that includes the motherboard and graphics card. The designer must keep in mind that the image may appear differently on different computers or different printers.

OpenGL. OpenGL is an API that facilitates writing programs across a wide variety of computer languages and hardware and software platforms. OpenGL consists of graphics libraries that provide the programmer with a basic set of commands to render images. OpenGL is implemented through a rendering pipeline (graphics pipeline). Both vector and raster data are accepted for processing but follow different steps. At the rasterization stage, all data are converted to pixels. At the final step, the pixels are written into a 2-D grid known as a framebuffer. A framebuffer is the memory portion of the computer allocated to hold the graphics information for a single frame or picture.

Maya. Maya (trade name Autodesk Maya) is computer-graphics software that has become the industry standard for generating 3-D models for game development and film. Maya is particularly effective in producing dazzling animation effects. Maya is imported into OpenGL or another API to display the models on the screen.

Video Games. Video-game development takes a specialized direction. The term "game engine" was coined to refer to software used to create and render video games on a computer screen. Many features need to work together to create a successful game, including 2-D and 3-D graphics, a "physics engine" to prepare realistic collision effects, sound, animation, and many other functions. Because of the highly competitive nature of the video-game industry, it is necessary to develop new games rapidly. This has led to reusing or adapting the same game engine to create new games. Some companies specialize in developing so-called middleware software suites that are conceived to contain basic elements on which the game programmer can build to create the complete game.

Television. To create 3-D images for television, individual objects are created in individual layers in computer memory. This way, the individual objects can move independently without affecting the other objects. TV graphics are normally produced a screen at a time and can be layered with different images, text, backgrounds, and other elements to produce rich graphic images. Editing of digital graphics is much faster and efficient than the traditional method of cutting and pasting film strips.

Film. The film *Avatar* (2009) illustrated how far 3-D animation had been developed. The production used a technique called performance or motion tracking. Video cameras were attached to computers and focused on the faces of human actors as they performed their parts. In this manner, subtle facial expressions could be transferred to their animated avatars. Backgrounds, props, and associated scenery moved in relation to the actors.

APPLICATIONS AND PRODUCTS

Game Development. Game development has become a major consumer industry, grossing $20 billion in 2010. Ninety-two percent of juveniles play video or computer games. Major players in the field include Sony PlayStation, Nintendo, and Microsoft Xbox. The demographics of video-game players are changing, resulting in a greater number of female and adult players. Whereas previously video games focused on a niche market (juvenile boys), in the future game developers will be increasingly directing their attention to the mass market. These market changes will result in games that are easier to play and broader in subject matter.

Film. The influence of computer graphics in the film industry is largely related to animation. Of course, animation predates computer graphics by many years, but it is realism that gives animation its force. Entire films can be animated, or animation can play a supplemental role. The fantasy of Hollywood was previously based on constructing models and miniatures, but now computer-generated imagery can be integrated into live action.

Television. The conversion of the broadcast signal for television from analogue to digital, and later to high definition, has made the role of the computer-graphics designer even more important. It is common for many shows to have a computer-graphics background instead of a natural background, such as when a weatherperson stands in front of a weather map. This development results in more economical productions, since there are no labor costs involved in preparing sets or a need to store them.

Computer graphics was used in television advertising before film, since it is more economical to produce. A combination of dazzling, animated graphics with a product or brand-name can leave a lasting impression on the viewer.

Medicine. Computer graphics typically works in concert with other advanced technologies, such as computed tomography, magnetic resonance, and positron emission tomography, to aid in diagnosis and treatment. The images obtained by these technologies are reconstructed by computer-graphics techniques to form 3-D models, which can then be visualized. The development of virtual human bodies has proven invaluable to illustrate anatomical structures in healthy and diseased subjects. These virtual images have found application in surgery simulations and medical education and training. The use of patient-specific imaging data guides surgeons to conduct minimally invasive interventions that are less traumatic for the patient and lead to faster healing. Augmented reality provides a larger view of the surgical field and allows the surgeon to view structures that are below the observed surface.

Science. Computer graphics has proven valuable to illustrate scientific principles and concepts that are not easily visible in the natural environment. By preparing virtual 3-D models of molecular structures or viruses moving through tissues, a student or scientist who is a visual learner is able to grasp these concepts.

Architecture and Engineering. Computer-aided design has greatly helped the fields of architecture and engineering. Computer graphics was initially used only as a replacement for drafting with pencil and paper. However, the profession has come to recognize its value in the early stages of a project to help designers check and reevaluate their proposed designs. Multimedia designs such as animations and panoramas are very useful in client presentations. The designs allow clients to walk through a building virtually and interactively look around interior spaces. Engineers can also test the effect of various inputs to a system, model, or circuit.

Business. Presentation of numeric data in graphs and charts is an important application of computer graphics in business. Market trends, production data, and other business information presented in graphic form are often more understandable to an audience or reader.

Education. Computer graphics has proven very useful in education because of the power of visualization in the learning process. There are many benefits to using computer graphics in education: Students learn at their own pace at their own time, the instruction is interactive rather than passive, the student is engaged in the learning process, and textual and graphic objects can be shared among applications via tools such as cutting and pasting, clipboards, and scrapbooks.

IMPACT ON INDUSTRY

The worldwide market for hardware, software, and services for 3-D visualization is expected to grow rapidly. Spending in the defense and government markets reached $16.5 billion in 2010 and is expected to increase to $20 billion by 2015. The popularity of games and the film industry has resulted in huge investments in developing 3-D technology. With the technology already in place, world governments can acquire 3-D visualization at a much more reasonable price than previously. The military sector is the largest market, followed by civil-aviation and civil-service sectors, including law-enforcement and emergency-response agencies.

Government and University Research. Although the U.S. government played a leading role in the early development of computer graphics, it has come to play a more collaborative role with universities. A typical example is the Graphics and Visualization

---

Fascinating Facts About Computer Graphics

- *Avatar* (2009) became the highest-grossing film of all time, earning $2.7 billion worldwide and demonstrating how compelling stories told using computer graphics had become.
- Creation of virtual persons allows surgeons to perform a trial run of their procedures before the actual surgery.
- Real-time visualization models prepared by architects allow buyers to walk through virtual homes and change the views they see.
- Scientists can study chemical reactions with molecules in real time to gain insight into the structural changes taking place among the molecules.
- Engineers can prepare a 3-D model of an engine under development and study the effects of various changes.
- The features visible in 3-D through Google Earth—topography, cities, street views, trees, and other details–dramatically demonstrate the uses of computer graphics when wedded to satellite technologies.
- 3-D visualization of atmosphere and terrain has resulted in improved weather forecasting and stunning presentations by news meteorologists.
- 3-D augmented reality is being studied to animate traffic-simulation models to aid in the planning and design of highway construction projects. In the 3-D simulation tests, the user has the opportunity to drive a virtual or real car.

Center, which was founded by the National Science Foundation to pursue interdisciplinary research at five universities (Brown, California Institute of Technology, Cornell, University of North Carolina, and University of Utah). The center pursues research in four main areas of computer graphics: modeling, rendering, user interfaces, and high-performance architectures. The Department of Defense also conducts research involving computer graphics.

On the local level, computer graphics has proven useful in preparing presentation material, such as visually appealing graphs, charts, and diagrams. Complex projects and issues can be presented in a manner that is more understandable than traditional speeches or handouts. County and community planners can explore "what if" scenarios for land-use projects and investigate the potential for change. Computers can enhance information used in the planning process or explain the scope of a project. The presentation material can be modified to incorporate community suggestions over a series of meetings.

Universities tend to offer courses and degree programs in the types of research they specialize in. Brown, Penn State, and several University of California campuses are working on scientific visualization that has applications in designing virtual humans and medical illustrations. Universities in Canada and Europe are also active in these areas. The University of Utah, a longtime leader in the field, conducts research in geometric design and scientific computing.

Industry and Business. The computer-graphics industry will continue to grow, resulting in an increased demand for programmers, artists, scientists, and designers. Pixar, DreamWorks, Disney, Warner Bros., Square Enix, Sony, and Nintendo are considered top animation producers in the film and video-game industries. Architectural and engineering consulting firms have been contracted to use virtual images in planning public buildings such as stadiums. Virtual people help to determine traffic flow through buildings. If congestion points are observed as virtual crowds travel to concession stands or restrooms, for example, changes can be made to provide more efficient flow. Virtual models were also used to study traffic flow in Hong Kong harbor.

Computer graphics are a big improvement over architectural scale models. They can readily portray a variety of alternative plans without incurring large costs.

CAREERS AND COURSE WORK

Computer-graphics specialists must have a unique combination of artistic and computer skills. They must have good math, programming, and design skills, and be able to visualize 3-D objects. The specialist must be creative, detail oriented, and able to work well individually and as part of a team.

Course offerings can vary considerably among universities, so the student must consider the specialties of the prospective schools in relation to the field in which he or she is most interested. For example, the student may want to focus more on software design than on graphics design. Essential courses can include advanced mathematics, programming, computer animation, geometric design and modeling, multimedia systems and applications, and software engineering.

Computer graphics can be applied to a vast number of fields, and its influence can only increase. In addition to animation in video games and film, computer graphics has proven valuable in architectural and engineering design, education, medicine, business, and cartography. Typical positions for a computer-graphics specialist include 3-D animator or modeler, special effects designer, video game designer, and Web designer.

Computer-graphics specialists can work in the public or private sector; they can also work independently. Although companies prefer to hire candidates with a bachelor's degree, many workers in the field have only an associate's degree or vocational certificate.

In the private sector, computer-graphics specialists can be employed by architectural, construction, and engineering consulting firms, electronics and software manufacturing companies, and petrochemical, food processing, and energy industries. In the public sector, they can work at all levels of government and in hospitals and universities.

SOCIAL CONTEXT AND FUTURE PROSPECTS

Computer graphics will continue to have a profound effect on the visual arts, freeing the artist from the need to master technical skills in order to focus on creativity and imagination in his or her work. The artist can experiment with unlimited variations in structures and designs in a single work to see which produces the desired effect. Continuing advances in producing virtual images and 3-D animation will enhance understanding of scientific principles and processes in education, medicine, and science.

David Olle, B.S., M.S.

FURTHER READING

McConnell, Jeffrey. *Computer Graphics: Theory into Practice.* Boston: Jones & Bartlett, 2006. The basic principles of graphic design are amply presented with reference to the human visual system. OpenGL is integrated with the material, and examples of 3-D graphics are illustrated.

Shirley, Peter, and Steve Marschner. *Fundamentals of Computer Graphics.* 3d ed. Natick, Mass.: A. K. Peters, 2009. This book emphasizes the underlying mathematical fundamentals of computer graphics rather than learning particular graphics programs.

Vidal, F. P., et al. "Principles and Applications of Computer Graphics in Medicine." *Computer Graphics Forum* 25, no. 1 (2006): 113-137. Excellent review of the state of the art. Discusses software development, diagnostic aids, educational tools, and computer-augmented reality.

WEB SITES
ACMSIGGRAPH (Association for Computing Machinery's Special Interest Group on Graphics and Interactive Techniques)
http://www.siggraph.org

Computer Society
http://www.computer.org

Institute of Electrical and Electronics Engineers
http://www.ieee.org

OpenGL Overview
http://www.opengl.org/about/overview

See also: Calculus; Computer-Aided Design and Manufacturing; Computer Engineering; Computer Networks; Computer Science; Human-Computer Interaction; Software Engineering; Video Game Design and Programming.

COMPUTER LANGUAGES, COMPILERS, AND TOOLS

FIELDS OF STUDY

Computer science; computer programming; information systems; information technology; software engineering.

SUMMARY

Computer languages are used to provide the instructions for computers and other digital devices based on formal protocols. Low-level languages, or machine code, were initially written using the binary digits needed by the computer hardware, but since the 1960's, languages have evolved from early procedural languages to object-oriented high-level languages, which are more similar to English. There are many of these high-level languages, with their own unique capabilities and limitations, and most require some type of compiler or other intermediate translator to communicate with the computer hardware. The popularity of the Internet has created the need to develop numerous applications and tools designed to share data across the Internet.

KEY TERMS AND CONCEPTS

- **Application:** Computer program that completes a specific task.
- **Basic Input-Output System (BIOS):** Computer program that allows the central processing unit of a computer to communicate with other computer hardware.
- **Browser:** Software program used to view Web pages on the Internet.
- **Compiler:** Program that converts source code in a text file into a format that can be executed by a computer.
- **Graphical User Interface (GUI):** Visual interface that allows a user to position a cursor over a displayed object or icon and click on that object or icon to make a selection.
- **Interpreter:** Computer tool that is much faster and more efficient than a compiler.
- **Mainframe:** Large stand-alone computer that completes batch processing (groups of computer instructions completed at once).

- **Operating System:** Computer program written in a language so that it can function as an interface between a computer user and the hardware that runs the computer by managing resources.
- **Portability:** Ability of a program to be downloaded from a remote location and executed on a variety of computers with different operating systems.
- **Protocol:** Set of rules to be followed to allow for communication.
- **Server:** Computer dedicated to managing network activities, such as printers and email, which are shared by many users (clients).
- **Structured Query Language (SQL):** Computer language used to access databases.

DEFINITION AND BASIC PRINCIPLES

The traditional process of using a computer language to write a program has generally involved the initial design of the program using a flowchart based on the purpose and desired output of the program, followed by typing the actual instructions for the computer (the code) into a file using a text editor, and then saving this code in a file (the source code file). A text editor is used because it does not have the formatting features of a word processor. An intermediate tool called a compiler then has been used to convert this source code into a format that can be run (executed) by a computer.

However, as of 2011, there are new tools that are much faster and more efficient than compilers. Therefore, many compilers have been replaced by these new tools, called interpreters. Larger, more complex programs have evolved that have required an additional step to link external files. This process is called linking and it joins the main, executable program created by the compiler to other necessary programs. Finally, the executable program is run and its output is displayed on the computer monitor, printed, or saved to another digital file. If errors are found, the process of debugging is followed to go back through the code to make corrections.

BACKGROUND AND HISTORY

Early computers such as ENIAC (Electronic Numerical Integrator and Computer), the first general-purpose computer, were based on the use of switches that could be turned on or off. Thus, the binary digits of 0 and 1 were used to write machine code. In addition to being tedious for a programmer, the code had to be rewritten if used on a different type of machine, and it certainly could not be used to transmit data across the Internet, where different computers all over the world require access to the same code.

Assembly language evolved by using mnemonics (alphabetic abbreviations) for code instead of the binary digits. Because these alphabetic abbreviations of assembly language no longer used the binary digits, additional programs were developed to act as intermediaries between the human programmers writing the code and the computer itself. These additional programs were called compilers, and this process was initially known as compiling the code. This compilation process was still machine and vendor dependent, however, meaning, for example, that there were several types of compilers that were used to compile code written in one language. This was expensive and made communication of computer applications difficult.

The evolution of computer languages from the 1950's has accompanied technological advances that have allowed languages to become increasingly powerful, yet easier for programmers to use. FORTRAN and COBOL languages led the way for programmers to develop scientific and business application programs, respectively, and were dependent on a command-line user interface, which required a user to type in a command to complete a specific task. Several other languages were developed, including Basic, Pascal, PL/I, Ada, Lisp, Prolog, and Smalltalk, but each of these had limited versatility and various problems. The C and C++ languages of the 1970's and 1980's, respectively, have emerged as being the most useful and powerful languages, are still in use, and have been followed by development tools written in the Java and Visual Basic languages, including integrated development environments with editors, designers, debuggers, and compilers all built into a single software package.

How It Works

BIOS and Operating System. The programs within the BIOS are the first and last programs to execute whenever a computer device is turned on or off. These programs interact directly with the operating system (OS). The early mainframe computers

that were used in the 1960's and 1970's depended on several different operating systems, most of which are no longer in usage, except for UNIX and DOS. DOS (Disk Operating System) was used on the initial microcomputers of the 1980's and early 1990's, and it is still used for certain command-line specific instructions.

Graphical User Interfaces (GUIs). Microsoft dominates the PC market with its many updated operating systems, which are very user-friendly with GUIs. These operating systems consist of computer programs and software that act as the management system for all of the computer's resources, including the various application programs most taken for granted, such as Word (for documents), Excel (for mathematical and spreadsheet operations), and Access (for database functions). Each of these applications is actually a program itself, and there are many more that are also available.

Since the 1980's, many programming innovations increasingly have been built to involve the client-server model, with less emphasis on large main

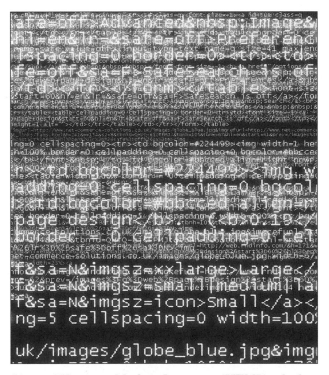

Lines of Hypertext Markup Language (HTML), the language used by programmers to develop Web sites. (Christian Darkin/Photo Researchers, Inc.)

413

frames and more emphasis on the GUIs for smaller microcomputers and handheld devices that allow consumers to have deep color displays with high resolution and voice and sound capabilities. However, these initial GUIs on client computers required additional upgrades and maintenance to be able to interact effectively with servers.

World Wide Web. The creation of the World Wide Web provided an improvement for clients to be able to access information, and this involvement of the Internet led to the creation of new programming languages and tools. The browser was developed to allow an end user (client) to be able to access Web information, and hypertext markup language (HTML) was developed to display Web pages. Because the client computer was manufactured by many different companies, the Java language was developed to include applets, which are mini-programs embedded into Web pages that could be displayed on any type of client computer. This was made possible by a special type of compiler-like tool called the Java Virtual Machine, which translated byte code. Java remains the primary computer language of the Internet.

APPLICATIONS AND PRODUCTS

FORTRAN and COBOL. FORTRAN was developed by a team of programmers at IBM and was first released in 1957 to be used primarily for highly numerical and scientific applications. It derived its name from formula translation. Initially, it used punched cards for input, because the text editors were not available in the 1950's. It has evolved but still continues to be used primarily in many engineering and scientific programs, including almost all programs written for geology research. Several updated versions have been released. FORTRAN77, released in 1980, had the most significant language improvements, and FORTRAN2000, released in 2002, is the most recent. COBOL (Common Business-Oriented Language) was released in 1959 with the goal of being used primarily for tracking retail sales, payroll, inventory control, and many other accounting-related activities. It is still used for most of these business-oriented tasks.

C and C++. The C computer language was the predecessor to the C++ language. Programs written in C were procedural and based on the usage of functions, which are small programming units. As programs grew in complexity, more functions were added to a

Fascinating Facts About Computer Languages, Compilers, and Tools

- Job prospects for software engineers are expected to be excellent, with salaries typically in the range of $70,000 to $80,000 per year.
- As of October, 2010, Java is the most widely used computer language, followed by C++.
- Using computer languages to write programs for video games is expected to be one of the most in-demand skills in the gaming area.
- Computer languages are necessary for a wide variety of everyday applications, ranging from cell phones, microwaves, and security systems to banking, online purchases, Web sites, auction sites (such as eBay), social-networking sites (such as Facebook), and Internet search engines (such as Google).
- Businesses, including banks, almost exclusively use COBOL, which uses two-digit rather than four-digit years. The use of two-digit years was the basis for the Y2K fear, in which it was predicted that on January 1, 2000, all the banking and other business-related applications would default to the year 1900, documents would be lost, and chaos would ensue. January 1, 2000, passed without incident.
- In 2010, Mattel introduced a Barbie doll called Computer Software Engineer Barbie. She has her own laptop with the Linux operating system and an iPhone.

C program. The problem was that eventually it became necessary to redesign the entire program, because trying to connect all of the functions, which added one right after the other, in a procedural way, was too difficult. C++ was created in the 1980's based on the idea of objects grouped into classes as the building blocks of the programs, which meant that the order did not have to be procedural anymore. Object-oriented programming made developing complex programs much more efficient and is the current standard.

Microsoft.NET. In June, 2000, Microsoft introduced a suite of languages and tools named Microsoft.NET along with its new language called Visual C#. Microsoft.NET is a software infrastructure that consists of many programs that allow a user to write programs for a range of new applications such as server components and Web applications by using new tools. Although programs written in Java can be run on any machine, as long as the entire program is written in Java, Microsoft.NET allows various programs to be run on the Windows OS. Additional advantages of Microsoft.NET involve its new language of Visual C#. Visual C# provides services to help Web pages already in existence, and C# can be integrated with the Visual Basic and Visual C++ languages, which facilitate the work of Web programmers by allowing them to update existing Web applications, rather than having to rewrite them.

The Microsoft.NET framework uses a common type system (CTS) tool to compile programs written in Cobol.NET, PerlNET, Visual Basic.NET, Jscript, Visual C++, Fortran.NET, and Visual C# into an intermediate language. This common intermediate language (CIL) can then be compiled to a common language runtime (CLR). The result is that the .NET programming environment promotes interoperability to allow programs originally written in different languages to be executed on a variety of operating systems and computer devices. This interoperability is crucial for sharing data and communication across the Internet.

IMPACT ON INDUSTRY

Microsoft was founded in 1975 by Bill Gates, who dropped out of Harvard, and Paul Allen. Their first software sold was in the BASIC language, which was the first language to be used on personal computers (PCs). As PC prices decreased, the number of consumers able to purchase a PC increased, as did the revenue and market dominance of Microsoft. Microsoft is the largest software company in the world, with annual revenues of more than $32 billion. Because it has gained such widespread dominance in the technology field, it offers its own certifications in order to maintain quality standards in its current and prospective employees. Among the most commonly held and useful certifications offered by Microsoft are the Microsoft Certified Applications Developer (MCAD), Microsoft Certified Solution Developer (MCSD), Microsoft Certified Professional Developer (MCPD), Microsoft Certified Systems Engineer (MCSE), Microsoft Certified Systems Administrator (MCSA), Microsoft Certified Database Administrator (MCDBA), and Microsoft Certified Trainer (MCT).

Sun Microsystems began with four employees in 1982 as a computer hardware manufacturer. With the rise of the Internet, the company seized the opportunity to create its computer language with its special tool, the Java Virtual Machine, which could be downloaded to any type of machine. This Java Virtual Machine contained a program called an interpreter, rather than a compiler, to act as an interface between the specific machine (platform) and the user. The result was a great enhancement in the interoperability of data on the Internet, and Java became the primary programming language on the Web.

Sun Microsystems also created Java servlets to allow more interaction with dynamic, graphical Web pages written in Java. However, although these servlets work very well with Java, they also must be compiled as classes before execution, which takes extra time.

Apple, one of the most successful American companies, has experienced phenomenal growth in the first decade of the twenty-first century. Its position as a market leader in consumer electronics is because of the success of its Macintosh computer accompanied by several Internet-enabled gadgets, such as the iPod, introduced in 2001, which uses various audio file formats, such as MP3 and WAV, to function as a portable media player. To further expand on the popularity of the iPod, Apple opened its own online media and applications stores, iTunes in 2003 and the App Store in 2008, to allow consumers to access music, video, and numerous applications via the Internet. In 2007, Apple expanded its market share by releasing the iPhone, which added to the phone the functions of text messaging, emailing of digital photos using its own camera capabilities, and a multi-touch screen for enhanced Internet access. In 2010, Apple further expanded with the release of the iPad, an electronic tablet that allows users access to the Internet to read, write, listen to, and view almost anything, including e-mail, movies, books, music, and magazines. The iPad also has a Global Positioning System (GPS) and a camera.

CAREERS AND COURSE WORK

Although it is becoming more common for entry-level job seekers to have a bachelor's degree in a computer-related field, either through a university computer-science or business-school department, it is possible to find rewarding employment as a programmer, software engineer, application developer, Web programmer or developer, database administrator, or software support specialist with just an associate's degree or relevant experience. This career field is unique in the large number of certifications that can be obtained to enhance job skills and increase the likelihood of finding employment, especially with just an associate's degree or related experience. Databases that can be accessed from within programming languages have also created the need for database administrators. There are many vendor-specific certifications for numerous job openings as database administrators who know computer languages such as SQL.

The future job prospects for software engineers are projected by the Bureau of Labor Statistics to be better than the job prospects for programmers for the period from 2008 to 2018. There is a great deal of similarity between these two careers, but generally programmers spend more time writing code in various languages, while the software engineers develop the overall design for the interaction of several programs to meet the needs of a customer. Software engineers are required more often to have a bachelor's or master's degree, usually in software engineering. The job growth for software engineers is expected to grow by 32 percent over the time period of 2008 to 2018, which is much faster than most other occupations.

SOCIAL CONTEXT AND FUTURE PROSPECTS

The Internet continues to bring the world together at a rapid pace, which has both positive and negative ramifications. Clearly, consumers have much easier access to many services, such as online education, telemedicine, and free search tools to locate doctors, learn more about any topic, comparison shop and purchase, and immediately access software, movies, pictures, and music. However, along with this increase in electronic commerce involving credit card purchases, bank accounts, and additional financial transactions has been the increase of cybercrime. Thousands of dollars are lost each year to various Internet scams and hackers being able to access private financial information. Some programmers even use computer languages to produce viruses and other destructive programs for purely malicious purposes, which have a negative impact on computer security.

Modern computer programs follow standard engineering principles to solve problems involving detail-driven applications ranging from radiation therapy and many medical devices to online banking, auctions, and stock trading. These online application programs require the special characteristics of Web-enabled software such as security and portability, which has given rise to the development of additional tools and will continue to produce the need for increased security features within computer languages and tools.

Jeanne L. Kuhler, B.S., M.S., Ph.D.

FURTHER READING

Das, Sumitabha. *Your UNIX: The Ultimate Guide.* 2d ed. Boston: McGraw-Hill, 2006. This textbook provides background and instructions for using the UNIX operating system.

Dhillon, Gupreet. "Dimensions of Power and IS Implementation." *Information and Management* 41, no. 5 (May, 2004): 635-644. This article describes some of the work done by managers when choosing which computer languages and tools to implement.

Guelich, Scott, Shishir Gundavaram, and Gunther Birznieks. *CGI Programming with Perl.* 2d ed. Cambridge, Mass.: O'Reilly, 2000. This introductory overview text describes the use of the common gateway interface programming and its interactions with the Perl programming language.

Horstmann, Cay. *Big Java.* 4th ed. Hoboken, N.J.: John Wiley & Sons, 2010. This book describes the most recent version of the Java language, along with its history and special features of the Java interpreter.

Snow, Colin. "Embrace the Role and Value of Master Data Management," *Manufacturing Business Technology,* 26, no. 2 (February, 2008): 92-95. This article, written by the VP and research director of Ventana Research, covers how to implement data management and avoid the errors that cost businesses millions of dollars each year.

WEB SITES

Association of Information Technology Professionals
http://www.aitp.org

The Computer Society
http://www.computer.org

Computing Technology Industry Association
http://www.comptia.org/home.aspx

Microsoft
Microsoft Certification Overview
http://www.microsoft.com/learning/en/us/
certification/cert-overview.aspx

See also: Computer Engineering; Computer Networks; Computer Science; Human-Computer Interaction; Internet and Web Engineering; Software Engineering.

COMPUTER NETWORKS

FIELDS OF STUDY

Computer science; computer engineering; information technology; telecommunications; engineering; communications; mathematics; physics.

SUMMARY

Computer networks consist of the hardware and software needed to support communications and the exchange of data between computing devices. The computing devices connected by computer networks include large servers, business workstations, home computers, and a wide array of smart mobile devices. The most popular computer network application is e-mail, followed by exchanging audio and video files. Computer networks provide an infrastructure for the Internet, which in turn provides support for the World Wide Web.

KEY TERMS AND CONCEPTS

- **Cloud Computing:** Software developed to execute on multiple servers, and delivered by a Web service.
- **Internet:** Another name for a set of networked computers.
- **Local Area Network (LAN):** Network that logically uses a shared media to connect its computers.
- **Media:** Term used to describe the physical components used to actually connect computers in a network.
- **Network Protocol:** Description of the rules and syntax used for computer networking software.
- **Packet:** Description of a block of data exchanged by two computers. For some protocols, the packet is the entire message, while for others, a message consists of many packets.
- **Routers:** Computing devices that receive and retransmit packets of information between networks.
- **Transmission Control Protocol/Internet Protocol (TCP/IP):** Primary network protocol used for wide area networks.

DEFINITION AND BASIC PRINCIPLES

A computer network is a collection of computer devices that are connected in order to facilitate the sharing of resources and information. Underlying the computer network is a communications network that establishes the basic connectivity of the computing devices. This communications network is often a wired system but can include radio and satellite paths as well. Devices used on the network include large computers, used to store files and execute applications; workstations, used to execute small applications and interact with the servers; home computers, connected to the network through an Internet service provider (ISP); and mobile devices, connected to the network by radio wave transmissions. Middleware is the software that operates on top of the communications network to provide a software layer for developers to add high-level applications, such as a search engine, to the basic network.

The high-level applications are what most people using the network see as the network. Some of the most important computer network applications provide communications for users. Older examples of this are e-mail, instant messaging, chat rooms, and videoconferencing. Newer examples of communications software are the multitude of social networks, such as Facebook. Other high-level applications allow users to share files. One of the oldest and still quite popular file-sharing programs is the file transfer protocol (FTP) program; the newer Flickr makes it easy to share photographs. Another way to use computer networks is to share computing power. The Telnet program (terminal emulation program for TCP/IP networks) allowed one to use an application on a remote mainframe in the early days of networking; and modern Web services allow one to run an application on a mobile device while getting most of the functionality from a remote server.

BACKGROUND AND HISTORY

The scientists who developed the early computers in the 1950's recognized the advantage of connecting computing devices. Teletype machines were in common use at that time, and many of the early computers were "networked" with these teletype machines over wired networks. By 1960, American Telephone and Telegraph (AT&T) had developed the first modem to allow terminal access to mainframes, and in 1964, IBM and American Airlines introduced the SABRE (Semi-Automated Business Research Environment) networked airline reservation system.

The Defense Department created ARPANET (Advanced Research Projects Agency Network) in 1966 to connect its research laboratories with college researchers. The early experience of ARPANET led the government to recognize the importance of being able to connect different networks so they could interoperate. One of the first efforts to promote interoperability was the addition of packet switching to ARPANET in 1969. In 1974, Robert Kahn and Vinton Cerf published a paper on packet-switching networks that defined the TCP/IP protocol, and in 1980, the U.S. government required all computers it purchased to support TCP/IP. When Microsoft added a TCP/IP stack to its Windows 95 operating system, TCP/IP became the standard wide area network in the world.

The development of the microcomputer led to the need to connect these devices to themselves and the wide area networks. In 1980, the Institute of Electrical and Electronics Engineers (IEEE) 802 standard was announced. It has provided most of the connectivity to networks since that time, although many wireless computing devices can connect with Bluetooth.

HOW IT WORKS

Computer networks consist of the hardware needed for networking, such as computers and routers; the software that provides the basic connectivity, such as the operating system; and middleware and applications, the programs that allow users to use the network. In understanding how these components work together, it is useful to look at the basic connectivity of the wide area network and contrast that to the way computers access the wide area network.

Wide Area Networks. A wide area network is one that generally covers a large geographic area and in which each pair of computers connect via a single line. The first wide area networks consisted of a number of connected mainframes with attached terminals. By connecting the mainframes, a user on one system could access applications on every networked computer. IBM developed the Systems Network Architecture (SNA), which was a popular early wide area network in the United States. The X.25 packet switching protocol standard provided a common architecture for the early wide area networks in Europe. Later,

as all computers provided support for the TCP/IP protocol, it became possible for different computer networks to work together as a single network. In the twenty-first century, any device on a single network, whether attached as a terminal by X.25 or as a part of a local area network (LAN), can access applications and files on any other network on the Internet. This complete connectivity allows browsers on any type of mobile device to access content over the World Wide Web because it runs transparently over the Internet.

Routing. A key to making the Internet operate efficiently is the large number of intermediate devices that route IP packets from the computer making a connection to the computer receiving the data. These devices are called routers, and they are the heart of the Internet. When a message is sent from a computer, it is decomposed into IP packets, each having the IP address of the receiver. Then each packet is forwarded to a border router of the sender. From the border router, the packet is routed through a set of intermediate routers, using an algorithm-like best path, and delivered to the border router of the receiver, and then the receiver. Once all the packets have arrived, the message is reassembled and used by the receiver.

Local Area Networks. A local area network (LAN) is a collection of computers, servers, printers, and the like that are logically connected over a shared media. A media access control protocol operates over the LAN to allow any two devices to communicate at the link level as if they were directly connected. There are a number of LAN architectures, but Ethernet is the most popular LAN protocol for connecting workplace computers on a LAN to other networks. Ethernet is also usually the final step in connecting home computers to other networks using regular, cable, and ADSL (Asymmetric Digital Subscriber Line) modems. The original coaxial cable Ethernet, developed by Robert Metcalfe in 1973, has been largely supplanted by the less expensive and easier to use twisted-pair networks. The IEEE 802.11 wireless LAN is a wireless protocol that is almost as popular as Ethernet.

Wireless Networks. A laptop computer's initial connection to a network is often through a wireless device. The most popular wireless network is the IEEE 802.11, which provides a reliable and secure connection to other networks by using an access point, a radio transmitter/receiver that usually includes border-routing capabilities. The Bluetooth network is another wireless network that is often used for peer-to-peer connection of cameras and cell phones to a computer and then through the computer to other networks. Cell phones also provide network connectivity of computing devices to other networks. They use a variety of techniques, but most of them include using a cell phone communications protocol, such as code division multiple access (CDMA) or Global System for Mobile Communications (GSM), and then a modem.

APPLICATIONS AND PRODUCTS

Computer networks have many applications. Some improve the operation of networks, while others provide services to people who use the network. The World Wide Web supplies many exciting new applications, and technologies such as cloud computing appear to be ready to revolutionize computing in the future.

Network Infrastructure. Computer networks are complex, and many applications have been developed to improve their operation and management. A typical example of the software developed to manage networks is Cisco's Internetwork Operating System (IOS) software, an operating system for routers that also provides full support for the routing functions. Another example of software developed for managing computer networks is the suite of software management programs provided by IBM's Tivoli division.

Communications and Social Networking. Communications programs are among the most popular computer applications in use. Early file-sharing applications, such as FTP, retain their popularity, and later applications such as Flickr, which shares photographs, and eDonkey2000, which shares all types of media files, are used by all. Teleconferencing is used to create virtual workplaces; and Voice over Internet Protocol (VoIP) allows businesses and home users to use the Internet for digital telephone service with services provided by companies such as Vonage.

The earliest computer network communications application, e-mail, is still the largest and most successful network application. One accesses e-mail services through an e-mail client. The client can be a stand-alone program, such as Outlook Express or Eudora, or it can be browser based, using Internet Explorer or Foxfire. The e-mail server can be a proprietary system such as Lotus Notes or one of the many Web-based e-mail programs, such as Hotmail, Gmail or Yahoo! Mail.

One of the most important types of applications being developed for computer networks is social networking sites. Facebook is probably the most widely used social network, although sites such as MySpace and Twitter are also very popular. Facebook was conceived by Mark Zuckerberg in 2004 while he was still a student at Harvard University. Facebook members can set up a profile that they can share with other members. The site supports a message service, event scheduling service, and news feed. By setting up a friend relationship with other members, one can quickly rekindle old friendships and acquaintances.

Fascinating Facts About Computer Networks

- Two students, Richard Bartle and Roy Trubshaw, went online in 1978 with Multi-User Dungeons (MUD), a game that allowed two users to compete against each other. Extensions of MUD in the 2010's have many of the features of social networks.

- Robert Metcalfe's doctoral thesis at Harvard University defined Ethernet, but it was initially rejected. In 1973, an extended version was accepted.

- In 1971, Ray Tomlinson sent the first e-mail over ARPANET, demonstrating that it was possible to exchange e-mail between different computer systems.

- Microsoft embedded a TCP/IP stack as a part of its Windows 95 operating system. When combined with an earlier government requirement that all government computers support TCP/IP, this made TCP/IP the only wide-area network.

- In 1990, Tim Berners-Lee demonstrated that a browser could access content on a Web server, thus creating the World Wide Web.

- In 1991, James Gosling created Java as a computer language to support embedded systems such as a home security system. Later, it was adapted to Web programming and became a big success.

- In the early 1980's, court decisions and changes in telecommunications laws eliminated the high tariffs charged on telecommunications, which facilitated the development of wide-area networking.

Although Zuckerberg's initial goal was to create a friendly and interactive environment for individuals, it has fast become a great way to promote businesses, organizations, and even politicians. As of 2010, Facebook did not charge a subscription fee and had a reasonably successful business model selling customer information and advertising as well as working with partners.

Cloud Computing. Cloud computing is a generic name for applications on the Web that are implemented as a Web service and used by desktop computers, wireless-connected laptops, and smartphones. A Web service application exposes its functionality by placing a description of itself on a Universal Description, Discovery, and Integration (UDDI) server. When a Web application, for example, a Web page in a browser, requests data from the Web service, the service sends the data to the user over the Internet, just as if the service was on the user's computer. The real appeal of the Web service is the promise of true interoperability of applications running on a wide variety of computing devices and requesting services from an equally wide variety of Web servers.

As an example of a useful Web service, assume that an individual has written a Web application to select the best automobile insurance to purchase. The application would return key information about each insurance company and its rates. Other programmers could develop similar applications, but there would be differences in some of the details and the cost for accessing the application. An enterprising smartphone developer could then create an application to assist people in buying a new car, and as a part of the interface, display the automobile insurance information. The smartphone developer could insert the individual's automobile insurance Web service information, or that of a competitor, in the automobile purchase interface. The smartphone user would see the auto insurance data displayed, with no idea that the information was coming from the first individual's application.

Next to e-mail, the most common application is word processing, and Microsoft Word is the dominant word processor. Microsoft has announced Azure Word, part of Azure Office, as a version of Word that will execute in the cloud. Because most of the work of Azure Word will be done on Microsoft servers and customers will pay for use at the time of use, Azure Word should run on smartphones almost as well as

on a desktop. Google has already released a word processor of its own to run in the cloud, and as more and more applications are produced to run on the cloud, it is predicted that this will become the dominant form of computing, replacing the desktops of the early twenty-first century.

IMPACT ON INDUSTRY

Although computer networks are important for the communications infrastructure in each country, internetworking (networking between countries) is almost as important. For example, when someone does a search using Google for the best air cleaner to purchase, results are returned from the United States, Sweden, Switzerland, and China. This would not be possible unless the underlying networks were able to work together. Although early networking hardware and software were largely produced in the United States, there were significant contributions from Europe as well. For example, the token-ring LAN was developed in England and the X.25 standard was developed in Europe. By the twenty-first century, computer network hardware and software were being produced all over the world.

Government, Industry, and University Research. Most governments recognize the importance of

Computer server room in a university. (Colin Cuthbert/ Photo Researchers, Inc.)

having a good national computer network and support research to develop networks and standards for networking. One of the best examples of this is ARPANET, a network developed by the United States government to connect the Department of Defense laboratories with university scientists. ARPANET started in 1969 with a few nodes; by 1970, it was growing at the rate of one node a month; and by 1975, it included nodes from nondefense organizations. In 1980, Larry Landweber produced a plan to expand the ARPANET to a network for everyone, including some individual access from telephone networks. When the U.S. government decided to become the world leader in supercomputers in 1985, it also supported the creation of the National Science Foundation Network (NSFNET) as a high-speed backbone for ARPANET and a number of feeder networks to complement ARPANET. By 1998, the government was allowing considerable commercial traffic on NSFNET, and the passage of the National Research and Education Network Act in 1991 led to the creation of the Internet.

The International Telecommunications Union (ITU), located in Geneva, Switzerland and founded in 1965, is an international organization that supports developing international standards for networks on behalf of all countries. It actually guides the development of networks in Europe. The X.25 protocol for wide area networks was approved by the ITU in 1976 and was used for commercial networks in Europe and the United States after its release. The ITU remains active in standards development for networks, including a green networking initiative.

In the early days of computer networking in the United States, most research done at universities and companies was government related. In spite of this, considerable computer network research was done, especially for local area networks. European universities did considerable research in basic network theory and applications, with the development of token-ring networks in English universities being a prime example.

Computer Network Companies. A great deal of the network research is done by industries. In 1979, Robert Metcalfe cofounded 3Com, a company to implement the Ethernet system he invented, and it remains active. In 1990, a twisted-pair implementation of Ethernet, called 10BaseT, was produced and soon became the most widely deployed implementation

of Ethernet. Novell, founded in Utah in 1979, developed a network operating system, Netware, that ran on top of Ethernet (and other LAN protocols as well) and was the dominant LAN during the 1980's and 1990's. IBM developed the Token-Ring network in the mid-1980's, and although it never achieved commercial acceptance, it remains in use. Digital Equipment Corporation's DecNet and IBM's SNA were mainstays for mainframe networks in the 1980's, but these companies, and many more, have come to produce a wide variety of TCP/IP networks.

Microsoft entered the LAN market in 1984 with its production of a LAN to support IBM personal computer networks, but Microsoft and IBM decided to develop separate products for networks with the introduction of the Personal System/2 (PS/2). Still, Microsoft's highly successful Window's New Technology (NT) network operating system and IBM's successful LAN Manager network operating system for the PS/2 had much in common. Both Microsoft and IBM continue to be leaders in research in computer networking.

Considerable research is done in computer networking by smaller companies. Cisco leads in router research and many other areas, such as wireless access points, video-conferencing equipment, and network management software. Symantec and MacAfee develop a great deal of security software for networks. Google, Yahoo!, and Microsoft spend millions every year on research and development to develop a better search engine.

CAREERS AND COURSE WORK

A major in computer science is the traditional way to prepare for a computer networking job. Students first need courses in ethics, mathematics, and physics to form the basis for a computer science degree. Then they take about thirty-six hours of computer hardware and software courses. Those getting a computer science degree often take jobs developing network software or managing a network.

A major in information systems is another way to prepare for a computer networking job. Students must take courses in mathematics and business as a background for this degree. Then they take about thirty hours of courses on information systems development. Those getting degrees in computer information systems often take jobs as network managers, especially in small businesses.

In addition to the traditional academic programs that prepare someone for a computer networking job, a large number of professional training programs result in certification. Novell was the first company to develop a certification program for its NetWare network operating system. Microsoft followed with its Microsoft Certified Professional program for Windows Server operating system, which includes networking. The Cisco Certified Network Professional program is a large program that covers all networking, not just routers.

SOCIAL CONTEXT AND FUTURE PROSPECTS

The development of computer networks and network applications has often resulted in some sticky legal issues. File exchange programs, such as Flickr, are very popular but can have real copyright issues. Napster developed a very successful peer-to-peer file exchange program but was forced to close after only two years of operation by a court decision in 2001. Legislation has played an important role in the development of computer networks in the United States. The initial deregulation of the communications system and the breakup of AT&T in 1982 reduced the cost of the base communications system for computer networks and thus greatly increased the number of networks in operation. Opening up NSFNET for commercial use in 1991 made the Internet possible.

Networking and access to the Web have been increasingly important to business development and improved social networking. The emergence of cloud computing as an easy way to do computing promises to be just as transformative. Although there are still security and privacy issues to be solved, most believe that in the future, people will be using mobile devices to access a wide variety of Web services through the cloud. For example, using a smartphone, people will be able to communicate with friends and associates, transact business, compose letters, pay bills, read books, and watch films.

George M. Whitson III, B.S., M.S., Ph.D.

FURTHER READING

Cerf, V. G., and R. E. Kahn. "A Protocol for Packet Network Interconnection." *IEEE Transactions on Communication Technology* 22 (May, 1974): 627-641. The article that defined the Internet and a TCP/IP network.

Dumas, M. Barry, and Morris Schwartz. *Principles of Computer Networks and Communications.* Upper Saddle River, N.J.: Pearson/Prentice Hall, 2009. Covers all aspects of computer networks, including signal carriers and fundamentals, local area networks, wide area networks, protocols, wireless networks, security, and network design and implementation.

Forouzan, Behrouz. *Data Communications and Networking.* 4th ed. New York: McGraw-Hill, 2007. A complete, readable text that provides excellent coverage of the physical implementation of networking components.

Metcalfe, Robert, and David Boggs. "Ethernet: Distributed Packet Switching for Local Computer Networks." *Communications of the ACM* 19, no. 7 (July, 1976): 395-404. Described the fundamentals of Ethernet and provided the foundation for local area networks.

Stallings, William. *Data and Computer Communications.* 9th ed. Upper Saddle River, N.J.: Prentice Hall, 2011. An excellent overview of the entire field of computer networks. Includes many diagrams illustrating network architectures.

Tanenbaum, Andrew. *Computer Networks.* 5th ed. Upper Saddle River, N.J.: Prentice Hall, 2010. An excellent theoretical coverage of computer networks.

WEB SITES

Computer Society
http://www.computer.org

Institute of Electrical and Electronics Engineers
http://www.ieee.org

Internet Society
http://www.isoc.org/internet/history/brief.shtml

See also: Communications; Computer Engineering; Computer Science; Information Technology; Internet and Web Engineering; Software Engineering; Telecommunications; Telephone Technology and Networks; Wireless Technologies and Communication.

COMPUTER SCIENCE

FIELDS OF STUDY

Computer science; computer engineering; electrical engineering; mathematics; artificial intelligence; computer programming; human-computer interaction; software engineering; databases and information management; bioinformatics; game theory; scheduling theory; computer networking; computer and data security; computer forensics; computer simulation and modeling; computation methods (including algorithms); ethics; computer graphics; multimedia.

SUMMARY

Computer science is the study of real and imagined computers, their hardware and software components, and their theoretical basis and application. Almost every aspect of modern life involves computers in some way. Computers allow people to communicate with people almost anywhere in the world. They control machines, from industrial assembly equipment to children's toys. Computers control worldwide stock market transactions and analyze and regulate those markets. They allow physicians to treat patients even when the physicians and patients cannot meet in the same location. Researchers endeavor to make the computers of science fiction everyday realities.

KEY TERMS AND CONCEPTS

- **Algorithm:** Step-by-step procedure for solving a problem; a program can be thought of as a set of algorithms written for a computer.
- **Artificial Intelligence:** Study of how machines can learn and mimic natural human and animal abilities; it can involve enhancing or going beyond human abilities.
- **Binary System:** Number system that has only two digits, 0 and 1; forms the basis of computer data representation.
- **Computer:** Real or theoretical device that accepts data input, has a running program, stores and manipulates data, and outputs results.
- **Computer Network:** Group of computers and devices linked together to allow users to communicate and share resources.
- **Database:** System that allows for the fast and efficient storage and retrieval of vast amounts of data.
- **Hardware:** Tangible part of a computer, consisting mostly of electronic and electrical components and their accessories.
- **Programming:** Using an artificial programming language to instruct a computer to perform tasks.
- **Software:** Programs running within a computer.
- **Theoretical Computer:** Imaginary computer made up by researchers to gain insights into computer science theory.

DEFINITION AND BASIC PRINCIPLES

Computer science is the study of all aspects of computers, applied and theoretical. However, considerable disagreement exists over the definition of such basic terms as computer science, computer, hardware, and software. This disagreement can be seen as a testament to the vitality and relative youth of this field. The Association for Computing Machinery's computing classification system, developed in 1998, is an attempt to define computer science.

The science part of computer science refers to the underlying theoretical basis of computers. Broadly speaking, computation theory, part of mathematics, looks at what mathematical problems are solvable. In computer science, it focuses on which problems can be solved using a computer. In 1936, English mathematician Alan Turing attempted to determine the limits of mechanical computation using a theoretical device, now called a Turing machine. Mathematics also forms the basis for research in programming languages, artificial languages developed for computers. Because computers do not have the ability to think like humans, these languages are very formal with strict rules on how programs using these languages can be written and used.

Another part of the underlying structure of computer science is engineering. The physical design of computers involves a number of disciplines, including electrical engineering and physics. The quest for smaller, faster, more powerful devices has led to research in fields such as quantum physics and nanotechnology.

Computer science is not just programming. Computer scientists view programming languages as tools to research such issues as how to create better programs, how information is represented and used by computers, and how to do away with programming languages altogether and instead use natural languages (those used by people).

BACKGROUND AND HISTORY

Computer science can be seen as the coming together of two separate strands of development. The older strand is technology and machines, and the newer is the theoretical one.

Technology. The twentieth century saw the explosive rise of computers. Computers began to incorporate electronic components instead of the mechanical components that had gone into earlier machines, and they made use of the binary system rather than the decimal system. During World II, major developments in computer science arose from the attempt to build a computer to control antiaircraft artillery. For this project, the Hungarian mathematician John von Neumann wrote a highly influential draft paper on the design of computers. He described a computer architecture in which a program was stored along with data. The decades after World War II saw the development of programming languages and more sophisticated systems, including networks, which allow computers to communicate with one another.

Theory. The theoretical development of computer science has been primarily through mathematics. One early issue was how to solve various mathematical equations. In the eighteenth and nineteenth centuries, this blossomed into research on computation theory. At a conference in Germany in 1936 to investigate these issues, Turing presented the Turing machine concept.

The World War II antiaircraft project resulted not only in the development of hardware but also in extensive research on the theory behind what was being done as well as what else could be done. Out of this ferment eventually came such work as Norbert Wiener's cybernetics and Claude Shannon's information theory. Modern computer science is an outgrowth of all this work, which continues all around the world in industry, government, academia, and various organizations.

HOW IT WORKS

Computer Organization and Architecture. The most familiar computers are stand-alone devices based on the architecture that von Neumann sketched out in his draft report. A main processor in the computer contains the central processing unit, which controls the device. The processor also has arithmetic-processing capabilities.

Electronic memory is used to store the operating system that controls the computer, numerous other computer programs, and the data needed for running programs. Although electronic memory takes several forms, the most common is random access memory (RAM), which, in terms of size, typically makes up most of a computer's electronic memory. Because electronic memory is cleared when the computer is turned off, some permanent storage devices, such as hard drives and flash drives, were developed to retain these data.

Computers have an input/output (I/O) system, which communicates with humans and with other devices, including other computers. I/O devices include keyboards, monitors, printers, and speakers.

The instructions that computers follow are programs written in an artificial programming language. Different kinds of programming languages are used in a computer. Machine language, the only language that computers understand, is used in the processor and is composed solely of the binary digits 0 and 1.

Computers often have subsidiary processors that take some of the processing burden away from the main processor. For example, when directed by the main processor, a video processor can handle the actual placing of images on a monitor. This is an example of parallel processing, in which more than one action is performed simultaneously by a computer. Parallel processing allows the main processor to do one task while the subsidiary processors handle others. A number of computers use more than one main processor. Although one might think that doubling the number of processors doubles the processing speed, various problems, such as contention for the same memory, considerably reduce this advantage. Individual processors also use parallel processing to speed up processing; for example, some processors use multiple cores that together act much like individual processors.

Modern computers do not actually think. They are best at performing simple, repetitious operations incredibly fast and accurately. Humans can get bored and make mistakes, but not computers (usually).

Mathematics. Mathematics underlies a number of areas of computer science, including programming languages. Researchers have long believed that programming languages incorporating rules that are mathematically based will lead to programs that contain fewer errors. Further, if programs are mathematically based, then it should be possible to test programs without running them. Logic could be used to deduce whether the program works. This process is referred to as program proving.

Mathematics also underlies a number of algorithms that are used in computers. For example, computer games have a large math and physics component. Game action is often expressed in mathematical formulas that must be calculated as the game progresses. Different algorithms that perform the same task can be evaluated through an analysis of the relative efficiencies of different approaches to a problem solution. This analysis is mathematically based and independent of an actual computer.

Software. Computer applications are typically large software systems composed of a number of different programs. For example, a word-processing program might have a core set of programs along with programs for dictionaries, formatting, and specialized tasks. Applications depend on other software for a number of tasks. Printing, for example, is usually handled by the operating system. This way, all applications do not have to create their own basic printing programs. The application notifies the operating system (OS), which can respond, for example, with a list of all available printers. If an application user wishes to modify printer settings, the OS contacts the printer software, which can then display the settings. This way, the same interface is always used. To print the application, the user sends the work to the OS, which sends the work to a program that acts as a translator between the OS and the printer. This translator program is called a driver. Drivers work with the OS but are not part of it, allowing for new drivers to be developed and installed after an OS is released.

Machine language, since it consists only of 0's and 1's, is difficult for people to readily interpret. For human convenience, assembly language, which uses mnemonics rather than digits, was developed. Because computers cannot understand this language, a program called an assembler is used to translate assembly language to machine language.

For higher-level languages such as C++ and Java, a compiler program is used to translate the language statements first to assembly language and then to machine language.

APPLICATIONS AND PRODUCTS

Computers have penetrated into nearly every area of modern life, and it is hard to think of an area in which computer technology is not used. Some of the most common areas in which computers are used include communications, education, digitalization, and security.

Communication. In the early twentieth century, when someone immigrated to the United States, they knew that communication with those whom they were leaving behind would probably be limited. In the modern world, people across the world can almost instantly communicate with anyone else as long as the infrastructure is available. Products such as Skype and Vonage allow people to both see and talk to other people throughout the world. Instead of traveling to the other person's location, people can hold meetings through software such as Cisco's WebEx or its TelePresence, which allows for face-to-face meetings.

One of the most far-reaching computer applications is the Internet, a computer network that includes the World Wide Web. People are increasingly relying on the Internet to provide them with news and information, and traditional sources, such as printed newspapers and magazines are declining in popularity. Newer technologies such as radio and television have also been affected. The Internet seems to be able to provide everything: entertainment, information, telephone service, mail, business services, and shopping.

Telephones and the way people think of telephones have also been revolutionized. Telephones have become portable computing devices. Apple iPhones and Research in Motion (RIM) Blackberry phones are smart phones, which provide access to e-mail and the Internet; offer Global Positioning Systems (GPS) that guide the user to a selected destination; download motion pictures, songs, and other entertainment; and shoot videos and take photographs. Applications (apps) are a burgeoning industry for the iPhone. Apps range from purely entertaining applications to those that provide useful information, such as weather and medical data. Some people no longer have traditional land-line telephones and rely instead on their cell phones.

Smart phones demonstrate convergence, a trend toward combining devices used for individual activities into a single device that performs various activities and services. For example, increasingly high-definition televisions are connected to the Internet. Devices such as Slingbox allow users, wherever they are, to control and watch television on their desktop computer, laptop computer, or mobile phone. Digital video recorders (DVRs) allow users to record a television program and watch it at a later time or to pause live broadcasts. DVRs can be programmed to record shows based on the owner's specifications. Televisions are becoming two-way interactive devices, and viewers are no longer passive observers. Although televisions and DVRs are not themselves computers, they contain microprocessors, which are miniature computers.

Networking is possible through a vast network infrastructure that is still being extended and improved. Companies such as Belkin provide the cable, Cisco the equipment, and Microsoft the software that supports this infrastructure for providers and users. It is common for homes to have wireless networks with a number of different devices connected through a modem, which pulls out the Internet signal from the internet service provider (ISP), and a router, which manages the network devices.

Education. Distance education through the Internet (online education) is becoming more and more commonplace. The 2010 Horizon Report noted that people expect anywhere, anytime learning. Learning (or course) management systems (LMS or CMS) are the means by which the courses are delivered. These can be proprietary products such as Blackboard or Desire2Learn, or nonproprietary applications such as Moodle. Through these management systems, students can take entire courses without ever meeting their instructor in person. Applications such as Wimba and Elluminate allow classes to be almost as interactive as traditional classes and record those sessions for future viewing. Those participating can be anywhere as long as they can connect to the Internet. Through such software, students can ask questions of an instructor, student groups can meet online, and an instructor can hold review sessions.

Digitalization. This area is concerned with how data are translated for computer usage. Real-world experience is usually thought of as continuous phenomena. Long-play vinyl records (LPs) and magnetic tapes captured a direct analogy for the original sound; therefore, any change to the LP or tape meant a change to the sound quality, usually introducing distortions. Each copy (generation) that was made of that analogue signal introduced further distortions. LPs were several generations removed from the original recording, and additional distortions were added to accommodate the sound to the LP medium.

Rather than record an analogue of the original sound, digitalization translates the sound into discrete samples that can be thought of as snapshots of the original. The more samples (the higher the sampling rate) and the more information captured by each sample (the greater the bit depth), the better the result. Digitalization enables the storage of data of all types. The samples are encoded into one of the many binary file formats, some of which are open and some proprietary. MPEG-1 Audio Layer 3 (MP3) is an open encoding standard that is used in a number of different devices. It is a lossy standard, which means that the result is of a lower quality than the original. This tradeoff is made to gain a smaller file size, enabling more files to be stored. The multimedia equivalent is MPEG-4 Part 14 (MP4), also a lossy format. Lossless formats such as Free Lossless Audio Codec (FLAC) and Apple Lossless have better sound quality but require more storage space.

These formats and others have resulted in a proliferation of devices that have become commonplace. iPods and MP3 players provide portable sound in a relatively small container and, in some cases, play videos. Some iPods and computers can become media centers, allowing their users to watch television shows and films.

In a relatively short span of time, digital cameras have made traditional film cameras obsolete. Camera users have choices as to whether they want better quality (higher resolution) photographs or lower resolution photos but the capacity to store more of them in the same memory space. Similarly, DVDs have made tape videos obsolete. Blu-ray, developed by Sony, allows for higher quality images and better sound by using a purple laser rather than the standard red laser. Purple light has a higher frequency than red light, which means that the size (wavelength) of purple light is smaller than that of red light. This can be visualized as allowing the purple

Fascinating Facts About Computer Science

- Konrad Zuse, working in Germany in the 1930's and 1940's, developed what many consider the first electronic computer. He also invented a programming language, Plankalkül, which is considered the first high-level computer language. His work was not generally known until after World War II.

- In 1973, a U.S. Federal District Court ruled that John Atanasoff, a professor at Iowa State University, was the inventor of the first electronic digital computer, the Atanasoff-Berry Computer, designed in 1939 and successfully tested in 1942.

- The 1952 U.S. presidential election was the first time that a computer was used to predict an outcome. The CBS News computer predicted that Dwight Eisenhower would defeat Adlai Stevenson, even though a number of polls predicted Stevenson's victory. Because of this discrepancy, the projection was not announced for some time.

- The UNIVAC, used to predict the outcome of the 1952 U.S. presidential election, was a warehouse-sized computer. Most modern laptop computers exceed the UNIVAC's computing power and capabilities.

- A Global Positioning System device with augmented reality capabilities can not only give directions to a restaurant but also provide its menu. If the device can access personal information from such sources as Facebook, it can be tailored to make restaurant recommendations based on an individual's preferences.

- The RoboBees project of the School of Engineering and Applied Sciences at Harvard University is attempting to create robotic bees. These bees are designed not only to fly but also to reproduce other bee activities, including colony behaviors. The project is funded by the National Science Foundation.

- Users of mobile devices use touch-screen technology, which requires a relatively large display for finger access. Patrick Baudisch's nanoTouch device has a touchpad on the back of the device, allowing it to be much smaller.

laser to get into smaller spaces than the red laser can; thus the 0's and 1's on a Blu-ray disc can be smaller than those on a standard DVD disc, so more data can be stored on the same size disc. Because the recordings are encoded in binary, they can be easily and exactly copied and manipulated without any distortion, unless that distortion is deliberately introduced, as with lossy file formats.

Security. The explosion of digital communication and products has caused some individuals to illegally, or at least unethically, exploit people's dependence on these technologies. These people write malware, programs designed to harm or compromise devices and send spam (digital junk mail). Malware includes viruses (programs that embed themselves in other programs and then make copies of themselves and spread), worms (programs that act like viruses but do not have to be embedded in another program), Trojan horses (malicious programs, often unknowingly downloaded with something else), and key loggers (programs that record all keystrokes and mouse movements, which might include user names and passwords). The key-stroke recorder is a type of spyware, programs that collect information

about the device that they are on and those using it without their knowledge. These are often associated with Internet programs that keep track of people's browsing habits.

Spam can include phishing schemes, in which a person sends an e-mail that appears to come from a well-known organization such as a bank. The e-mail typically states that there is a problem with the recipient's account and to rectify it, the recipient must verify his or her identity by providing sensitive information such as his or her user name, password, and account number. The information, if given, is used to commit fraud using the recipient's identity.

Another type of spam is an advance-fee fraud, in which the sender of an e-mail asks the recipient to help him or her claim a substantial sum of money that is due the sender. The catch is that the recipient must first send money to "show good faith." The money that the recipient is to get in return never arrives. Another asks the recipient to cash a check and send the scammer part of the money. By the time the check is determined to be worthless, the recipient has sent the scammer a considerable sum of money. These scams are a form of social engineering, in which the scammer first obtains the trust of the e-mail recipient so that he or she will help the scammer.

These problems are countered by companies such as Symantec and McAfee, which produce security suites that contain programs to manage these threats. These companies keep databases of threats and require that their programs be regularly updated to be able to combat the latest threats. Part of these suites is a firewall, which is designed to keep the malware and spam from reaching the protected network.

IMPACT ON INDUSTRY

It is hard to think of any industry that has not felt the impact of computers in a major, perhaps transforming, way. The United States, the United Kingdom, and Japan led research and development in computer science, but Israel, the European Union, China, and Taiwan are increasingly engaging in research. Although India is not yet a major center of research, it has been very successful in setting up areas in Mumbai and elsewhere that support technology companies from around the world. Other countries, such as Pakistan, have tried to duplicate India's success. Indian and Chinese computer science graduates have been gaining a reputation worldwide as being very well prepared.

Industry and Business Sectors. Computer science has led to the creation of electronic retailing (e-tailing). Online sales rose 15.5 percent in 2009, while brick and mortar (traditional) sales were up 3.6 percent from the previous year. Online-only outlets such as Amazon.com have been carving out a significant part of the retail market. Traditional businesses, such as Macy's and Home Depot, are finding that they must have an online presence. Online sales for the 2009 Christmas season were estimated at $27 billion.

Some traditional industries are finding that survival has become difficult. Newspapers and magazines have been experiencing a circulation decline, as more and more people read the free online versions of these publications or get their news and information from other Internet sites. Some newspapers, such as *The Wall Street Journal*, sell subscriptions to their Web sites or offer subscriptions to enhanced Web sites. Travel agencies are losing business, as their clients instead use online travel sites such as Expedia. com or the Web sites of airlines and hotels to make their own travel arrangements. DVD rental outlets such as Blockbuster (which declared Chapter 11

bankruptcy in September, 2010) have been declining in the face of competition from Netflix, which initially offered a subscription service and sent DVDs through the U.S. Postal Service. As consumers have turned to streaming television programs and films, Netflix has entered this business, but faces competition from Amazon.com, Google, Apple, and cable and broadband television providers.

CD and DVD sales have been affected not only by online competition but also by piracy. Since digital sources can be copied without error, copying of CDs and DVDs has become rife. Online file sharing through such applications as BitTorrent has caused the industry to fight back through highly publicized lawsuits and legislation. The Digital Millennium Copyright Act of 1998 authorized, among many other measures, the use of Digital Rights Management (DRM) to restrict copying. Many DVDs and downloads use DRM protection. However, a thriving industry has sprung up to block the DRM features and allow copying. This is a highly contentious area, which engendered more controversy with the passage of the 2010 Digital Economy Act in the United Kingdom.

A thriving industry has been developed using virtual reality. Linden Research's Second Life is one of the most popular virtual reality offerings. Members immerse themselves in a world where they have the opportunity to be someone that they might never be in real life. A number of organizations use Second Life for various purposes, including education.

Professional Organizations. Probably the two largest computer organizations are the Association for Computer Machinery (ACM), based in New York, and Computer Society of the Institute of Electrical and Electronic Engineers (IEEE), based in Washington, D.C. The ACM Special Interest Groups explore all areas of computer science. When computer science higher education was a new frontier and the course requirements for a degree varied widely, it was the ACM that put forward a curriculum that has become the standard for computer science education in colleges and universities in the United States. Its annual Turing Award is one of the most prestigious awards for computing professionals.

The IEEE partners with the American National Standards Institute (ANSI) and the International Organization for Standardization on many of its wireless specifications, such as 1394 (FireWire) and Ethernet (802.3). Both computer organizations sponsor various contests for students from middle school to college.

Research. The National Science Foundation, a federal agency created in 1950, is responsible for about 20 percent of all federally funded basic research at colleges and universities in the United States. An example project is the California Regional Consortium for Engineering Advances in Technological Excellence (CREATE), a consortium of seven California community colleges, which works toward innovative technical education in community colleges. The consortium has funded a number of computer-networking initiatives.

The Defense Advanced Research Projects Agency (DARPA), part of the U.S. Department of Defense, funds research into a number of areas, including networks, cognitive systems, robotics, and high-priority computing. This agency initially funded the project that became the basis of the Internet.

CAREERS AND COURSE WORK

Computer science degrees require courses in mathematics (including calculus, differential equations, and discrete mathematics), physics, and chemistry as well as the usual set of liberal arts courses. Lower-division computer science courses are usually heavily weighted toward programming. These might include basic programming courses in C++ or Java, data structures, and assembly language. Other courses usually include computer architecture, networks, operating systems, and software engineering. A number of specialty areas, such as computer engineering, affect the exact mix of course work.

Careers in software engineering typically require a bachelor's degree in computer science. The U.S. Bureau of Labor Statistics sees probable growth in the need for software engineers. Computer skills are not all that employers want, however. They also want engineers to have an understanding of how technology fits in with their organization and its mission.

Those who wish to do research as computer scientists will generally require a doctorate in computer science or some branch of computer science. Computer scientists are often employed by universities, government agencies, and private industries such as IBM. AT&T Labs also has a tradition of Nobel Prize-winning research and was where the UNIX operating system and the C and C++ programming languages were developed.

SOCIAL CONTEXT AND FUTURE PROSPECTS

Computers are considered an essential part of modern life, and their influence continues to revolutionize society. The advantages and disadvantages of an always-wired society are being keenly debated. People not only are connected but also can be located by computer devices with great accuracy. Because people are doing more online, they are increasingly creating an online record of what they have done at various times in their lives. Many employers routinely search online for information on job candidates, and people are finding that comments they made online and the pictures they posted some years ago showing them partying are coming back to haunt them. People are starting to become aware of these pitfalls and are taking actions such as deleting these possibly embarrassing items and changing their settings on social network sites such as Facebook to limit access.

These issues bring up privacy concerns, including what an individual's privacy rights are on the Internet and who owns the data that are being produced. For example, in 1976, the U.S. Supreme Court ruled that financial records are the property of the financial institution not the customer. This would seem to suggest that a company that collects information about an individual's online browsing habits—not the individual whose habits are being recorded—owns that information. Most of these companies state that individuals are not associated with the data that they sell and all data are used in aggregate, but the capacity to link data with specific individuals and sell the information exists. Other questions include whether an employer has a right to know whether an employee visits a questionable Internet site. Certainly the technology to accomplish this is available.

These concerns lead to visions of an all-powerful and all-knowing government such as that portrayed in George Orwell's *Nineteen Eighty-Four* (1949). With the Internet a worldwide phenomenon, perhaps no single government can dictate regulations governing the World Wide Web and enforce them. The power of technology will only grow, and theses issues and many others will only become more pressing.

Martin Chetlen, B.S.E.E, M.C.S.

FURTHER READING

Brooks, Frederick P., Jr. *The Mythical Man-Month: Essays on Software Engineering.* Anniversary ed.

Boston: Addison-Wesley, 2008. Brooks, who was the leader of the IBM 360 project, discusses computer and software development. The first edition of this book was published in 1975, and this 2008 edition adds four chapters.

Gaddis, Tony. *Starting Out With C++*. 6th ed. Boston: Addison-Wesley, 2009. A good introduction to a popular programming language.

Goldberg, Jan, and Mark Rowh. *Great Jobs for Computer Science Majors*. Chicago: VGM, 2003. Designed for a young and general audience, this career guide includes tools for self-assessment, researching computer careers, networking, choosing schools, and understanding a variety of career paths.

Henderson, Harry. *Encyclopedia of Computer Science and Technology*. Rev. ed. New York: Facts On File, 2009. An alphabetical collection of information about computer science. Contains bibliography, chronology, list of awards, and a list of computer organizations.

Kidder, Tracy. *The Soul of the New Machine*. 1981. Reprint. New York: Back Bay Books, 2000. Chronicles the development of a new computer and the very human drama and comedy that surrounded it.

Schneider, G. Michael, and Judith L. Gersting. *Invitation to Computer Science*. 5th ed. Boston: Course Technology, 2010. Gives a view of the breadth of computer science, including social and ethical issues.

WEB SITES
Association for Computer Machinery
http://www.acm.org

Computer History Museum
http://www.computerhistory.org

Computer Society
http://www.computer.org

Institute of Electrical and Electronic Engineers
http://www.ieee.org

See also: Communications; Computer Engineering; Computer Languages, Compilers, and Tools; Computer Networks; Information Technology; Internet and Web Engineering; Parallel Computing; Robotics; Software Engineering; Virtual Reality.

CRACKING

FIELDS OF STUDY

Chemistry; engineering; chemical engineering; chemical process modeling; fluid dynamics; heat transfer; distillation design; mechanical engineering; environmental engineering; control engineering; process engineering; industrial engineering; electrical engineering; safety engineering; plastics engineering; physics; thermodynamics; mathematics; materials science; metallurgy; business administration.

SUMMARY

In the petroleum industry, cracking refers to the chemical conversion process following the distillation of crude oil, by which fractions and residue with long-chain hydrocarbon molecules are broken down into short-chain hydrocarbons. Cracking is accomplished under pressure by thermal or catalytic means and by injecting extra hydrogen. Cracking is done because short-chain hydrocarbons, such as gasoline, diesel, and jet fuel, are more commercially valuable than long-chain hydrocarbons, such as fuel and bunker oil. Steam cracking of light gases or light naphtha is used in the petrochemical industry to obtain lighter alkenes, which are important petrochemical raw products.

KEY TERMS AND CONCEPTS

- **Catalytic Cracking:** Use of a catalyst to enhance cracking.
- **Coking:** Most severe form of thermal cracking.
- **Crude Oil:** Liquid part of petroleum; has a wide mix of different hydrocarbons.
- **Distillation:** Process of physically separating mixed components with different volatilities by heating them.
- **Fluid Catalytic Cracker (FCC):** Cracking equipment that uses fluid catalysts.
- **Fractionator:** Distillation unit in which product streams are separated and taken away; also known as fractionating tower, fractionating column, and bubble tower.
- **Fractions:** Product streams obtained after each distillation; also known as cuts.

- **Hydrocracking:** Special form of catalytic cracking that uses extra hydrogen to obtain the end products with the highest values.
- **Regenerator:** Catalytic cracker in which accumulated carbon is burned off the catalyst.
- **Residue:** Accumulated elements of crude oil that remain solid after distillation.
- **Steam Cracking:** Petrochemical process used to obtain lighter alkenes.
- **Thermal Cracking:** Oldest form of cracking; uses heat and pressure.

DEFINITION AND BASIC PRINCIPLES

Cracking is a key chemical conversion process in the petroleum and petrochemical industries. The process breaks down long-chain hydrocarbon molecules with high molecular weights and recombines them to form short-chain hydrocarbon molecules with lower molecular weights. This breaking apart, or cracking, is done by the application of heat and pressure and can be enhanced by catalysts and the addition of hydrogen. In general, products with short-chain hydrocarbons are more valuable. Cracking is a key process in obtaining the most valuable products from crude oil.

At a refinery, cracking follows the distillation of crude oil into fractions of hydrocarbons with different molecule chain lengths and the collection of the heavy residue. The heaviest fractions with the longest molecule chains and the residue are submitted to cracking. For petrochemical processes, steam cracking is used to convert light naphtha, light gases, or gas oil into short-chain hydrocarbons such as ethylene and propylene, crucial raw materials in the petrochemical industry.

Cracking may be done by various technological means, and the hydrocarbons cracked at a particular plant may differ. In general, the more sophisticated the cracking plant, the more valuable its end products will be but the more costly it will be to build. Being able to control and change a cracker's output to conform to changes in market demand has substantial economic benefits.

BACKGROUND AND HISTORY

By the end of the nineteenth century, demand rose for petroleum products with shorter hydrogen molecule chains, in particular, diesel and gasoline to fuel the new internal combustion engines. In Russia, engineer Vladimir Shukhov invented the thermal cracking process for hydrocarbon molecules and patented it on November 27, 1891. In the United States, the thermal cracking process was further developed and patented by William Merriam Burton on January 7, 1913. This doubled gasoline production at American refineries.

To enhance thermal cracking, engineers experimented with catalysts. American Almer McAfee was the first to demonstrate catalytic cracking in 1915. However, the catalyst he used was too expensive to justify industrial use. French mechanical engineer Eugene Jules Houdry is generally credited with inventing economically viable catalytic cracking in a process that started in a Paris laboratory in 1922 and ended in the Sun Oil refinery in Pennsylvania in 1937. Visbreaking, a noncatalytic thermal process that reduces fuel oil viscosity, was invented in 1939. On May 25, 1942, the first industrial-sized fluid catalytic cracker started operation at Standard Oil's Baton Rouge, Louisiana, refinery.

Research into hydrocracking began in the 1920's, and the process became commercially viable in the early 1960's because of cheaper catalysts such as zeolite and increased demand for the high-octane gasoline that hydrocracking could yield. By 2010, engineers and scientists worldwide sought to improve cracking processes by optimizing catalysts, using less energy and feedstock, and reducing pollution.

HOW IT WORKS

Thermal Cracking. If hydrocarbon molecules are heated above 360 degrees Celsius, they begin to swing so vigorously that the bonds between the carbon atoms of the hydrocarbon molecule start to break apart. The higher the temperature and pressure, the more severe this breaking is. Breaking, or cracking, the molecules, creates short-chain hydrocarbon molecules as well as some new molecule combinations. Thermal cracking—cracking by heat and pressure only—is the oldest form of cracking at a refinery. In modern refineries, it is usually used on the heaviest residues from distillation and to obtain petrochemical raw materials.

Modern thermal crackers can operate at temperatures between 440 and 870 degrees Celsius. Pressure can be set from 10 to about 750 pound-force per square inch (psi). Heat and pressure inside different thermal crackers and the exact design of the crackers vary considerably.

Typically, thermal crackers are fed with residues from the two distillation processes of the crude oil. Steam crackers are primarily fed with light naphtha and other light hydrocarbons. After their preheating, often feedstocks are sent through a rising tube into the cracker's furnace area. Furnace temperature and feedstock retention time is carefully set, depending on the desired outcome of the cracking process. Retention time varies from fractions of a second to some minutes.

After hydrocarbon molecules are cracked, the resulting vapor is either first cooled in a soaker or sent directly into the fractionator. There, the different fractions, or different products, are separated by distillation and extracted.

In cokers, severe thermal cracking creates an unwanted by-product. It is a solid mass of pure carbon, called coke. It is collected in coke drums, one of which supports the cracking process while the other is emptied of coke.

Catalytic Cracking. Because it yields more of the desired short-chain hydrocarbon products with less use of energy than thermal cracking, catalytic cracking has become the most common method of cracking. Feedstocks are relatively heavy vacuum distillate hydrocarbon fractions from crude oil. They are preheated before being injected into the catalytic cracking reactor, usually at the bottom of a riser pipe. There they react with the hot catalyst, commonly synthetic aluminum silicates called zeolites, which are typically kept in fluid form. In the reactor, temperatures are about 480 to 566 degrees Celsius and pressure is between 10 and 30 psi.

As feedstock vaporizes and cracks during its seconds-long journey through the riser pipe, feedstock vapors are separated from the catalyst at cyclones at the top of the reactor and fed into a fractionator. There, different fractions condense and are extracted as separate products. The catalyst becomes inactive as coke builds on its surface. Spent catalyst is collected and fed into the regenerator, where coke deposits are burned off. The regenerated catalyst is recycled into the reactor at temperatures of 650 to 815 degrees Celsius.

Hydrocracking. The most sophisticated and flexible cracking process, hydrocracking delivers the

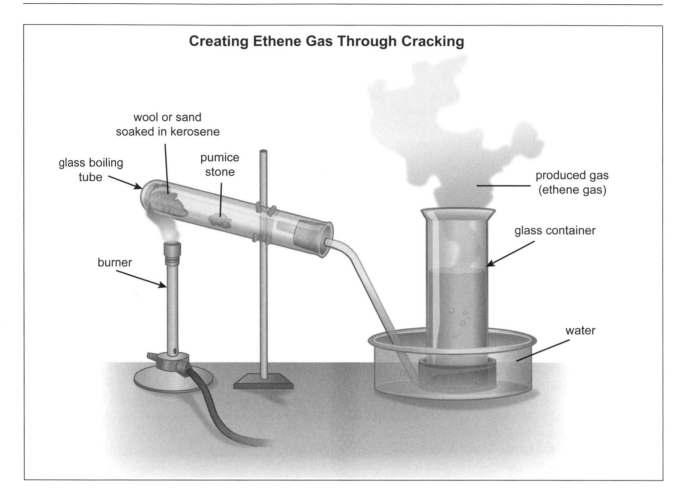

Creating Ethene Gas Through Cracking

wool or sand soaked in kerosene

glass boiling tube

pumice stone

burner

produced gas (ethene gas)

glass container

water

highest value products. It combines catalytic cracking with the insertion of hydrogen. Extra hydrogen is needed to form valuable hydrocarbon molecules with a low boiling point that have more hydrogen atoms per carbon atom than the less valuable, higher boiling point hydrocarbon molecules that are cracked. All hydrocracking involves high temperatures from 400 to 815 degrees Celsius and extremely high pressure from 1,000 to 2,000 psi, but each hydrocracker is basically custom designed.

In general, hydrocracking feedstock consists of middle distillates (gas oils), light and heavy cycle oils from the catalytic cracker, and coker distillates. Feedstock may also be contaminated by sulfur and nitrogen. Feedstocks are preheated and mixed with hydrogen in the first stage reactor. There, excess hydrogen and catalysts convert the contaminants sulfur and nitrogen into hydrogen sulfide and ammonia.

Some initial hydrogenating cracking occurs before the hydrocarbon vapors leave the reactor. Vapors are cooled, and liquefied products are separated from gaseous ones and hydrogen in the hydrocarbon separator. Liquefied products are sent to the fractionator, where desired products can be extracted.

The remaining feedstock at the bottom of the fractionator is sent into a second-stage hydrocracking reactor with even higher temperatures and pressures. Desired hydrocracked products are extracted through repetition of the hydrocracking process. Unwanted residues can go through the second stage again.

APPLICATIONS AND PRODUCTS

There are three modern applications of thermal cracking: visbreaking, steam cracking, and coking.

Visbreaking. Visbreaking is the mildest form of thermal cracking. It is applied to lower the viscosity (increasing the fluidity) of the heavy residue, usually from the first distillation of crude oil. In a visbreaker, the residue is heated no higher than 430 degrees Celsius. Visbreaking yields about 2 percent light gases such as butane, 5 percent gasoline products, 15 percent gas oil that can be catalytically cracked, and 78 percent tar.

Steam Cracking. Steam cracking is used to turn light naphtha and other light gases into valuable petrochemical raw materials such as ethene, propene, or butane. These are raw materials for solvents, detergents, plastics, and synthetic fibers. Because light feed gases are cracked at very high temperatures between 730 and 900 degrees Celsius, steam is added before they enter the furnace to prevent their coking. The mix remains in the furnace for only 0.2 to 0.4 second before being cooled and fractionated to extract the cracked products.

Coking. Coking is the hottest form of cracking distillation residue. Because it leaves no residue, it has all but replaced conventional thermal cracking. Cracking at about 500 degrees Celsius also forms coke. Delayed coking moves completion of the cracking process, during which coke is created, out of the furnace area. To start cracking, feedstock stays in the furnace for only a few seconds before flowing into a coking drum, where cracking can take as long as one day.

In the coke drum, about 30 percent of feedstock is turned into coke deposits, while the valuable rest is sent to the fractionator. In addition to coke, delayed coking yields about 7 percent light gas, 20 percent light and heavy coker naphtha from which gasoline and gasoline products can be created, and 50 percent light and heavy gas oils. Gas oils are further processed through hydrocracking, hydrotreating, or subsequent fluid catalytic cracking, or used as heavy fuel oil. The coke drum has to be emptied of coke every half or full day. To ensure uninterrupted cracking, at least two are used. The coke comes in three kinds, in descending order of value: needle coke, used in electrodes; sponge coke, for part of anodes; and shot coke, primarily used as fuel in power plants.

Flexicoking. Flexicoking, continuous coking, and fluid coking are technological attempts to recycle coke as fuel in the cracking process. Although these cokers are more efficient, they are more expensive to build.

Fluid Catalytic Cracking. Because of its high conversion rate of vacuum wax distillate into far more valuable gasoline and lighter, olefinic gases, fluid catalytic crackers (FCCs) are the most important crackers at a refinery. By 2007, there were about four hundred FCCs worldwide working continuously at refineries. Together, they cracked about 10.6 million barrels of feedstock each day, about half of which was cracked in the United States. FCCs are essential to meet the global demand for gasoline.

Design of individual FCCs, while following the basic principle of fluid catalytic cracking, varies considerably, as engineers and scientists continuously seek to improve efficiency and lessen the environmental impact. By 2010, there were five major patents for FCCs that arranged the reactor, where cracking occurs, and the regenerator for the spent catalyst side by side. One major patent places the reactor atop the generator.

The typical products derived from vacuum distillate conversion in a FCC are about 21 percent olefinic gases, often called liquefied petroleum gas; 47 percent gasoline of high quality; 20 percent light cycle (or gas) oil, often called middle distillate; 7 percent heavy cycle (gas) oil, often called heavy fuel oil; and 5 percent coke. Gasoline is generally the most valuable. Light cycle oil is blended into heating oil and diesel, with highest demand for these blends in winter. The more a FCC can change the percentages of its outcome, the higher its economic advantage.

Hydrocracking. Hydrocrackers are the most flexible and efficient cracker, but they have high building and operation costs. High temperatures and pressure require significant energy, and the steel wall of a hydrocracker reactor can be as thick as 15 centimeters. Its use of hydrogen in the conversion process often requires a separate hydrogen generation plant. Hydrocrackers can accept a wide variety of feedstock ranging from crude oil distillates (gas oils or middle distillate) to light and heavy cycle oils from the FCC to coker distillates.

Hydrocracker output typically falls into flexible ranges for each product. Liquefied petroleum gas and other gases can make up 7 to 18 percent. Gasoline products, particularly jet fuel, one of the prime products of the hydrocracker, can be from 28 to 55 percent. Middle distillates, especially diesel and kerosene, can make up from 15 to 56 percent.

Heavy distillates and residuum can range from 11 to 12 percent.

Hydrocracking produces no coke but does have high carbon dioxide emissions. Its products have very low nitrogen and sulfur content.

IMPACT ON INDUSTRY

Cracking is a core activity at any modern refinery. Without modern cracking proesses, the world's demand for gasoline, diesel, kerosene, and jet fuel, as well as for basic petrochemicals, could not easily be met from processing crude oil. About 45 percent of the world's gasoline, for example, comes from FCCs or related units in 2006. Significant public and private research is focusing on improving cracking efficiency.

Government Agencies. Throughout the industrialized and industrializing world, national governments and their agencies, as well as international entities, have promoted more environmentally friendly patterns of hydrocarbon production, processing, and consumption. This has increased the importance of and demand for cracking, as most of the heavy oil distillates and residues once used as heavy fuel oil in power plants are increasingly being replaced by cleaner burning natural gas. National legislation such as the United States Clean Air Act of 1963 with its crucial 1990 amendment imposes strict limits on emissions including those generated during the cracking process, as well as quality specifications for refined products such as cleaner burning gasoline and diesel. Crackers have come under governmental pressure to reduce their often high emissions of carbon dioxide and other gases through innovations or process changes, which means more costs.

Universities and Research Institutes. Particularly in industrialized societies, there is considerable public research into improving existing and developing new cracking processes. Of special interest are catalysts used by FCCs and hydrocrackers. Catalyst research looks into properties of and new materials for catalysts and experiments with optimizing their placement within the reactor, enhancing their recovery, and lengthening their life cycle.

However, a critical gap exists between what works in the laboratory or in small pilot plants and what can be implemented in an industrial setting. Occasionally, promising new catalysts have not risen to their perceived potential, and new reactor designs have proved commercially impractical. University

and institute sciences tend to have a certain bias toward energy-saving and environmentally friendly discoveries, while industrial research tends to be more concerned with raising feedstock economy.

Industry and Business Sectors. Cracking provides a key process for the oil and gas and petrochemical industries. For example, in January, 2010, in the first step of atmospheric distillation, U.S. refineries processed a total of 17.8 million barrels of crude oil per stream day (almost identical to a calendar day). Of this, 8.5 million barrels were subjected to vacuum distillation. Among the distillates and residue from both steps, 2.5 million barrels (15 percent of the original crude oil entering the refinery) were processed in the thermal cracking process of delayed coking (fluid coking and visbreaking were negligible at U.S. refineries). Another 6.2 million barrels (35 percent

---

Fascinating Facts About Cracking

- If engineers had not invented cracking hydrocarbons, the world would need to produce almost double the amount of crude oil to meet the global demand of gasoline, diesel, and jet fuel.

- Crackers at refineries and petrochemical plants operate continuously for twenty-four hours and are typically staffed in three shifts.

- During World War II, newly invented catalytic cracking provided plenty of powerful high-octane gasoline for aircraft engines of the United States and Great Britain, giving their air forces an edge over German and Japanese air forces.

- Crackers are expensive chemical processing plants. The ethane steam cracker and high olefin fluid catalytic cracker constructed in 2009 at Rabigh Refining and Petrochemical Complex in Saudi Arabia cost $850 million to build.

- Reactors of catalytic and steam crackers are incredibly hot, reaching up to 900 degrees Celsius. At these temperatures, oil vapors are blown through the reactor in fractions of a second.

- New pharmaceuticals can contain some ingredients made from raw materials that came out of a steam cracker.

- Catalysts fundamentally support cracking processes. If operating staff mistreats or abuses catalysts, the whole cracker may have to be shut down for catalyst replacement.

of all products from the original crude oil) were submitted to an FCC for catalytic cracking, and 1.8 million barrels (10 percent of the base) were processed in a hydrocracker. Although the balance between FCC and hydrocracker processing is somewhat different in Europe and Asia, where demand for diesel and kerosene is higher than in the United States, the example is representative for the petroleum industry. Overall, about 60 percent of the crude oil that enters a refinery undergoes some form of cracking.

Major Corporations. Because crackers are integral parts of a refinery, require a very large initial investment, and have relatively high operating costs, they are typically owned by the large international or national oil and gas companies that own the refinery. Their construction, however, may be executed by specialized engineering companies, and their design can involve independent consultants. In addition, many cracking facilities are built under license from the company that holds the patent to their design.

Innovative or improved crackers can be added to existing refineries or built together with green field refineries that are increasingly erected in the Middle East, the source of much crude oil, or Asia, the source of much demand. Crackers follow economies of scale. In 2005, to build an FCC with a capacity of 10 million barrels of feedstock per stream day in a Gulf Coast refinery cost from $80 million to $120 million. An FCC with ten times more capacity, however, cost only between $230 million and $280 million. The economy-of-scale factor favors world-scale crackers at equally large refineries.

CAREERS AND COURSE WORK

Cracking is a key refining process, so good job opportunities should continue in this field. Students interested in designing and constructing or operating and optimizing a cracker at a refinery or petrochemical plant should take science courses, particularly chemistry and physics, in high school. The same is true for those who want to pursue research in the field. There also are many opportunities for technicians in building, operationing, and maintaining a cracker.

A bachelor of science degree in an engineering discipline, particularly chemical, electrical, computer, or mechanical engineering, is excellent job preparation in this field. A bachelor of science or arts degree in a major such as chemistry, physics, computer science, environmental science, or mathematics is also a good foundation. An additional science minor is useful.

For an advanced career, any master of engineering degree is a suitable preparation. A doctorate in chemistry or chemical or mechanical engineering is needed if the student wants a top research position, either with a corporation or at a research facility. Postdoctoral work in materials science (engaged in activities such as searching for new catalysts) is also advantageous.

Because cracking is closely related to selecting crude oil for purchase by the refinery, there are also positions for graduates in business or business administration. The same is true for members of the medical profession with an emphasis in occupational health and safety. Technical writers with an undergraduate degree in English or communication may also find employment at oil and engineering companies. As cracking is a global business, career advancement in the industry often requires willingness to work abroad.

SOCIAL CONTEXT AND FUTURE PROSPECTS

As fossilized hydrocarbons are a finite resource, they must be used as efficiently as possible. To this end, cracking at a refinery seeks to create the most valuable products out of its feedstocks derived from crude oil. This is not limited to fuels such as gasoline. The steam crackers of the petrochemical industry create raw materials for many extremely valuable and useful products such as pharmaceuticals, plastics, solvents, detergents, and adhesives, stretching the use of hydrocarbons for consumers.

At the same time, international concern with the negative environmental side effects of some hydrocarbon processes is increasing. Traditionally, crackers such as cokers or even hydrocrackers released a large amount of carbon dioxide or, in the case of cokers, other airborne pollutants as well. To make cracking more environmentally friendly, to save energy, and to convert feedstock efficiently are concerns shared by the public and the petroleum and petrochemical industries. This is especially so for companies when there are commercial rewards for efficient, clean operations.

The possible rise of alternative fuels, replacing some hydrocarbon-based fuels such as gasoline and diesel, would challenge refineries to adjust the output of their crackers. The more flexible hydro-crackers are best suited to meet this challenge.

R. C. Lutz, B.A., M.A., Ph.D.

FURTHER READING

Burdick, Donald. *Petrochemicals in Nontechnical Language.* 3d ed. Tulsa, Okla.: Penn Well, 2010. Accessible introduction. Covers use of petrochemical materials gained from steam cracking. Figures and tables.

Conaway, Charles. *The Petroleum Industry: A Nontechnical Guide.* Tulsa, Okla.: Penn Well, 1999. Chapter 13 provides a short summary of cracking processes.

Gary, James, et al. *Petroleum Refining: Technology and Economics.* 5th ed. New York: CRC Press, 2007. Well-written textbook. Good presentation of all types of cracking in the petrochemical industry. Includes economic aspects of cracking. Five appendixes, index, and photographs.

Leffler, William. *Petroleum Refining in Nontechnical Language.* Tulsa, Okla.: Penn Well, 2008. Covers all major aspects of cracking at a refinery, from chemical foundations to catalytic cracking and hydrocracking.

Meyers, Robert, ed. *Handbook of Petroleum Refining Processes.* 3d ed. New York: McGraw-Hill, 2004. Advanced-level technical compendium covers various industry methods for catalytic cracking, hydro-cracking, visbreaking, and coking. In-depth information for those considering a career in this field.

Sadeghbeigi, Reza. *Fluid Catalytic Cracking Handbook.* 2d ed. Houston, Tex.: Gulf Publishing, 2000. Detailed, technical look at this key cracking process. Figures, conversion table, glossary.

WEB SITES

American Association of Petroleum Geologists
http://www.aapg.org

American Petroleum Institute
http://www.api.org

National Petrochemical and Refiners Association
http://www.npra.org

National Petroleum Council
http://www.npc.org

See also: Chemical Engineering; Coal Liquefaction; Detergents; Distillation; Gasoline Processing and Production; Petroleum Extraction and Processing; Plastics Engineering.

CRIMINOLOGY

FIELDS OF STUDY

Criminal justice; law enforcement; law; psychology; philosophy; sociology; anthropology; political science; social work; forensic science; psychiatry.

SUMMARY

Criminology is the study of crime causation and control. Criminologists attempt to elucidate the characteristics and motivation of criminals—from the most mundane and innocuous (scofflaws and petty thieves) to the most obscure and heinous (mass murderers and serial killers)—and how they differ from noncriminals. Why do people commit crimes? Are people born criminals, or do their experiences dictate whether they will break the law? Can crime be controlled or eliminated? Why are some communities safer than others? What causes changes in crime rates? Criminologists search for answers to these and related questions.

Understanding who commits crimes and why can directly affect the passage of laws and the operations and practices of the criminal justice system, which comprises the agencies authorized to respond to criminal acts (law enforcement, courts, and corrections). Criminological theories can provide the basis for the creation of new and more effective programs and interventions designed to help lower crime rates, thereby making communities safer.

KEY TERMS AND CONCEPTS

- **Biological Theory:** Explanations of crime that focus on genetically based, inherited, or physical characteristics.
- **Crime:** Act that constitutes the violation of a law or criminal statute.
- **Crime Typology:** Classification of crimes according to different types of offenses and criminal motivations as well as offender and victim characteristics.
- **Criminal Career:** Frequency, type, and duration of criminal activities committed by an individual offender over a period of time.
- **Criminality:** Strong psychological propensity or tendency to commit crime.
- **Desistance:** Cessation of criminal activity after a period of offending.
- **Deterrence:** Attempt to prevent crime by making punishments match the severity of the crime committed.
- **Positivism:** Application of empirical scientific methods to study crime and criminals.
- **Psychological Theory:** Explanations of crime that focus on early-childhood experiences and personality traits.
- **Sentence:** Punishment that follows a conviction for a crime, such as a fine, community supervision, incarceration, or the death penalty.
- **Social Theory:** Explanation of crime that focuses on environmental processes or structures and the relationships between and among social groups.
- **Theory:** Set of testable, interrelated propositions intended to describe, explain, and predict an event or activity.

DEFINITION AND BASIC PRINCIPLES

Criminology draws from writings in the areas of law and crime control as well as from research in the social and behavioral sciences, such as sociology (particularly the subspecialty of the sociology of social-anthropology deviance), medicine (particularly the subspecialty of psychiatry), and psychology (particularly the subspecialty of clinical psychology).

BACKGROUND AND HISTORY

Edwin H. Sutherland, who came to be known as the "dean of American criminology," wrote the first modern-day textbook on criminology in 1924. *Criminology* paved the way for twentieth century criminological academic pursuits, and his contributions to the discipline are still relevant. Sutherland recognized that crime was a complicated phenomenon affected by political, social, economic, and geographic variables, and he rejected the notion that criminals were simpleminded. In two of his classic textbooks, *The Professional Thief* and *White Collar Crime*, Sutherland presented in-depth ethnographic analyses of a professional thief and white-collar criminals, demonstrating the humanity and complexity of criminals.

HOW IT WORKS

Criminologists are theoreticians and researchers. In studying the distribution and causes of crime, criminologists conduct research and other quantitative studies, analyzing large public surveys and data sets and asking offenders to report their histories, experiences, and decisions to pursue criminal activities. Criminologists also use qualitative research, such as ethnographic studies, to explore firsthand how people move into and out of criminal lifestyles and develop their skills and expertise in criminal specialties.

Criminologists develop theories using a deductive or an inductive approach. The deductive method of theory development starts with general propositions that are used to generate hypotheses, that is, testable questions. These questions are examined in carefully conducted studies, and the data are tested to see if they support the researcher's theory. The inductive method of theory development starts with data used to generate propositions, which are the building blocks of theory. As the theory becomes increasingly elaborate, more data are collected in a process of theory refinement. As both of these approaches suggest, the key elements of a good criminological theory are testability and empirical support.

Criminological theories are primarily derived from schools of thought that provide basic frameworks for formulating theories and their propositions as well as testable hypotheses. The schools also provide a general perspective and level of analysis for studying crime. Theories of crime are built on the concepts, variables, methodologies, and traditions of a particular discipline. The major disciplines in criminology include sociology, psychology, and biology. Sociology-based theories examine social structures and processes as well as the relationship between social groups. One sociological theory, strain theory, suggests that people without access to legitimate means to achieve success turn to crime. These people are feeling a certain amount of strain (economic, social, emotional, mental) in their lives, and they turn to crime to alleviate it. If members of society are unable to attain their goals because of lack of resources (money, education) those members may engage in illegitimate actions to achieve their goals. Some will retreat from mainstream society to join deviant subcultures, such as gangs and communities of drug users.

Psychology-based theories examine individual differences in characteristics and traits. Michael Gottfredson and Travis Hirschi, authors of *A General Theory of Crime*, posit that people commit crimes because of poorly developed or low self-control, which explains all delinquent and criminal behaviors and is the single, most important cause of crime at the individual level. People who are in control of themselves are able to consider the consequences of their behavior and are careful and deliberate in their decision-making. Such individuals self-monitor effectively and conform in socially desirable ways. In contrast, people with low self-control are present-oriented, impulsive, reckless, and lack empathy. They prefer physical rather than mental activities and engage in antisocial behaviors to meet their selfish needs. Self-control is learned during childhood and once learned it is difficult to change.

Biology-based theories examine the genetic contribution to crime, known as heritability, by employing several different methods of research. For example, twin studies compare identical and fraternal twins raised in the same household to determine the degree to which they are similar regarding crime—that is, their concordance rates. If both twins in identical pairs, who share all the same genes, are more likely to engage in crime than those in fraternal pairs, who share half of their genes, the evidence suggests that crime has a genetic component.

Another type of biology-based investigation strategy is the adoption study. In this type of research, investigators compare the criminality of parents and children who were separated at birth. The children in these investigations were reared by adoptive parents and had no contact with their biological parents. If the criminal involvement of the adoptive children was more like that of their biological parents than their adoptive parents, the data suggest that crime is affected more by nature (biology) than by nurture (environment).

Criminological theories are published in books. Even more important, the work of criminologists is published in journals, which are repositories for the body of knowledge in the field. Publication in a journal lends a degree of prestige and respectability to a study because it has been scrutinized by the author's peers prior to publication. Journals that publish the research of criminologists include *Criminology, Justice Quarterly, Crime and Delinquency, Criminal Justice and*

Behavior, Journal of Research in Crime and Delinquency, Journal of Criminal Law and Criminology, International Journal of Offender Therapy and Comparative Criminology, and *Journal of Quantitative Criminology.*

APPLICATIONS AND PRODUCTS

Several schools of thought are responsible for most of the major theories of crime. These include the Classical School, the Positivist School, and the Chicago (Ecological) School.

Classical School. This school of thought came about during the age of Enlightenment and was fueled by major reforms in penology and law in which imprisonment replaced corporal punishment as the predominant sanction. Cesare Beccaria (who argued against the death penalty in *On Crimes and Punishments,* published in 1764), Jeremy Bentham (English philosopher and the inventor of the panopticon, an early, revolutionary prison design), and other Classical School theorists argued that all people—including criminals—act on the basis of free will. Classical criminologists advocated for a humane penology.

In the Classical framework, people are considered fundamentally rationale beings who maximize pleasure and minimize pain as well as weigh the costs and benefits of each action in a process of rational choice. Thus, people decide to commit crimes using a straightforward, cost-benefit analysis. Beccaria wrote that threat of sufficient punishment can deter people from committing crimes, and that the most effective punishments are swift and consistent, and the severity is commensurate with the seriousness of the crime.

In addition to being against the death penalty, Beccaria also stood against torture and the inhumane treatment of prisoners—punishments he considered to be nonrational (ineffective) deterrents. Bentham similarly considered punishment for crimes only as useful as they served as deterrents for future offenses.

Positivist School. Formed in the late nineteenth century, the Positivist School maintained that criminal behavior stemmed from factors beyond a person's control, both internal (biological and psychological) and external (sociological). Positivists believed that crime and criminals could be best understood through the application of scientific techniques, and that biological, personal, and environmental factors determine criminal behaviors in a cause-and-effect relationship. Biological positivism was first proposed by the Italian physician Cesare Lombroso. Lombroso described criminals as "atavistic throwbacks," who acted on their primitive urges because they had failed to evolve fully. Their brains were underdeveloped, rendering them incapable of comporting their behaviors to the rules and regulations of society. They could also be differentiated from noncriminals by their physical characteristics. For example, murderers had glassy eyes, aquiline noses, and thin lips. Enrico Ferri, a student of Lombroso, believed that social as well as biological factors played a role in criminality. He argued that criminals should not be held responsible for their crimes when the factors causing their criminality were patently deterministic.

As an example of psychological positivism, Hans Eysenck, a British psychologist and author of *Crime and Personality* (1964), contended that certain psychological traits were at the root of what drove one to crime. His model of personality also contained the dimension of psychoticism, which consists of traits similar to those found in profiles of people with psychopathy, a set of behaviors and characteristics (the lack of empathy, conscience, and impulse control) that predispose people to commit serious crimes. Eysenck's model also acknowledged the influence of early parental socialization on childhood and adult tendencies to engage in criminal behaviors. His approach bridged the gaps among biological (William H. Sheldon), environmental (B.F. Skinner), and social learning-based (Albert Bandura) explanations of criminal behavior.

One of the tenets of sociological positivism is that societal factors (poverty, membership in subcultures, low levels of education) create a predisposition to crime. Belgian mathematician Adolphe Quetelet was one of the first to explore the relationship between crime and societal factors. He reported that poverty and low educational levels were important components in crime. British statistician Rawson W. Rawson linked population density and crime rates with statistics. He theorized that crowded cities create an environment conducive to crime and violence. French sociologist Émile Durkheim viewed crime as an inevitable consequence of the uneven distribution of wealth among the social classes.

Chicago (Ecological) School. Between 1915 and the early 1940's, sociological research in the United States was dominated by various academic disciplines at the University of Chicago—most notably, political science

Fascinating Facts About Criminology

- Early criminologists meticulously measured the facial and bodily features of criminals and non-criminals to identify distinguishable differences between the two groups. Based on so-called somato-typing criteria, men were classified along certain physical dimensions (endomorph, mesomorph, and ectomorph) that purportedly made them more or less susceptible to criminal behavior.

- Spurious research in the 1960's suggested that violent male criminals had an extra Y chromosome, which made them hypermasculine and contributed to their criminality.

- Crime rates have been declining steadily since the early 1990's, especially in big cities. However, the public remains highly fearful of crime.

- The incarceration rate in the United States is the highest in the world among industrialized countries. The explosion in the prison population in the past thirty years is not due to an increase in the number of crimes committed but rather to changes in laws and crime-control policies.

- The United States has 5 percent of the world's population but nearly 25 percent of the number of people incarcerated worldwide.

- No theory of crime has ever fully explained criminal behavior. Each theory has limitations, and no criminologist has ever formulated a fully integrated theory of criminal behavior. The causes of crime are simply too multifarious and complex to elucidate within a single theoretical framework.

- The relationship between the economy and crime is complicated and confounding. Crime has gone up during periods of economic prosperity, down during periods of economic woe, and vice versa.

- Called the "immigrant paradox," data suggest that immigrants (both legal and illegal) are less likely to commit crimes than are native-born Americans.

and sociology. To journalists, social reformers, and sociologists, the everyday struggles associated with living in Chicago became a microcosm of the human condition and an encapsulation of human suffering. In this atmosphere of urban despair and blight, many creative scholars combined their talents and applied their intellects to examine the harsh sequelae of urbanism, particularly those problems generated by inner-city living. The Chicago School brought to its research on urbanism many innovative, trenchant, and eclectic methods of social scientific analyses. Members of the Chicago School used a wide array of methodologies in their research, which was conducted in the field (streets, housing developments, opium dens, brothels, alleys, and parks) rather than in the sterility of a library, laboratory, or faculty office.

The Chicago School is exemplified in the work of urban sociologists at the University of Chicago, notably Robert E. Park and Ernest Burgess. In their 1925 book, *The City*, Park and Burgess identified five zones that often form as cities develop. The business district is in the center of the Park-Burgess model, and the zones that appear concentrically include the factory zone, the zone of transition (which was often crime prone), followed by the working-class, residential, and commuter zones. While researching juvenile delinquency in the 1940's, Chicago School sociologists Henry McKay and Clifford R. Shaw discovered that most troubled and troublesome adolescents were concentrated in the zone of transition.

Chicago School sociologists adopted a social-ecology approach in their study of cities, postulating that urban neighborhoods with high levels of poverty often experience a breakdown in social structure and institutions (families and schools). In the ecological model, such areas are hotbeds of social pathology, including disorganization, disorder, and decline—all of which render social institutions unable to control behavior, resulting in a downward spiral of neighborhood decay and creating an environment ripe for criminal and other deviant behaviors. Such neighborhoods tend to experience high rates of population turnover, which does not allow informal social structures to develop adequately and, in turn, makes it difficult to maintain social order in the community.

IMPACT ON INDUSTRY

Criminology has greatly affected crime-control policies and practices. Theories of crime have implications for punishments and crime-prevention techniques. The adoption of utilitarian-based (Classical) approaches to punishment led to the creation of determinate sentencing structures in which sanctions—in particular, lengths of incarceration—are pre-established and based on the seriousness of the offense. The punishment fits the crime in terms of severity, while characteristics of the offender are given little weight in the sentencing decision.

Another application of the Classical School of criminology stems from rational-choice theory, which argues that criminals, like everyone else, weigh costs and benefits when deciding whether to commit a crime. An economic calculus is their primary decision-making modality. Rational-choice theories also suggest that increasing the likelihood of being caught through added surveillance (cameras), the visible presence of police officers or security guards, more voluminous street lighting, and other environmental measures is effective in reducing crime.

Based on the Positivist School of thought, social-disorganization theory suggests strategies to reduce crime by strengthening communities. The Chicago Area Project (CAP) exemplifies this strategy and has achieved legendary status in the annals of criminology. Founded in 1934 by Chicago School sociologist Clifford R. Shaw, CAP has a long history of community building in which low-income residents assume responsibility for addressing critical neighborhood problems, such as delinquency, gang violence, substance abuse, and unemployment. Skeptical of psychological explanations of delinquency and programs aimed solely at reforming individuals, Shaw created CAP as a new form of grassroots community organization. Its goal was to prevent delinquency by encouraging local residents' active participation in community self-renewal. CAP rests on a powerful network of organizations and special projects that promote positive youth development and prevent juvenile delinquency through community building.

CAREERS AND COURSE WORK

Educational programs in criminology focus largely on crime and deviant behavior and include courses in criminal law and procedures, psychology, sociology, research methodology, and statistics. Criminology students also study the components and operations of the criminal justice system. Academic programs in criminology differ from those in criminal justice, which focus more on the criminal justice system itself and often provide training and job placement for particular careers in the field.

Criminologists are mostly doctoral-level academicians and policy analysts. The individuals who work in the criminal justice system are most properly described as criminal justice professionals or practitioners. These include a wide variety of personnel: judges, state's attorneys, public defenders, police officers, probation officers, parole agents, correctional officers, and victims' advocates. Criminologists can also be confused with criminalists, who specialize in the collection and analyses of the physical evidence deposited at a crime scene (also known as ballistic, fingerprint, shoe-print experts; crime laboratory technicians; and crime scene investigators and photographers).

Primarily involved in theory construction, research, teaching, writing, and policy analysis, criminologists contribute a great deal of expertise to the study of policing, police administration and policies, juvenile justice and delinquency, corrections, correctional administration and policies, drug addiction and enforcement, criminal subcultures, typologies of criminals, and victimology. In addition, they examine the various biological, sociological, and psychological factors related to criminal trajectories, which are pathways into and out of criminal behavior. Some criminologists also engage directly in community initiatives as well as in evaluation and policy projects with local, state, and federal criminal justice agencies.

Criminologists conduct their own research while teaching courses in psychology, legal studies, criminal justice, criminology, sociology, and pre-law at two- and four-year colleges and universities. Others work for state and federal justice agencies as policy advisers or researchers. These agencies include the Bureau of Justice Statistics, the Bureau of Justice Assistance, the National Institute of Justice, the National Institute of Corrections, the Office of National Drug Control Policy, the National Criminal Justice Reference Service, the Office of Juvenile Justice and Delinquency Prevention, and the Federal Bureau of Investigation. Criminologists can also be found in the private sector, where they provide consulting services on various issues such as crime statistics, juvenile and adult correctional programming, crime prevention and security protocols, legal reform, and justice initiatives, or they can work for large think tanks such as RAND or Abt Associates, pursuing policy-oriented research and evaluations.

Becoming a criminologist requires a minimum of a master's degree in criminology. However, criminologists who work in university settings typically possess doctoral degrees and postdoctoral training. In addition, criminologists can receive their doctoral degrees in other disciplines, such as sociology, psychology, political science, or public policy, and specialize in crime and criminals during the course of their advanced studies.

SOCIAL CONTEXT AND FUTURE PROSPECTS

The precipitous increases in crime that accompanied the creation of the urban ghetto and the alienation of the immigrant populations during the turbulent 1960's, as well as the crime-accelerating effects of unstable drug markets in the 1990's, provided further impetus for criminological theory and research and the continual search for solutions to Americans' crime problems. Despite the admonitions of European criminologist Hermanus Bianchi, who warned criminologists to ply their trade away from politics, modern criminologists are interested in informing political agendas and public policies relating to crime control and justice issues. Indeed, many practitioners increasingly seek to incorporate into their everyday activities evidence-based practices grounded in scientific knowledge and solid theorizing. Nonetheless, as historian Lawrence Friedman notes, enduring cultural taboos have become obstacles to the implementation of lasting and effective crime-control efforts; these include widespread resistance to adopting strict gun-control laws, legalizing/decriminalizing drugs, and increasing taxes to pay for social and rehabilitative programs.

Arthur J. Lurigio, Ph.D.

FURTHER READING

Beccaria, Cesare. *On Crimes and Punishments.* Translated by David Young. Indianapolis, Ind.: Hackett, 1986. This is the definitive, seminal work that launched the Classical School of criminology and remains relevant to the field despite its origins nearly 250 years ago.

Gaines, Larry K., and Roger Leroy Miller. *Criminal Justice in Action: The Core.* Belmont, Calif.: Wadsworth, 2010. An engaging textbook that covers the criminal justice process from arrest to incarceration and describes its three major components of operation: law enforcement, courts, and corrections. Contains text boxes with "fast facts," landmark criminal cases, and information about careers in criminal justice. Many illustrations and photographs enliven the text.

Gottfredson, Michael R., and Travis Hirschi. *A General Theory of Crime.* Stanford, Calif.: Stanford University Press, 1990. The authors explore the real reasons people commit crimes and bring together classic and modern theories on criminology in an effort to provide answers to the questions criminologists have been asking for centuries.

Hayward, Keith, Shadd Maruna, and Jayne Mooney, eds. *Fifty Key Thinkers in Criminology.* Abingdon, England: Routledge, 2010. This comprehensive collection includes essays on the earliest proponents of the science (Cesare Beccaria and Jeremy Bentham) as well as contemporary practitioners (Frances Heidensohn and Travis Hirschi).

Jacoby, Joseph, ed. *Classics of Criminology.* 3d ed. Long Grove, Ill.: Waveland Press, 2004. An anthology of the most influential papers published in the field from the 1700's through the 1990's, exploring numerous criminological theories and featuring the most prominent theoreticians in the discipline.

Schmalleger, Frank. *Criminology Today.* 6th ed. Upper Saddle River, N.J.: Prentice Hall, 2011. A highly readable tome that contains a wealth of colorful illustrations and boxed text that provides interesting case studies and contemporary takes on criminology

Shoemaker, Donald J. *Theories of Delinquency: An Examination of Explanations of Delinquent Behavior.* 6th ed. New York: Oxford University Press, 2010. This well-researched book presents the best known sociological theories of crime, which are each described in a separate chapter, as well as chapters that summarize psychological and biological theories of crime and issues in the field, such as female delinquency and radical criminology.

Sutherland, Edwin H., and Donald R. Cressey. *Criminology.* Philadelphia: Lippincott, 1978. This is an updated edition of the first recognized textbook in the field, which defined the discipline and articulated basic concepts and methodologies for the study of crime and criminals.

WEB SITES

American Society of Criminology
http://www.asc41.com

American Sociological Association
http://www.asanet.org

Federal Bureau of Investigation
http://www.fbi.gov

U.S. Department of Justice
http://www.justice.gov

See also: Anthropology; Forensic Science; Penology; Psychiatry.

CRYOGENICS

FIELDS OF STUDY

Astrophysics; cryogenic engineering; cryogenic electronics; nuclear physics; cryosurgery; cryobiology; high-energy physics; mechanical engineering; chemical engineering; electrical engineering; cryotronics; materials science; biotechnology; medical engineering; astronomy.

SUMMARY

Cryogenics is the branch of physics concerned with creation of extremely low temperatures and involves the observation and interpretation of natural phenomena resulting from subjecting various substances to those temperatures. At temperatures near absolute zero, the electric, magnetic, and thermal properties of most substances are greatly altered, allowing useful industrial, automotive, engineering, and medical applications.

KEY TERMS AND CONCEPTS

- **Absolute Zero:** Temperature measured 0 Kelvin (−273 degrees Celsius), where molecules and atoms oscillate at the lowest rate possible.
- **Cryocooler:** Device that uses cycling gases to produce temperatures necessary for cryogenic work.
- **Cryogenic Processing:** Deep cooling of matter using cryogenic temperatures so the molecules and atoms of the matter slow or almost stop movement.
- **Cryogenic Tempering:** Onetime process using sensitive computerization to cool metal to cryogenic temperatures then tempering the metal to enhance performance, strength, and durability.
- **Cryopreservation:** Cooling cells or tissues to subzero temperatures to preserve for future use.
- **Evaporative Cooling:** Process that allows heat in a liquid to change surface particles from a liquid to a gas.
- **Heat Conduction:** Technique where a substance is cooled by passing heat from matter of higher temperature to matter of lower temperature.
- **Joule-Thomson Effect:** Technique where a substance is cooled by rapid expansion, which drops the temperature. So named for its discoverers, British physicists James Prescott Joule and William Thomson (Lord Kelvin).
- **Kelvin Temperature Scale (K):** Used to study extremely cold temperatures. On the Kelvin scale, water freezes at 273 K and boils at 373 K.
- **Superconducting Device:** Device known for its electrical properties and magnetic fields, such as magnetic resonance imaging (MRI) in medicine.
- **Superconducting Magnet:** Electromagnet with a superconducting coil where a magnetic field is maintained without any continuing power source.
- **Superconductivity:** Absence of electrical resistance in metals, ceramics, and compounds when cooled to extremely low temperatures.
- **Superfluidity:** Phase of matter, such as liquid helium, that is absent of viscosity and flows freely without friction at very low temperatures.

DEFINITION AND BASIC PRINCIPLES

Cryogenics comes from two Greek words: *kryo*, meaning "frost," and *genic*, "to produce." This science studies the implications of producing extremely cold temperatures and how these temperatures affect substances such as gases and metal. Cryogenic temperature levels are not found naturally on Earth. The usefulness of cryogenics is based on scientific principles. The three basic states of matter are gas, liquid, and solid. Matter moves from one state to another by the addition or subtraction of heat (energy). The molecules or atoms in matter move or vibrate at different rates depending on the level of heat. Extremely low temperatures, as achieved through cryogenics, slow the vibration of atoms and can change the state of matter. For example, cryogenic temperatures are used in the liquefaction of atmospheric gases such as oxygen, nitrogen, hydrogen, and methane for diverse industrial, engineering, automotive, and medical applications.

Sometimes cryogenics and cryonics are mistakenly linked, but use of subzero temperatures is the only thing these practices share. Cryonics is the practice of freezing a body right after death to preserve it for a future time when a cure for fatal illness or remedy for fatal injury may be available. The practice of cryonics is based on the belief that technology from cryobiology can be applied to cryonics.

If cells, tissues, and organs can be preserved by cryogenic temperatures, then perhaps whole body can be preserved for future thawing and life restoration. Facilities exist for interested persons or families, although the cryonic process is not considered reversible as of this writing.

BACKGROUND AND HISTORY

The history of cryogenics follows the evolution of low-temperature techniques and technology. Principles of cryogenics can be traced to 2500 B.C.E., when Egyptians and Indians evaporated water through porous earthen containers to produce cooling. The Chinese, Romans, and Greeks collected ice and snow from the mountains and stored it in cellars to preserve food. In the early 1800's, American inventor Jacob Perkins created a sulfuric-ether ice machine, a precursor to the refrigerator. By the mid-1800's, William Thomson, a British physicist known as Lord

Technician placing blood stem cells collected from a donated human placenta and umbilical cord into a cryostorage tank. (Tek Image/Photo Researchers, Inc.)

Kelvin, theorized that extremely cold temperatures could stop the motion of atoms and molecules. This became known as absolute zero, and the Kelvin scale of temperature measurement emerged.

Scientists of the time focused on liquefaction of permanent gases. By 1845, the work of British physicist Michael Faraday accomplished liquefaction of permanent gases by cooling immersion baths of ether and dry ice followed by pressurization. Six permanent gases—oxygen, hydrogen, nitrogen, methane, nitric oxide, and carbon monoxide—still resisted liquefaction. In 1877, French physicist Louis-Paul Cailletet and Swiss physicist Raoul Pictet produced drops of liquid oxygen, working separately and using completely different methods. In 1883, S. F. von Wroblewski at the University of Krakow in Poland, discovered oxygen would liquefy at 90 Kelvin (K) and nitrogen at 77 K. In 1898, Scottish chemist James Dewar discovered the boiling point of hydrogen at 20 K and its freezing point at 14 K.

Helium, with the lowest boiling point of all known substances, was liquefied in 1908 by Dutch physicist Heike Kamerlingh Onnes at the University of Leiden. Onnes was the first person to use the word "cryogenics." In 1892, Scottish physicist James Dewar invented the Dewar flask, a vacuum flask designed to maintain temperatures necessary for liquefying gases, which was the precursor to the Thermos. The liquefaction of gases had many important commercial applications, and many industries use Dewar's concept in applying cryogenics to their processes and products.

The usefulness of cryogenics continued to evolve, and by 1934 the concept was well established. During World War II, scientists discovered that metals became resistant to wear when frozen. In the 1950's, the Dewar flask was improved with the multilayer insulation (MLI) technique for insulating cryogenic propellants used in rockets. Over the next thirty years, Dewar's concept led to the development of small cryocoolers, useful to the military in national defense. The National Aeronautics and Space Administration (NASA) space program applies cryogenics to its programs. Cryogenics can be used to preserve food for long periods—this is especially helpful during natural disasters. Cryogenics continues to grow globally and serve a wide variety of industries.

HOW IT WORKS

Cryogenics is an ever-expanding science. The basic principle of cryogenics that the creation of extremely low temperatures will affect the properties of matter so the changed matter can be used for a number of applications. Four techniques can create the conditions necessary for cryogenics: heat conduction, evaporative cooling, rapid-expansion cooling (Joule-Thomson effect), and adiabatic demagnetization.

Creating Low Temperatures. With heat conduction, heat flows from matter of higher temperature to matter of lower temperature in what amounts, basically, to a transfer of thermal energy. As the process is repeated, the matter cools. This principle is used in cryogenics by allowing substances to be immersed in liquids with cryogenic temperatures or in an environment such as a cryogenic refrigerator for cooling.

Evaporative cooling is another technique employed in cryogenics. Evaporative cooling is demonstrated in the human body when heat is lost through liquid (perspiration) to cool the body via the skin. Perspiration absorbs heat from the body, which evaporates after it is expelled. In the early 1920's in Arizona during the summers, people hung wet sheets inside screened sleeping porches. Electric fans pulled air through the sheets to cool the sleeping space. In the same way, a container of liquid can evaporate, so the heat is removed as gas; the repetitive process drops the temperature of the liquid. An example is reducing the temperature of liquid nitrogen to its freezing point.

The Joule-Thomson effect occurs without the transfer of heat. Temperature is affected by the relationship between volume, mass, pressure, and temperature. Rapid expansion of a gas from high to low pressure results in a temperature drop. This principle was employed by Dutch physicist Heike Kamerlingh Onnes to liquefy helium in 1908 and is useful in home refrigerators and air conditioners.

Adiabatic demagnetization uses paramagnetic salts to absorb energy from liquid, resulting in a temperature drop. The principle in adiabatic demagnetization is the removal of the isothermal magnetized field from matter to lower the temperature. This principle is useful in application to refrigeration systems, which may include a superconducting magnet.

Cryogenic Refrigeration. Cryogenic refrigeration, used by the military, laboratories, and commercial businesses, employs gases such as helium (valued for its low boiling point), nitrogen, and hydrogen to cool equipment and related components at temperatures lower than 150 K. The selected gas is cooled through

---

Fascinating Facts About Cryogenics

- American businessman Clarence Birdseye revolutionized the food industry when he discovered that deep-frozen food tasted better than regular frozen food. In 1923, he developed the flash-freezing method of preserving food at below-freezing temperatures under pressure. The "Father of Frozen Food" first sold small-packaged foods to the public in 1930 under the name Birds Eye Frosted Foods.

- In cryosurgery, super-freezing temperatures as low as -200 Celsius are introduced through a probe of circulating liquid nitrogen to treat malignant tumors, destroy damaged brain tissue in Parkinson's patients, control pain, halt bleeding, and repair detached retinas.

- Cryogenics can be used to save endangered species from extinction. Smithsonian researcher Mary Hagedorn is using cryogenics to establish the first coral seed banks: She's collecting thousands of sample species and freezing them for the future. Hagedorn refers to this as an insurance policy for natural resources.

- The Joule-Thomson effect, discovered in 1852 by James Prescott Joule and William Thomson (Lord Kelvin), is responsible for the cooling used in home refrigerators and air conditioners.

- Helium's boiling point, 173 Kelvin, is the lowest of all known substances.

- Surgical tools and implants used by surgeons and dentists have increased strength and resistance to wear because of cryogenic processing.

- Cryogenic processing is 100 percent environmentally friendly with no use of harmful chemicals and no waste products.

- In 1988, microbiologist Curt Jones, who studied freezing techniques to preserve bacteria and enzymes for commercial use, created Dippin' Dots, a popular ice cream treat, using a quick-freeze process with liquid nitrogen.

---

pressurization to liquid or solid forms (dry ice used in the food industry is solidified carbon dioxide). The cold liquid may be stored in insulated containers until used in a cold station to cool equipment in an immersion bath or with sprayer.

Cryogenic Processing and Tempering. Cryogenic processing or treatment increases the length of wear of many metals and some plastics using a deep-freezing process. Metal objects are introduced to cooled liquid gases such as liquid nitrogen. The computer-controlled process takes about seventy-two hours to affect the molecular structure of the metal. The next step is cryogenic or heat tempering to improve the strength and durability of the metal object. There are about forty companies in the United States that provide cryogenic processing.

APPLICATIONS AND PRODUCTS

Early applications of cryogenics targeted the need to liquefy gases. The success of this process in the late 1800's paved the way for more study and research to apply cryogenics to developing life needs and products. Examples include applications in the auto and health care industries and in development of rocket fuels and methods of food preservation. Cryogenic engineering has applications related to commercial, industrial, aerospace, medical, domestic, and defense ventures.

Superconductivity Applications. One property of cryogenics is superconductivity. This occurs when the temperature is dropped so low that the electrical current experiences no resistance. An example is electrical appliances, such as toasters, televisions, radios, or ovens, where energy is wasted trying to overcome electrical resistance. Another is with magnetic resonance imaging (MRI), which uses a powerful magnetic field generated by electromagnets to diagnosis certain medical conditions. High magnetic field strength occurs with superconducting magnets. Liquid helium, which becomes a free-flowing superfluid, cools the superconducting coils; liquid nitrogen cools the superconducting compounds, making cryogenics an integral part of this process. Another application is the use of liquefied gases that are sprayed on buried electrical cables to minimize wasted power and energy and to maintain cool cables with decreased electrical resistance.

Health Care Applications. The health care industry recognizes the value of cryogenics. Medical applications using cryogenics include preservation of cells or tissues, blood products, semen, corneas, embryos, vaccines, and skin for grafting. Cryotubes with liquid nitrogen are useful in storing strains of bacteria at low temperatures. Chemical reactions needed to release active ingredients in statin drugs, used for cholesterol control, must be completed at very low temperatures (−100 degrees Celsius). High-resolution imaging, like MRI, depends on cryogenic principles for the diagnosis of disease and medical conditions. Dermatologists uses cryotherapy to treat warts or skin lesions.

Food and Beverage Applications. The food industry uses cryogenic gases to preserve and transport mass amounts of food without spoilage. This is also useful in supplying food to war zones or natural-disaster areas. Deep-frozen food retains color, taste, and nutrient content while increasing shelf life. Certain fruits and vegetables can be deep frozen for consumption out of season. Freeze-dried foods and beverages, such as coffee, soups, and military rations, can be safely stored for long periods without spoilage. Restaurants and bars use liquid gases to store beverages while maintaining the taste and look of the drink.

Automotive Applications. The automotive industry employs cryogenics in diverse ways. One is through the use of thermal contraction. Because materials will contract when cooled, the valve seals of automobiles are treated with liquid nitrogen, which shrinks to allow insertion and then expands as it warms up, resulting in a tight fit. The automotive industry also uses cryogenics to increase strength and minimize wear of metal engine parts, pistons, cranks, rods, spark plugs, gears, axles, brake rotors and pads, valves, rings, rockers, and clutches. Cryogenic-treated spark plugs can increase an automobile's horsepower as well as its gasoline mileage. The use of cryogenics allows a race car to race as many as thirty times without a major rebuild on the motor compared with racing twice on an untreated car.

Aerospace Industry Applications. NASA's space program utilizes cryogenic liquids to propel rockets. Rockets carry liquid hydrogen for fuel and liquid oxygen for combustion. Cryogenic hydrogen fuel is what enables NASA's workhorse space shuttle to get into orbit. Another application is using liquid helium to cool the infrared telescopes on rockets.

Tools, Equipment, and Instrument Applications. Metal tools can be treated with cryogenic applications that provide wear resistance. In surgery or dentistry, tools can be expensive, and cryogenic treatment can prolong usage. Sports equipment, such as golf clubs, benefits from cryogenics as it provides increased wear resistance and better performance. Another is the ability of a scuba diver to stay submerged for hours with an insulated Dewar flask of cryogenically cooled nitrogen and oxygen. Some claim musical instruments receive benefits from cryogenic treatment; in brass instruments, a crisper and cleaner sound is allegedly produced with cryogenic enhancement.

Other Applications. Other applications are evolving as industries recognize the benefits of cryogenics to their products and programs. The military have used cryogenics in various ways, including infrared tracking systems, unmanned vehicles, and missile warning receivers. Companies can immerse discarded recyclables in liquid nitrogen to make them brittle, then these recyclables can be pulverized or grinded down to a more eco-friendly form. No doubt with continued research, many more applications will emerge.

IMPACT ON INDUSTRY

The science of cryogenics continues to impact the quality and effectiveness of products of many industries. Various groups have initiated research studies to support the expanded use of cryogenics.

Government and University Research. Government agencies and universities have conducted research on various aspects of cryogenics. The military have investigated the applications of cryogenics to national defense. The Air Force Research Laboratory (AFRL) Space Vehicles Directorate addressed applications of cryogenic refrigeration in the area of ballistic missile defense. The study looked at ground-based radars and space-based infrared sensors requiring cryogenic refrigeration. Future research targets the availability of flexible technology such as field cryocoolers. Such studies can be significant to a cost-effective national defense for the United States.

The health care industry has many possible applications for cryogenics. In 2005, Texas A&M University Health Science Center-Baylor College of Dentistry investigated the effect of cryogenic treatment of nickel-titanium on instruments used by dentists. Past research had been conducted on stainless-steel endodontic instruments with no significant increase in wear resistance. This research demonstrated an increase in microhardness but not in cutting efficiency.

Industry and Business. In 2006, the Cryogenic Institute of New England (CINE) recognized that although cryogenic processing was useful in many industries, research validating its technologic advantages and business potential was scant. CINE located forty commercial companies that provided cryogenic processing services and conducted telephone surveys with thirty of them. The survey found that some $8 million were generated by these deep cryogenic services in the United States, with 75 percent coming from the services and 25 percent from the sales of equipment.

The survey asked participants to identify the list of top metals they worked with in cryogenic processing. These were given as cast irons, various steels (carbon, stainless, tool, alloy, mold), aluminum, copper, and others. The revenue was documented by market application. Some 42 percent of the cryogenic-processing-plant market was in the motor sports and automotive industry, where the goal was treatment of engine components to improve performance, extend life wear, or treat brake rotors. Thirty percent of the application market fell into the category of heavy metals, tooling, and cutting; examples of these include manufacturing machine tools, dies, piping, grinders, knives, food processing, paper and pulp processing, and printing. Ten percent were listed in heavy components such as construction, in-ground drilling, and mining, while 18 percent of the market was in areas such as recreational, firearms, electronics, gears, copper electrodes, and grinding wheels.

The National Institute of Standards and Technology (NIST) initiated the Cryogenic Technologies Project, a collaborative research effort between industry and government agencies to improve cryogenic processes and products. One goal is the investigation of cryogenic refrigeration.

The nonprofit Cryogenic Society of America (CSA) offers conferences on related work areas such as superconductivity, space cryogenics, cryocoolers, refrigeration, and magnet technology. It also has continuing-education courses and lists job postings on its Web site.

CAREERS AND COURSE WORK

Careers in cryogenics are as diverse as the applications of cryogenics. Interested persons can enter the profession in various ways, depending on their field of interest. Some secure jobs through additional education, while others learn on the job. In general, the jobs include engineers, technologists or technicians, and researchers.

A primary career track for those interested in working in cryogenics is cryogenic engineering. To become a cryogenic engineer requires a bachelor's or master's degree in engineering. Course work may include thermodynamics, production of low temperatures, refrigeration, liquefaction, solid and fluid properties, and cryogenic systems and safety.

In the United States, some four hundred academic institutions offer graduate programs in engineering with about forty committed to research and academic opportunities in cryogenics. Schools with graduate programs in cryogenics include the University of California, Los Angeles; University of Colorado; Cornell University; Georgia Institute of Technology; Illinois Institute of Technology; Florida International University; Iowa State University; Massachusetts Institute of Technology; Ohio State University; University of Wisconsin-Madison; Florida State University; Northwestern University; and University of Southampton in the United Kingdom.

SOCIAL CONTEXT AND FUTURE PROSPECTS

The economic and ecological impact of cryogenic research and applications holds global promise for the future. In 2009, Netherlands firm Stirling Cryogenics built a cooling system with liquid argon for the ICARUS project, which is being carried out by Italy's National Institute of Nuclear Physics. In China, the Cryogenic and Refrigeration Engineering Research Centre (CRERC) focuses on new innovations and technology in cryogenic engineering. Both private industry and government agencies in the United States are pursuing innovative ways to utilize existing applications and define future implications of cryogenics. Although cryogenics has proved useful to many industries, its full potential as a science has not yet been realized.

Marylane Wade Koch, M.S.N., R.N.

FURTHER READING

Hayes, Allyson E., ed. *Cryogenics: Theory, Processes and Applications.* Hauppauge, N.Y.: Nova Science Publishers, 2010. Details global research on cryogenics and applications such as genetic engineering and cryopreservation.

Jha, A. R. *Cryogenic Technology and Applications.* Burlington, Mass.: Elsevier, 2006. Deals with most aspects of cryogenics and cryogenic engineering, including historical development and various laws, such as heat transfer, that make cryogenics possible.

Schwadron, Terry. "Hot Sounds From a Cold Trumpet? Cryogenic Theory Falls Flat." *New York Times,* November 18, 2003. Explains how two Tufts University researchers studied cryogenic freezing of trumpets and determined the cold did not improve the sound.

Ventura, Gugliemo, and Lara Risegari. *The Art of Cryogenics: Low-Temperature Experimental Techniques.* Burlington, Mass.: Elsevier, 2008. Comprehensive discussion of various aspects of cryogenics from heat transfer and thermal isolation to cryoliquids and instrumentation for cryogenics, such as the use of magnets.

WEB SITES

Cryogenic Society of America
http://www.cryogenicsociety.org

Help Mary Save Coral
http://www.helpmarysavecoral.org/obe

National Aeronautics and Space Administration
Cryogenic Fluid Management
http://www.nasa.gov/centers/ames/research/technology-onepagers/cryogenic-fluid-management.html

National Institute of Standards and Technology
Cryogenic Technologies Project
http://www.nist.gov/mml/properties/cryogenics/index.cfm

See also: Chemical Engineering; Computer Science; Electrical Engineering; Electromagnet Technologies; Magnetic Resonance Imaging; Mechanical Engineering.

CRYONICS

FIELDS OF STUDY

Cardiopulmonary resuscitation; cardiothoracic surgery; cryopreservation; nanotechnology; surgery.

SUMMARY

Cryonics is a theoretical life support technology, which involves stabilizing the condition of a terminally ill patient via freezing until a future date when technology will be able to revive that person and hopefully return him or her to a normal life. Storing a person at the temperature of liquid nitrogen (–196 degrees Celsius) can prevent further tissue damage indefinitely; however, the freezing process inflicts a degree of tissue damage that is not reversible by existing technology. Modern technology is successful in freezing sperm, ova (eggs), and embryos, which can later be thawed and restored to life. Human embryos have been thawed, implanted into a uterus, and have ultimately developed into a healthy newborn.

KEY TERMS AND CONCEPTS

- **Cryopreservation:** Process whereby cells and tissues are preserved by cooling to subzero temperatures.
- **Cryoprotectant:** Substance that protects biological tissue from freezing damage.
- **Ischemia:** Inadequate supply of oxygenated blood to a bodily organ or tissue.
- **Liquid Nitrogen:** Nitrogen, which is a gas at room temperature, can be maintained in a liquid state at a very low temperature (–196 degrees Celsius). It is used for the cryopreservation of sperm, ova (eggs), embryos, and adult human bodies.
- **Nanotechnology:** Technology that entails the manipulation of individual atoms or molecules, eventually to build or repair any physical object, including human cells and biological tissue.
- **Necrosis:** Cell death, which can be caused by factors such as ischemia or lack of oxygen.
- **Vitrification:** Process of converting something into a glass-like solid that is free of any crystal formation, which can cause tissue damage; vitrification lowers the freezing point of a solution.

DEFINITION AND BASIC PRINCIPLES

Cryonics (from the Greek *kryos*, which means "icy cold") involves the freezing, or cryopreservation, of an entire body or the brain until a future date when technology will be able to thaw the tissue and restore life. If only the brain is frozen, the ultimate goal is to grow a new body around the head. The brain is customarily frozen with the head to provide protection to the structure. In the United States, the process is begun when a patient's heart stops beating. In addition to humans, cryonics facilities preserve pets. Once frozen in liquid nitrogen, and with proper storage, a frozen body will not deteriorate further over time. As of 2011, technology is not available to thaw a cryopreserved individual and restore life. Therefore, any future thawing methods are purely hypothetical. As the technology improves, so will the chances of life restoration. A number of biological specimens have been cryopreserved, stored in liquid nitrogen, and revived. They include insects, eels, and organs. For example, in 2005, a cryopreserved rabbit kidney was thawed and transplanted to a rabbit; the organ functioned normally. The best examples of restoration of viability after cryopreservation of human tissue are ova (eggs), sperm, embryos, and ovarian tissue. A frozen ovum can be thawed and fertilized with previously frozen sperm, and a normal infant can develop. On a larger scale, human embryos can be cryopreserved, later thawed, and implanted into the uterus for growth. Infants derived from this process are no different in growth or intelligence than those who came into being under natural conditions.

BACKGROUND AND HISTORY

In 1866, Italian physician Paolo Mantegazza suggested that before soldiers departed for the battlefield, they should leave behind frozen sperm. He also suggested collection of sperm for freezing just before a mortally wounded soldier would die. The first successful human pregnancy using frozen sperm occurred in 1953; thirty years later, cryopreservation of oocytes was accomplished. In 1986, Australian physician Christopher Chen reported the world's first pregnancy using previously frozen oocytes. The rate of success of pregnancies using frozen oocytes, embryos, and fresh embryos are comparable.

The first cryonics organization, the Life Extension Society, was founded by Evan Cooper in 1963. Growth was slow over the next two years and most of the people interested in the procedure were wealthy celebrities. The first person to be cryogenically frozen was James Bedford, a seventy-three-year-old psychologist. He was cryopreserved in 1967, and his body is reportedly still in good condition at Alcor Life Extension Foundation. By the late 1970's, there were six cryonics companies operating in the United States. However, the preservation and subsequent maintenance of bodies indefinitely proved to be too expensive, and many of the companies ceased operations in the 1980's. Currently, seven companies offer cryonics services: Alcor Life Extension Foundation (Scottsdale, Arizona), American Cryonics Society (Cupertino, California), Cryonics Institute (Clinton Township, Michigan), Eucrio (Braga, Portugal), KrioRus (Alabychevo, Russia), Suspended Animation (Boynton Beach, Florida), and Trans Time (San Leandro, California). A facility in Australia is in the planning stages.

HOW IT WORKS

Embryo Cryopreservation. Embryo cryopreservation has evolved to a technology with a high success rate. Embryos can be frozen at the pronuclear stage (one cell) up to the blastocyst stage (five to seven days after fertilization; 75 to 100 cells). Freezing and thawing of embryos is overseen by embryologists who are assisted by laboratory technicians. The embryos are mixed with a cryoprotectant and placed in straws before freezing. This allows for vitrification during the freezing process. The scientific definition of vitrification is the conversion of a liquid to a glass-like solid, which is free of any ice-crystal formation. The straw is placed in a cooling chamber for freezing. After freezing, the straw is placed into a carefully labeled metal cane, which is lowered into a liquid nitrogen tank with other frozen embryos. The thawing process involves warming the embryos to room temperature in 35 seconds. Over the next half hour, the embryos are incubated in decreasing concentrations of the cryoprotectant and increasing concentrations of water. These embryos are then transferred to a woman's uterus for growth. In the past, several embryos were transferred in the hope that at least one would survive. Existing technology has improved to the point that a single embryo has a high survival rate.

The trend is to transfer a single embryo to prevent multiple births. Multiple births—even twins—have a much higher incidence of complications. Embryos stored for ten to twelve years have been thawed with subsequent development of a live infant. The cost for freezing an embryo is around $700. Storage costs vary depending on the length of storage.

Sperm and Ovary Cryopreservation. For men planning to have a vasectomy or men with testicular cancer, sperm can be cryopreserved, which will allow the future fathering of children. In addition to cryopreservation of embryos, women can have a portion of an ovary cryopreserved. This ovarian tissue can be thawed and transplanted at a later date. For women who must undergo chemotherapy for a malignancy, which can compromise their fertility, this is an extremely attractive option. The ovarian tissue contains thousands of eggs; furthermore, this tissue will produce female hormones.

Comparison of Embryo Cryopreservation to Cryonics. In contrast to the high success rate of human embryo transfer, cryopreservation of a fully formed individual is imperfect and theoretical. Embryos consist of a small number of cells, each of which has the potential to develop into any type of tissue or organ. Embryos are preselected and, if available, only high-quality embryos are used. The embryos are in good health and well oxygenated when frozen. The whole process from freezing to storage and ultimate thawing is done in an orderly process. An adult who wishes to be cryopreserved is usually of advanced age with multiple health problems, such as an advanced malignancy, severe heart disease, or Alzheimer's disease; the individual also is near death when the process is begun. Often, as death approaches, multiple organ failure occurs. The patient may have previously suffered a stroke or suffer one as death approaches. Unless the patient dies at a location and can be promptly instituted, a significant delay in the process can occur. Cryoprotectants must be infused into the body to circulate throughout it. With current technology, despite the use of cryoprotectants, the blood vessels are damaged. Even in cases where only the brain is frozen, blood-vessel damage occurs. Once frozen, no further damage will occur; however, all would agree, a cryopreserved human is significantly damaged. When thawed, the individual must be restored to life as well as good physical and mental health.

Cryonics. According to Alcor, when a patient who has enrolled in a cryonics program becomes critically ill, cryonics personnel will be placed on standby. When the heart stops beating, the patient is placed in an ice-water bath, and cardiopulmonary resuscitation is begun. Intravenous lines are established to infuse cryoprotectants, medications, and anesthetics. The medications maintain blood pressure. The cooling and the anesthetic lower the oxygen consumption to protect the brain and vital organs. If the patient is in a hospital that does not allow cryonics procedures, the patient is transferred to another facility; resuscitation and cooling are maintained during this process. The subsequent process includes surgically assessing the femoral (upper leg) arteries and veins and the patient is placed on cardiopulmonary bypass with a portable heart-lung machine. External resuscitation is discontinued. In the heart-lung machine, a heat exchanger works to lower the body temperature to a few degrees above the freezing point of water (0 degrees Celsius), and then blood is replaced with a cryoprotectant. The patient is then transferred to the cryonics facility (if not already there). Major blood vessels are accessed by thoracic surgery and attached to the perfusion circuit. Cryoprotectant is infused at nearly the freezing point of water for several minutes, which removes any residual blood. Then the cryoprotectant concentration is increased over a two-hour period to half the final target concentration. A rapid increase to the final concentration is then made. After cryoprotective perfusion, the patient is rapidly cooled under computer control by fans circulating nitrogen gas at a temperature of –125 degrees Celsius. The patient is then further cooled to –196 degrees Celsius over the next two weeks. The cost of the procedure and storage can exceed $160,000 for whole-body cryopreservation and $80,000 for the head-only option. The fee includes a basic cost at enrollment and usually requires an annual membership fee until death. Often, arrangements can be made to have these costs covered by life insurance.

APPLICATIONS AND PRODUCTS

Cryoprotectants. A variety of cryoprotectants are available for cryopreservation of embryos and adults. Research is ongoing to improve these products. Conventional cryoprotectants in use include dimethyl sulfoxide (DMSO), ethylene glycol, glycerol, propylene glycol, propanediol (PROH), and sucrose. DMSO and glycerol have been used for decades by embryologists to cold-preserve embryos and sperm. Usually, a combination of ingredients is used because they have less toxicity and increased efficacy compared to a single substance. Cryoprotectants lower the freezing point of water; in addition, many cryoprotectants displace water molecules with hydrogen bonds in biological materials. This hydrogen bonding maintains proper protein and DNA function.

Cryonics Procedures. Cryonics companies employ specially designed resuscitation equipment to perform immediate cardiopulmonary resuscitation (CPR) immediately after death. Two such devices are the LUCAS Chest Compression System and Michigan Instruments' Thumper. These devices are powered by pressurized oxygen and restore blood flow much more efficiently than manual CPR. Later, a portable heart-lung machine is used to continue perfusion. After instillation of cryoprotectants, transport, ranging from a few miles to thousands, is needed. Cryonics facilities contain an operating room with a heart-lung machine as well as a variety of specialized equipment for instilling cryoprotectants and controlled cooling. The use of the operating room is sporadic at best. Months and sometimes years transpire between patients. Nevertheless, it must be maintained in a standby status that can be made fully operational in a matter of hours. Technicians and medical personnel, including doctors and nurses, are obviously not part of the regular staff. They are alerted and assembled when the death of a cryonics applicant is imminent. In some cases, sudden death occurs, and the facility must have a system in place to obtain the necessary personnel on short notice.

Embryo Culture. Specially designed media and incubators are used for the culture of embryos. Research is ongoing to improve both the media and incubators. Around day five of development, the embryo will be in the blastocyst stage, and it must hatch out from its "shell," a glycoprotein membrane called the zona pellucida, in order to implant in the uterus. If the embryo does not hatch, it cannot implant. Since the mid-1990's, assisted hatching has been performed to encourage embryo to come out. This is done by an embryologist, who makes a small hole in the zona of each embryo. The zona is not a living part of the embryo and making the hole does not harm the embryo. A laser beam is often used to make the hole.

Nanotechnology. Nanotechnology is focused on the manipulation of individual atoms or molecules with the ultimate goal of building or repairing any physical object, including human cells and biological tissue. Proponents of cryonics feel that, in the future, damaged tissue can be restored to a healthy state. This technology, if perfected, could repair damage from cryoprotectant toxicity, lack of oxygen, thermal stress (fracturing), ice-crystal formation in tissues that were not successfully vitrified, and reversal of the effects that caused the patient's death in the first place (heart attack, kidney failure, etc.). Advocates of

---

Fascinating Facts About Cryonics

- The most well-known cryopreserved patient is the baseball player Ted Williams.
- Science-fiction author Robert Heinlein wrote enthusiastically about cryonics; however, he was cremated after his death.
- The psychologist Timothy Leary, well known for his advocacy of psychedelic drugs such as LSD, was a cryonics advocate. However, shortly before his death, he changed his mind and was not cryopreserved.
- In a 1773 letter, Benjamin Franklin expressed regret that he lived "in a century too little advanced, and too near the infancy of science" that he could not be preserved and revived to fulfill his "very ardent desire to see and observe the state of America a hundred years hence."
- Arctic and Antarctic insects, fish, amphibians, and reptiles naturally produce cryoprotectants to minimize freezing damage during cold spells. Insects commonly use sugars or polyols (alcohols) as cryoprotectants. Arctic frogs use glucose, and Arctic salamanders produce glycerol in their livers for use as cryoprotectant.
- Woolly bear caterpillars can survive for ten months of the year frozen solid at temperatures as low as −50 degrees Celsius.
- Several novels based on contemporary cryonics have been published, including Bill Clem's *Immortal* (2008), which presents "a chilling look at the science behind cryonics," and James Halperin's *The First Immortal: A Novel of the Future* (1998), in which the protagonist suffers a massive coronary, is cryopreserved, and revived in 2072 by his great-grandson.

---

freezing only the head feel that nanotechnology can regrow an entire body from the head. Theoretical revival scenarios describe repairs being accomplished via large numbers of devices or microscopic organisms. Repairs would be done at the molecular level before thawing.

Preservation of Mental Function and Memory. A successful cryonics outcome entails not only restoration of a healthy body but also a healthy mind, which is fully functional with an intact memory. It is well known that if the brain is deprived of oxygen for more than four to six minutes, ischemic changes occur, which result in brain damage and brain death. Proponents of cryonics claim that these ischemic changes may be reversible with future technology. They also claim that personality, identity, and long-term memory persist for some time after death because they are stored in resilient cell structures and patterns within the brain; thus, these features do not require continuous brain activity to survive. Another more radical cryonics concept is mind transfer. This entails a future technology that could scan the memory contents of a preserved brain.

Storage Facilities. Storage facilities for embryos are rooms containing tanks of liquid nitrogen, each containing many embryos. The tanks are, in essence, large thermoses. Cryonic storage tanks are obviously much larger; some are designed to contain a single individual while others can hold more than a dozen. Cryopreservation requires a constant source of liquid nitrogen because it evaporates, even in specially designed storage tanks. The tanks contain monitoring devices with alarms to indicate a drop in the nitrogen level. Some tanks have automated refilling devices attached. The nitrogen level in the air surrounding the tanks is also monitored. An increase in the nitrogen level would indicate leakage. The tanks are also inspected on a daily basis for any sign of leakage.

IMPACT ON INDUSTRY

Cryonics has minimal impact on industry because of the small number of patients involved. As of 2010, about 200 individuals in the United States have undergone the procedure since it was first offered in 1962. Cryogenics research is also limited and is ongoing on a small scale at cryonics facilities such as the Cryonics Institute and Alcor. In 2006, the Cryonics Society, a nonprofit corporation, was formed. It is focused solely on the promotion and public education of the concept of cryonics. In the much broader field of cryogenics, considerable research is ongoing at facilities such as the National Institutes of Health (NIH) and universities. Furthermore, these facilities conduct research on nanotechnology and aging.

In contrast to cryonics, embryo cryopreservation and the associated areas of assisted reproduction technology (ART) have a significant impact on industry. ART facilities can be found in many metropolitan areas throughout the United States. These facilities are major consumers of medical equipment and supplies such as culture media, ultrasound equipment, and surgical instruments.

CAREERS AND COURSE WORK

Career opportunities are extremely limited in cryonics. The handful of cryonics facilities currently in existence employs a small staff of technicians to maintain the cryopreservation tanks and sales personnel. During a cryopreservation procedure, the staff expands tremendously for a brief period and includes physicians, nurses, and technicians. All temporary employees have full-time positions elsewhere in their areas of expertise.

SOCIAL CONTEXT AND FUTURE PROSPECTS

The field of cryonics is speculative and controversial on scientific, sociological, and religious grounds. Proponents claim that it provides a possibility of a future existence to the finality of death. They suggest the possibility of immortality if revival is successful as well as the associated benefits that postponing or avoiding dying would bring. They tout that the cost of the procedure pales in comparison to the possible benefits. One of the arguments against the procedure is that it would change the concept of death. If life is restored at some time in the future, no friends or family would be left alive. Also, from a religious perspective, the soul would leave the body at the time of cryopreservation. Furthermore, religious beliefs often include a spiritual afterlife. Opponents state that the funds spent for cryopreservation could better be spent on worthwhile causes such as charities or providing funds for education of family members. They also claim that cryonics could lead to premature euthanasia to maximize the chances for revival. In fact, Alcor became involved in a lawsuit involving that topic. In 1987, an eighty-three-year-old woman who had opted for cryonics developed pneumonia after several years of poor health. Alcor personnel deemed that death was imminent and transferred the woman to their facility. She subsequently underwent a "neuro," a head preservation. After the local coroner received a headless body, he demanded the head, which Alcor refused. The coroner launched an investigation and accused Alcor of murder. According to Alcor, the case was settled out of court in the company's favor in 1991.

In its short history, several cryonics facilities have ceased operation. An economic downturn or a large lawsuit could bankrupt a company. If a suitable buyer could not be located to accept the ongoing maintenance costs, cryopreservation would not be maintained for the patients. Continuing the operations of a floundering company would occur only if the projected revenues from living individuals enrolled in the program would justify the costs. If, at some future date, technology exists to resuscitate cryopreserved patients and incorporate nanotechnology, the cost might be prohibitive.

Wide scientific acceptance of cryonics requires both cryopreservation and successful revival. This can be accomplished with animal experimentation. For example, dogs that are placed in a pound and scheduled for euthanasia after expiration of the adoption period could become subjects for cryonics experimentation.

Robin L. Wulffson, M.D., F.A.C.O.G.

FURTHER READING

Foster, Lynn E. *Nanotechnology: Science, Innovation, and Opportunity*. Upper Saddle River, N.J.: Prentice Hall, 2006. Experts in the field present information on where the industry currently stands, how it will evolve, and how it will impact the individual.

Immortality Institute. *The Scientific Conquest of Death: Essays on Infinite Lifespans*. Buenos Aires: Libros en Red, 2004. A collection of essays by nineteen doctors, scientists, and philosophers describing a positive perspective on cryonics.

Johnson, Larry, and Scott Baldyga. *Frozen: My Journey into the World of Cryonics, Deception, and Death*. New York: Vanguard Press, 2009. After a period of employment at Alcor, veteran paramedic Johnson became a whistleblower and described "horrific discoveries" that took place at the facility.

Perry, R. Michael. *Forever for All: Moral Philosophy, Cryonics, and the Scientific Prospects for Immortality*. Boca Raton, Fla.: Universal Publishers, 2000. The book follows recent immortalist thinking that places hope in future advances in humankind's understanding and technology.

Pommer, R. W., III. "Donaldson v. Van de Kamp: Cryonics, Assisted Suicide, and the Challenges of Medical Science." *Journal of Contemporary Health Law and Policy* Spring (1993): 589-603.Very interesting paper chronicling California mathematician Thomas Donaldson's constitutional right to premortem cryogenic suspension. Donaldson was diagnosed with an inoperable brain tumor.

Romain, T. "Extreme Life Extension: Investing in Cryonics for the Long, Long Term." *Medical Anthropology* 29 no. 2 (April, 2010): 194-215. Explores the possibility of biotechnology through cryonics; social aspect of anxiety about aging is also covered.

Shaw D. "Cryoethics: Seeking Life After Death." *Bioethics* 23 (November, 2009): 515-521. This article examines the ethical considerations of cryonic preservation.

WEB SITES
Alcor Life Extension Foundation
http://www.alcor.org

Cryonics Institute
http://www.cryonics.org

The Cryonics Society
http://www.cryonicssociety.org

Sperm Bank, Inc.
http://www.spermbankcalifornia.com/embryo-egg-banking.html

See also: Nanotechnology; Obstetrics and Gynecology; Surgery.

CRYPTOLOGY AND CRYPTOGRAPHY

FIELDS OF STUDY

History; computer science; mathematics; systems engineering; computer programming; communications; cryptographic engineering; security engineering; statistics.

SUMMARY

Cryptography is the use of a cipher to hide a message by replacing it with letters, numbers, or characters. Traditionally, cryptography has been a tool for hiding communications from the enemy during times of war. Although it is still used for this purpose, it is more often used to encrypt confidential data, messages, passwords, and digital signatures on computers. Although computer ciphers are based on manually applied ciphers, they are programmed into the computer using complex algorithms that include algebraic equations to encrypt the information.

KEY TERMS AND CONCEPTS

- **Algorithm:** Mathematical steps to solve a problem with a computer.
- **Ciphertext:** Meaningless text produced by applying a cipher to a message.
- **Cleartext:** Original message; also known as plaintext.
- **Cryptosystem:** Computer program or group of programs that encrypt and decrypt a data file, message, password, or digital signature.
- **Key:** Number that is used both in accessing encrypted data and in encrypting it.
- **Polyalphabetic Cipher:** Cipher that uses multiple alphabets or alphabetic arrangements.
- **Prime Number:** Number that can be divided only by itself and the numeral one.
- **Proprietary Information:** Information that is the property of a specific company.

DEFINITION AND BASIC PRINCIPLES

Cryptography is the use of a cipher to represent hidden letters, words, or messages. Cryptology is the study of ciphers. Ciphers are not codes. Codes are used to represent a word or concept, and they do not have a hidden meaning. An example is the international maritime signal flags, part of the International Code of Signals, known by all sailors and available to anyone else. Ciphers are schemes for changing the letters and numbers of a message with the intention of hiding the message from others. An example is a substitution cipher, in which the order of letters of the alphabet is rearranged and then used to represent other letters of the alphabet. Cryptography has been used since ancient times for communicating military plans or information about the enemy. In modern times, it is most commonly thought of in regard to computer security. Cryptography is critical to storing and sharing computer data files and using passwords to access information either on a computer or on the Internet.

BACKGROUND AND HISTORY

The most common type of cipher in ancient times was the substitution cipher. This was the type of cipher employed by Julius Caesar during the Gallic Wars, by the Italian Leon Battista Alberti in a device called the Alberti Cipher Disk described in a treatise in 1467, and by Sir Francis Bacon of Great Britain. In the 1400's, the Egyptians discovered a way to decrypt substitution ciphers by analyzing the frequency of the letters of the alphabet. Knowing the frequency of a letter made it easy to decipher a message.

German abbot Johannes Trithemius devised polyalphabetic ciphers in 1499, and French diplomat Blaise de Vigenère did the same in 1586. Both used the tableau, which consists of a series of alphabets written one below the other. Each alphabet is shifted one position to the left. De Vigenère added a key to his tableau. The key was used to determine the order in which the alphabets were used.

The Greeks developed the first encryption device, the scytale, which consisted of a wooden staff and a strip of parchment or leather. The strip was wrapped around the staff and the message was written with one letter on each wrap. When wrapped around another staff of the same size, the message appeared. In the 1780's, Thomas Jefferson invented a wheel cipher that used wooden disks with the alphabet printed around the outside. They were arranged side by side on a spindle and were turned to create a huge number of ciphers. In 1922, Jefferson's wheel cipher was adopted by the U.S. Army. It used metal disks and was named the M-94.

Four inventors—Edward H. Hebern, United States, in 1917; Arthur Scherbius, Germany, in 1918; Hugo Alexander Koch, Netherlands, in 1919; and Arvid Gerhard Damm, Sweden, in 1919—independently developed cipher machines that scrambled letters using a rotor or a wired code wheel. Scherbius, an electrical engineer, called his machine the Enigma, which looked like a small, narrow typewriter. The device changed the ciphertext with each letter that was input. The German navy and other armed forces tried the machine, and the German army redesigned the machine, developing the Enigma I in 1930 and using various models during World War II. The Japanese also developed an Enigma-like device for use during the war. The Allies were unable to decrypt Enigma ciphers, until the Polish built an Enigma and sold it to the British. The German military became careless in key choice and about the order of words in sentences, so the cipher was cracked. This was a factor in the defeat of Germany in World War II.

In the 1940's, British intelligence created the first computer, Colossus. After World War II, the British destroyed Colossus. Two Americans at the University of Pennsylvania are credited with creating the first American computer in 1945. It was named the Electronic Numerical Integrator and Calculator (ENIAC). It was able to easily decrypt manual and Enigma ciphers.

How It Works

The data, message, or password starts out as cleartext or plaintext. Once it is input into the computer, a computer algebra system (CAS) performs the actual encryption, using the key and selected algebraic equations, based on the cryptographic algorithm. It stores the data, message, or password as ciphertext.

Most encryption systems require a key. The length of the key increases the complexity of the cryptographic algorithm and decreases the likelihood that the cipher will be cracked. Modern key lengths range from 128 to 2,048 bits. There are two key types. The first is a symmetric or secret key that is used both to encrypt and to decrypt the data. A different key generates different ciphertext. It is critical to keep the key secret and to use a sufficiently complex cryptographic algorithm. The algebraic equations used with a symmetric key are two-way equations.

Symmetric key cryptography uses a polyalphabetic encryption algorithm, such as a block cipher, a stream cipher, or a hash function to convert the data into binary code. A block cipher applies the same key to each block of data. Stream ciphers encrypt the data bit by bit. They can operate in several ways, but two common methods are self-synchronizing and synchronous. The self-synchronizing cipher encrypts the data bit by bit, using an algebraic function applied to the previous bit. Synchronous ciphers apply a succession of functions independent of the data.

Public key or asymmetric cryptography uses two different keys: a public key and a private key. The public key can be distributed to all users, whereas the private key is unknown. The private key cannot be calculated from the public key, although there is a relationship between the two numbers. The public key is used for encryption and the private key is used for decryption. The algebraic equations used in public key cryptography can be calculated only one way, and different equations are used to encrypt and decrypt. The calculations are complex and often involve factorization of a large number or determining the specific logarithm of a large number. Public key algorithms use block ciphers and the blocks may be variable in size.

Digital signatures can be linked to the public key. A digital signature provides a way to identify the creator of the data and to authenticate the source. A digital signature is difficult to replicate. The typical components of a digital signature are the public key, the user's name and e-mail address, the expiration date of the public key, the name of the company that owns the signature, the serial number of the digital identification number (a random number), and the digital signature of the certification authority.

A hash function, or message digest, is an encryption algorithm that uses no key. A hash function takes a variable length record and uses a one-way function to calculate a shorter, fixed-length record. Hash functions are difficult to reverse, and so they are used to verify a digital signature or password. If the new hash is the same as the encrypted version, then the password, or digital signature, is accepted. There are a number of hash algorithms. Typically, they break the file into even-sized blocks and apply either a random number or a prime number to each block. Some hash functions act on all the values in the block, and others work on selected values. Sometimes they generate duplicate hashes, which are called collisions. If there is any change in the data, the hash changes.

There are specific security standards that cryptographic data must meet. They are authentication, privacy/confidentiality, integrity, and nonrepudiation. These standards not only protect the security of the data but also verify the identity of the user, the validity of the data, and that the sender provided the data.

APPLICATIONS AND PRODUCTS

Cryptographic software is created so that it can interact with a variety of computer systems. Some of this software is integrated into other computer programs, and some interfaces with other computer systems. Most cryptographic programs are written in the computer languages of Java, JavaScript, C, C+, or C++. The types of software used for cryptography include computer algebra systems, symmetric key algorithms, public key algorithms, hash functions schemes, and digital signature algorithms.

CAS Software. A computer algebra system (CAS) is a software package that performs mathematical functions. Basic CAS software supports functions such as linear algebra, algebra of polynomials, calculus, nonlinear equations, functions, and graphics. More complex CAS software also supports command lines, animation, statistics, number theory, differential equations, networking, geometric functions, graphing, mathematical maximization, and a programming language. Some of the CAS software that is used for encryption of data includes FriCAS 1.0.3, Maple 12, Mathematica 6.0, Matlab 7.2.0.283, Maxima 5.19.1, and Sage 3.4.

Symmetric Key Algorithms. Symmetric key algorithms work best for storing data that will not be shared with others. This is because of the need to communicate the secret key to the receiver, which can compromise its secrecy. There are two types of secret key encryption algorithms: stream ciphers and block ciphers. The standard for secret key encryption is the Data Encryption Standard (DES) software that the National Bureau of Standards has adopted. Other secret key encryption algorithms include: IDEA (International Data Encryption Algorithm), Rivest Ciphers (RC1 through RC6), Blowfish, Twofish, Camellia, MISTY1, SAFER (Secure and Fast Encryption Routine), KASUMI, SEED, ARIA, and Skipjack. They are block ciphers, except for RC4, which is a stream cipher.

Public Key Algorithms. Asymmetric or public key algorithms support the review of digital signatures and variable length keys. They are used for data that are sent to another businesses or accessed by users, because there is no need to keep the public key secure. The U.S. National Institute of Standards and Technology (NIST) has adopted a new encryption cipher, called Advanced Encryption Standard (AES). AES uses a public key and a block algorithm and is considered to be more secure than DES. Some examples of public key algorithms are: RSA1, Diffie-Hellman, DSA, ElGamal, PGP (Pretty Good Privacy), ECC (Elliptic Curve Cryptography), Public-Key Cryptography Standards (PKCS 1 through 15), Cramer-Shoup, and LUC.

Hash Function Algorithms. Hash function algorithms are not considered actual encryption, although they are often used with encryption algorithms. Hash functions are used to verify passwords or digital signatures, to look up a file or a table of data, to store data, to verify e-mail users, to verify data integrity, and to verify the parties in electronic funds transfer (EFT). Some examples of actual hash algorithms are: SHA-1, SHA-2, MD2, MD4, MD5, RIPEMD, Haval, Whirlpool, and Tiger.

Digital Signatures. A digital signature scheme is used with a public key system and may be verified by a hash. It can be incorporated into the algorithms or just interface with them. If a digital signature algorithm is used, it requires that the user know both the public and the private keys. For hash functions, there may be no key. The digital key also verifies that the message or data were not altered during transmission. Digital signature schemes are used to verify a student's identity for access to an academic record, to verify the identity of a credit card user or a banking account owner, to verify the identity of an e-mail user, to verify the company identities in electronic funds transfers, to verify the source of data that is being transferred, and to verify the identity of the user who is storing data. Some digital signature algorithms include RSA, DSA, the elliptical curve variant ElGamel, variants of Schnorr and Pointcheval-Stern, Rabin, pairing-based schemes, undeniable signature, and aggregate signature.

IMPACT ON INDUSTRY

As society becomes more dependent on computers, data security by encryption becomes more important. Data need to be available to those who use them to perform their jobs, but they must be safe from hacking and unregistered users. Encryption of computerized data and passwords affects all types of businesses, including health care providers, the government, and schools.

Government. The U.S. government has standards for computer cryptography that are used to encrypt some of its own data. IBM designed Data Encryption Standard (DES) in the 1970's, and it has become the most commonly used secret key algorithm. DES has some shortcomings, but modifications have been made to strengthen it. In 2001, the National Institute of Standards and Technology adopted Advanced Encryption Standard, a more secure cryptosystem, for its applications. The U.S. government has to ensure the security of its own data, such as Internal Revenue records, personnel data, and confidential military data. In addition, it provides computers for its personnel. These require security of work data, messages, passwords, and e-mail.

Enigma machine, a German electromechanical device used in World War II to encrypt messages. (James King-Holmes / Bletchley Park Trust / Photo Researchers, Inc.)

Some government agencies are responsible for gathering intelligence about foreign nations. This may involve attempting to decrypt messages sent by these countries.

Universities and Schools. Universities require a secure computer system to store student demographic and health data; student grades and course work; computer access information; personnel information, including salaries; research data; and Internet use. Most students have their own personal computers and are able to access some areas of the university's computer system. Their digital signatures must be linked to a table that reflects their limited access. Faculty members, both on and off campus, are able to access the university's system, and their access also has limits. Researchers use the university system to store data for their projects and to perform analyses. Much of the research on encryption of data has been performed by university researchers. University staff have access to additional data based on their responsibilities. Even elementary and secondary schools have computerized data that include student records and grades, as well as personnel information. Some school systems permit parents to access the grades of their children online. Security is important to protect this data.

Industry and Business. All big businesses and most small businesses have computers and use encryption algorithms. Computers have made it possible to store business data in a small space, to aggregate and analyze information, and to protect proprietary data through encryption. Banking has been changed to a primarily electronic business with the use of automated teller machines (ATMs), debit cards, electronic funds transfer, and credit cards. A key or password within the debit and credit cards is verified by a hash function. Access to an ATM is granted by a public key or credit card number, along with the PIN (personal identification number). Some banking transactions can even be done at home on a personal computer. Encryption keeps this information confidential. Health insurance companies record customer transactions on a computer, process claims, list participating health care providers, link employees to their employer's individual health insurance plans, perform aggregate reporting for income taxes (1099s), and report on employee claims. Actuaries use aggregate claims data to determine insurance premiums. All these functions require that only employees access computerized data. Similar types of data use are performed by most businesses.

CAREERS AND COURSE WORK

Most cryptographers have at least a bachelor's degree in computer science, mathematics, or engineering. Often they have either a master's degree or a doctorate. A number of universities and information technology institutes offer nondegree programs in cryptography. Cryptographers are persons who are knowledgeable in encryption programming, computer security software, data compression, and number theory, as well as firewalls and Windows and Internet security. People with this background may be employed in computer software firms, criminology, universities, or information technology.

There is a voluntary accreditation examination for cryptographers, which is administered by the International Information Systems Security Certification Consortium. The examination is called the Certified Information Systems Security Professional (CISSP) examination. To be certified, a cryptographer must have five years of relevant job experience and pass the CISSP examination. Recertification must be done every three years, either by retaking the examination or by earning 120 continuing professional education credits.

Position titles include cryptoanalysts, cryptosystem designers, cryptographic engineers, and digital rights professionals. A cryptoanalyst is involved in the examination and testing of cryptographic security systems. Cryptosystem designers develop the complex cryptographic algorithms. Cryptographic engineers work with the hardware of cryptographic computer systems. Digital rights professionals are responsible for the security of encrypted data, passwords, and digital signatures. They may be certificate administrators and be responsible for accepting new users and approving their digital signatures.

SOCIAL CONTEXT AND FUTURE PROSPECTS

Computer technology is an important part of contemporary life. Encryption of data, messages, passwords and digital signatures makes this possible. Without data and password security, both personal and professional users would find that anything that they loaded onto their computers would be available to hackers and spies, and they would be vulnerable to computer viruses that could damage their computers and corrupt their files. It would be difficult to use a computer under these circumstances because of the lack of dependability of the data and of the computer system. Many workers perform their jobs on a computer, and some are able to work from home. Wireless systems are increasingly being used. These capabilities require adequate security.

Despite the complexity of modern cryptography, there still are risks of attacks on computer data, messages, and passwords. No cryptography program is without its weak point. The likelihood of breaking a cryptographic algorithm is assessed by how long it would take to break the cipher with a high-speed computer. A common attack on key cryptography is brute force. This involves trying all the possible number combinations in order to crack the key. The longer the key is, the longer this will take. The only way to create an unbreakable cipher is to use a one-time pad, in which the secret key is encrypted using an input number that is used only once. Each time the data are accessed, another random number of the same length is used as the secret key.

Cryptosystems are not totally secure. As computer technology and cryptography knowledge advances, any particular encryption algorithm is increasingly likely to be broken. Cryptoanalysts must constantly evaluate and modify or abandon encryption algorithms as needed. A future risk to cryptographic systems is the development of a quantum computer, based on quantum theory. It is thought that a quantum computer could crack all the cryptographic ciphers in existence; however, no quantum computer has been created yet.

Christine M. Carroll, B.S.N., R.N., M.B.A.

FURTHER READING

Blackwood, Gary. *Mysterious Messages: A History of Codes and Ciphers.* New York: Penguin Books, 2009. Provides a comprehensive history of the development of cryptography from ancient history through modern times.

Kahate, Atul. *Cryptography and Network Security.* 2d ed. Boston: McGraw-Hill Higher Education, 2009. Describes the concepts involved with the use of cryptography for network security.

Katz, Jonathan, and Yehuda Lindell. *Introduction to Modern Cryptography.* Boca Raton, Fla.: Chapman & Hall/CRC, 2008. Explores both public and private key cryptography. Looks at various models and systems.

Lunde, Paul, ed. *The Book of Codes: Understanding the World of Hidden Messages.* Los Angeles: University of California Press, 2009. Covers codes and ciphers used in society and the history of cryptography. Chapter 13 discusses the development of the computer and computer cryptography.

Seife, Charles. *Decoding the Universe.* New York: Penguin Group, 2006. Discusses information theory, which is the basis of computer cryptography.

WEB SITES

American Cryptogram Association
http://cryptogram.org

International Association for Cryptographic Research
http://www.iacr.org

International Financial Cryptography Association
http://ifca.ai

International Information Systems Security Certification Consortium
https://www.isc2.org

See also: Computer Networks; Computer Science; Criminology; Information Technology; Internet and Web Engineering; Software Engineering.

D

DEMOGRAPHY AND DEMOGRAPHICS

FIELDS OF STUDY

Sociology; political science; social psychology; statistics; probability; data collection; data modeling; economics; anthropology; biology; population dynamics; urban studies; public policy.

SUMMARY

Demography and demographics both involve the study of human populations. Demography refers to the discipline of measuring and analyzing factors about groups of people, such as group size, composition, density, and growth. These factors may be studied at a single point in time or followed over years. Demographics commonly refers to the factors themselves but can also mean the study of populations as a discipline. Many demographers work in professional fields related to marketing. Others seek to understand populations to predict political activity such as voting patterns, public health trends, or the expansion and decline of cities.

KEY TERMS AND CONCEPTS

- **Census:** Gathering of data on all members of a population.
- **Cohort:** Population that experienced the same type of life-shaping event, such as birth or marriage, within a defined time period.
- **Migration:** Movement of a population from one place to another to establish new homes.
- **Morbidity:** Rate at which disease and injuries affect a population.
- **Mortality:** Rate at which deaths occur within a population.
- **Natality:** Rate at which births occur within a population.
- **Population:** Group of people studied by demographers that share a characteristic, such as living in a region.
- **Survey:** Gathering of data on selected, but not all, members of a population.

DEFINITION AND BASIC PRINCIPLES

Demography is the study of human populations and the ways in which they grow and change over time. Although a population might be a group of people as small as a single rural village, most demographic studies look at groups living in a metropolitan area, a region, or even a country or continent. Demographers are also likely to use data categories such as age, family status, ethnicity, household size, education, and income level in answering questions about population change.

Three primary factors govern changes in population size. Fertility is the rate at which new members of the population are born. Migration shows the rate at which members move in and out of the population during their lifetimes. Mortality is the rate at which population members die. Demography is the social science in which these factors, along with many others, are analyzed to better understand why a population is growing or declining.

Demographics is a term that refers to the data gathered by demographic analyses. Government agencies are some of the leading users of demographic data as they create and implement public policy. Companies also rely on demographic data as they design products and services for growing markets.

BACKGROUND AND HISTORY

Philosophers and scholars have discussed ideas about population growth for thousands of years. In ancient Greece, Plato and Aristotle advocated the concept that civilizations should strive to reach certain population levels determined by the maximum quality of life that could be achieved under such numbers. Both Plato and Aristotle examined questions about migration, population and the environment, and fertility control. Similar issues were explored by writers in China and India.

Demography was further developed in Europe in the sixteenth through eighteenth centuries under a school of thought known as mercantilism. Under mercantilism, population growth was seen as a necessary force behind the increase of a country's trading power and wealth. In 1798, this philosophy was challenged by British professor Thomas Malthus, who anonymously published a pamphlet in which he argued that human biology would, by nature, create a future society that would be overpopulated and struggle with heightened problems such as poverty and famine.

Many nineteenth century demographers such as William Farr and Louis-Adolphe Bertillon made advancements in the field through their work in public health. By the late 1800's, statistical analysis had become a key component of demographic study. Actuarial science, led by insurance providers to better understand mortality and risk, also contributed to the discipline.

How It Works

Demographic studies seek to answer questions about human populations. To conduct a study, demographers must define their questions and design a survey or census to collect the necessary data. Once the data are gathered, demographers apply statistical tools to measure and analyze the information. The results allow demographers to determine whether their question has been answered or whether further research is needed.

Question Definition. Every demographic study begins with a question about a group of people that can be answered by data analysis. A well-designed study needs a question that is thoroughly defined. For example, a city's board of education might ask the question, "How many students are expected to attend public school in the city each year over the next ten years?" Demographers working for the board would choose the items to be measured in the study and define each item in detail, such as the number of days in a year that a student would need to be present at school in order to be counted.

Study Design and Data Collection. Once the question is defined, demographers decide what sources of data to use and how to collect the information. Many demographic studies rely on census data, which involves gathering information on each member of a population. Census data are thorough but difficult

As the global population hits seven billion, experts are warning that skewed gender ratios could fuel the emergence of volatile "bachelor nations" driven by an aggressive competition for brides. (AFP/Getty Images)

and expensive to compile. Many demographers instead rely on surveys, which collect information from people chosen as a representative sample of the population. The survey data are analyzed statistically to infer conclusions about the population as a whole. In the example of the school board, the demographers might gather data through a census or survey of households with at least one child below the age of nine.

Measurement and Analysis. Demographers use tools that store the information gathered in surveys or censuses and assist with statistical analysis. The demographers in this example might transfer their survey data into a database or other application that would allow them to create multiyear growth forecasts. Additional information might come from population growth projections for the city or region. The demographers would use this information to compensate for areas not covered by the survey, such as households where there are not yet any children or households living outside the city limits at the time of the survey's mailing. Once these sources of data are combined and analyzed, demographers document their findings and make recommendations about additional research, if needed.

The types of data gathered by demographers often fall into categories. Fertility measures the rate at which a population grows because of the number of babies being born to population members. Migration is the rate at which people join or leave a population by moving to new homes inside or outside its borders. Immigration refers to people moving into an area, while emigration describes people leaving the area. Some studies follow patterns known as domestic migration, in which people move from one part of an area to another. Mortality measures a population's rate of death. Aside from these categories, demographers seek to understand factors such as the distribution of age and sex across a population. These factors are particularly important when using population data to build forecasts.

APPLICATIONS AND PRODUCTS

Demographic data are used in a wide range of contexts. Some of the most common tools for demographic research are population census data, social surveys, and commercial and marketing surveys.

Population Census Data. Countries throughout the world track the number of people living within their borders. Most national governments have a department or agency that uses a tool such as a census to gather information. Census data are used by other government agencies to track population growth, decline, and migration. Aggregate data from censuses are often published by governments for use by businesses and the public. These data are released in the form of reports and electronic databases. In many countries, a full population census is conducted roughly once every five to fifteen years. Countries such as France gather population data on a yearly basis. Other countries such as Germany do not conduct a census of all residents but use statistical sampling to estimate information about their populations.

Social Surveys. Social surveys are conducted on selected members of a population to gather data that

will be used to understand the population as a whole. Although they present a higher rate of error than a census, they cost less and can be carried out more quickly. Many large-scale social surveys focus on specific issues such as education, housing, employment, or health. Others are focused by geography. The American Community Survey from the Census Bureau is one of the best known social surveys in the United States. The United Nations conducts a large number of social surveys that are focused by both topic and region, such as the economic and social survey of Asia and the Pacific.

Commercial and Marketing Surveys. Demographic information about groups of people is useful to companies as well as government agencies. Commercial and marketing surveys gather some of the same types

---

Fascinating Facts About Demography and Demographics

- The terms "demography" and "demographics" come from the Greek words for "people" and "writing."
- Greek philosophers Plato and Aristotle were in favor of giving rewards such as medals to parents for increasing a city's population.
- Mercantilism, a popular school of thought in sixteenth- through eighteenth-century Europe, linked population growth to economic strength. It supported the idea that more people—and therefore more labor—enabled a country to arm for war more quickly than its enemies could.
- Characters in Aldous Huxley's novel *Brave New World* (1932) wear contraceptive Malthusian belts, named after Malthus, as a form of population control.
- The U.S. POPClock and World POPClock are maintained by the U.S. Census Bureau on the agency's Web site. The clocks estimate the U.S. and global population and are updated once a minute.
- Aside from counting people, the U.S. Census Bureau studies many things about the way they live, work, and talk. The bureau found that the number of U.S. residents who speak a language other than English at home rose 140 percent from 1980 to 2007. The population as a whole grew 34 percent.

of data as social surveys. Companies focusing on a new target market or expanding their hold on an existing group of customers are likely to track factors such as average age, ethnic background, family structure, education, and income levels. When analyzed, these factors are used to guide the company's decisions about which products and services to offer and how to communicate them to the customers most likely to buy them. Market data publisher Nielsen's Pop-Facts reports are an example of demographic data available commercially.

Areas of Focus. Professionals working within the field of demography often hold jobs that require them to focus by subject area. Common areas in which demographers are employed are public health and epidemiology, immigration and emigration policy, and urban and environmental planning.

Public Health and Epidemiology. Demographers working in the field of public health and epidemiology seek to answer questions such as which public health problems pose the greatest risk to a population, based on its age structure and sex ratio, and what factors will define a population's public health needs in ten to twenty years. They conduct surveys to gather data, which they may combine with other sources, such as statistics from hospitals, to assess which health issues are most critical for a selected group of people.

One of the largest subfields of demography in public health concerns fertility and family planning. In many of the world's most developed economies, the overall birth rate is high enough to support stable growth without posing unusual challenges to a country's public resources such as health care. However, developing economies, such as those in many African and Asian countries, have been experiencing rapid growth in their populations because of high fertility levels. Demographers studying these trends are likely to be involved with health issues concerning mothers and children. They may also be employed by government agencies as advisers on the development of public policies shaping long-term population growth, such as economic incentives for having more or fewer offspring.

Immigration and Emigration Policy. Demographers advising on immigration and emigration policy examine issues such as how much of a country's population growth is caused by people moving into the country and why a country's most educated residents are leaving for other countries. Although questions of migration are important in a wide range of contexts, public policy is developed by government agencies and specialized research firms, where demographers focusing on the issues are most likely to work. Migration policy has become a topic of great public interest in the twenty-first century in areas such as the European Union, where residents may live and work not only in the country of which they are a citizen but also in member countries. Regions such as the Caribbean and the countries that were formerly Soviet republics or satellite states are studying emigration to better understand how to retain their most skilled citizens, who have moved in large numbers to countries with stronger economies in a trend known as brain drain.

Urban and Environmental Planning. Population change has a profound effect on the environment, especially where housing and employment are concerned. As people live and work in an area, they influence its natural resources and the types of infrastructure needed to sustain further growth, such as houses, roads, schools, stores, and offices. Demographers who work in the field of urban and environmental planning hold a variety of jobs. They may work for city planning agencies or provide research for national offices such as the U.S. Department of Housing and Urban Development. They also may specialize in specific categories of growth, such as families with young children, to advise boards and departments of education in planning school systems. Many demographers choose to work in private industry for companies such as homebuilders, where they support the design of new types of houses or make forecasts about where and when population growth is likely to be highest.

IMPACT ON INDUSTRY

Demography is, by nature, a multidisciplinary field. Few areas of government and industry are not affected by population change in some way. Changes in age ranges, family types, sex ratios, and migration patterns have an impact on public services of all kinds as well as the types of products and services that can be sold to the members of a population.

The Population Association of America, the leading professional organization for demographic specialists in the United States, provides a forum in which demographers can share their expertise.

Although the association does not release statistics on the types of jobs held by its members or the sectors in which they work, the organization has active connections in a number of areas. These relationships provide insight into the influence held by demographers in many professional spheres. They include the Ad Hoc Group for Medical Research, a group that provided $35 billion in federal funding for the National Institutes of Health in fiscal year 2011; the Census Project, a team of public- and private-sector leaders in demography that were active in lobbying for $14 billion to support the work of the 2010 U.S. Census; the Coalition for Health Funding, made up of fifty national professional associations in the health care field that seek funding for the U.S. Public Health Service and its many agencies; and the Council of Professional Associations on Federal Statistics, a group of associations, businesses, research institutions, and individual members who foster discussion and collaboration on uses of the statistics released by agencies of the federal government.

Demographic data, while useful, can be difficult and expensive to compile. Information must be mined from existing resources such as government reports, or it must be collected directly from population members through censuses and surveys. Because the same demographic data can be valuable in multiple applications, an industry of demographic research firms has evolved to serve the needs of companies. These research firms gather information and sell it to clients in the form of raw data, published studies, or customized consulting services.

Some providers of demographic data are large and multinational in scope. Leading commercial firms of this type include Nielsen, Information Resources, Kantar, and Ipsos. These firms concentrate on data about not only consumers themselves but also their behavior, specifically in purchasing goods and services. Nielsen and several of its competitors also monitor the ways in which consumers use media such as television and the Internet. Firms such as Nielsen invite consumers, known as members, to participate in surveys. Information about buying habits and media use is gathered through a variety of channels. These channels range from the voluntary answering of questions posed to consumers to the automated collection of data through monitoring devices. In 2009, Nielsen, the largest consumer demographics research firm worldwide, reported $4.8 billion in revenues.

Other demographic research providers are smaller and concentrate on specific populations. These populations might be defined by age, such as those studied by Teenage Research Unlimited or Childwise, or by ethnic background, such as Multicultural Marketing Resources. Specialist firms provide data to companies that allow products and services to be tailored to the needs of a more focused market.

CAREERS AND COURSE WORK

Demographers have a wide range of career options in government, industry, and nonprofit settings. Because there is no association that certifies college and university programs in demography, few numbers are available when it comes to students or full-time employees in the field. However, the professional demands of studying population change make a bachelor's degree or graduate degree an asset when looking for a job.

Students interested in demography—also known as population studies at many institutions—are likely to approach it as a topic within a major in sociology, anthropology, political science, economics, geography, or biology. Course work varies widely, depending on the student's focus and department. Most students interested in demography take classes on topics such as population theory and human development. Multiple courses on research methods, study design, and statistical analysis are required. Issue-specific classes cover topics such as poverty, reproductive policy and health, and the environmental impact of human populations. Students who earn graduate degrees often take managerial positions or work as professors or researchers at academic institutions.

Government agencies and nonprofit organizations employ many demographers. The U.S. Census Bureau is one of the best-known agencies, but population studies play a key role in aspects of government ranging from health care policy to land use. In industry, jobs in marketing and new product development often call for experience in gathering and interpreting demographic data. This experience can come through college-level course work as well as internships with consumer products manufacturers.

SOCIAL CONTEXT AND FUTURE PROSPECTS

The demand for professionals with a background in demography and population studies is expected to increase significantly in the 2010's. Job candidates with bachelor's or graduate degrees and years of focused experience in demographic analysis are likely to benefit most. There will also be many opportunities for new college graduates with course work in demography.

The U.S. Bureau of Labor Statistics reports that sociologists and political scientists, a group to which demographers belong, will see job growth of 21 percent from 2008 to 2018—a rate considered much faster than average. Sociologists earned a median annual income of $68,570 in 2008, and political scientists reported a median income of $104,130. Many of the opportunities open to demographic specialists are not called "demographer" but instead carry titles such as "policy analyst" or "research analyst."

Government agencies will continue to hire demographers as elected officials look for more efficient ways in which to use tax funds. Demographers can help evaluate whether a particular program is reaching its target audience most effectively. Consumer products manufacturers and marketing firms will hire more demographic specialists as new product development and advertising focus on smaller customer niches. Research centers and management consulting firms are likely to be interested in demographers for their experience working with complex data on population change.

Julia A. Rosenthal, B.A., M.S.

FURTHER READING

Buchholz, Todd G., and Martin Feldstein. "Malthus: Prophet of Doom and Population Boom." Chapter 3 in *New Ideas from Dead Economists: An Introduction to Modern Economic Thought.* New York: Penguin, 2007. A brief overview of the life of Thomas Malthus and his landmark contributions to demographics.

Magnus, George. *The Age of Aging: How Demographics Are Changing the Global Economy and Our World.* New York: John Wiley & Sons, 2008. A discussion of the effects of decreasing birth rates and increasing average ages worldwide.

Malthus, Thomas. *An Essay on the Principle of Population.* New York: Oxford University Press, 2008. A scholarly edition of Malthus's most famous work.

Poston, Dudley L., Jr., and Leon F. Bouvier. *Population and Society: An Introduction to Demography.* New York: Cambridge University Press, 2010. A college-level textbook that covers demography as a science, written for a nonspecialist audience.

Siegel, Jacob S., and David A. Swanson, eds. *The Methods and Materials of Demography.* 2d ed. San Diego, Calif.: Elsevier, 2004. Provides the tools and techniques used by demographers.

Yaukey, David, Douglas L. Anderton, and Jennifer Hickes Lundquist. *Demography: The Study of Human Population.* 3d ed. Long Grove, Ill.: Waveland Press, 2007. Examines the causes of changes in the human population and the effects of these changes. Features an international focus and long-term projections.

WEB SITES

The Census Project
http://www.thecensusproject.org

Council of Professional Associations on Federal Statistics
http://www.copafs.org

Population Association of America
http://www.populationassociation.org

Population Reference Bureau
http://www.prb.org

U.S. Bureau of the Census
http://www.census.gov

See also: Land-Use Management; Noise Control; Probability and Statistics; Reproductive Science and Engineering; Urban Planning and Engineering.

DENTISTRY

FIELDS OF STUDY

Biology; health science; physiology; anatomy; pharmacology; biochemistry; chemistry; mathematics; microbiology; physics.

SUMMARY

Dentistry involves the diagnosis, treatment, and prevention of disorders and diseases of the teeth, mouth, jaw, and face. Dentistry includes instruction on proper dental care, removal of tooth decay, teeth straightening, cavity filling, and corrective and reconstructive work on teeth and gums. Dentistry is recognized as an important component of overall health. Practitioners of dentistry are called dentists. Dental hygienists, technicians, and assistants aid dentists in the provision of dental care.

KEY TERMS AND CONCEPTS

- **Appliance:** Removable restorative or corrective dental or orthodontic device.
- **Bite:** Contact of the upper and lower teeth; also known as occlusion.
- **Caries:** Tooth decay; also known as cavities.
- **Cleaning:** Removal of plaque and tartar from the teeth, generally above the gum line.
- **Enamel:** Hard ceramic that covers and protects the exposed part of the tooth.
- **Gingiva:** Soft, pink tissue surrounding the base of the teeth; also known as gums.
- **Permanent Teeth:** The thirty-two teeth that appear after the loss of the primary teeth, beginning around the age of six years; also known as adult teeth.
- **Plaque:** Sticky film of food particles, bacteria, and saliva that forms on the teeth and can eventually turn into tartar.
- **Primary Teeth:** The first set of teeth that appear between the ages of six months and one year that help children learn to speak and chew; also known as baby teeth or deciduous teeth.
- **Pulp:** Soft, inner part of the tooth that contains nerves and blood vessels.
- **Root:** Part of the tooth that is embedded in the gums.
- **Tartar:** Hard deposit that adheres to teeth and attracts plaque; also known as dental calculus.

DEFINITION AND BASIC PRINCIPLES

Dentistry is a branch of medicine that focuses on diseases and disorders of the teeth, mouth, face, and oral cavity. Dentistry includes examining the teeth, gums, mouth, head, and neck to evaluate dental health. The examination may include a variety of dental instruments, imaging techniques, and other diagnostic equipment. Dentistry involves diagnosing oral or dental diseases or disorders and formulating treatment plans. Dentistry is instrumental in teaching patients about the importance of maintaining oral health and instructing patients on proper oral hygiene techniques.

Although dentistry is an independent health care field, it is not entirely detached from other health care services and collaboration between dentistry and other health care providers ensures positive outcomes for patients. Dentists often see patients more often than physicians and may be the first to diagnose systemic diseases, including inflammatory conditions, autoimmune diseases, and cardiovascular risk factors. Dentists also work with pharmacists to prescribe the best antibiotics or anesthetics for dental patients, as well as to understand how certain medications affect dental care and oral health.

Dentistry not only prevents and treats serious oral health disorders but also provides cosmetic services to enhance facial features and correct signs of aging. Dentistry strives to promote oral health as a part of overall health and applies principles of basic medicine, pharmacology, and psychology to dental care.

BACKGROUND AND HISTORY

Before the seventeenth century, dental care was crude, unrefined, and most often provided by physicians. Through the eighteenth and nineteenth centuries, dentistry emerged as its own medical discipline, and most dentists trained through apprenticeships.

Pierre Fauchard is credited as the founder of modern dentistry. In 1728, the French surgeon published *Le Chirurgien Dentiste: Ou, Traité des dents* (*The Surgeon Dentist: Or, Treatise on the Teeth*, 1946), which summarized all available knowledge of dental anatomy, diseases of the teeth, and the construction of dentures.

In 1840, Horace Hayden and Chapin Harris established the world's first dental school, the Baltimore College of Dental Surgery, in Baltimore, Maryland. In 1867, Harvard University became the first university to establish a university-affiliated dental school.

Several scientific milestones transformed dentistry in the nineteenth century. In 1844, Horace Wells administered nitrous oxide to a patient before a tooth extraction, becoming the first dentist to use anesthesia. In 1890, dentist Willoughby Dayton Miller connected microbes to the decay process, extending the germ theory to dental disease. In 1898, William Hunter introduced the term "oral sepsis" to the profession of dentistry and called attention to the contaminated practices and instruments used by dentists. In 1918, radiology was added to dental school curricula, and by the 1930's, most dentists in the United States were using X rays as part of routine dental diagnostics.

Advances in science and technology, including the sequencing of the human genome and the arrival of the digital age, have revolutionized dentistry, rendering it nearly unrecognizable when compared with nineteenth-century dentistry and improving the diagnostic and treatment capabilities within the field.

How It Works

Dental Tools. Many common dental tools are available for home use as part of a daily oral care routine. The most basic of dental tools is the toothbrush. Toothbrushes come in a variety of sizes, shapes, and stiffness. Patient age and oral condition determine the best toothbrush for each individual. Toothbrushes usually consist of a plastic handle with nylon bristles that remove food, bacteria, and plaque that can lead to tartar and dental caries. Toothpaste is usually added to a toothbrush to aid in cleaning the teeth and freshening the mouth. Toothpaste is available in a variety of flavors and compositions and may contain polishing or bleaching agents. Dental floss is another basic tool used to remove food and debris from between the teeth. Floss is available in waxed and unwaxed formulations and in a variety of widths and thicknesses. Mouthwash is a rinse that prevents gum disease. Mouthwash is available in many flavors, but all types reduce the number of germs in the mouth that cause gingivitis.

More sophisticated dental tools are used by dentists during dental examinations and procedures. A routine dental cleaning removes stains on the teeth, as well as tartar that brushing and flossing cannot remove. Polishing the teeth aids the dentist in visualizing the teeth and makes it more difficult for plaque to accumulate on the surface of the teeth. Mirrors, scrapers, scalers, and probes are essential in-office dental tools.

Dental Therapy and Devices. Countless therapies and devices are available to diagnose, prevent, and treat disorders and diseases of the teeth and mouth. Extraction, previously the mainstay of dentistry, involves simply removing the affected tooth. Fillings are used to replace a portion of a tooth that is missing or decayed. Fillings are often made of gold or silver but may also be made of composite resins or amalgam depending on the size and location of the filling. A dental implant is the extension or replacement of a tooth or its root by inserting a post made of metal or other material into the bone to support a new artificial tooth. Crownwork involves covering a damaged tooth with porcelain or other alloy to restore the tooth's original size and shape. A denture is a removable prosthetic appliance that replaces missing teeth. Dentures may replace all or just some teeth. In contrast, a bridge is a tooth-replacement device that cannot be removed. A bridge is made of one or more artificial crowns that are cemented to adjacent teeth.

Orthodontic appliances are necessary to correct and prevent irregularities in the alignment of the teeth, face, and jaw. Braces are among the most common orthodontic appliances, along with headgear and retainers. Conventional braces have metal brackets that are attached to the outer surfaces of the teeth. Wires are attached to the brackets, and manipulation of the wire allows movement and rotation of the teeth into the desired position. The braces may be attached to headgear to help move teeth or secure them into position. Retainers are often worn after braces are removed to maintain the new position of the teeth. Retainers may be permanent or removable. Removable retainers consist of a wire attached to a resin base that is worn at all times (except during meals) to hold the teeth in place for up to several years after braces are removed. A permanent retainer is a metal wire attached to the tongue side of the lower teeth that can maintain the desired position of severely crowded or rotated teeth.

Various scalpels are seen at the University of Nebraska school of dentistry in Lincoln, Nebraska. (AP Photo)

APPLICATIONS AND PRODUCTS

Most dentists practice dentistry as general practitioners. In addition, the American Dental Association recognizes nine specialties within the field of dentistry, each of which requires additional education or training beyond dental school.

General Dentistry. A general practitioner of dentistry deals with the overall maintenance of patients' teeth, gum, and mouth health. Ideally, general dentistry is preventive in nature, focusing on the maintenance of oral health and hygiene to avoid the occurrence of disorders and diseases of the mouth. Dentists who practice general dentistry encourage regular checkups to ensure proper functioning of the mouth and teeth. A general dentist will provide individualized treatment plans that include dental examinations, tooth cleanings, and X rays or other diagnostic tests to prevent or treat disorders of the mouth as early as possible. General dentists also repair and restore injuries of the teeth and mouth that result from decay, disease, or trauma. All dentists are able to prescribe medicines and treatments to diagnose, prevent, or treat diseases of the mouth and teeth.

Orthodontics. The largest specialty within dentistry is orthodontics. Orthodontics focuses on straightening teeth and correcting misalignment of the bite, usually using braces and retainers. Misalignment of the teeth or bite can cause eating or speaking disorders, making orthodontics an important part of overall health. Also, orthodontics may be aesthetic in nature, focusing on improving the structure and appearance of the teeth, mouth, and face to improve a patient's self-esteem. Most orthodontic patients are children because corrective procedures of the teeth are most effective when started early. However, an increasing number of adults are seeking orthodontic care, owing to the development of new methods and techniques in orthodontics that allow minimal discomfort and improved healing.

Oral and Maxillofacial Surgery. Commonly referred to as oral surgery, oral and maxillofacial surgery is the application of surgical techniques to the diagnosis and treatment of disorders of the teeth, mouth, face, and jaw. An oral surgeon may remove damaged or decayed teeth under intravenous sedation or general anesthesia; place dental implants to replace missing or damaged teeth; repair facial trauma, including injuries to soft tissues, nerves, and bones; evaluate and treat head and neck cancers; alleviate facial pain; perform cosmetic surgery of the face; perform corrective and reconstructive surgery of the face and jaw; and correct sleep apnea.

Pedodontics. Also known as pediatric dentistry, pedodontics focuses on dental care and oral hygiene of children and adolescents. Pediatric dentists apply the principles of dentistry to the growth and development of young patients, oral disease prevention, and child psychology. Some pediatric dentists also specialize in the treatment of patients with developmental or physical disabilities. Pediatric dentists emphasize proper oral hygiene, beginning with baby teeth, because healthy teeth allow for proper chewing and correct speech. Pediatric dentists also stress the importance of proper nutrition for its role in oral health, as well as overall growth and development. Early dental care facilitates lifelong oral health.

Periodontics. The field of dentistry called periodontics studies the bone and connective tissues that surround the teeth. Periodontics also involves the placement of dental implants. Periodontists prevent, diagnose, and treat periodontal disorders and infections, including gingivitis and periodontitis. Most periodontal diseases are inflammatory in nature, as are some cardiovascular diseases, and a connection between these two disease states has prompted physicians and periodontists to work together to treat patients at risk for either condition.

Prosthodontics. Also known as prosthetic dentistry, prosthodontics is the specialized field of dentistry that focuses on restoring and replacing teeth with dental implants, bridges, dentures, and crowns. Although general dentists can perform simple restoration or replacement of teeth, prostodontists handle severe or extreme cases of tooth loss because of trauma, disease, congenital defects, and age.

Endodontics. The field of dentistry that studies abnormal tooth pulp and focuses on the prevention, diagnosis, and treatment of diseases of the tooth pulp is called endodontics. Endodontic treatment is also known as root canal therapy. Endodontic therapy

may also include surgery necessary to save a diseased tooth. Endodontists are often able to treat the diseased or damaged inside of a tooth instead of extracting it completely.

Oral and Maxillofacial Pathology. In oral and maxillofacial pathology, the principles of dentistry are applied to investigating the causes and effects of diseases of the mouth, head, and neck. Oral pathologists are trained to diagnose and treat such diseases, as well as to expose the connection between oral disease and systemic disease.

Oral and Maxillofacial Radiology. The use of advanced imaging techniques to diagnose and treat disorders of the mouth, teeth, head, and neck is known as oral and maxillofacial radiology. An oral and maxillofacial radiologist is a dentist who uses radiographic images to diagnose disease and guide treatment plans. Radiologists may use X rays, computed tomography (CT) scans, magnetic resonance imaging (MRI), ultrasound, and positron emission tomography (PET) to visualize the oral cavity or maxillofacial regions. Specialized sialography images the salivary glands. Intraoral radiographs are used routinely by general dentists as part of regular dental checkups.

Dental Public Health. The field of dental public health is involved in the epidemiology of dental diseases and applies the principles of dentistry to populations rather than individuals. Dental public health specialists have been involved in promoting fluoridation of drinking water and examining the links between commercial mouthwash and cancer. Dental health specialists assess the oral health needs of communities, develop programs to teach and promote oral health, and implement policies and regulations to address oral health issues.

IMPACT ON INDUSTRY

Globally, demand for dental care and consumer oral health care products, including toothbrushes, toothpaste, dental floss, and mouth rinse, is growing at an annual rate of at least 2 percent, with stronger growth in Latin America and Europe. In 2004, sales of oral care products in the United States—the world's strongest oral health care market—reached almost $3 billion. Globally, the dentistry industry reaches revenues of nearly $20 billion annually, dominated by markets in the United States, Europe, and Japan. The industry is highly segmented, with primarily small office practices and clinics and no suppliers of dental technology or products attaining a majority of market share.

Government Initiatives. In the United States, the government is increasingly involved in the provision and regulation of health care and medical services, including dentistry, in order to maintain the safety and health of patients, consumers, and workers within the industry.

The Occupational Safety and Health Administration (OSHA), a division of the United States Department of Labor, maintains workplace safety standards for the dentistry industry. Primarily, dental professionals may be exposed to toxic or harmful chemicals, materials, infectious substances, or medications as part of their professional duties. OSHA establishes guidelines to recognize and prevent situations that place workers in the dentistry industry at risk for exposure to harmful substances or pathogens. OSHA also provides safety training for all health care workers.

The United States Department of Health and Human Services established the Health Insurance Portability and Accountability Act (HIPAA) in 1996, which, in part, protects the privacy and security of private health information. Dental care and history is considered private health information and is, therefore, protected by HIPAA regulations. Dental practices must adhere to strict guidelines to maintain patient confidentiality and improve patient safety.

The National Institute of Dental and Craniofacial Research (NIDCR), part of the National Institutes of Health, is the government-sponsored research arm within the field of dentistry. The NIDCR conducts research covering the entire spectrum of dental-related diseases and disorders and encourages interdisciplinary approaches to oral health. The NIDCR is a major contributor to the government's Healthy People 2010 program, along with the Centers for Disease Control and Prevention, the Indian Health Service, and the Health Resources and Services Administration. Overall, the goal of the oral health arm of the program is to prevent oral and craniofacial diseases and injuries by promoting oral hygiene and health maintenance and improving access to oral health care for underserved populations.

Consumer Initiatives. Consumer demand for oral health care products is increasing as people recognize the importance of oral health in overall health and well-being. There are several major corporations that

---

Fascinating Facts About Dentistry

- In the mid-1850's, a textbook on dental surgery taught that asbestos could be placed under the filling of a sensitive tooth because asbestos is unable to conduct heat or electricity.

- The earliest evidence of dental caries dates back 100 million years to the Cretaceous period; dinosaur and fish fossils from this period show signs of dental decay.

- Toothpicks were in use at least 3,000 years ago and were made of wood, metal, thorns, or porcupine quills; ornate metal toothpicks were a sign of wealth in ancient Egypt.

- Saint Appollonia, a Christian martyr whose teeth were removed by her Roman captors, is the patron saint of dental pain sufferers, and her intercession is thought to bring healing to all oral pain and afflictions.

- The structure and anatomy of the teeth and mouth are unique to each individual, and dental records and examinations are used to identify victims of accidents, terrorism, or natural disasters. Paul Revere was the first dentist to suggest using bridgework to identify remains—namely, a Revolutionary War general—and he became a pioneer of forensic dentistry in the United States.

- More than 90 percent of systemic diseases exhibit oral manifestations; oral signs and symptoms may come before, after, or at the same time as signs and symptoms elsewhere in the body but are often the first signs of systemic illness.

---

contribute to the worldwide market for dental products, but no single company dominates the landscape. The largest players include Colgate, Procter & Gamble, and Johnson & Johnson. Many consumer oral health product manufacturers are involved in sponsoring dentistry research and education. In addition to educating dentistry scientists and professionals, these large corporations sponsor education programs for school-age children and the community at large to promote awareness of oral health and teach proper oral hygiene. Industry-supported programs also bring dental care to underserved patients.

Consumers demand safe and effective oral health products and rely on the Seal of Acceptance of the American Dental Association (ADA) to choose goods.

All major manufacturers strive to receive the ADA seal, which denotes extensive clinical and laboratory research to ensure product safety and effectiveness. More than three hundred manufacturers of dental care products volunteer their products for scrutiny by the ADA. Although the Food and Drug Administration establishes and enforces safety and effectiveness guidelines for the manufacture and use of health care products, the ADA surpasses these guidelines to offer reassurance in product choice for consumers.

CAREERS AND COURSE WORK

All fifty states in the United States, plus the District of Columbia, require a license to practice dentistry. To obtain a license, applicants must graduate from an accredited school of dentistry and pass written and practical exams. There are more than fifty dental schools in the United States, offering either the doctor of dental surgery (D.D.S.) or the doctor of dental medicine (D.M.D.) degree, which are equivalent degrees.

To apply to dental school, students should be proficient in basic sciences, including biology, chemistry, physics, health science, and mathematics. A minimum of two years of college education is necessary to apply to dental school, but most applicants have completed an undergraduate degree in a science discipline. The Dental Admission Test is also required for applicants to dental school.

Dental school is a four-year program consisting of two years of didactic learning and two years of clinical training. Education in anatomy, microbiology, biochemistry, physiology, and pharmacology is essential to training in dentistry. Clinical practice experience takes place under the supervision of licensed dentists.

Dentists must possess superb diagnostic skills, supreme visual memory, and excellent manual dexterity. Most dentists work in private practice, either alone or with partners. Therefore, dentists also need business management talents, self-discipline, and communication skills.

Dental hygienists, assistants, and laboratory technicians are related professions within the field of dentistry. They each work closely with a dentist to perform the technical duties associated with oral care and teach patients about proper hygiene and good nutrition. The educational and licensing requirements for these dental careers vary by state, although formal education is encouraged and favorable in a competitive job market.

SOCIAL CONTEXT AND FUTURE PROSPECTS

The connection between oral health and overall health has led to an increase in oral home care as well as in professional dentistry services. Patients seek dental care for routine maintenance of oral health and cosmetic procedures to improve the appearance of the face and mouth. In the future, dentistry will increasingly play a fundamental role in people's overall health and wellness. From preventing childhood tooth decay and age-related tooth loss to improving self-esteem through a brighter, straighter smile, dentistry has evolved from a fearful, painful process of tooth extraction to a respected field of medicine that is associated with comfortable care and daily hygiene.

Dentistry of the future will emphasize less painful therapy and disease prevention. It will seek to identify at-risk groups and to provide services to underserved populations to improve dental public health, which will have lasting benefits in education and overall disease morbidity and mortality. Emerging research is focused on mouthwashes that prevent the buildup of plaque on teeth, vaccines that prevent decay and dental caries, and long-lasting pellets that deliver a continuous dose of fluoride to the teeth. Braces may soon be replaced or aided by small, battery-operated paddles that deliver an undetectable electric current to the gums to rearrange bone and tissue structures of the mouth. Lasers will replace existing surgical techniques, allowing for painfree treatment of dental disease. Further, new enzymes and plastics are emerging as options for tooth restoration and dental diagnostics.

Dentistry will continue to be a collaborative and interdisciplinary practice that meets the growing and changing needs of dental health.

Jennifer L. Gibson, B.S., D.P.

FURTHER READING

Kendall, Bonnie. *Opportunities in Dental Care Careers.* Rev. ed. New York: McGraw-Hill, 2006. A review of the educational requirements and professional expectations for all specialties of dentistry and dental-related careers.

Picard, Alyssa. *Making the American Mouth: Dentists and Public Health in the Twentieth Century.* New Brunswick, N.J.: Rutgers University Press, 2009. Presents a history of dentistry as well as essays on issues such as dental hygiene, dental economics, and the American diet.

Pyle, Marsha, et al. "The Case for Change in Dental Education." *Journal of Dental Education* 70, no. 9 (September, 2006): 921-924. The American Dental Education Association's Commission on Change and Innovation in Dental Education examines the need for change in dental education. It takes into account the financial expense of a dental education and the professional responsibilities of meeting all individual and public health needs.

Rossomando, Edward F., and Mathew Moura. "The Role of Science and Technology in Shaping the Dental Curriculum." *Journal of Dental Education* 72, no. 1 (January, 2008): 19-25. Offers a history of the changing dental school curricula in the United States and offers perspectives for the future of dentistry education.

Wynbrandt, James. *The Excruciating History of Dentistry: Toothsome Tales and Oral Oddities from Babylon to Braces.* New York: St. Martin's Press, 1998. An entertaining history of the development of the dental profession, offering humorous anecdotes and macabre tales of the profession.

WEB SITES
American Dental Association
http://www.ada.org

American Dental Education Association
http://www.adea.org

See also: Anesthesiology; Cardiology; Pediatric Medicine and Surgery; Radiology and Medical Imaging; Surgery.

DERMATOLOGY AND DERMATOPATHOLOGY

FIELDS OF STUDY

Medicine; pathology; surgery; surgical pathology; biology; histology; chemistry; physics; immunodermatology; pediatric dermatology; cosmetic dermatology; surgical dermatology; veterinary dermatology; Mohs micrographic surgery.

SUMMARY

Dermatologists diagnose and treat medical conditions of the skin, including acne, rosacea, psoriasis, warts, hair loss, and various forms of skin cancer. Dermatopathologists analyze the mechanisms of skin diseases and perform microscopic diagnoses based on the tissue samples submitted by dermatologists. Skin disorders have a high prevalence and can affect patients of all ages, from neonates to elderly people. Because of the great variety and dynamic nature of the lesions, specialties focusing on skin are among the most complex in medicine.

KEY TERMS AND CONCEPTS

- **Botulinum Toxin:** Neurotoxin produced by the bacterium *Clostridium botulinum;* commonly known as Botox, its trade name.
- **Epidermis:** Upper (outer) skin layer.
- **Flow Cytometry:** Technique for separating and counting cells or chromosomes by suspending them in fluid and passing them by a focused light.
- **Immunohistochemistry:** Antibody-based method of detecting a specific protein in a tissue sample.
- **Keratinocyte:** Common epidermal cell that synthesizes keratin and changes while moving upward from basal to superficial layers.
- **Macule:** Flat, colored skin area that measures less than 10 millimeters in diameter.
- **Melanocyte:** Epidermal cell that produces the skin pigment melanin.
- **Papule:** Solid, raised spot on the skin that measures less than 10 millimeters in diameter.
- **Plaque:** Broad, raised area of skin.
- **Pustule:** Small skin swelling filled with pus.
- **Retinoids:** Class of compounds chemically related to vitamin A.

DEFINITION AND BASIC PRINCIPLES

Dermatology is the branch of medicine dedicated to the diagnosis, treatment, and prevention of diseases and conditions of the skin, the hair, the nails, and mucous membranes. A subspecialty of pathology and dermatology, dermatopathology focuses on studying the mechanisms of skin diseases and on the microscopic examination of cutaneous tissue.

Dermatologists assess the appearance and distribution of any abnormalities in the skin, identifying primary and secondary lesions. These lesions can manifest in numerous forms, including macules, papules, plaques, nodules, pustules, vesicles, wheals (hives), scales, fissures, and scars. The patient may complain of itchiness (pruritus), pain, or hair loss, or may be uncomfortable with the appearance of a skin area. If a diagnosis is not readily apparent, the dermatologist performs a skin biopsy. A dermatopathologist examines the tissue under a microscope and renders a pathological diagnosis.

The skin is the largest and most visible organ of the human body, with essential functions in storage, absorption, thermoregulation, vitamin D synthesis, and protection against pathogens. It is readily accessible to the examiner; however, the potential abnormalities are numerous and the differential diagnoses extensive, rendering dermatology one of the most complex medical disciplines. Although the field has been morphologically oriented for centuries, advances in molecular medicine and genetics have opened new opportunities for understanding the pathogenesis of skin diseases and for improved diagnosis strategies. An evolving interrelationship with other disciplines such as plastic surgery and endocrinology has been expanding the frontiers of this medical specialty.

BACKGROUND AND HISTORY

People have been concerned with the health and appearance of their skin throughout history. Egyptian physicians used arsenic applications to treat skin cancer and sandpaper to smooth scars. Queen Cleopatra was known for her cosmetic knowledge. Geoffrey Chaucer's *The Canterbury Tales* (1387-1400) and William Shakespeare's plays contain numerous references to unsightly skin afflictions, such as boils, carbuncles, and scabs. Not surprisingly, their

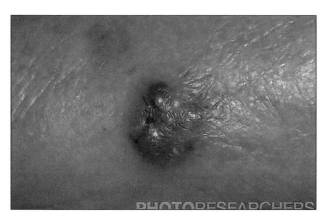

Close-up of basal cell carcinoma (BCC), or rodent ulcer, on the forearm in a 100-year-old female patient. This is the most common form of skin cancer among fair-skinned people. It usually appears on the face and is caused by sun damage. (Dr. P. Marazzi/Photo Researchers, Inc.)

appearance is frequently a metaphor for character flaws.

Some of the first skin treatments were undoubtedly borrowed from the plant world, making use of leaves, flowers, and roots. The juice of the aloe vera, for example, is an ancient and effective remedy that continues to be used for some skin conditions. For centuries, physicians treated a wide range of afflictions, from rashes to wounds, using oils, powders, and salves they mixed themselves. Sunlight was used by European physicians in the eighteenth and nineteenth centuries to treat psoriasis and eczema.

Starting in the nineteenth century, a true revolution in biology galvanized the progress of skin sciences. The terms "dermatology" and "dermatosis" were introduced. In the late 1800's, dermatologists began using a variety of chemicals to smooth facial wrinkles and scars. Cryosurgery and electrosurgery came into use. Soon after the development of the laser in the 1950's, dermatologists used it to treat skin conditions. The surge of innovations has continued, making dermatology an exciting and rapidly evolving specialty.

HOW IT WORKS

Skin diseases and conditions affect patients of all ages and ethnicities. Physicians may specialize in a specific age group, such as children, or a category of conditions. Some dermatologists focus on cosmetic disorders of the skin and may be certified to perform

procedures such as injections of botulinum toxin, chemical peels, and laser therapy. Others concentrate on skin cancers or immunological conditions. Regardless of the focus of a dermatologist's practice, the day-to-day work can be divided into three main areas: diagnosis, treatment, and management.

Diagnosis. Dermatologists obtain the patient's medical history and assess his or her status. They examine the affected skin and adjacent areas to determine the nature and extent of the lesions. A frequently used method is dermoscopy (or epiluminescent microscopy), which employs a quality magnifying lens and a powerful lighting system to allow a close examination of the skin's structure. It is useful in evaluating pigmented skin lesions and can facilitate the diagnosis of melanoma.

Some skin conditions are more readily diagnosable than others. Acne and psoriasis, for example, often do not necessitate further tests. The lesions, however, may be of an ambiguous nature or potentially malignant. In these cases, the physician takes a tissue sample (for example, a biopsy or nail clippings) and submits it, usually with a differential diagnosis, to a laboratory. There, the sample undergoes a dermatopathological evaluation.

Dermatopathologists interpret tissue samples on specially prepared slides using light, fluorescent, and sometimes electron microscopy. They first determine how the specimen was obtained (for example, a punch or shave biopsy), then establish if the condition appears to be infectious, inflammatory, degenerative, or neoplastic (benign or malignant). Often, consultation with other dermatopathologists and the attending dermatologist or primary care physician is necessary. Additional sections of the specimen may be required before a diagnosis can be rendered and the report sent to the clinician. The work needs to be extremely thorough; no part of the microscopy slide can be left unexamined. Ancillary methods used by dermatopathologists include immunohistochemistry and flow cytometry.

Additional tests that may be undertaken in the dermatologist's office include a potassium hydroxide examination for fungi, bacterial stains, fungal and bacterial cultures, skin scrapings for scabies, patch tests (for contact allergies), and blood tests.

Treatment. Once the diagnosis has been made, treatment options are considered and discussed at length with the patient or caregiver. Dermatopathologists often play an active role in this process. Treatment may involve medications to be administered externally or internally, injections, or surgical procedures. Punch biopsy, shave biopsy, electrodesiccation and curettage, blunt dissection, and simple excision and suture closure are the basic techniques that dermatologists master. They are also familiar with more sophisticated techniques, such as Mohs micrographic surgery and, if appropriate, may refer patients to physicians who perform these techniques.

Management. Skin conditions can be lifelong problems. Eczema, acne, and psoriasis are only a few of many conditions that require regular visits to the dermatologist. Managing the patient's condition often takes the form of control rather than cure.

APPLICATIONS AND PRODUCTS

Dermatologist diagnose and treat many disorders and diseases. The most common examples of disorders treated are infections, inflammatory diseases, papulosquamous diseases, and tumors.

Infections. Several categories of pathogens cause infections with cutaneous manifestations. *Staphylococcus aureus* and group A beta-hemolytic streptococci account for most skin and soft tissue infections, such as impetigo, folliculitis, cellulitis, and furuncles. Syphilis is an infectious disease caused by the bacterium *Treponema pallidum*. Primary syphilis, acquired by direct contact with a skin or mucosal lesion, manifests with a cutaneous ulcer (chancre). Warts are benign epidermal tumors caused by numerous types of human papillomaviruses (HPVs). These viruses infect epithelial cells of the skin, mouth, and other areas, causing both benign and malignant lesions.

Herpesvirus infections are caused by herpes simplex virus 1 (HSV1) and herpes simplex virus 2 (HSV2), distinguishable by laboratory tests. HSV1 is generally associated with oral infections, and HSV2 causes genital infections. The lesions appear as grouped vesicles on a red base.

The agents that induce superficial fungal infections include dermatophytes (responsible for tinea, or ringworm) and *Candida* species yeasts.

Inflammatory Diseases. Eczema is the most common inflammatory disorder. It manifests with itchiness and exhibits three clinical stages: acute (redness and vesicles), subacute (redness, scaling, fissuring, and scalded appearance), and chronic (thickened skin). There are numerous types of eczemas, including atopic dermatitis (in patients with personal or family history of allergies) and contact dermatitis (allergy to a common material such as nickel or poison oak).

Acne is a common disorder with important psychosocial effects. It occurs in predisposed individuals when sebum production increases. Proliferation of the microorganism *Propionibacterium acnes* in the sebum alters it and causes pore clogging. Lesions are noninflammatory (comedones, also known as blackheads and whiteheads) or inflammatory (papules, pustules, or nodules). The extent and severity of the lesions varies, from a few comedones to the strongly inflammatory acne conglobata.

Papulosquamous Diseases. The group of disorders known as papulosquamous diseases are characterized by scaly papules and plaques. Psoriasis, an immune-mediated skin and joint inflammatory disease, develops when inflammation primes basal stem keratinocytes to proliferate excessively. Initial red, scaling papules coalesce to form round-oval plaques. The scales are adherent, silvery white, and show bleeding points when removed (Auspitz sign). Inflammatory arthritis is present in some patients.

Tumors. The two most common skin cancers are basal cell carcinoma (BCC) and squamous cell carcinoma (SCC). Approximately 80 percent of nonmelanoma skin cancers are the basal cell type, and 20 percent are the squamous cell type.

Basal cell cancer, the most common invasive malignant skin tumor in humans, represents more than 90 percent of skin cancers in the United States. The patient typically has a bleeding or scabbing sore that heals and subsequently recurs. The tumor advances by direct extension and destroys normal tissue but rarely metastasizes. The cells of basal cell carcinoma resemble those of the basal epidermal layer. They have a large nucleus and develop an orderly line around the periphery of tumor nests (palisading).

Squamous cell carcinoma is the second most common cancer among light-skinned individuals. The relationship to ultraviolet radiation is stronger and the chances of metastasis much higher than for basal cell carcinoma. Actinic keratosis, the most common precursor of squamous cell carcinoma, begins on sun-exposed skin as isolated or multiple flat, pink-brown, rough lesions. Abnormal squamous cells originate in the epidermis from keratinocytes and proliferate indefinitely.

Malignant melanoma originates from melanocytes. Skin melanoma either begins on its own or develops from a preexisting lesion, such as a mole (nevus). One of the most aggressive tumors, melanoma can metastasize to any organ, including the brain and heart. Individuals who sunburn easily or who experienced multiple or severe sunburns have a twofold to threefold increased risk for developing skin melanoma. The goal of specialists and patients alike is to recognize melanomas as early as possible in their development. Compared with common acquired melanocytic nevi, malignant melanoma tend to have four characteristics: asymmetry, border irregularity, color variation, and diameter enlargement (ABCD). These four characteristics are the primary criteria for clinical melanoma recognition. Changes in the shape and color of a mole are important early signs and should always arouse suspicion. Ulceration and bleeding are late signs; at this stage, the chance of cure diminishes greatly.

Important Treatment Modalities. Common ways of dealing with dermatological problems are topical treatments (such as ointments and creams) and oral treatments (drugs taken by mouth). Any bodily injury, irritation, or trauma that eliminates water, lipids, or protein from the epidermis compromises its function. Restoration of the normal epidermal barrier can often be accomplished using mild soaps and emollient creams or lotions. The often-cited dermatologic adage is "If it is dry, wet it; if it is wet, dry it." Consequently, wet compresses are a frequently used remedy. A multitude of other topical treatments are available, from antibiotic, antiviral, or steroid ointments applied to treat infectious diseases or eczema to vitamin D derivative creams for psoriasis and retinoid creams for acne. Drugs can also be taken orally to treat a variety of conditions such as acne and autoimmune disorders.

Surgical and Cosmetic Procedures. Dermatologists use several techniques to obtain skin biopsies. Most procedures are done in the doctor's office, and each technique has specific indications. Punch biopsies are employed for most superficial inflammatory diseases and skin tumors (except melanoma). Shave biopsies are used for superficial benign and malignant tumors. Deep inflammatory diseases and malignant melanoma benefit from excisions.

Electrodesiccation and curettage (ED&C; also known as scrape and burn) is an important technique

---

Fascinating Facts About Dermatology and Dermatopathology

- Under normal conditions, the top layer of skin on an adult human sheds every twenty-four hours, and the skin completely renews itself in three to four weeks.

- Throughout the centuries, people with leprosy have been ostracized by their communities. Although modern medicine has made diagnosis and treatment of leprosy easy, the stigma associated with the disease remains and presents an obstacle to self-reporting. About one hundred patients are diagnosed each year in the United States.

- The use of botulinum toxin is not limited to the treatment of wrinkles. It has also been used as a remedy for muscle spasms, migraines, strabismus (lazy eye), and other conditions.

- Vitiligo, a skin disorder that affects one in every two hundred people, causes patches of skin that lack pigment and are prone to sun damage but not to skin cancer. A gene mutation responsible for increasing the risk of developing vitiligo also decreases the risk of skin malignancy.

- Researchers are studying noninvasive techniques for removing adipose tissue that could help eliminate localized fat deposits in individuals of average weight. These include exposing fat cells beneath the skin to ultrasound waves or low temperatures.

- Scientists have created artificial skin with biomechanical properties similar to real skin using biomaterials such as fibrin (from blood), agarose (from seaweed), chitosan (from crustacean shells), and collagen.

---

for removing a variety of superficial skin lesions, such as cancerous growths and genital warts. The physician uses a sharp dermal curette to cut away the growth and a needle-shaped electrode that delivers an electric current to remove any remaining material and to stop the bleeding.

Blunt dissection is a fast, elementary, usually nonscarring surgical procedure used to remove warts and other epidermal tumors. Unlike ED&C and excision, it does not disturb normal tissue.

Small, superficial, nonmalignant lesions may be quickly and efficiently frozen with liquid nitrogen, administered with a spray or sterile contact probe. Cryosurgery for malignant lesions, however, requires experience and sophisticated equipment with thermocouples that measure the depth of freeze. This minimally invasive technique is also successfully employed for common lesions, including genital warts, actinic keratoses, and certain infectious conditions.

An important surgical breakthrough occurred in the 1930's, when physician Frederic Mohs developed a microscope-guided method of tracing and removing basal cell carcinomas. These—and other tumors—may not grow in a well- circumscribed fashion but instead extend in fingerlike projections. Thin layers of tissue are removed, and all margins of the specimen are mapped to determine whether any tumor remains. This tissue-sparing technique has high cure rates.

Chemical peeling of facial skin uses a caustic agent to achieve a controlled, chemical burn of the epidermis and the outer dermis. Skin regeneration results in a fresh and orderly epidermis with ablation of fine wrinkles and pigmentation reduction.

In liposuction surgery, fat is removed through half-inch incisions using small- diameter cannulae. Multiple to-and-fro movements mechanically disrupt the fat and create tunnels. The loosened fat is removed by strong suction.

Photothermolysis is based on the property of a chromophore (melanin, hemoglobin, tattoo ink) in a target tissue to strongly absorb a selected laser wavelength and generate heat. It removes the target tissue while producing only a local thermal injury, resulting in less injury to the surrounding tissue and lowered risk of scarring. Vascular lesions, for example, can be treated in this manner, including port-wine stains, benign tumors, and spider veins in legs. In vascular lesions, the targeted chromophore is hemoglobin.

Specific types of lasers can be used to treat benign pigmented lesions with a predominant epidermal component such as freckles and tattoos. In addition, numerous laser-based devices can remove unwanted hair.

Other common techniques and devices include the use of intense pulsed light for resurfacing (to treat vascular lesions and acne) and light-activated drugs in photodynamic therapy (for precancerous and cancerous cells, acne, rosacea, or skin enhancement).

One of the most popular nonsurgical cosmetic procedures is injections of botulinum toxin (Botox).

This neurotoxin blocks the release of the chemical messenger acetylcholine, effectively causing chemical denervation. The injections reduce facial lines caused by hyperfunctional muscles.

IMPACT ON INDUSTRY

Government and University Research. The etiology, pathogenesis, and optimal therapeutic strategies for many skin disorders are still unclear. Understanding the biology of skin tumors, especially melanoma, has become imperative. The collaboration between academic dermatologists and basic scientists—geneticists, immunologists, and molecular biologists—in the United States and abroad is growing stronger.

Sources of research funds include grants and scholarships from governmental institutions, foundations, associations, and corporate partners. In the United States, at least two branches of the National Institutes of Health (NIH)—the National Cancer Institute and the National Institute of Arthritis and Musculoskeletal and Skin Diseases—are involved in patient education and research regarding cutaneous disorders. The American Academy of Dermatology and the Society for Investigative Dermatology also are committed to supporting the development of a strong skin research environment.

The mission of the European Academy of Dermatology and Venereology includes promoting excellence in research, education, and training. The European Society for Dermatological Research has focused on advancing basic and clinical science and has facilitated the exchange of information relevant to investigative dermatology among clinicians and scientists worldwide.

Industry and Business. Cosmetic dermatology has become vastly popular. More and more dermatologists devote a significant part of their time to nonsurgical and surgical cosmetic procedures. The financial gains are undeniable. The global market for cosmetic surgery is predicted to exceed $40 billion in revenues by 2013. A survey conducted by the American Academy of Cosmetic Surgery (AACS) found that more than 17 million cosmetic surgery procedures were performed in the United States in 2009. Nonsurgical procedures, such as injections of botulism toxin and acid peels, and prescription drugs that are used to treat conditions associated with perceived deficits in physical appearance, such as retinoids, are also in great demand. The aging baby boomers are strongly driving the demand for facial rejuvenation products. Although for dermatologists, cosmetic procedures are generally more lucrative than medical treatments, pharmaceutical companies find drugs that act on the skin to be a significant source of revenue. The psoriasis market is forecast to grow the most aggressively, doubling from $3.4 billion in 2009 to $6.8 billion in 2019 in the United States, France, Germany, Italy, Spain, the United Kingdom, and Japan.

Teledermatology and teledermatopathology use telecommunication technologies to exchange medical information for diagnostic, therapeutic, and educational purposes. Digital dermoscopy is an evolving computer-based version of traditional dermoscopy. It provides a reliable way to capture and store images, send them to pathologists, and compare them over time. The technique is quickly becoming an essential tool in dermatology practice. Numerous companies are developing hardware and software products for these advanced technologies; even more are in the business of providing cosmetic surgery products (lasers, instruments), phototherapy equipment, and other devices essential to modern dermatologic practices.

CAREERS AND COURSE WORK

Those considering a career in dermatology will require a strong foundation in biology, chemistry, and physics, followed by a decade of training and a lifetime of learning. Medical school requires four years of intense preparation in an accredited school. Subsequent specialization in dermatology requires a one-year internship, followed by a three-year residency in an accredited program. A one- or two-year fellowship can be undertaken after residency; examples of possible paths are dermatopathology, pediatric dermatology, immunodermatology, and dermatologic surgery. Veterinary colleges may also offer a three-year residency program in dermatology.

To pursue a career in dermatopathology, a physician also can specialize in surgical pathology via a three-year residency, then undergo further training in a dermatopathology fellowship program. Competency in dermatopathology is of utmost importance for both pathology and dermatology residents.

Opportunities to practice exist all over the United States and worldwide. Dermatopathologists and dermatologists can be self-employed, partner with others, or be employed by hospitals, clinics, and governmental agencies. Both categories of specialists teach in universities and colleges that have degree programs in these disciplines. Research opportunities are available in universities and in laboratories operated by corporations, pharmaceutical companies, or governmental agencies.

SOCIAL CONTEXT AND FUTURE PROSPECTS

The burden of skin diseases on society is significant. According to a 2004 study by the American Academy of Dermatology Association and the Society for Investigative Dermatology, the annual cost of skin diseases in the United States is about $39 billion; direct medical costs account for $29 billion and indirect costs related to lost productivity make up the remaining $10 billion. At any given time, one in three people in the United States suffers from an active cutaneous condition. The most prevalent disorders are herpes simplex, shingles, sun damage, eczema, warts, and hair and nail conditions. The incidence of melanoma is on the rise. The main reasons for this high level of skin disease are increased exposure to the sun during recreational activities and the atmospheric changes brought on by pollutants that result in increased radiation. Understanding the biology of skin tumors, especially melanoma, has become a priority of research efforts worldwide.

New therapeutic agents such as antibodies and immunomodulators offer hope for stubborn medical conditions such as psoriasis, still an incurable disease in need of good long-term therapeutic approaches. Biological treatments are on their way to bringing relief. Stem cells hold promise for tissue regeneration.

Advances in understanding the pathogenesis of various disorders have led to improved management and to a reduced risk of incorporating nonevidence-based components into dermatological practice. The close cooperation between dermatologists, pathologists, rheumatologists, and surgeons enhances the quality and efficiency of care.

The ever-increasing preoccupation with young, healthy skin has fueled an unprecedented explosion in the popularity of cosmetic procedures. More important, skin diseases with significant aesthetic, psychological, and social consequences have prompted dermatologists to implement and refine numerous cosmetic techniques involving peeling, botulinum toxin, hyaluronic acid, and lasers. These techniques have enabled many categories of patients with skin disorders to lead a normal social life.

Mihaela Avramut, M.D., Ph.D.

FURTHER READING

Bickers, D. R., et al. "The Burden of Skin Diseases, 2004: A Joint Project of the American Academy of Dermatology Association and the Society for Investigative Dermatology." *Journal of the American Academy of Dermatology* 55, no. 3 (September, 2006): 490-500. Summary of the well-documented study assessing the prevalence and economic burden of skin diseases and how they effect quality of life.

Bolognia, Jean, et al., eds. *Dermatology.* 2d ed. 2 vols. St. Louis, Mo.: Mosby Elsevier, 2008. A basic textbook that covers nearly all aspects of dermatology, from cancers to cosmetic procedures.

Ferri, Fred. *Ferri's Fast Facts in Dermatology: A Practical Guide to Skin Diseases and Disorders.* Philadelphia: Saunders/Elsevier, 2011. A handbook for the diagnosis of dermatological disorders.

Habif, Thomas P. *Clinical Dermatology.* 5th ed. St. Louis, Mo.: Mosby Elsevier, 2010. Leading manual with excellent photographs, online access, multiple appendixes, and an online differential diagnoses (DDX) mannequin for lesion localization.

Hall, Brian J., and John C. Hall. *Sauer's Manual of Skin Diseases.* 10th ed. Philadelphia: Lippincott, Williams & Wilkins, 2010. Accessible textbook includes numerous color photographs, diagnostic algorithms, and a dictionary-index. Has an accompanying Web site.

Pilla, Louis. "Cosmetic Versus Medical Dermatology: A Widening Gap?" *Skin and Aging* 11, no. 6 (June, 2003). Analysis of the interplay between medical and cosmetic dermatology in modern practices.

WEB SITES

American Academy of Dermatology
http://www.aad.org

European Academy of Dermatology and Venereology
http://www.eadv.org

European Society for Dermatological Research
http://www.esdr.org

Society for Investigative Dermatology
http://www.sidnet.org

See also: Geriatrics and Gerontology; Pathology; Pharmacology; Surgery.

DETERGENTS

FIELDS OF STUDY

Biochemistry; biology; biotechnology; engineering; chemistry; environmental science; chemistry; materials science; pharmacy; physics and kinematics; polymer chemistry; process design; systems engineering; textile design.

SUMMARY

Synthetic detergents enable otherwise immiscible materials (water and oil) to form homogeneous dispersions. In addition to facilitating the breakdown and removal of stains or soil from textiles, hard surfaces, and human skin, detergents have found widespread application in food technology, oil-spill cleanup, and other industrial processes, such as the separation of minerals from their ores, the recovery of oil from natural deposits, and the fabrication of ceramic materials from powders. Detergents are used in the manufacture of thousands of products and have applications in the household, personal care, pharmaceutical, agrochemical, oil and mining, and automotive industries, as well as in the processing of paints, paper coatings, inks, and ceramics.

KEY TERMS AND CONCEPTS

- **Critical Micelle Concentration (CMC):** Surfactant concentration at which appreciable micelle formation occurs, enabling the removal of soils.
- **Hydrophilic-Lipophilic Balance (HLB):** Value expressing hydrophilic (polar/water-loving) and lipophilic (nonpolar/oil-loving) character of surfactants.
- **Interface:** Surface that forms the boundary between two bodies of matter (liquid, solid, gas) or the area where two immiscible phases come in contact.
- **Micelle:** Aggregate or cluster of individual surfactant molecules formed in solution, whereby polar ends are on the outside (toward the solvent) and nonpolar ends are in the middle.
- **Surface Tension:** Property of liquids to preferentially contract or expand at the surface, depending on the strength of molecular association and attraction forces.
- **Surfactant:** Surface active agent containing both hydrophilic and hydrophobic groups, enabling it to lower surface tension and solubilize or disperse immiscible substances in water.

DEFINITION AND BASIC PRINCIPLES

Detergents comprise a group of synthetic water-soluble or liquid organic preparations that contain a mix of surfactants, builders, boosters, fillers, and other auxiliary constituents, the formulation of which is specially designed to promote cleansing action or detergency. Liquid laundry detergents are by far the largest single product category: Sales in the United States were reported at $3.6 billion for 2010. As of 2011, products are largely mixtures of surfactants, water softeners, optical brighteners and bleach substitutes, stain removers, and enzymes. They must be formulated with ingredients in the right proportion to provide optimum detergency without damaging the fabrics being washed.

Before the advent of synthetic detergents, soaps were used. Soaps are salts of fatty acids and made by alkaline hydrolysis of fats and oils. They consist of a long hydrocarbon chain with the carboxylic acid end group bonded to a metal ion. The hydrocarbon end is soluble in fats and oils, and the ionic end (carboxylic acid salt) is soluble in water. This structure gives soap surface activity, allowing it to emulsify or disperse oils and other water-insoluble substances. Because soaps are alkaline, they react with metal ions in hard water and form insoluble precipitates, decreasing their cleaning effectiveness. These precipitates became known as soap scum—the "gunk" that builds up and surrounds the bathtub and causes graying or yellowing in fabrics.

Once synthetic detergents were developed, this problem could be avoided. Detergents are structurally similar to soap and work much the same to emulsify oils and hold dirt in suspension, but they differ in their water-soluble portion in that their calcium, magnesium, and iron forms are more water soluble and do not precipitate out. This allowed detergents to work well in hard water, and thus, reduce the discoloration of clothes.

BACKGROUND AND HISTORY

Soap is the oldest cleaning agent and has been in use for 4,500 years. Ancient Egyptians were known to bathe regularly and Israelites had laws governing personal cleanliness. Records document that soap-like material was being manufactured as far back as 1500 B.C.E. Before soaps and detergents, clothes were cleansed by beating them on rocks by a stream. Plants such as soapwort and soapbark that contained saponins were known to produce soapy lather and probably served as the first crude detergent. By the 1800's, soap making was widespread throughout Europe and North America, and by the 1900's, its manufacture had grown into an industry. The chemistry of soap manufacturing remained primarily the same until 1907, when the first synthetic detergent was developed by Henkel. By the end of World War I, detergents had grown in popularity as people learned that they did not leave soap scum like their earlier counterparts.

The earliest synthetic detergents were short-chain alkyl naphthalene sulfonates. By the 1930's sulfonated straight-chain alcohols and long-chain alkyl and aryl sulfonates with benzene were being made. By the end of World War II, alkyl aryl sulfonates dominated the detergent market. In 1946, the use of phosphate compounds in combination with surfactants was a breakthrough in product development and spawned the first of the "built" detergents that would prove to be much better-performing products. Sodium tripolyphosphate (STPP) was the main cleaning agent in many detergents and household cleaners for decades. By 1953, U.S. sales of detergents had surpassed those of soap. In the mid-1960's, it was discovered that lakes and streams were being polluted, and the blame was laid on phosphate compounds; however, the actual cause was found to be branching in their molecular structure, which prevented them from being degraded by bacteria. Detergent manufacturers then switched from commonly used compounds such as propylene tetramer benzene sulphonate to a linear alkyl version. Detergent manufacturers are still grappling with sustainability issues and are focusing much attention on developing products that are safe for the environment as well as consumers.

HOW IT WORKS

Just as forces exist between an ordered and disordered universe, so too do they between soil and

Phosphate-free dishwasher detergent is on display at a Rosauers supermarket in Spokane, Washington. (AP Photo)

cleanliness. People have been conditioned to believe that soil on the surface of an object is unwanted matter, but in reality, soil is being deposited continuously on all surfaces around us, and cleanliness itself is an unnatural, albeit desirable, state. In order to rid any surface of soil, one must work against nature and have an understanding of the concept of detergency—the act of cleaning soil from a surface (substrate).

The Function of Detergency. The cleaning action of detergents is based on their ability to emulsify or disperse different types of soil and hold it in suspension in water. The workhorse involved in this job is the surfactant, a compound used in all soaps and detergents. This ability comes from the surfactant's molecular structure and surface activity. When a soap or detergent product is added to water that contains insoluble materials like dirt, oil, or grease, surfactant molecules adsorb onto the substrate (clothes) and form clusters called micelles, which surround the immiscible droplets. The micelle itself is water soluble and allows the trapped oil droplets to be dispersed throughout the water and rinsed away. While this is a simplified explanation, detergency is a complex set of interrelated functions that relies on the diverse properties of surfactants, their interactions in solution, and their unique ability to disrupt the surface tension of water.

Surface Tension. The internal attraction or association of molecules in a liquid is called surface tension. However simple this may seem, it is a complex phenomenon, and for many students can be hard to grasp. Examining the properties of water and the action of surfactants may dispel any confusion.

Water is polar in nature and very strongly associated, such that the surface tension is high. This is because of its nonsymmetrical structure, in which the double-atom oxygen end is more negative than the single-atom hydrogen end is positive. As a result, water molecules associate so strongly that they are relatively stable, with only a slight tendency to ionize or split into oppositely charged particles. This is why their boiling point and heat of vaporization are very high in comparison to their low molecular weight.

The surface tension of water can be explained by how the molecules associate. Water molecules in the liquid state, such as those in the center of a beaker full of water, are very strongly attracted to their neighboring molecules, and the pull is equal in all directions. The molecules on the surface, however, have no neighboring molecules in the air above; hence, they are directed inward and pulled into the bulk of the liquid where the attraction is greater. The result is a force applied across the surface, which contracts as the water seeks the minimum surface area per unit of volume. An illustration of this is the fact that one can spin a pail of water around without spilling the contents.

Surfactant compounds are amphiphilic, meaning their backbone contains at least one hydrophilic group attached to a hydrophobic group (called the hydrophobe), usually consisting of an 8 to 18 carbon-hydrocarbon chain. All surfactants possess the common property of lowering surface tension when added to water in small amounts, at which point the surfactant molecules are loosely integrated into the water structure. As they disperse, the hydrophilic portion of the surfactant causes an increased attraction of water molecules at the surface, leaving fewer sides of the molecule oriented toward the bulk of the liquid and lessening the forces of attraction that would otherwise pull them into solution.

Micelle Formation and Critical Micelle Concentration (CMC). As surface active agents, surfactants not only have the ability to lower surface tension but also to form micelles in solution, unique behavior that is at the core of detergent action. Micelles are aggregate or droplet-like clusters of individual surfactant molecules, whose polar ends are on the outside, oriented toward the solvent (usually water), with the nonpolar ends in the middle. The driving force for micelle formation is the reduction of contact between the hydrocarbon chain and water, thereby reducing the free energy of the system. The micelles are in dynamic equilibrium, but the rate of exchange between surfactant molecule and micelle increase exponentially, depending on the structure of each individual surfactant. As surfactant concentration increases, surface tension decreases rapidly, and micelles proliferate and form larger units.

The concentration at which this phenomenon occurs is known as the critical micelle concentration (CMC). The most common technique for measuring this is to plot surface tension against surfactant concentration and determine the break point, after which surface tension remains virtually constant with additional increases in concentration. The corresponding surfactant concentration at this discontinuity point corresponds to the CMC. Every surfactant has its own characteristic CMC at a given temperature and electrolyte concentration.

Fascinating Facts About Detergents

- According to biblical accounts, the ancient Israelites created a hair gel by mixing ashes and oil.
- Persil was the name of the first detergent made in 1907 by Germany's Henkel.
- Soap scum became immortalized with the ad campaign that coined the terms "ring around the tub" and "ring around the collar."
- Automatic dishwasher powders and liquid fabric softeners were invented in the 1950's, laundry powders with enzymes in the 1960's, and fabric softener sheets in the 1970's.
- Dawn dishwashing liquid was used to clean wildlife during the Exxon Valdez and BP oil spills in 1989 and 2010, respectively.
- The human lung contains a surface-active material called pulmonary surfactant that helps prevent the lung from collapsing after expiration.
- Enzymes used in detergents work much like they do in the body, since each has a personalized target soil it breaks down.

APPLICATIONS AND PRODUCTS

The workhorse of detergents is the surfactant or more commonly, a mix of surfactants. The most important categories are the carboxylates (fatty acid soaps), the sulfates (alkyl sulfates, alkyl ether sulfates, and alkyl aryl sulfonates), and the phosphates.

Laundry Products. The primary purpose of laundry products is the removal of soil from fabrics. As the cleaning agent, the detergent must fulfill three functions: wet the substrate, remove the soil, and prevent it from redepositing. This usually requires a mix of surfactants. For example, a good wetting agent is not necessarily a good detergent. For best wetting, the surface tension need only be lowered a little, but it must be done quickly. That requires surfactants with short alkyl chain lengths of 8 carbons and surfactants with an HLB of 7 to 9. For best detergency, the surface tension needs to be substantially lowered and that requires surfactants with higher chain lengths of 12 to 14 carbons and an HLB of 13 to 15. To prevent particles from redepositing, the particles must be stabilized in a solution, and that is done best by nonionic surfactants of the polyethylene oxide type. In general, nonionics are not as effective in removing dirt as anionic surfactants, which is one reason why a mixture of anionic and nonionic surfactants is used. However, nonionics are more effective in liquid dirt removal because they lower the oil-water interfacial tension without reducing the oil-substrate tension.

Skin-Cleansing Bars. As of 2011, the bar soap being manufactured is called superfatted soap and is made by incomplete saponification, an improved process over the traditional method in which superfatting agents are added during saponification, which prevents all of the oil or fat from being processed. Superfatting increases the moisturizing properties and makes the product less irritating. Transparent soaps are like traditional soap bars but have had glycerin added. Glycerin is a humectant (similar to a moisturizer) and makes the bar transparent and much milder.

Syndet bars are made using synthetic surfactants. Since they are not made by saponification, they are actually not soap. Syndet bars are very mild on the skin and provide moisturizing and other benefits. Dove was the first syndet bar on the market.

Mining and Mineral Processing. Because minerals are rarely found pure in nature, the desired material, called values, needs to be separated from the rocky, unwanted material, called gangue. Detergents are used to extract metals from their ores by a process called froth flotation. The ore is first crushed and then treated with a fatty material (usually an oil) which binds to the particles of the metal (the values), but not to the unwanted gangue. The treated ore is submerged in a water bath containing a detergent and then air is pumped over the sides. The detergent's foaming action produces bubbles, which pick up the water-repellant coated particles or values, letting them float to the top, flow over the sides, and be recovered. The gangue stays in the water bath.

Enhanced Oil Recovery (EOR). This process refers to the recovery of oil that is left behind after primary and secondary recovery methods have either been exhausted or have become uneconomical. Enhanced oil recovery is the tertiary recovery phase in which surfactant-polymer (SP) flooding is used. SP flooding is similar to waterflooding, but the water is mixed with a surfactant-polymer compound. The surfactant literally cleans the oil off the rock and the polymer spreads the flow through more of the rock. An additional 15 to 25 percent of original oil in place (OOIP) can be recovered. Before this method is used, there is a great deal of evaluation and laboratory testing involved, but it has become a reliable and cost-effective method of oil recovery.

Ceramic Dispersions. Ceramic is a nonmetallic inorganic material. Ceramic dispersions are the starting material for many applications. The use of detergents or surfactants enhances the wetting ability of the binder onto the ceramic particles and aids in the dispersion of ceramic powders in liquids. As dispersants, they reduce bulk viscosity of high-solid slurries and maintain stability in finely divided particle dispersions. Bi-block surfactants help agglomeration of the ceramic particles. In wastewater treatment, detergents are used in ceramic dispersions to reduce the amount of flocculents.

IMPACT ON INDUSTRY

Detergents are big business, not only in the United States, but around the world, and the industry and business sectors catering to it are expansive. These include the laundry and household products market; the industrial and institutional (I&I) market, which makes heavy-duty disinfectants, sanitizers, and cleaners; pharmaceuticals; the oil industry; food service and other service industries, such as hospitals and hospitality (hotels), as well as the personal-care industry.

The Global Picture. Worldwide, the detergent industry is estimated to be around $65 billion, $52 billion of which is the laundry-care category. Despite global cleaning products being a multibillion-dollar business, the global household products industry has seen slow growth over the last few years. The global I&I market is reported to be $30 billion.

Worldwide, the most growth for the detergent, household, and I&I sectors is expected to be in emerging markets, with China, India, and Brazil leading the pack. The three countries have been in the midst of a construction boom. China, for instance, has a new alkyl polyglucoside plant that is in full production. In contrast to the domestic market's slow-growth mode, experts say companies in these countries are seeing double-digit sales figures, while sales in other countries outside U.S. borders are rising by 5 to 10 percent. The consensus is that growth in the United States, Canada, Japan, and Western Europe will be slow compared with world averages. Some analysts predict that the growth in emerging markets will be explosive.

The Local Picture. The domestic market for detergents is not so rosy, say analysts. Market data for 2010 was mixed, as many companies are still feeling the pinch from the 2007-2009 recession. U.S. sales of liquid laundry detergents in food, drug, and mass merchant (FDMx) outlets (excluding Walmart) fell to $3.6 billion, a 3.14 percent drop from the previous year. However, growth in I&I as a whole rose 1.7 percent, amounting to about $11 billion. Projection data indicate that demand for disinfectants and sanitizers through 2014 will increase by more than 6 percent, outpacing the industry's overall growth for the same period. In the same vein, the personal-care sector saw FDMx sales of personal cleansers rise 22.22 percent, an impressive jump to nearly $143 million. A 2010 observational study sponsored by the American Society for Microbiology and the American Cleaning Institute (formerly the Soap and Detergent Association), indicates that 85 percent of adults now wash their hands in public restrooms—the highest percentage observed since studies began in 1996. FDMx soap sales reached $2.08 billion in 2010, which represents a 5 percent increase over the previous year.

The Supplier Side. Surfactants are the primary ingredient in detergents and are made from two different types of raw materials. Those derived from agricultural feedstocks are called oleochemicals, and those derived from petroleum (crude oil), synthetic surfactants. The global market for surfactants by volume size is more than 18 million tons per year, with 80 percent of the demand represented by just ten different types of compounds. The global share of surfactant raw materials represents only about 0.1 percent of crude-oil consumption.

Major Manufacturers. The top laundry detergent manufacturers in the United States are Procter & Gamble (P&G), Henkel, and Unilever, until 2008, when it sold its North American detergent business to Vestar Capital Partners, saying it could not compete with P&G. The former All, Wisk, Sunlight, Surf, and Snuggle brands are now part of a new company, the Sun Products Corporation. Vestar said Sun Products will have annual sales of more than $2 billion. Unilever NV of the Netherlands is still the largest detergent manufacturer in the United Kingdom.

P&G's detergent sales totaled $79 billion in 2010 and represent 70 percent of the company's annual sales. Tide continues to dominate the laundry detergent market, commanding more than 40 percent of sales and is the standout product of P&G's $1 Billion Club. Purex Complete was a big growth product for Henkel in the United States, but not in Europe. Besides the Purex brands, Henkel also markets Dial soaps, a brand it acquired in 2004. Henkel's worldwide sales grew 11 percent to $20.6 billion in 2010, but had stagnant sales in its U.S. division.

In the liquid laundry detergent category, the top five products by sales volume for the year ending October 31, 2010, were Tide (P&G), All (Sun Products), Arm & Hammer (Church & Dwight), Gain (P&G), and Purex (Henkel). Although ranked number eight, Cheer Brightclean (P&G) had the biggest jump in sales, reported at 41 percent. Reacting to the recession, P&G reduced the price of its line of laundry detergents by about 4.5 percent during 2010.

In the United States, Ecolab dominates the I&I sector. With a market share of more than 30 percent, it easily outdoes the number-two player, Diversey. Spartan Chemical is another big contender. Dial (Henkel), Purell (Gojo), and Gold Bond are top-selling brands in the personal cleanser and sanitizer market. Overall, in the personal-care market, private-label brands are by far the leader at mass merchandisers, with sales of $73.9 million in 2009.

CAREERS AND COURSE WORK

Science courses in the organic, inorganic, and physical chemistries, biology, and biochemistry, plus courses in calculus, physics, materials science, polymer technology, and analytical chemistry are typical requirements for students interested in pursuing careers in detergents. Other pertinent courses may include differential equations, instrumental analysis, statistics, thermodynamics, fluid mechanics, process design, quantitative analysis, and instrumental methods. Earning a bachelor of science degree in chemistry or chemical engineering is usually sufficient for entering the field or doing graduate work in a related area.

There are few degree programs in formulation chemistry, but students need to understand the chemistry involved, such as thermodynamics of mixing, phase equilibria, solutions, surface chemistry, colloids, emulsions, and suspensions. Even more important is how their dynamics affect such properties as adhesion, weather resistance, texture, shelf life, biodegradability, and allergenic response.

Students in undergraduate chemistry programs are encouraged to select a specialized degree track but are also being advised to take substantial course work in more than one of the primary fields of study related to detergent chemistry because product development requires skills drawn from multiple disciplines. Often a master's degree or doctorate is necessary for research and development.

There are a number of career paths for students interested in the detergent industry. Manufacturers of laundry detergents, household cleaning products, industrial and institutional (I&I) products, and personal-care products, as well as the ingredients suppliers of all these products, are the biggest employers of formulation chemists, technicians, and chemical operators. There are also career opportunities in research and development, marketing, or sales. Other areas where detergent chemists and technicians are needed include food technology, pharmaceuticals, oil drilling and recovery, mining and mineral processing, and ceramic powder production.

SOCIAL CONTEXT AND FUTURE PROSPECTS

Around the world, sustainability and environmental protection remain the buzzwords for detergents in the twenty-first century. The industry is very dependent on the price and availability of fats and oils, since these materials are needed to make the fatty acids and alcohols used in the manufacture of surfactants. The turmoil in the oil industry is not only impacting how much is paid at the gas pump but is also directly related to the price paid to keep the environment clean.

While the detergents industry, on the whole, has traditionally been relatively recession-proof, the economic slowdown and conflicts in the Middle East have taken a toll on this huge market. Robert McDonald, chairman of P&G, alluded to changes in consumer habits, volatility in commodity costs, and increasing complexity of the regulatory environment as rationales for the slowdown. A spokesperson for Kline and Company, a market research firm reporting on the industry, stated: "The industry's mantra is greener, cleaner, safer," but goes on to say that the challenge is a double-edge sword, as the industry is battling to hold down costs while trying to produce environmentally sustainable products.

Sustainability and environmental protection are global issues, but insights are becoming more astute and action is being taken. Henkel published its twentieth sustainability report in conjunction with its annual report for 2010, saying the company met its targets for 2012 early, namely to commit to principles and objectives relating to occupational health and safety, resource conservation, and emissions reduction. Important emerging markets such as China are no exception in this battle. The Chinese government puts great emphasis on environmental regulations. Analysts say this is a clear message that to expand their sales to other parts of the world, China needs to hone in on the green demand and manufacture products that offer innovative and sustainable solutions without compromising on performance.

Barbara Woldin, B.S.

FURTHER READING

Carter, C. Barry, and M. Grant Norton. *Ceramic Materials: Science and Engineering.* New York: Springer, 2007. Covers ceramic science, defects, and the mechanical properties of ceramic materials and how these materials are processed. Provides many examples and illustrations relating theory to practical applications; suitable for advanced undergraduate and graduate study.

Myers, Drew. *Surfactant Science and Technology.* 3d ed. Hoboken, N.J.: John Wiley & Sons, 2006. Written with the beginner in mind, this text clearly illustrates the basic concepts of surfactant action and application.

Rosen, Milton J. *Surfactants and Interfacial Phenomena.* 3d ed. Hoboken, N.J.: Wiley-Interscience, 2004. Easy-to-understand text on properties and applications of surfactants; covers many topics, including dynamic surface tension and other interfacial processes.

Tadros, Tharwat F. *Applied Surfactants: Principles and Applications.* Weinheim, Germany: Wiley-VCH Verlag, 2005. Author covers a wide range of topics on the preparation and stabilization of emulsion systems and highlights the importance of emulsion science in many modern-day industrial applications; discusses physical chemistry of emulsion systems, adsorption of surfactants at liquid/liquid interfaces, emulsifier selection, polymeric surfactants, and more.

Zoller, Uri, and Paul Sosis, eds. *Handbook of Detergents, Part F: Production.* Boca Raton, Fla.: CRC Press, 2009. One of seven in the Surfactant Science Series, the book discusses state of the art in the industrial production of the main players in detergent formulation—surfactants, builders, auxiliaries, bleaching ingredients, chelating agents, and enzymes.

WEB SITES
American Chemical Society
http://portal.acs.org/portal/acs/corg/content

American Chemistry Council
http://www.americanchemistry.com

American Oil Chemists' Society
http://www.aocs.org

Household and Personal Products Industry
http://www.happi.com

See also: Chemical Engineering; Environmental Chemistry; Pharmacology.

DIFFRACTION ANALYSIS

FIELDS OF STUDY

Physics; chemistry; geology; X-ray diffraction; electron diffraction; neutron diffraction; materials science; crystallography; mechanical engineering; physical chemistry; quantum mechanics; organic chemistry; molecular biology; fiber diffraction; mineralogy; metallurgy; differential equations; partial differential equations; Fourier analysis; optics; spectroscopy.

SUMMARY

Diffraction analysis is a general term used to describe various methods of scattering beams of X rays, electrons, or neutrons from targeted materials to generate diffraction patterns from which atomic arrangements in gases, liquids, and solids can be precisely determined. Diffraction techniques are useful in identifying and characterizing both natural materials, such as minerals, and engineered materials, such as ceramics. X-ray diffraction played an important role in the discoveries of the three-dimensional structures of molecules such as proteins and deoxyribonucleic acid (DNA), and neutron diffraction has allowed researchers to investigate stresses in automobile and aerospace constituents.

KEY TERMS AND CONCEPTS

- **Bragg Diffraction:** Scattering of X rays from a three-dimensional periodic structure such as a crystal, caused by waves reflecting from different crystal planes.
- **Crystal Lattice:** Three-dimensional array of points, each of which represents a unit cell.
- **Crystalline Material:** Chemical substance in which the atoms or ions are arranged in an orderly pattern.
- **Diffraction:** Directional change of a wave group after it encounters an obstacle or passes through an aperture.
- **Electron Diffraction:** Technique used to study the atomic structures of various substances through the interference patterns resulting when electrons are directed at a sample.

- **Fiber Analysis:** Study of biological or artificial filaments by X-ray-, electron-, or neutron-diffraction techniques.
- **Neutron Diffraction:** Analytic technique in which a beam of neutrons is scattered from gaseous, liquid, crystalline, or amorphous materials to determine their atomic arrangements.
- **Powder-Diffraction Analysis:** Technique in which a coherent beam of monochromatic X rays is directed at a powdered sample to obtain a diffraction pattern used to determine the structure of a particular substance.
- **Texture Analysis:** Use of diffraction analysis to determine the nature of components in such materials as textiles, food ingredients, or soil samples.
- **Unit Cell:** Basic building block, or the smallest assemblage of atoms or ions, the repetition of which generates the overall pattern of a crystal.
- **X-Ray Diffraction:** Analytic method in which X rays are reflected from the lattice of a crystal to determine the crystal's structural arrangement of atoms or ions.

DEFINITION AND BASIC PRINCIPLES

In general, diffraction denotes a change in the directions and intensities of waves when they encounter obstacles. All waves, be they water, sound, or light, or even such particles as electrons and neutrons that exhibit wavelike properties, are subject to diffraction. When X rays, the wavelengths of which range from 0.001 nanometer to 10 nanometers (a nanometer is one billionth of a meter), interact with matter such as a crystal, a pattern is generated that can be photographed and analyzed. This X-ray diffraction analysis has led to the discovery of the precise atomic arrangements in an enormous number of substances, from a simple crystal such as sodium chloride (whose sodium and chloride ions are cubically arranged) to such complex crystals as vitamin B_{12} (cyanocobalamin), the gigantic molecular structure of which was worked out by Dorothy Crowfoot Hodgkin, helping her to win the 1964 Nobel Prize in Chemistry.

The discovery that moving electrons exhibit wave properties led to a new kind of diffraction analysis

English physicist Sir William Henry Bragg (right) and his son Australian-born British physicist Sir Lawrence Bragg. (Getty Images)

employing electron beams. This technique proved to be particularly valuable in studies of the structures of gaseous substances, adsorbed gases, and surface layers. It could also be used to study liquids and solids, and it was often used in tandem with X-ray diffraction analyses. For example, X-ray diffraction studies confirmed that liquid cyclohexane existed in a chair form, and electron diffraction showed that cyclohexane oscillated among two chair forms and a boat form. This led scientists to the discovery of the basic principles of conformational analysis, which proved essential to an understanding of a variety of molecules.

Although X-ray and electron-diffraction methods are powerful in elucidating the structures of various substances, they are limited in studying elements that are close together in the periodic table. However, such elements vary greatly in their neutron-scattering ability, and so neutron-diffraction analysis can easily distinguish such elements as carbon, nitrogen, and oxygen. Certain details of molecular structure such as hydrogen bonding can be more precisely observed by neutron diffraction than by X-ray or electron techniques. Because of this, neutron scattering has proved advantageous in structural studies of colloids, membranes, dissolved proteins, and viruses.

BACKGROUND AND HISTORY

The term "diffraction" owes its origin to Francesco Maria Grimaldi, who derived it from the Latin

root *diffringere*, meaning "to break apart," in the seventeenth century. Others had observed diffraction patterns when a light beam was broken up by a bird feather, and in the early nineteenth century the English physician and physicist Thomas Young performed his famous double-slit experiment in which he proved that light is a wave phenomenon, since interfering waves constituted the sole explanation of the striped light and dark bands that he observed when light managed to traverse the two slits.

The discovery that X rays also produced interference patterns was made by the German physicist Max von Laue in 1912, when he showed that X rays passing through a crystal produced a diffraction pattern on a photographic plate. The next year the British father-and-son team of William Henry Bragg and William Lawrence Bragg used the X-ray diffraction technique to determine the structures of such crystals as sodium chloride and diamond, and William Lawrence Bragg formulated a mathematical equation relating the wavelength of the X rays, the angle between the incident X rays and the crystal's parallel atomic layers, and its interplanar spacing.

The Braggs' work led to the flourishing new field of X-ray crystallography, in which the three-dimensional arrangements of a wide variety of materials were determined, for example, fluorite (calcium fluoride) and calcite (calcium carbonate). In 1916, the Dutch physical chemist Peter Debye showed that X-ray diffraction could be extended to powdered solids, and he used this technique to figure out the structure of graphite. During the 1920's and 1930's, the American physical chemist Linus Pauling used X-ray diffraction to determine the structures of more than thirty minerals, including some important silicates. Pauling learned about the electron diffraction of molecules in the gaseous state from its discoverer, Herman Mark, and Pauling as well as others determined the structures of many substances that could be studied in the gaseous and liquid states. The techniques of X-ray and electron diffraction were also helpful throughout the second half of the twentieth century in elucidating the structures of proteins and DNA.

HOW IT WORKS

X-Ray Diffraction. Crystals are orderly arrangements of atoms or ions, and crystals constitute about 95 percent of all solid materials. The scattering of X rays from a crystal is actually due to its electron densities (atomic nuclei have a negligible contribution). Every crystal bombarded by X rays gives a unique pattern, and so each pattern acts as a distinctive identifier of the substance. This pattern is a result of secondary waves emanating from the electrons, which are called scatterers, and the phenomenon itself is known as elastic scattering. In a crystal these scatterers are regularly arrayed, but the reflected X rays can interfere destructively as well as constructively. The Bragg equation describes how these waves add constructively in certain directions. The reason X rays can do this is that their wavelength is generally similar to the spacings between crystal planes. When a researcher changes the angle between the X-ray beam and a crystal face, the diffraction pattern changes, and a series of diffraction photographs taken at different angles allows the investigator to formulate the three-dimensional atomic structure of the crystal.

Powder X-Ray Diffraction. In this technique experimenters use fine grains of a crystalline substance instead of a single crystal. It is extensively employed for identifying such materials as minerals and chemical compounds, or such engineered materials as ceramics. It is also used for characterizing materials, for studying particles in liquid suspensions or thin films, and for elucidating the structures of components of polycrystalline materials. Because the tiny crystals in the powder sample are randomly oriented, researchers can collect diffraction data either by reflecting X rays from the sample or by transmitting them through it. Powder X-ray diffractometer systems allow scientists to obtain a diffraction pattern for the substance under investigation, which can then be used to calculate the unit cell of the substance.

Electron Diffraction. This technique depends on the wave nature of electrons, but because electrons are negatively charged they interact much more intensely with the electromagnetic environment of samples than neutral X rays or neutrons. Like X rays, electron beams can be scattered by atoms in a sample, producing patterns that can be registered on a photographic plate or fluorescent screen. To determine the positions of atoms or ions in solids, electron-diffraction instruments that can generate high-energy electrons in a thin beam are required. Nevertheless, electrons do not have the penetrating power of X rays, which leads researchers to use thin slices of solids. Relative to solids, molecules in a gas are far apart, which means that electron diffraction readily generates patterns that allow molecular dimensions to be determined. Because air molecules scatter electrons, these measurements have to be made in a vacuum. The electron-diffraction technique has enabled scientists to study films on solid surfaces. The new field of quasicrystals, which exhibit fivefold symmetry in violation of the traditional principles of crystallography, has benefited from electron-diffraction analyses.

Despite its usefulness, electron diffraction has its limitations. In electron-diffraction experiments, interplanar spacings can be discovered to accuracies in the range of one to one hundred parts per thousand, whereas in powder X-ray diffraction, interplanar spacings can be found to precisions of one to one hundred parts per million.

Neutron Diffraction. For many scientists and engineers, neutron diffraction, when compared with X-ray and electron diffraction, is the least utilized technique for structural determinations and other applications. Nevertheless, as instruments become more sophisticated and techniques more refined, this method has been increasing in popularity. Neutron diffraction differs from X-ray diffraction in terms of scattering points—electrons for X rays and nuclei for neutrons. This makes neutron diffraction useful in distinguishing certain isotopes and helpful in studies of such compounds as the metal hydrides (hydrogen is difficult to determine by X-ray diffraction). Disadvantages of this technique include the requirement of large crystals and an efficient neutron source. Large crystals can often be difficult to grow, and neutron sources such as nuclear reactors are not as generally available as diffractometers.

Despite these problems, neutron diffraction studies have been successfully performed to measure precisely the carbon-carbon distances in graphite as well as to determine the absolute configuration of atoms in several complex chiral structures (those in which it is impossible to superimpose a configuration on its mirror image). Chemists have also found neutron diffraction studies helpful in illuminating the function of hydrogen bonds in inorganic and organic compounds, the role of magnetism in such classes of compounds as the ferrites and rare-earth nitrides, and the nature of such condensed inert gases as helium, argon, neon, and krypton.

APPLICATIONS AND PRODUCTS

X-Ray Diffraction. German physicist Wilhelm Conrad Röntgen discovered X rays in 1895. The power of X rays to reveal structures within the human body was quickly recognized, leading to many applications. With the discovery of the power of X rays to uncover the previously hidden atomic structures of crystals in the second decade of the twentieth century, physicists, chemists, and geologists enthusiastically embraced the X-ray diffraction technique as an essential tool in their disciplines. Starting with such simple substances as sodium chloride and diamond, scientists began working out the structures of increasingly complex inorganic materials such as the silicates, and the technique was also applied to organic crystals (hexamethylenetetramine was the first to be determined). As instrumentation and methods became more sophisticated, X-ray diffraction, in both its single-crystal and powder forms, was most notably applied to specify the molecular structures in living things, such as proteins and DNA. Once the three-dimensional structure of DNA was determined by James Watson and Francis Crick in 1953, an explosion of applications followed, leading to the flourishing field of biotechnology as well as to applications in criminology, genetically modified foods, and medicine.

At the beginning of diffraction analysis the instruments and other products associated with this technique (X-ray diffractometers, photographic film, and various crystals) were relatively simple and inexpensive. As the field evolved, instruments became more complex and expensive. In the early years analyses of diffraction photographs were time-consuming and labor-intensive (some complex structures took years to figure out). With the development of efficient mathematical techniques and ever-more-powerful computers, complex structures could be determined in days or even hours.

A major application of X-ray diffraction analysis has been the study of defects in metals, alloys, ceramics, and other materials. These studies can be done more quickly and efficiently, and methods have become so refined that they can be applied to defects in nanocrystalline materials. Although the use of diffraction techniques in such industries as aerospace, iron, and steel is well known, not-so-well known is their use in the manufacture of such common household products as cleansers. Diffraction techniques enable researchers to monitor the effectiveness of the abrasive minerals these cleansers.

Fascinating Facts About Diffraction Analysis

- By the early twenty-first century, the Cambridge Structural Database contained precise information for more than one-quarter million small-molecule crystal structures, and most of the new data had arrived electronically.

- Nicolaus Steno, a Danish anatomist and the father of modern geology, and René Just Haüy, a French mineralogist, made important contributions to crystallography. Steno discovered the first law of crystallography, which states that every crystal of a particular kind has angles between the faces that are the same. Haüy found that every crystal face can be explained by the stacking of basic building blocks having constant angles and sides interrelated by simple integral ratios, now known as unit cells.

- Max von Laue was a Privatdozent and Paul Peter Ewald was a doctoral student at the University of Munich in 1912 when Ewald told Laue about his thesis on crystal models. In what he later called the biggest mistake of his scientific life, Ewald did not realize that the spacings between layers of his model suggested the possibility of X-ray diffraction, which Laue quickly went on to discover and prove.

- Laue won the Nobel Prize in Physics in 1914 for his discovery of X-ray diffraction, but during World War II he decided to have his gold medal dissolved in aqua regia to prevent its falling into the hands of the Nazis. After the war, the Nobel Society recast his medal in its original gold.

- The young English physicist Henry Moseley, influenced by the discoveries of Laue and the Braggs, used X-ray diffraction to discover atomic number, which led to important improvements in the periodic table, but his promising scientific career was cut short when he was killed in action in the Gallipoli campaign during World War I.

- In 1952, the English physical chemist Rosalind Franklin took what has been called the most famous X-ray diffraction photograph ever made. It was of the B form of DNA, and, without her permission, it was shown to James Watson and Francis Crick, who used it in making their monumental discovery of the double-helix structure of DNA.

Electron- and Neutron-Diffraction Analysis. Like X-ray diffraction, electron and neutron analyses have applications in the areas of characterizations of materials, substance identification, measurement of purity, texture description, and so on. Electron diffraction is better than X-ray analysis in studying such substances as membrane proteins, because, unfortunately, X rays pass through these thin layers without forming a diffraction pattern. Electron crystallographic analysis has been used to determine the atomic arrangements in certain proteins. More common is determining structures of inorganic crystals, including complex materials such as zeolites. As with X-ray analysis, the instruments and other products associated with electron-diffraction methods have become more advanced, efficient, and expensive, but certain applications, such as the use of electron diffraction in electron microscopy, have led many scientists, from mineralogists to biologists, to expand and deepen their discoveries.

Applications of neutron analysis developed much later than those found in X-ray and electron diffraction. Like electron diffraction, neutron analysis has often been used in conjunction with X-ray techniques. For example, researchers were able, in such a combined study, to elucidate the internal dynamics of protein molecules. Independently, neutron diffraction has allowed investigators to study the details of atomic movements in substances. This basic scientific knowledge has helped others to develop better products, including window glass, semiconductors, and other electronic devices. In industry, neutron diffraction has been used to study the stresses, strains, and textures of various building materials. For example, metal alloys and welds often exhibit cracks or expansion as well as shrinkage, which limits the value of the respective products. Indeed, this practice is so prevalent that it has been named "engineering diffraction."

IMPACT ON INDUSTRY

Many thousands of structure determinations being done at a variety of institutions—governmental, academic, and industrial—together with a rough estimate of the average cost per determination indicate that this scientific and technical field's valuation is most likely in the billions of dollars. Data from various professional organizations reveal that the growth or decline of different segments of this field,

which is often tied to the market value of various materials, is sensitive to the economic state of particular countries. Viewed from a historical perspective, though, the growth of this field since the discovery of X-ray diffraction in 1912 has been truly remarkable.

Government and University Research. Diffraction analysis thrives in most advanced industrialized societies, and government agencies have played an important role in funding structural studies and the development of new equipment. Governmental agencies in the United States such as Sandia National Laboratories, part of the Department of Energy's National Nuclear Security Administration, and the U.S. Patent Office either fund their own projects or provide contracts for proposals and programs at academic institutions and industrial research centers. A representative example is the X-ray Diffraction Center in the Department of Chemistry at the University of Missouri-Columbia.

Industry and Business Sectors. The mining, metallurgical, biochemical, and pharmaceutical industries make extensive use of diffraction-analysis technologies. As chemists create new compounds, alloys, and drugs, knowledge of the three-dimensional structures of these substances is essential in understanding how they function in nature, the human body, and machines. Diffraction techniques are important in new materials development, especially in the growing field of nanomaterials. Forensic scientists use these techniques for identifying and characterizing crime-scene materials. X-ray diffraction even plays a role in the imaging industries, for example, in evaluating existing and future photographic materials. Large companies such as Eastman Kodak have supported the research and development of new imaging chemicals, sensors, and hybrid substances. Other companies, such as Matco Services, use diffraction techniques in their corrosion investigations.

CAREERS AND COURSE WORK

Associations such as the American Crystallographic Association provide detailed information on the education needed to pursue a career in diffraction analysis. Courses in physics, chemistry, and higher mathematics form the basis for later study in specialized diffraction techniques, mineralogy, or metallography. If, for example, a person wishes to pursue a career in failure analysis, courses in fracture mechanics, fractography, and corrosion testing and engineering form part of the curriculum. On the other hand, a student desiring a career in diffraction analysis as it applies to biotechnology would take courses in molecular biology, molecular biophysics, molecular genetics, and biostatistics.

Many careers are possible for students who complete programs in diffraction analysis, from being an X-ray diffraction technician to becoming the head of an integrated imaging facility. Some companies want applicants with interdisciplinary expertise, for example, someone with a mastery of mineralogy, X-ray diffraction, and modern computer-modeling methods. Many universities have X-ray diffraction facilities and will hire research assistants as well as tenured professors. The increasing number of forensic science laboratories hire large numbers of scientists and technicians, many of whom should have expertise in diffraction techniques.

SOCIAL CONTEXT AND FUTURE PROSPECTS

If the past is a prelude to the future, then the twenty-first century will be a time when the determination of structures by diffraction techniques will become faster, more efficient, and more accurate than determinations in the twentieth century. Another established trend has been the application of these techniques to more complex structures, and this, too, should continue. Many prognosticators predict that nanotechnology will flourish in the twenty-first century, and diffraction techniques have already been applied to smaller and smaller crystals. The environmental concerns that intensified in the latter part of the twentieth century will most likely continue throughout the twenty-first, and diffraction techniques provide powerful tools for identifying, characterizing, and developing detailed structural understanding of pollutants and also of the chemical compounds that make up the life-forms that are increasingly threatened by these pollutants.

Another trend that is likely to continue is the improvement of traditional techniques and instruments and the discovery of new ones. In the late twentieth century computer-controlled diffractometers came into wide use because they facilitated the collection and processing of digitized diffraction patterns. According to Moore's law, the computing power of integrated circuits doubles about every eighteen months, and, if this law remains valid, the computerization of diffraction technologies should become more powerful and less expensive than ever before. As more and more structures are determined by X-ray, electron, and neutron diffraction, larger and larger databases will be created, and more sophisticated computer software will be written to mine this treasure trove for useful information. Diffraction analyses have been largely concentrated in educationally and industrially well-developed societies, but as the less developed countries evolve, diffraction techniques will undoubtedly spread to these nations.

Robert J. Paradowski, M.S., Ph.D.

FURTHER READING

Bacon, G. E., ed. *X-Ray and Neutron Diffraction.* New York: Pergamon Press, 1966. This book, part of a series intended to introduce undergraduates to important topics in physics through original sources, treats diffraction techniques as the culmination of a long tradition in crystallography.

Chung, Frank H., and Deane K. Smith, eds. *Industrial Applications of X-Ray Diffraction.* New York: Marcel Dekker, 2000. This well-illustrated book features extensive coverage of X-ray diffraction applications for a large number of industries.

Hammond, Christopher. *The Basics of Crystallography and Diffraction.* 3d ed. New York: Oxford University Press, 2009. The author's intention is to provide the essentials of crystallography and diffraction to beginning students from a wide variety of fields. He emphasizes making difficult topics crystal clear and using history and biography to bring out the human element in his subjects.

McPherson, Alexander. *Introduction to Macromolecular Crystallography.* 2d ed. Hoboken, N.J.: Wiley-Blackwell, 2009. Intended for students interested in pursuing careers in such areas as pharmacology, protein engineering, bioinformatics, and nanotechnology, this well-illustrated book provides an easily understood introduction for readers without a background in science and mathematics.

Scott, Robert A., and Charles M. Lukehart, eds. *Applications of Physical Methods to Inorganic and Bioinorganic Chemistry.* Hoboken, N.J.: John Wiley & Sons, 2007. The editors intend this book as a practical introduction to those physical methods, including diffraction techniques, that have been used to gather relevant structural information about inorganic and organic materials. Includes an extensive list of abbreviations.

Warren, B. E. *X-Ray Diffraction*. Mineola, N.Y.: Dover Publications, 1990. This inexpensive reprint is a well-illustrated introductory text that is an excellent resource for learning X-ray diffraction from the ground up.

WEB SITES
American Crystallographic Association
http://www.amercrystalassn.org

American Physical Society
http://www.aps.org

Elmer O. Schlemper X-Ray Diffraction Center
http://www.chem.missouri.edu/x-ray

International Union of Crystallography
http://www.iucr.org

Pittsburgh Diffraction Society
http://www.pittdifsoc.org

See also: Applied Physics; Biophysics; Calculus; Ceramics; Electrometallurgy; Metallurgy; Mineralogy; Nanotechnology; Optics.

DIGITAL LOGIC

FIELDS OF STUDY

Mathematics; electronics; physics; analytical chemistry; chemical engineering; mechanical engineering; computer science; biomedical technology; cryptography; communications; information systems technology; integrated-circuit design; nanotechnology; electronic materials

SUMMARY

Digital logic is electronic technology constructed using the discrete mathematical principles of Boolean algebra, which is based on binary calculation, or the "base 2" counting system. The underlying principle is the relationship between two opposite states, represented by the numerals 0 and 1. The various combinations of inputs utilizing these states in integrated circuits permit the construction and operation of many devices, from simple on-off switches to the most advanced computers.

KEY TERMS AND CONCEPTS

- **Boolean Algebra:** A branch of mathematics used to represent basic logic statements.
- **Clock Speed:** A specific frequency that controls the rate at which data bits are changed in a digital logic circuit.
- **Gate:** A transistor assembly that combines input signals according to Boolean logic to produce a specific output signal.
- **Interleaf:** To incorporate sections of different processes within the same body in such a way that they do not interfere with each other.
- **Karnaugh Map:** A tabular representation of the possible states resulting from combinations of a specific number of binary inputs.
- **Sample:** To take discrete measurements of a specific quantity or property at a specified rate.

DEFINITION AND BASIC PRINCIPLES

Digital logic is built upon the result of combining two signals that can have either the same or opposite states, according to the principles of Boolean algebra. The mathematical logic is based on binary calculation, or the base 2 counting system. The underlying principle is the relationship between two opposite states, represented by the numerals 0 and 1.

The states are defined in modern electronic devices as the presence or absence of an electrical signal, such as a voltage or a current. In modern computers and other devices, digital logic is used to control the flow of electrical current in an assembly of transistor structures called gates. These gates accept the input signals and transform them into an output signal. An inverter transforms the input signal into an output signal of exactly opposite value.

An AND gate transforms two or more input signals to produce a corresponding output signal only when all input signals are present. An OR gate transforms two or more input signals to produce an output signal if any of the input signals are present. Combinations of these three basic gate structures in integrated circuits are used to construct NAND (or not-AND) and NOR (or not-OR) gates, accumulators, flip-flops, and numerous other digital devices that make up the functioning structures of integrated circuits and computer chips.

BACKGROUND AND HISTORY

Boolean algebra is named for George Boole (1815-1864), a self-taught English scientist. This form of algebra was developed from Boole's desire to express concrete logic in mathematical terms; it is based entirely on the concepts of true and false. The intrinsically opposite nature of these concepts allows the logic to be applied to any pair of conditions that are related as opposites.

The modern idea of computing engines began with the work of Charles Babbage (1791-1871), who envisioned a mechanical "difference engine" that would calculate results from starting values. Babbage did not see his idea materialize, though others using his ideas were able to construct mechanical difference engines.

The development of the semiconductor junction transistor in 1947, attributed to William Shockley, John Bardeen, and Walter Brattain, provided the means to produce extremely small on-off switches that could be used to build complex Boolean circuits. This permitted electrical signals to carry out Boolean algebraic calculations and marked the beginning of what has come to be known as the digital revolution. These circuits helped produce the many modern-day devices that employ digital technology.

HOW IT WORKS

Boolean Algebra. The principles of Boolean algebra apply to the combination of input signals rather than to the input signals themselves. If one associates one line of a conducting circuit with each digit in a binary number, it becomes easy to see how the presence or absence of a signal in that line can be combined to produce cumulative results. The series of signals in a set of lines provides ever larger numerical representations, according to the number of lines in the series. Because the representation is binary, each additional line in the series doubles the amount of information that can be carried in the series.

Bits and Bytes. Digital logic circuits are controlled by a clock signal that turns on and off at a specific frequency. A computer operating with a CPU (central processing unit) speed of 1 gigahertz (10^9 cycles per second) is using a clock control that turns on and off 1 billion times each second. Each clock cycle transmits a new set of signals to the CPU in accord with the digital logic circuitry. Each individual signal is called a bit (plural byte) of data, and a series of 8 bits is termed one byte of data. CPUs operating on a 16-bit system pass two bytes of data with each cycle, 32-bit systems pass four bytes, 64-bit systems pass eight bytes, and 128-bit systems pass sixteen bytes with each clock cycle.

Because the system is binary, each bit represents two different states (system high or system low). Thus, two bits can represent four (or 2^2) different states, three bits represents eight (or 2^3) different states, four bits represents sixteen (or 2^4) different states, and so on. A 128-bit system can therefore represent 2^{128} or more than 3.40×10^{38} different system states.

Digital Devices. All digital devices are constructed from semiconductor junction transistor circuits. This technology has progressed from individual transistors to the present technology in which millions of transistors can be etched onto a small silicon chip. Digital electronic circuits are produced in "packages" called integrated circuit, or IC, chips. The simplest digital logic device is the inverter, which converts an input signal to an output signal of the opposite value. The AND gate accepts two or more input signals such that the output signal will be high only if all of the input signals are high. The OR gate produces an output high signal if any one or the other of the input signals is high.

All other digital logic devices are constructed from these basic components. They include NAND gates, NOR gates, X-OR gates, flip-flops that produce two simultaneous outputs of opposite value, and counters and shift registers, which are constructed from series of flip-flops. Combinations of these devices are used to assemble accumulators, adders, and other components of digital logic circuits.

One other important set of digital devices is the converters that convert a digital or analogue input signal to an analogue or digital output signal, respectively. These find extensive use in equipment that relies on analogue, electromagnetic, signal processing.

Karnaugh Maps. Karnaugh maps are essential tools in designing and constructing digital logic circuits. A Karnaugh map is a tabular representation of all possible states of the system according to Boolean algebra, given the desired operating characteristics of the system. By using a Karnaugh map to identify the allowed system states, the circuit designer can select the proper combination of logic gates that will then produce those desired output states.

APPLICATIONS AND PRODUCTS

Digital logic has become the standard structural operating principle of most modern electronic devices, from the cheapest wristwatch to the most advanced supercomputer. Applications can be identified as programmable and nonprogrammable.

Nonprogrammable applications are those in which the device is designed to carry out a specific set of operations as automated processes. Common examples include timepieces, CD and DVD players, cellular telephones, and various household appliances. Programmable applications are those in which an operator can alter existing instruction sets or provide new ones for the particular device to carry out. Typical examples include programmable logic controllers, computers, and other hybrid devices into which some degree of programmability has been incorporated, such as gaming consoles, GPS (global positioning system) devices, and even some modern automobiles.

Digital logic is utilized for several reasons. First, the technology provides precise control over the

John Bardeen, William Shockley, and Walter Brattain shared the Nobel Prize for Physics in 1956 for their invention of the transistor. Transistors are solid state switches found in nearly every electronic component today. (Emilio Segrè Visual Archives/American Institute of Physics/Photo Researchers, Inc.)

processes to which it is applied. Digital logic circuits function on a very precise clock frequency and with a rigorously defined data set in which each individual bit of information represents a different system state that can be precisely defined millions of times per second, depending on the clock speed of the system. Second, compared with their analogue counterparts, which are constructed of physical switches and relays, digital circuits require a much lower amount of energy to function. A third reason is the reduced costs of materials and components in digital technology. In a typical household appliance, all of the individual switches and additional wiring that would be required of an analogue device are replaced by a single small printed circuit containing a small number of IC chips, typically connected to a touchpad and LCD (liquid crystal display) screen.

Programmable Logic Controller. One of the most essential devices associated with modern production methods is the programmable logic controller, or PLC. These devices contain the instruction set for the automated operation of various machines, such as CNC (computer numerical control) lathes and milling machines, and all industrial robotics. In operation, the PLC replaces a human machinist or operator, eliminating the effects of human error and

fatigue that result in undesirable output variability. As industrial technology has developed, the precision with which automated machinery can meet demand has far exceeded the ability of human operators.

PLCs, first specified by General Motors Corporation (now Company) in 1968, are small computer systems programmed using a reduced-instruction-set programming language. The languages often use a ladder-like structure in which specific modules of instructions are stacked into memory. Each module consists of the instructions for the performance of a specific machine function. More recent developments of PLC systems utilize the same processors and digital logic peripherals as personal computers, and they can be programmed using advanced computer programming languages.

Digital Communications. Digital signal processing is essential to the function of digital communications. As telecommunication devices work through the use of various wavelengths of the electromagnetic spectrum, the process is analogue in nature. Transmission of an analogue signal requires the continuous, uninterrupted occupation of the specific carrier frequency by that signal, for which analogue radio and television transmission frequencies are strictly regulated.

A digital transmission, however, is not continuous, being transmitted as discrete bits or packets of bits of data rather than as a continuous signal. When the signal is received, the bits are reassembled for audio or visual display. Encryption codes can be included in the data structure so that multiple signals can utilize the same frequency simultaneously without interfering with each other. Data reconstruction occurs at a rate that exceeds human perception so that the displayed signal is perceived as a continuous image or sound. The ability to interleaf signals in this way increases both the amount of data that can be transmitted in a limited frequency range and the efficiency with which the data is transmitted.

A longstanding application of digital logic in telecommunications is the conversion of analogue source signals into digital signals for transmission, and then the conversion of the digital signal back into an analogue signal. This is the function of digital-to-analogue converters (DACs), and analogue-to-digital converters (ACDs). A DAC uses sampling to measure the magnitude of the analogue signal, perhaps many millions of times per second depending upon the clock speed of the system. The greater the sampling rate, the closer its digital representation will be to the real nature of the analogue signal. The ACD accepts the digital representation and uses it to reconstruct the analogue signal as its output.

One important problem that exists with this method, however, is what is called aliasing, in which the DAC analogue output correctly matches the digital representation, but at the wrong frequencies. Present-day telecommunications technology is eliminating these steps by switching to an all-digital format that does not use analogue signals.

Servomechanisms. Automated processes controlled by digital logic require other devices by which the function of the machine can be measured. Typically, a measurement of some output property is automatically fed back into the controlling system and used to adjust functions so that desired output parameters are maintained. The adjustment is carried out through the action of a servomechanism, a mechanical device that performs a specific action in the operation of the machine. Positions and rotational speeds are the principal properties used to gauge machine function.

In both cases, it is common to link the output property to a digital representation such as Gray code, which is then interpreted by the logic controller of the machine. The specific code value read precisely describes either the position or the rotational speed of the component, and is compared by the controller to the parameters specified in its operating program. Any variance can then be corrected and proper operation maintained.

IMPACT ON INDUSTRY

Digital logic devices in industry, often referred to as embedded technology, represent an incalculable value, particularly in industrial production and fabrication, where automated processes are now the rule rather than the exception.

Machine shops once required the services of highly skilled machinists to control the operation of machinery for the production of quality parts. Typically, each machine required one master machinist per working shift, and while the work pieces produced were generally of excellent quality, dimensional tolerance was variable. In addition, human error and inconsistency caused by fatigue and other factors resulted in significant waste, especially given that master machinists represented the high end of the wage scale for skilled tradespersons.

Compare this earlier shop scenario to a modern machine shop, in which a single millwright ensures the proper operating condition of several machines, each of which is operated by a digital logic controller. The machines are monitored by unskilled laborers who must ensure that raw parts are supplied to the machine as needed and who must report any discrepancies to the millwright for needed maintenance. The actual function of the machine is automated through a PLC or similar device, providing output products that are dimensionally consistent, with minimal waste, and are produced at the same rate throughout the entire operating period.

It is easy to realize the economic value to industry of even just this one example of change in production methods. Given that almost all industrial production methods that can be automated with digital logic have been, the effects on productivity and

Fascinating Facts About Digital Logic

- Graphene, a one-atom-thick form of carbon that may allow transistors the size of single molecules, is one of the most exotic materials ever discovered, even though people have unknowingly been writing and drawing with it for centuries.
- The first functional transistor used two pieces of gold leaf that had been split in half with a razor blade.
- The term "transistor" was devised to indicate a resistor that could transfer electrical signals.
- Digital logic devices have automated nearly all modern production processes.
- An X-OR (Exclusive-OR) gate outputs a positive signal when just one or the other of its input signals is positive.
- Quantum dots can be thought of as artificial atoms containing a single electron with a precisely defined energy.
- Moore's law, stated by Gordon Moore of Intel Corporation in a 1965 paper, predicts that the number of transistors that can be put on a computer chip doubles about every eighteen months.
- The most sophisticated of digital logic devices, the CPU chips of advanced computers, are constructed from only three basic transistor gate designs, the NOT, AND, and OR gates. All other logic gates are constructed from these three.

profitability are fairly obvious. Automation, however, reduces the requirement for skilled tradespersons. Machine operators now represent a much lower position on the wage scale than did their skilled predecessors. The reduced production costs in this area have led to the relocation of many company facilities to parts of the world where daily wages are typically much lower and willing workers are plentiful.

PLCs and other digital logic controls automate almost every industrial process. In many cases the function is entirely automatic, not even requiring the presence of a human overseer. Sophisticated machine design in these cases replaces each action that a human would otherwise be required to carry out. In other cases, human intervention is required, as some aspect of the process cannot be left to automatic control. For example, die-casting of parts from magnesium alloys involves a level of art over science that calls for continuous human attention to produce an acceptable cast and prevent a piece from jamming in the die. Between these two more extreme situations are those common ones in which digital logic replaces the need for an operator to carry out several minor actions with the press of a single button. Once that button is pressed, the operation proceeds automatically, and the operator can only stand and watch while preparing for the next cycle of the machinery.

CAREERS AND COURSEWORK

An understanding of digital logic and its applications is obtained through courses of study in electronics engineering technology. This will necessarily include the study of mathematics and physics as foundation courses. Because digital logic is used to control physical devices, students can also expect to study mechanics and mechanical engineering, control systems and feedback, and perhaps hydraulic and pneumatic systems technology.

The development of new electronic materials and production methods will draw students to the study of materials science and chemistry, while others will specialize in circuit design and layout. Graphic methods are extremely important in this area, and specialized design programs are essential components of careers involving logic circuit design.

Introductory and college-level courses will focus on providing a comprehensive understanding of logic gates and their functions in the design of relatively simple logic circuits. Special attention may be given to machine language programming, because this is the level of computer programming that works directly with the logical hardware. Typically, study progresses in these two areas, as students will design and build more complex logic circuits with the goal of interfacing a functional device to a controlling computer.

Postgraduate electronics engineering programs build on the foundation material covered in college-level programs and are highly specialized fields of study. This level will take the student into quantum mechanics and nanotechnology, as research continues to extend the capabilities of digital logic in the form of computer chips and other integrated circuits. This involves the development and study of new materials and methods, such as graphenes and fullerenes, superconducting organic polymers, and quantum dots, with exotic properties.

SOCIAL CONTEXT AND FUTURE PROSPECTS

At the heart of every consumer electronic device is an embedded digital logic device. The transistor quickly became the single most important feature of electronic technology, and it has facilitated the rapid development of everything from the transistor radio to space travel. Digital logic, as embedded devices, is becoming an ever more pervasive feature of modern technology; an entire generation has now grown up not knowing anything but digital computers and technology. Even the accoutrements of this generation are rapidly being displaced by newer versions, as tablet computers and smartphones displace more traditional desktop personal computers, laptops, and cell phones. The telecommunications industry in North America is in the process of making a government-regulated switch-over to digital format, making analogue transmissions a relic of the past.

Research to produce new materials for digital logic circuits and practical quantum computers is ongoing. The eventual successful result of these efforts, especially in conjunction with the development of nanotechnology, will represent an unparalleled advance in technology that may usher in an entirely new age for human society.

Richard M. Renneboog, M.Sc.

FURTHER READING

Brown, Julian. *Minds, Machines, and the Multiverse: The Quest for the Quantum Computer.* New York: Simon & Schuster, 2000. Discusses much of the history of digital logic systems, in the context of ongoing research toward the development of a true quantum computer.

Bryan, L. A., and E. A. Bryan. *Programmable Controllers: Theory and Application.* Atlanta: Industrial Text Company, 1988. Provides a detailed review of basic digital logic theory and circuits, and discusses the theory and applications of PLCs.

Holdsworth, Brian, and R. Clive Woods. *Digital Logic Design.* 4th ed. Woburn, Mass.: Newnes/Elsevier Science, 2002. A complete text providing a thorough treatment of the principles of digital logic and digital logic devices.

Jonscher, Charles. *Wired Life: Who Are We in the Digital Age?* London: Bantam Press, 1999. Discusses the real and imagined effects that the digital revolution will have on society in the twenty-first century.

Marks, Myles H. *Basic Integrated Circuits.* Blue Ridge Summit, Pa.: TAB Books, 1986. A very readable book that provides a concise overview of digital logic gates and circuits, then discusses their application in the construction and use of integrated circuits.

Miczo, Alexander. *Digital Logic Testing and Simulation.* 2d ed. Hoboken, N.J.: John Wiley & Sons, 2003. An advanced text that reviews basic principles of digital logic circuits, then guides the reader through various methods of testing and fault simulations.

WEB SITES
ASIC World
http://www.asic-world.com/

NobelPrize.org
http://www.nobelprize.org

See also: Algebra; Applied Mathematics; Applied Physics; Automated Processes And Servomechanisms; Communication; Computer Engineering; Computer Science; Electronics and Electronic Engineering; Engineering Mathematics; Human-Computer Interaction; Information Technology; Integrated-Circuit Design; Liquid Crystal Technology; Mechanical Engineering; Nanotechnology; Telecommunications; Transistor Technologies; Zone Refining.

DIODE TECHNOLOGY

FIELDS OF STUDY

Electrical engineering; materials science; semiconductor technology; semiconductor manufacturing; electronics; physics; chemistry; nanotechnology; mathematics.

SUMMARY

Diodes act as one-way valves in electrical circuits, permitting electrical current to flow in only one direction and blocking current flow in the opposite direction. The original diodes used in circuits were constructed using vacuum tubes, but these diodes have been almost completely replaced by semiconductor-based diodes. Solid-state diodes, the most commonly used, are perhaps the simplest and most fundamental solid-state semiconductor devices, formed by joining two different types of semiconductors. Diodes have many applications, such as safety circuits to prevent damage by inadvertently putting batteries backward into devices and in rectifier circuits to produce direct current (DC) voltage output from an alternating current (AC) input.

KEY TERMS AND CONCEPTS

- **Anode:** More positive side of the diode, through which current can flow into a diode while forward biased.
- **Cathode:** More negative side of the diode, from which current flows out of the diode while forward biased.
- **Forward Bias:** Orientation of the diode in which current most easily flows through the diode.
- **Knee Voltage:** Minimum forward bias voltage required for current flow, also sometimes called the threshold voltage, cut-in voltage, or forward voltage drop.
- **Light-Emitting Diode (LED):** Diode that emits light when current passes through it in forward bias configuration.
- **P-N Junction:** Junction between positive type (p-type) and negative type (n-type) doped semiconductors.
- **Power Dissipation:** Amount of energy per unit time dissipated in the diode.

- **Rectifiers:** Diode used to make alternating current, or current flowing in two directions, into direct current, or current flowing in only one direction.
- **Reverse Bias:** Orientation of the diode in which current does not easily flow through the diode.
- **Reverse Breakdown Voltage:** Maximum reverse biased voltage that can be applied to the diode without it conducting electrical current, sometimes just called the breakdown voltage.
- **Thermionic Diode:** Vacuum tube that permits current to flow in only one direction, also called a vacuum tube diode.
- **Zener Diode:** Diode designed to operate in reverse bias mode, conducting current at a controlled breakdown voltage called the Zener voltage.

DEFINITION AND BASIC PRINCIPLES

A diode is perhaps the first semiconductor circuit element that a student learns about in electronics courses, though most early diodes were constructed using vacuum tubes. It is very simplistic in structure, and basic diodes are very simple to connect in circuits. They have only two terminals, a cathode and an anode. The very name diode was created by British physicist William Henry Eccles in 1919 to describe the circuit element as having only the two terminals, one in and one out.

Classic diode behavior, that for which most diodes are used, is to permit electric current to flow in only one direction. If voltage is applied in one direction across the diode, then current flows. This is called forward bias. The terminal on the diode into which the current flows is called the anode, and the terminal out of which current flows is called the cathode. However, if voltage is applied in the opposite direction, called reversed bias, then the diode prevents current flow. A theoretical ideal diode permits current to flow without loss in forward bias orientation for any voltage and prohibits current flow in reverse bias orientation for any voltage. Real diodes require a very small forward bias voltage in order for current to flow, called the knee voltage. The terms threshold voltage or cut-in voltage are also sometimes used in place of the term knee voltage. The electronic symbol for the diode signifies the classic diode behavior, with an arrow pointing in the direction of permitted current flow, and a bar on the other side of the diode signifying a block to current flow from the other direction.

Though most diodes are used to control the direction of current flow, there are many subtypes of diodes that have been developed with other useful properties, such as light-emitting diodes and even diodes designed to operate in reverse bias mode to provide a regulated voltage.

BACKGROUND AND HISTORY

Diode-like behavior was first observed in the nineteenth century. Working independently of each other in the 1870's, American inventors Thomas Alva Edison and Frederick Guthrie discovered that heating a negatively charged electrode in a vacuum permits current to flow through the vacuum but that heating a positively charged electrode does not produce the same behavior. Such behavior was only a scientific curiosity at the time, since there was no practical use for such a device.

At about the same time, German physicist Karl Ferdinand Braun discovered that certain naturally occurring electrically conducting crystals would conduct electricity in only one direction if they were connected to an electrical circuit by a tiny electrode connected to the crystal in just the right spot. By 1903, American electrical engineer Greenleaf Whittier Pickard had developed a method of detecting radio signals using the one-way crystals. By the middle of

the twentieth century, homemade radio receivers using galena crystals had become quite popular among hobbyists.

As the electronics and the radio communication industries developed, it became apparent that there would be a need for human-made diodes to replace the natural crystals that were used in a trial-and-error manner. Two development paths were followed: solid-state diodes and vacuum tube diodes. By the middle of the twentieth century, inexpensive germanium-based diodes had been developed as solid-state devices. The problem with solid-state diodes was that they lacked the ability to handle large currents, so for high-current applications, vacuum tube diodes, or thermionic diodes, were developed. In the twenty-first century, most diodes are semiconductor devices, with thermionic diodes existing only for the rare very high-power applications.

HOW IT WORKS

Thermionic Diodes. Though not used as frequently as they once were, thermionic diodes are the simplest type of diode to understand. Two electrodes are enclosed in an evacuated glass tube. Because the thermionic diode is a type of vacuum tube, it is often called a vacuum tube diode. The geometry of the electrodes in the tube depends on the manufacturer and the intended use of the tube. Heating one of the electrodes in some fashion permits electrons on that electrode to be thermally excited. If the electrode is heated past the work function of the material of which the electrode is fabricated, the electrons can come free of the electrode. If the heated electrode has a more negative voltage than the other electrode, then the electrons cross the space between the electrodes. More electrons flow into the negative electrode to replace the missing ones, and the electrons flow out of the positive electrode. Current flow is defined opposite to electron flow, so current would be defined as flowing into the positive electrode (labeled as the anode) and out of the negative electrode (labeled as the cathode). However, if the voltage is reversed, and the heated electrode is more positive than the other electrode, electrons liberated from the anode do not flow to the cathode, so no current flows, making the diode a one-way device for current flow.

Solid-State Diodes. Thermionic diodes, or vacuum tube diodes, tend to be large and consume a lot of electricity. However, paralleling the development of

Diodes are electronic components that allow current to pass in only one direction. (Andrew Lambert Photography/ Photo Researchers, Inc.)

vacuum tube diodes was the development of diodes based on the crystal structure of solids. The most important type of solid-state diodes are based on semi-conductor technology.

Semiconductors are neither good conductors nor good insulators. The purity of the semiconductor determines, in part, its electrical properties. Extremely pure semiconductors tend to be poor conductors. However, all semiconductors have some impurities in them, and some of those impurities tend to improve conductivity of the semiconductor. Purposely adding impurities of the proper type and concentration into the semiconductor during the manufacturing process is called doping the semiconductor. If the impurity has one more outer shell electron than the number of electrons in atoms of the semiconductor, then extra electrons are available to move and conduct electricity. This is called a negative doped or n-type semiconductor. If the impurity has one fewer electrons than the atoms of the semiconductor, then electrons can move from one atom to another in the semiconductor. This acts as a positive charge moving

in the semiconductor, though it is really a missing electron moving from atom to atom. Electrical engineers refer to this as a hole moving in the semiconductor. Semiconductors with this type of impurity are called positive doped or p-type semiconductor.

What makes a semiconductor diode is fabricating a device in which a p-type semiconductor is in contact with an n-type semiconductor. This is called a p-n junction. At the junction, the electrons from the n-type region combine with the holes of the p-type region, resulting in a depletion of charge carriers in the vicinity of the p-n junction. However, if a small positive voltage is applied across the junction, with the p-type region having the higher voltage, then additional electrons are pulled from the n-type region and additional holes are pulled from the p-type region into the depletion region, with electrons flowing into the n-type region from outside the device to make up the difference and out of the device from the p-type region to produce more holes. As with the thermionic device, current flows through the device, with the p-type side of the device being the anode and the n-type side of the device being the cathode. This is the forward bias orientation. When the voltage is reversed on the device, the depletion region simply grows larger and no current flows, so the device acts as a one-way valve for the flow of electricity. This is the reverse bias orientation. Though reverse bias diodes do not normally conduct electricity, a sufficiently high reverse voltage can create electric fields within the diode capable of moving charges through the depletion region and creating a large current through the diode. Because diodes act much like resistors in reverse bias mode, such a large current through the diode can damage or destroy the diode. However, two types of diodes, avalanche diodes and Zener diodes, are designed to be safely operated in reverse bias mode.

APPLICATIONS AND PRODUCTS

P-n junction devices, such as diodes, have a plethora of uses in modern technology.

Rectifiers. The classic application for a diode was to act as a one-way valve for electric current. Such a property makes diodes ideal for use in converting alternating current into direct-current circuits or circuits in which the current flows in only one direction. In fact, the devices were originally called rectifiers before the term diode was created to describe the

---

Fascinating Facts About Diode Technology

- The term diode comes from the Greek *di* (two) and *ode* (paths) signifying the two possible ways of connecting diodes in circuits.
- Naturally occurring crystals with diode-like properties were used in amateur crystal radio sets purchased by millions of hobbyists in the middle of the twentieth century.
- Despite their widely known property of allowing electricity to pass in only one direction, some diodes, such as Zener diodes, are made to be operated in reverse bias mode.
- Light-emitting diodes (LEDs) emit light only when current passes through them in the right direction.
- The laser light produced in laser pointers is made using laser diodes.
- Diodes are often used in battery-operated devices to protect the electronics in the device from users accidentally inserting batteries backward.
- Diodes are used with automobile alternators to produce the DC voltage required for automobile electrical systems.

function of these one-way current devices. Modern rectifier circuits consist of more than just a single diode, but they still rely heavily on diode properties.

Solid-state diodes, like most electronic components, are not 100 percent efficient, and so some energy is lost in their operation. This energy is typically dissipated in the diode as heat. However, semiconductor devices are designed to operate at only certain temperatures, and increasing the temperature beyond a specified range changes the electrical properties of the device. The more current that passed through the device, the hotter it gets. Thus, there is a limiting current that a solid-state diode can handle before it is damaged. Though solid-state diodes have been developed to handle higher currents, for the highest current and power situations, thermionic, or vacuum tube diodes, are still sometimes used, particularly in radio and television broadcasting.

Shottky Diodes. All diodes require at least a small forward bias voltage in order to work. Shottky diodes are fabricated by using a metal-to-semiconductor junction rather than the traditional dual semiconductor p-n junction used with other diodes. Such a construction allows Shottky diodes to operate with extremely low forward bias.

Zener Diodes. Though most diodes are designed to operate only in the forward bias orientation, Zener diodes are designed to operate in reverse bias mode. In such an orientation, they undergo a breakdown and conduct electric current in the reverse direction with a well-defined reverse voltage. Zener diodes are used to provide a stable and well-defined reference voltage.

Photodiodes. Operated in reverse bias mode, some p-n junctions conduct electricity when light shines on them. Such diodes can be used to detect and measure light intensity, since the more light that strikes the diodes, the more they conduct electricity.

Circuit Protection. In most applications of diodes, they are used to take advantage of the properties of the p-n junction on a regular basis in circuits. For some applications, though, diodes are included in circuits in the hope that they will never be needed. One such application is for DC circuits, which are typically designed for current to flow in only one direction. This is automatically accomplished through a power supply with a particular voltage orientation such as a DC source, power converter, or battery. However, if the power supply were connected in reverse or if the batteries were inserted backward, then damage to the circuit could result. Diodes are often used to prevent current flow in such situations where voltage is applied in reverse, acting as a simple but effective reverse voltage protection system.

Light-Emitting Diodes (LEDs). For diodes with just the right kind of semiconductor and doping, the combination of holes and electrons at the p-n junction releases energy equal to that carried by photons of light. Thus, when current flows through these diodes in forward bias mode, the diodes emit light. Unlike most lighting sources, which produce a great deal of waste heat in addition to light (with incandescent lights often using energy to produce more heat than visible light), most of the energy dissipated in LEDs goes into light, making them far more energy-efficient light sources than most other forms of artificial lighting. Unfortunately, large high-power applications of light-emitting diodes are somewhat expensive, limiting them to uses where their small size and long life characteristics offset the cost associated with other forms of lighting.

Laser Diodes. Very similar to light-emitting diodes are laser diodes, where the recombination of holes and electrons also produces light. However, with the laser diode, the p-n junction is placed inside a resonant cavity and the light produced stimulates more light, producing coherent laser light. Laser diodes typically have much shorter operational lifetimes than other diodes, including LEDs, and are generally much more expensive. However, laser diodes cost much less than other methods of producing laser light, so they have become more common. Most lasers not requiring high-power application are based on laser diodes.

IMPACT ON INDUSTRY

The control of electric current is fundamental to electronics. Diodes, acting like one-way valves for electric current are, therefore, very important circuit elements in circuits. They can act as rectifiers, converting alternating current into direct current, but they have many other uses.

Government and University Research. Though the basic concept of the p-n junction is understood, research continues into new applications for the junction. This research is conducted in both public and private laboratories. Much research goes into an effort to make diodes smaller, cheaper, and capable of higher power applications.

Industry and Business. Most diodes are used in fabrication of devices. They are used in circuits and are part of almost every electronic device and will likely continue to play an important role in electronics. Specialized diodes, such as LEDs or laser diodes, have come to be far more important for their specialized properties than for the their current directionality. Laser diodes produce laser light inexpensively enough to permit the widespread use of lasers in many devices that would otherwise not be practical, such as DVD players. Laser diodes are easily modulated, so a great deal of fiber-optic communications use laser diodes. LEDs produce light much more efficiently than most other means of producing light. Many companies are working on developing LED lighting systems to replace more conventional lights.

Many uses of diodes exist outside of the traditional electronics industry. Diodes are, of course, used in electronic devices, but diodes are also used in automobiles, washing machines, and clocks. Almost all electronic devices use diodes, from televisions and radios to microwave ovens and DVD players.

Thermionic diodes are used in only a few industries, such as the power industry and radio and television broadcasting, where the extremely high-power applications would burn out semiconductor diodes. Semiconductor diodes are used in all other applications, since they are more efficient to operate and less expensive to produce.

CAREERS AND COURSE WORK

The electronics field is vast and encompasses a wide variety of careers. Diodes exist in some form in most electronic devices. Thus, a wide range of careers come into contact with diodes, and therefore a wide range of background knowledge and preparation exists for the different careers.

Development of new types of diodes requires considerable knowledge of solid-state physics, materials science, and semiconductor manufacturing. Often advanced degrees in these fields would be required for research, necessitating students studying physics, electrical engineering, mathematics, and chemistry. However, diode technology is quite well evolved, so there are limited job prospects for developing new diodes or diode-like devices other than academic curiosity. Most of this area of study is simply determining how to manufacture or include smaller diodes in integrated circuits.

Electronics technicians repair electrical circuits containing diodes. So, knowledge of diodes and diode behavior is important in diagnosing failures in electronic circuits and circuit boards. Sufficient knowledge can be gained in basic electronics courses. A two-year degree in electronic technology is sufficient for many such jobs, though some jobs may require a bachelor's degree. Likewise, technicians designing and building circuits often do not need to know much about the physics of diodes—just the nature of diode behavior in circuits. Such knowledge can be gained through basic electronics courses or an associate's or bachelor's degree in electronics.

Manufacturing diodes does not actually require much knowledge about diodes for technicians who are actually making semiconductor devices. Such technicians need course work and training in operating the equipment used to manufacture semiconductors and semiconductor devices, and they must be able to follow directions meticulously in operating the machines. An associate's degree in semiconductor manufacturing is often sufficient for many such jobs. Manufacturing circuit boards with diodes, or any other circuit element, does not really require much knowledge of the circuit elements themselves, save for the ability to identify them by sight, though it would be helpful to understand basic diode behavior. Basic course work in circuits would be needed for such jobs.

SOCIAL CONTEXT AND FUTURE PROSPECTS

Diodes exist in almost every electronic device, though most people do not realize that they are using diodes. Because electronics have been increasing in use in everyday life, diodes and diode technology will continue to play an important role in everyday devices. Diodes are very simple devices, however, and it is unlikely that the field will advance further in the development of basic diodes. Specialized devices using the properties of p-n junctions, such as laser diodes, continue to be important. It can be anticipated that additional uses of p-n junctions may be discovered and new types of diodes developed accordingly. Because the p-n junction is the basis of diode behavior and is the basis of semiconductor technology, diodes will continue to play an important role in electronics for the foreseeable future. LEDs produce light very efficiently, and work is proceeding to investigate the possibility of such devices replacing many other forms of lighting.

Raymond D. Benge, Jr., B.S., M.S.

FURTHER READING

Gibilisco, Stan. *Teach Yourself Electricity and Electronics.* 5th ed. New York: McGraw-Hill, 2011. Comprehensive introduction to electronics, with diagrams. A chapter on semiconductors includes a good description of the physics and use of diodes.

Held, Gilbert. *Introduction to Light Emitting Diode Technology and Applications.* Boca Raton, Fla.: Auerbach, 2009. A thorough overview of light-emitting diodes and their uses. The book also includes a good description of how diodes in general work.

Paynter, Robert T. *Introductory Electronic Devices and Circuits.* 7th ed. Upper Saddle River, N.J.: Prentice Hall, 2006. An excellent and frequently used introductory electronics textbook, with an excellent description of diodes, different diode types, and their use in circuits.

Razeghi, Manijeh. *Fundamentals of Solid State Engineering.* 3d ed. New York: Springer, 2009. An advanced undergraduate textbook on the physics of semiconductors, with a very detailed explanation of the physics of the p-n junction.

Schubert, E. Fred. *Light Emitting Diodes.* 2d ed. New York: Cambridge University Press, 2006. A very good and thorough overview of light-emitting diodes and their uses.

Turley, Jim. *The Essential Guide to Semiconductors.* Upper Saddle River, N.J.: Prentice Hall, 2003. A brief overview of the semiconductor industry and semiconductor manufacturing for the beginner.

WEB SITES

The Photonics Society
http://photonicssociety.org

Schottkey Diode Flash Tutorial
http://cleanroom.byu.edu/schottky_animation.phtml

Semiconductor Industry Association
http://www.sia-online.org

University of Cambridge
Interactive Explanation of Semiconductor Diode
http://www-g.eng.cam.ac.uk/mmg/teaching/linearcircuits/diode.html

See also: Electrical Engineering; Electronic Materials Production; Light-Emitting Diodes; Nanotechnology.

DISTILLATION

FIELDS OF STUDY

Chemistry; chemical engineering; industrial studies; chemical hygiene; environmental chemistry.

SUMMARY

Distillation is a process for purifying liquid mixtures by collecting vapors from the boiling substance and condensing them back into the original liquid. Various forms of this technique, practiced since antiquity, continue to be used extensively in the petroleum, petrochemical, coal tar, chemical, and pharmaceutical industries to separate mixtures of mostly organic compounds as well as to isolate individual components in chemically pure form. Distillation has also been employed to acquire chemically pure water, including potable water through the desalination of seawater.

KEY TERMS AND CONCEPTS

- **Condensation:** Phase transition in which gas is converted into liquid.
- **Distilland:** Liquid mixture being distilled.
- **Distillate:** Product collected during distillation.
- **Forerun:** Small amount of low-boiling material discarded at the beginning of a distillation.
- **Fraction:** Portion of distillate with a particular boiling range.
- **Miscible:** Able to be mixed in any proportion without separation of phases.
- **Petrochemical:** Chemical product derived from petroleum.
- **Pot Residue:** Oily material remaining in the boiling flask after distillation.
- **Reflux:** To return condensed vapors (partially or totally) back to the original boiling flask.
- **Reflux Ratio:** Ratio of descending liquid to rising vapor during fractional distillation.
- **Theoretical Plate:** Efficiency of a fractionating column, being equal to the number of vapor-liquid equilibrium stages encountered by the distillate on passing through the column; often expressed as the height equivalent to a theoretical plate (HETP).
- **Vaporization:** Phase transition in which liquid is converted into gas.
- **Vapor Pressure:** Pressure exerted by gas in equilibrium with its liquid phase.

DEFINITION AND BASIC PRINCIPLES

Matter commonly exists in one of three physical states: solid, liquid, or gas. Any phase of matter can be changed reversibly into another at a temperature and pressure characteristic of that particular sample. When a liquid is heated to a temperature called the boiling point, it begins to boil and is transformed into a gas. Unlike the melting point of a solid, the boiling point of a liquid is proportional to the applied pressure, increasing at high pressures and decreasing at low pressures.

When a mixture of several miscible liquids is heated, the component with the lowest boiling point is converted to the gaseous phase preferentially over those with higher boiling points, which enriches the vapor with the more volatile component. The distillation operation removes this vapor and condenses it back to the liquid phase in a different receiving flask. Thus, liquids with unequal boiling points can be separated by collecting the condensed vapors sequentially as fractions. Distillation also removes nonvolatile components, which remain behind as a residue.

BACKGROUND AND HISTORY

Applications of fundamental concepts such as evaporation, sublimation, and condensation were mentioned by Aristotle and others in antiquity; however, many historians consider distillation to be a discovery of Alexandrian alchemists (300 B.C.E. to 200 C.E.), who added a lid (called the head) to the still and prepared oil of turpentine by distilling pine resin. The Arabians improved the apparatus by cooling the head (now called the alembic) with water, which allowed the isolation of a number of essential oils by distilling plant material and, by 800 C.E., permitted the Islamic scholar Jbir ibn Hayyn to obtain acetic acid from vinegar. Alembic stills and retorts were widely employed by alchemists of medieval Europe. The first fractional distillation was developed by Taddeo Alderotti in the thirteenth century. The first comprehensive manual of distillation techniques was *Liber de arte distillandi, de simplicibus,* by Hieronymus Brunschwig, published in 1500 in Strasbourg, France.

The first account of the destructive distillation of coal was published in 1726. Large-scale continuous stills with fractionating towers similar to modern industrial stills were devised for the distillation of alcoholic beverages in the first half of the nineteenth century and later adapted to coal and oil refining. Laboratory distillation similarly advanced with the introduction of the Liebig condenser around 1850. The modern theory of distillation was developed by Ernest Sorel and reduced to engineering terms in his *Distillation et rectification industrielles* (1899).

HOW IT WORKS

Simple Distillation. A difference in boiling point of at least 25 degrees Celsius is generally required for successful separations with simple distillation. The glass apparatus for laboratory-scale distillations consists of a round-bottomed boiling flask, a condenser, and a receiving flask. Vapors from the boiling liquid are returned to the liquid state by the cooling action of the condenser and are collected as distillate in the receiving flask. For high-boiling liquids, an air condenser may be sufficient, but often a jacketed condenser—in which a cooling liquid such as cold water is circulated—is required. The design of many styles of condensers (such as Liebig and Wes) enhances the cooling effect of the circulating liquid. An adapter called a still head connects the condenser to the boiling flask at a 45-degree angle and is topped with a fitting in which a thermometer is inserted to measure the temperature of the vapor (the boiling point). A second take-off adapter is often used to attach the receiving flask to the condenser at a 45-degree angle so that it is vertical and parallel to the boiling flask. One should never heat a closed system, so the take-off adapter contains a side-arm for connection to either a drying tube or a source of vacuum for distillations under reduced pressure. The apparatus was formerly assembled by connecting individual pieces with cork or rubber stoppers, but the ground-glass joints of modern glassware make these stoppers unnecessary.

Fractional Distillation. When the boiling points of miscible liquids are within about 25 degrees Celsius, simple distillation does not yield separate fractions. Instead, the process produces a distillate whose composition contains varying amounts of the components, being initially enriched in the lower-boiling and more volatile one. The still assembly is modified to improve efficiency by placing a distilling column

Huge distillation towers separate oil into its components for Pemex in Mexico. (Chris Sharp/Photo Researchers, Inc.)

between the still head and boiling flask. This promotes multiple cycles of condensation and revaporization. Each of these steps is an equilibration of the liquid and gaseous phases and is, therefore, equivalent to a simple distillation. Thus, the distillate from a single fractional distillation has the composition of one obtained from numerous successive simple distillations. Still heads allowing higher reflux ratios and distilling columns having greater surface area permit more contact between vapor and liquid, which increases the number of equilibrations. Thus, a Vigreux column having a series of protruding fingers is more efficient than a smooth column. Even more efficient are columns packed with glass beads, single- or multiple-turn glass or wire helices, ceramic pieces, copper mesh, or stainless-steel wool. The limit of efficiency is approached by a spinning-band column that contains a very rapidly rotating spiral of metal or Teflon over its entire length.

APPLICATIONS AND PRODUCTS

Batch and Continuous Distillation. Distilling very large quantities of liquids as a single batch is impractical, so industrial-scale distillations are often conducted by continuously introducing the material to be distilled. Continuous distillation is practiced in petrochemical and coal tar processing and can also be used for the low-temperature separation and purification of liquefied gases such as hydrogen, oxygen, nitrogen, and helium.

Vacuum Distillation. Heating liquids to temperatures above about 150 degrees Celsius is generally avoided to conserve energy, to minimize difficulties of insulating the still head and distilling column, and to prevent the thermal decomposition of heat-sensitive organic compounds. A vacuum distillation takes advantage of the fact that a liquid boils when its vapor pressure equals the external pressure, which causes the boiling point to be lowered when the pressure decreases. For example, the boiling point of water is 100 degrees Celsius at a pressure of 760 millimeters of mercury (mmHg), but this drops to 79 degrees Celsius at 341 mmHg and rises to 120 degrees Celsius at 1,489 mmHg. Vacuum distillation can be applied to solids that melt when heated in the boiling flask; however, higher temperatures may be required in the condenser to prevent the distillate from crystallizing. The term "vacuum distillation" is actually a misnomer, for these distillations are conducted at a reduced pressure rather than under an absolute vacuum. A pressure of about 20 mmHg is obtainable with ordinary water aspirators and down to about 1 mmHg with a laboratory vacuum pump.

Molecular Distillation. When the pressure of residual air in the still is lower than about 0.01 mmHg, the vapor can easily travel from the boiling liquid to the condenser, and distillate is collected at the lowest possible temperature. Distillation under high-vacuum conditions permits the purification of thermally unstable compounds of high molecular weight (such as glyceride fats and natural oils and waxes) that would otherwise decompose at temperatures encountered in an ordinary vacuum distillation. Molecular stills often have a simple design that minimizes refluxing and accelerates condensation. For example, the high-vacuum short-path still consists of two plates, one heated and one cooled, that are separated by a very short distance. Industrially, the distillate can be condensed on a rapidly rotating cone and removed quickly by centrifugal force.

Steam Distillation. Another method to lower boiling temperature is steam distillation. When a homogeneous mixture of two miscible liquids is distilled, the vapor pressure of each liquid is lowered according to Raoult's and Dalton's laws; however, when a heterogeneous mixture of two immiscible liquids is distilled, the boiling point of the mixture is lower than that of its most volatile component because the vapor pressure of each liquid is now independent of the other liquid. A steam distillation occurs when one of these components is water and the other an immiscible organic compound. The steam may be introduced into the boiling flask from an external source or may be generated internally by mixing water with the material to be distilled. Steam distillation is especially useful in isolating the volatile oils of plants.

Azeotropic Extractive Distillation. Certain nonideal solutions of two or more liquids form an azeotrope, which is a constant-boiling mixture whose composition does not change during distillation. Water (boiling point of 100.0 degrees Celsius) and ethanol (boiling point of 78.3 degrees Celsius) form a binary azeotrope (boiling point of 78.2 degrees Celsius) consisting of 4 percent water and 96 percent ethanol. No amount of further distillation will remove the remaining water; however, addition of benzene (boiling point 80.2 degrees Celsius) to this distillate forms a tertiary benzene-water-ethanol azeotrope (boiling point of 64.9 degrees Celsius) that leaves pure ethanol behind when the water is removed. This is an example of azeotropic drying, which is a special case of azeotropic extractive distillation.

Microscale Distillation. Microscale organic chemistry, with a history that spans more than a century, is not a new concept to research scientists; however, the traditional 5- to 100-gram macroscale of student laboratories was reduced one hundred to one thousand times by the introduction of microscale glassware in the 1980's to reduce the risk of fire and explosion, limit exposure to toxic substances, and minimize hazardous waste. Microscale glassware comes in a variety of configurations, such as Mayo-Pike or Williamson styles. Distillation procedures are especially troublesome in microscale because the ratio of wetted-glass

Fascinating Facts About Distillation

- The word "distill" in the late fourteenth century was originally applied to the separation of alcoholic liquors from fermented materials and comes from the Middle English "distillen," which comes from the Old French *distiller*, which comes from the Late Latin *distillare*, which is an alteration of *destillare* (*de* and *stillare*) meaning "to drip."

- In *Aristotelous peri geneses kai phthoras* (335-323 B.C.E.; *Meteoroligica*, 1812), Aristotle described how potable water was obtained as condensation in a sponge suspended in the neck of a bronze vessel containing boiling seawater.

- Distilled alcoholic beverages first appeared in Europe in the late twelfth century, developed by alchemists.

- The Celsius scale of temperature was originally defined in the eighteenth century by the freezing and boiling points of water at 0 and 100 degrees Celsius, respectively, at a pressure of one standard atmosphere (760 millimeters of mercury).

- The first vertical continuous distillation column was patented in France by Jean-Baptiste Cellier Blumenthal in 1813 for use in alcohol distilleries.

- The first modern book on the fundamentals of distillation, *La Rectification de l'alcool* (*The Rectification of Alcohol*), was published by Ernest Sorel in 1894.

- The gas lights of Sherlock Holmes's London burned coal gas from the destructive distillation of coal.

- In 1931, M. R. Fenske separated two isomeric hydrocarbons that boiled at only 3 degrees Celsius apart using a 52-foot fractionating column mounted in an airshaft at Pond Laboratory of Pennsylvania State College.

surface area to the volume of distillate increases as the sample size is reduced, thereby causing significant loss of product. Specialized microscale glassware such as the Hickman still head has been designed to overcome this difficulty.

Analytical Distillation. The composition of liquid mixtures can be quantitatively determined by weighing the individual fractions collected during a carefully conducted fractional distillation; however, this technique has been largely replaced by instrumental methods such as gas and liquid chromatography.

IMPACT ON INDUSTRY

Chemical and Petrochemical Industries. Distillation is one of the fundamental unit operations of chemical engineering and is an integral part of many chemical manufacturing processes. Modern industrial chemistry in the twentieth century was based on the numerous products obtainable from petrochemicals, especially when thermal and catalytic cracking is applied. Industrial distillations are performed in large, vertical distillation towers that are a common sight at chemical and petrochemical plants and petroleum refineries. These range from about 2 to 36 feet in diameter and 20 to 200 feet or more in height. Chemical reaction and separation can be combined in a process called reactive distillation, where the removal of a volatile product is used to shift the equilibrium toward completion.

Petroleum Industry. Distillation is extensively used in the petroleum industry to separate the hydrocarbon components of petroleum. Crude oil is a complex mixture of a great many compounds, so initial refining yields groups of substances within a range of boiling points: natural gas below 20 degrees Celsius (C_1 to C_4 hydrocarbons), petroleum ether from 20 to 60 degrees Celsius (C_5 to C_6 hydrocarbons), naphtha (or ligroin) from 60 to 100 degrees Celsius (C_5 to C_9 hydrocarbons), gasoline from 40 to 205 degrees Celsius (C_5 to C_{12} hydrocarbons and cycloalkanes), kerosene from 175 to 325 degrees Celsius (C_{10} to C_{18} hydrocarbons and aromatics), gas oil (or diesel) from 250 to 400 degrees Celsius (C_{12} and higher hydrocarbons). Lubricating oil is distilled at reduced pressure, leaving asphalt behind as a residue

Destructive Distillation. When bituminous or soft coal is heated to high temperatures in an oven that excludes air, destructive distillation converts the coal into coke as gases and liquids form and are distilled over. The coke is chiefly employed in the iron and steel industry, and coal gas is used for heating. The liquid fraction, called coal tar, contains some compounds originally in the coal and others produced by chemical reactions during heating. Coal tar yields numerous organic compounds on repeated distilling and redistilling of the crude fractions. It provided the raw material for the early synthetic organic chemical industry in the nineteenth century. In the early twentieth century, coal-derived chemicals provided gas to light street lamps and were the main source of phenol, toluene, ammonia, and naphthalene during World

War I. When wood is similarly heated in a closed vessel, destructive distillation converts the wood into charcoal and yields methyl alcohol (called wood alcohol), together with acetic acid and acetone as liquid distillates.

Essential Oils. Essential oils are generally obtained by steam distillation of plant materials (flowers, leaves, wood, bark, roots, seeds, and peel) and are used in perfumes, flavorings, cosmetics, soap, cleaning products, pharmaceuticals, and solvents. Examples include turpentine and oils of cloves, eucalyptus, lavender, and wintergreen.

Distilled Water and Desalination. Tap water commonly contains dissolved salts that contribute to its overall hardness. These are removed through distillation to provide distilled water for use in automobile batteries and radiators, steam irons, and other applications where pure water is beneficial. Specialized stills fed with continuously flowing tap water are used to prepare distilled water in chemistry laboratories; however, deionized water is becoming an increasingly popular and more convenient alternative. Seawater can be similarly distilled to provide potable drinking water. Desalination of seawater is especially important in arid and semi-arid regions of the Middle East, where the abundant sunshine is used in solar distillation facilities.

CAREERS AND COURSE WORK

The art of distillation is most commonly practiced by chemists whose college majors were chemistry or chemical engineering, the former distilling samples on a laboratory scale and the latter conducting distillations on larger pilot plant and considerably larger industrial scales. The difference between chemistry and chemical engineering majors occurs in advanced and elective course work. Chemistry majors concentrate on molecular structure to better understand the chemical and physical properties of matter, whereas chemical engineering students focus more on the properties of bulk matter involved in the large-scale, economical, and safe manufacture of useful products. Advanced graduate study can be pursued in both disciplines at the master's, and doctoral levels, the latter degree being very common among professors of chemistry and chemical engineering at colleges and universities.

Both chemistry and chemical engineering majors must have strong backgrounds in physics and mathematics and take a year of general chemistry followed by a year of organic chemistry and another year of physical chemistry. The ability to infer the relative boiling points of chemical substances by predicting the strength of intermolecular forces from molecular structure begins in general chemistry and is pursued in detail in organic chemistry. Student laboratories employ distillation to isolate and purify products of synthesis, often with very small amounts using microscale glassware. The theoretical and quantitative aspects of phase transitions are studied in depth in the physical chemistry lecture and laboratory. Participation in research by working with an established research group is an important way for students to gain practical experience in laboratory procedures, methodologies, and protocols. Knowledge of correct procedures for handling toxic materials and the disposal of hazardous wastes are vital for all practicing chemists.

SOCIAL CONTEXT AND FUTURE PROSPECTS

The process and apparatus of distillation, more than any other technology, gave birth to modern chemical industry because of the numerous chemical products derived first from coal tar and later from petroleum. The continued role played by distillation in modern technology will depend on several factors including the sustainability of raw materials and energy conservation. Increased demand and diminishing supplies of raw materials, together with the accumulation of increasing amounts of hazardous waste, have made recycling economically feasible on an industrial scale, and distillation has a role to play in many of these processes. Likewise, the increasing cost of crude oil because of diminishing supplies of this finite resource encourages the use of alternate sources of oil such as coal (nearly 75 percent of total fossil fuel reserves), and distillation would be expected to play the same central role as it does in refining petroleum. However, distillation is also an energy-intensive technology in the requirements of both heating liquids to boiling and cooling the resulting vapors so that they condense back to liquid products. Thus, one would expect that the chemical industry of the future would seek alternate energy sources such as solar power as well as ways to conserve energy through improvements in distillation efficiency.

Martin V. Stewart, B.S., Ph.D.

FURTHER READING

Donahue, Craig J. "Fractional Distillation and GC Analysis of Hydrocarbon Mixtures." *Journal of Chemical Education* 79, no. 6 (June, 2002): 721-723. Demonstrates analytical and physical methods employed in petroleum refining.

El-Nashar, Ali M. *Multiple Effect Distillation of Seawater Using Solar Energy.* New York: Nova Science, 2008. Tells the story of the Abu Dhabi, United Arab Emirates, desalinization plant that uses solar energy to distill seawater.

Forbes, R. J. *Short History of the Art of Distillation from the Beginnings Up to the Death of Cellier Blumenthal.* 1948. Reprint. Leiden, Netherlands: E. J. Brill, 1970. Discusses the history of distillation, focusing on earlier periods.

Kister, Henry Z. *Distillation Troubleshooting.* Hoboken, N.J.: Wiley-Interscience, 2006. Examines distillation operations and how they are maintained and repaired.

Owens, Bill, and Alan Dikty, eds. *The Art of Distilling Whiskey and Other Spirits: An Enthusiast's Guide to the Artisan Distilling of Potent Potables.* Beverly, Mass.: Quarry Books, 2009. Photographs from two cross-country road trips illustrate this paperback guide to the small-scale distillation of whiskey, vodka, gin, rum, brandy, tequila, and liquors.

Stichlmair, Johann G., and James R. Fair. *Distillation: Principles and Practices.* New York: Wiley-VCH, 1998. This 544-page comprehensive treatise for distillation technicians contains chapters on modern industrial distillation processes and energy savings during distillation.

Towler, Gavin P., and R. K. Sinnott. *Chemical Engineering Design: Principles, Practice, and Economics of Plant and Process Design.* Boston: Elsevier/Butterworth-Heinemann, 2008. Chapter 11 examines continuous and multicomponent distillation, looking at the principles involved and design variations. Also discusses other distillation systems and components of a system.

WEB SITES

American Institute of Chemical Engineers
http://www.aiche.org

American Petroleum Institute
http://www.api.org

Fractionation Research
http://www.fri.org

Institution of Chemical Engineers
http://unified.icheme.org

See also: Chemical Engineering; Coal Gasification; Coal Liquefaction; Cracking; Petroleum Extraction and Processing.

DNA ANALYSIS

FIELDS OF STUDY

Biochemistry; biotechnology; molecular biology; molecular genetics; population genetics; forensic science; statistics.

SUMMARY

DNA analysis involves the use of scientific tools to access the information found in DNA to identify its source, whether some infectious agent, another organism of interest, or a particular individual, such as in forensic applications. Medical applications of this technology include the search for mutations associated with genetic disorders and the design of probes that are able to diagnose these disorders in a timely fashion.

KEY TERMS AND CONCEPTS

- **Base Pair:** Single unit of double-stranded DNA.
- **Combined DNA Index System (CODIS):** Federal Bureau of Investigation (FBI) database that contains the DNA profiles of more than 8 million convicted violent offenders.
- **DNA (Deoxyribonucleic Acid):** Nucleic acid that is the genetic component of all living cells. Nucleic acids are polymers of nucleotides.
- **Nucleotide:** Chemical composed of a nucleoside (a nitrogen-containing purine or pyrimidine base linked to a five-carbon sugar such as ribose or deoxyribose) and a phosphate or polyphosphate group.
- **Polymerase Chain Reaction (PCR):** Technique used to produce a large amount of DNA from very small quantities.
- **Restriction Fragment Length Polymorphism (RFLP):** Early method of DNA analysis involving the cleavage and separation of DNA.
- **Short Tandem Repeat (STR):** Repeating element in DNA that is from one to six base pairs long; also known as microsatellite.
- **Single Nucleotide Polymorphism (SNP):** Point mutation in DNA that differs among individuals.
- **Southern Blotting:** Process used to detect the presence of a particular DNA sequence in a sample. DNA fragments are separated using gel electrophoresis, then transferred, or blotted, onto a membrane, where they are exposed to a hybridization probe (a fragment of DNA that will hybridize to a complementary sequence).
- **Variable Number of Tandem Repeats (VNTR):** Repeating element in DNA that is tens to hundreds of base pairs long; also known as minisatellite or long tandem repeat.

DEFINITION AND BASIC PRINCIPLES

DNA analysis is, in the strictest sense of the term, an actual observation of the length or sequence of a portion of DNA. The length of a fragment of DNA can be determined using gel electrophoresis. This technique involves placing DNA onto a semisolid support, or gel, and applying an electric current to the gel so that DNA migrates toward the positive pole. The migration of DNA in gel electrophoresis is proportional to its mass, which is, in turn, proportional to its length. Determining the actual sequence of bases in a strand of DNA is much more complicated.

Although DNA analysis has at times been equated with genetic testing, the two are not always the same. Certain types of genetic testing developed in the 1960's did not technically involve DNA analysis. Amniocentesis, which allowed for Down syndrome testing, actually involved chromosomal analysis following the creation of a karyotype (organized profile of a person's chromosomes). Similarly, genetic testing for phenylketonuria (PKU) and Tay-Sachs disease originally involved enzyme assays, not an analysis of the defective genes themselves. Thus, actual DNA analysis did not begin in earnest until the mid-1970's.

BACKGROUND AND HISTORY

Although the double helical structure of DNA was first described in 1953 by American molecular biologist James D. Watson and British biophysicist Francis Crick, more than twenty years passed before scientists developed methods of comparing DNA for the purpose of identification. In 1974, British molecular biologist Joseph Sambrook described the differentiation of human tumor viruses following cleavage by a restriction endonuclease (an enzyme that cleaves DNA at a specific nucleotide sequence). He noticed that different-sized bands of DNA were

Researcher supervising an automated probe that is distributing DNA (deoxyribonucleic acid) samples among an array of microplates. This is part of an automated polymerase chain reaction (PCR) process, carried out to amplify the amount of DNA present in the samples. (Pascal Goetgheluck/Photo Researchers, Inc.)

visible following their separation on a gel. This discovery formed the basis for what has become known as restriction fragment length polymorphism (RFLP). Although the DNA of viruses and even bacteria could be analyzed directly by RFLP, the restriction enzyme cleavage patterns of higher organisms were of sufficient complexity that only a subset of the bands that were produced could be analyzed. DNA analysis of higher organisms was made possible in 1975 by the development of the Southern blotting technique by British biochemist Edwin Southern.

The following decade saw two ideas that would revolutionize the field of DNA analysis. In 1983, American biochemist Kary Mullis developed the polymerase chain reaction (PCR), a process that enabled the amplification across several orders of magnitude of small amounts of starting sample DNA in the laboratory. Then, in 1985, British geneticist Alec Jeffreys realized that human DNA was peppered with regions of repeating sequences, or variable number tandem repeats (VNTRs), and that comparisons of these regions could create a unique DNA fingerprint for any given individual.

HOW IT WORKS

Probes and Primers. Most of the methods of DNA analysis take advantage of DNA's natural tendency to form a double helix. RFLP analysis has long been paired with Southern blotting. This procedure involves binding a small synthetic fragment of DNA to a region of interest that is contained within one or more of the bands of DNA that have been separated by gel electrophoresis and then blotted onto some type of membrane. This binding is made possible by the complementary nature of the DNA bases, the fact that adenine forms hydrogen bonds to pair with thymine and that cytosine pairs with guanine, a concept called Watson-Crick base pairing after the discoverers of DNA structure. The synthetic DNA, called a probe, is designed to contain about twenty complementary nucleotides to the sequence of interest. Binding of this probe to the blotted DNA is called hybridization. Originally, DNA probes were labeled with a radioactive marker to enable the detection of their position on a membrane, but later probe labeling has included nonradioactive alternatives such as fluorescent dyes.

The polymerase chain reaction (PCR) also involves the binding of small synthetic fragments of DNA to a region of interest, but in this process, two such fragments bind to opposite strands of the target DNA. These fragments, although identical in structure to the probes described earlier, are called primers because they are used to prime a DNA synthesis reaction. Also, the binding reaction, which involves the same process of complementary bases coming together to form hydrogen bonds, is referred to as annealing. It had been previously discovered that DNA could be made in the laboratory by taking a given single strand of DNA and adding a specific primer, the four types of nucleotides, and purified DNA polymerase (the enzyme normally involved in the polymerization process), but Mullis's insight in the early 1980's was that this process could be converted into a chain reaction that produced large amounts of DNA. By adding two primers instead of one and by using double-stranded DNA as a target, twice as many molecules of DNA could be created, but this necessitated a step in which the DNA had to be heated to near-boiling temperatures to separate, or melt, the two strands of the double helix. Mullis reasoned that a DNA polymerase that had been purified from a thermophilic microbe would be able to survive this heating step. This would allow the stringing together of a number of cycles with three different temperatures—one each for annealing, polymerization, and DNA melting—without having to add more DNA polymerase enzyme. Thus, after each cycle, the amount of DNA double helix would be doubled, resulting in more than a billion molecules of DNA after thirty cycles, even if the cycle started with only one strand of DNA.

DNA Polymorphism. DNA analysis takes advantage of the intrinsic variability that exists among organisms as well as among members of the same species. Polymorphism, a word derived from the Greek for "many forms," is used to describe this variability, the simplest form of which is a single nucleotide polymorphism (SNP). Single nucleotide polymorphisms, which are also referred to as point mutations in genetics, are detectable by RFLP analysis only when they occur within the recognition sequence for a particular restriction enzyme because the enzyme fails to cleave the altered sequence. RFLP analysis also readily detects deletions or insertions of DNA sequences that have occurred between restriction enzyme cleavage sites. What Jeffreys realized in the 1980's was that most restriction fragment length variation in humans was not caused by large insertions or deletions of unique DNA sequences but by a variation in the number of repetitive DNA elements that were found in tandem with one another. He did not, however, use the PCR method that was being developed at the time because a practitioner of PCR must know the precise sequences that flank a site of interest to design the primers used in this procedure. Instead, Jeffreys performed Southern blotting using a probe designed to hybridize with the about fifteen-nucleotide-long sequence that he was studying. This probe specifically labeled the regions of the membrane that contained these variable number tandem repeats (VNTRs). For this contribution, Jeffreys has been called the father of DNA fingerprinting.

Subsequent analysis of various regions in human DNA has taken advantage of PCR to produce results, focusing on even smaller tandem repeats with repeating units that are only one to six nucleotides in length. Discovered in 1989, these short tandem repeats (STRs) were eventually found to outnumber variable number tandem repeats by nearly one hundredfold, being found at more than 100,000 sites in human DNA. As more and more of these STR sites were characterized over time, primers that annealed to their flanking sequences were designed to amplify the repeat area in question.

APPLICATIONS AND PRODUCTS

Of Microbes and Man. Although the tools involved in DNA analysis are often used in basic research such as determining the evolutionary relationships between organisms, much of the application of this technology involves analysis for the purposes of identification. Although identification could potentially include any organism of interest, the primary focus of DNA analysis has been disease-causing viruses and microorganisms along with humans. Ever since Sambrook and colleagues first applied RFLP analysis to differentiate between two strains of viruses, viral epidemiology has remained an important application for tools such as PCR. For example, around the beginning of the twenty-first century, nucleic acid amplification testing (NAAT) was developed to detect the viral load of the human immunodeficiency virus (HIV). The procedure is a faster and more effective way to test for the presence of HIV in a person. NAAT has also been applied as a diagnostic test for certain bacterial infections. Other PCR-based methods have been adapted to test for bacterial contamination of foods as well as of hospital areas and supplies. In most cases, the identification of the precise strain of virus or microbe present is unnecessary because the physical presence of an infectious agent, not its detailed classification, is of interest. Tandem-repeat-based methods of identification are largely useless when analyzing such infectious agents because these agents tend to lack such repetitive DNA sequences. Because the DNA sequence of the entire genome (the complete set of DNA found in a particular organism) of most known infectious agents has been determined, it is possible to design primers that will specifically amplify DNA from a given target species.

In some cases, as in life-threatening illnesses, potential epidemics, and acts of bioterrorism, the speed at which an infectious agent is identified is critical to saving lives. For such applications, a type of PCR called real-time PCR has been developed. Rather than waiting to run gel electrophoresis after the full thirty or so cycles of a traditional PCR reaction have been completed, real-time PCR measures the production of a fluorescent-tagged product in real time, during the early phases of the reaction. This allows for an agent to be detected in minutes rather than hours.

Crime Scenes and Beyond. The best-known use of DNA analysis is probably in the area of forensics.

The first case in which Jeffreys applied DNA fingerprinting was an immigration dispute. In 1983, British authorities had denied a thirteen-year-old boy entry into the country, claiming that his passport was forged and that his stated mother, a British subject, was not his biological mother. The dispute continued until 1985, when Jeffreys was able to apply his new technique to prove that the maternal relationship stated on the passport was indeed correct. Since that time, maternity tests have been vastly outnumbered by paternity tests, but the principle used in both types of parental testing remains the same.

The first use of DNA fingerprinting in a criminal case occurred in 1986, when it was used to exonerate a suspect accused of the rape and murder of a teenage girl near Leicester, England. Later, the same technique was used to identify the real killer. Since this early case, evidence from DNA fingerprinting has helped convict thousands of criminals. The source of DNA is blood in about half of all cases; other common sources are semen and hair. DNA analysis also plays an important role in the identification of human remains following disasters, acts of terrorism, and war.

Limitations of PCR. Following the advent of PCR, the amount of forensic sample required for analysis was reduced significantly. The original DNA fingerprinting procedure developed by Jeffreys required a blood sample about the size of a quarter, but later methods needed only a few cells swabbed from a person's cheeks to perform an analysis. Although PCR requires much less starting material than RFLP analysis and is also a more rapid procedure to perform, it does have a number of limitations. The first limitation, that flanking DNA sequences must be known ahead of time, was largely overcome as more and more human short tandem repeats were characterized along with the DNA that surrounded them. A second limitation is that the method is so sensitive that it is prone to contamination by outside sources. Because even a single fragment of DNA can be amplified into large amounts on a gel, care must be taken not to introduce foreign DNA from an investigator's hair or fingertips. A third limitation is that only a single area, or locus, of DNA can be analyzed at one time. To overcome this limitation, a procedure called multiplex PCR has been developed. This method simultaneously employs a number of primers that have been labeled with fluorescent tags. These can be

identified during the subsequent gel electrophoresis step based on their specific labels.

Medical Applications. Besides using DNA analysis to identify infectious agents, the medical community has begun to use this technique to study genetic disorders. However, common methods of DNA analysis cannot identify most genetic disorders, with the exception of a class of disorders called trinucleotide repeat expansion disorders. This rare class of disorders, which includes Huntington's disease as well as fragile X syndrome, is readily detectable using PCR amplification of the short tandem repeats that contribute to the disorders in question. A more common class of genetic disorders results from point mutations in genes and can therefore be linked to particular single nucleotide polymorphisms in the human genome. Unfortunately, single nucleotide polymorphisms are not detectable by PCR and will show up in RFLP analysis only if they occur in the restriction enzyme recognition site itself, which is a rare occurrence. The identification of genetic disorders is therefore largely dependent on determining the actual sequence of the DNA, still a technically challenging and expensive undertaking despite progress that has been made since the inception of the Human Genome Project DNA sequencing program in the 1990's.

Methods involved in DNA sequencing include many of the same principles as other forms of DNA analysis. A single primer is labeled with a fluorescent dye and mixed with a target sequence in the presence of a thermostable DNA polymerase. This procedure does not amplify the DNA as in PCR but results in primer extension for a certain length along the target sequence. Another difference from PCR is that modified nucleotides are added to this mixture so that the primer extension is halted whenever these particular nucleotides are incorporated into a growing DNA strand. Four separate tubes are used in this method, one for each of the four DNA bases. Once these four reactions are separated by electrophoresis, the order of bases can be determined using computer software that monitors the relative migration of the bands that occurs from each of the four reaction tubes.

IMPACT ON INDUSTRY

Perhaps because the United States and the United Kingdom rank first and second, respectively, for the most crimes committed within an industrialized nation, or perhaps because they are world leaders in biomedical research, these two nations have played a leading role in the development of tools for use in DNA analysis from the very beginning. Soon after Jeffreys developed the DNA fingerprinting technique, the Cellmark Diagnostics division of the British company Imperial Chemical Industries (ICI) began offering the first commercially available DNA testing kit. Since then, molecular diagnostics/genetic testing has become a billion dollar industry, with many of the companies based in the United States. In addition to producing reagents required to perform genetic testing, some companies have focused on providing kits for the extraction of DNA from a variety of sources, including bacteria, soil, water, blood, tissue, and bone. Because the basic steps in DNA extraction usually include cell lysis, ribonucleic acid (RNA) removal, and the separation of DNA and proteins based on their differing solubilities in salt- and alcohol-containing solutions, many of these kits have a number of reagents in common, although the procedure used in extraction can be adapted to the specific source of the DNA.

The governments of the United States and the United Kingdom also assumed leading roles in DNA analysis. In 1990, the Federal Bureau of Investigation established the combined DNA index system (CODIS), a database to contain the information gained from DNA fingerprinting of convicted offenders. The numerical data that correspond to the migration of bands obtained from DNA fingerprinting are also referred to as an individual's DNA profile. Early on, the FBI chose thirteen specific STR loci to include in its system. As of 2010, CODIS contained more than 8 million DNA profiles. In 1995, the United Kingdom's Forensic Science Services established a similar system to catalog DNA profiles, including many of the same STR loci. In addition to law enforcement, the governments of both the United States and the United Kingdom were significantly involved in the Human Genome Project. On June 26, 2000, President Bill Clinton and Prime Minister Tony Blair made a simultaneous announcement that a first draft of the human genome had been completed ten years into the project.

CAREERS AND COURSE WORK

Scientists in general have traditionally chosen from three career paths, industry, academics, and government, with industry providing the most jobs

---

Fascinating Facts About DNA Analysis

- In addition to Southern blotting, two other forms of blotting are performed in molecular biology. These involve blotting of either RNA or protein for analysis and are named Northern blotting and Western blotting, respectively, as a humorous homage to Edwin Southern.

- Ironically, Kary Mullis, who developed the method used to screen for the viral load of HIV in humans, is among the handful of scientists who reject the scientific evidence that HIV is the cause of acquired immunodeficiency syndrome (AIDS).

- Colin Pitchfork, the first criminal convicted of murder based on DNA fingerprinting evidence, initially evaded arrest by telling a friend that he was terrified of needles and paying that friend to submit a blood sample for him.

- DNA extracted from blood actually comes from the white blood cells, not the red, even though the latter outnumber the former by a ratio of 700:1. Human red blood cells lose their nuclei, and therefore their DNA, during development.

- The first human genome sequence produced was actually a mosaic of DNA sequences from various anonymous donors. In 2007, J. Craig Venter, the head of a private company involved in the Human Genome Project, became the first individual to have his entire genome sequenced. Venter said his company had largely used his own DNA in the sequencing efforts that they had contributed to the project.

- According to the latest estimates, the Bureau of Justice Statistics at the U.S. Department of Justice reports that tens of thousands of requests for DNA analysis are backlogged at any given time because of the high demand for this service. This represents the highest percentage of backlogged requests for any type of analysis performed by crime laboratories under their jurisdiction.

---

and government the fewest. This general principle is true for those interested in the science of DNA analysis. Industry leads the way in the design of diagnostic tests and DNA extraction kits, academics is the domain of basic research, and government dominates the field of forensic science. Most of of the forensic analysis performed in the United States occurs at the governmental level. About half of the 400 publicly owned crime laboratories at the municipal, county, state, and federal level are capable of performing DNA analyses; the rest rely on a government laboratory higher up in their jurisdiction or on a private laboratory for this service.

Those who are interested in a career in forensic science should bear in mind a few facts. First, the hybrid job of police officer/detective/forensic scientist depicted on television dramas does not exist in reality. Forensic scientists either work in the field or in the laboratory but rarely in both. In large laboratories, the forensic analyst will tend to specialize in a particular area, and in small laboratories, the analyst will be more of a generalist but also will lack most of the resources and equipment shown on television. Second, although dozens of colleges have introduced degree programs in forensics to keep up with the increased demand, any bachelor's degree that gives its holder a solid background in science and mathematics and excellent communication skills is sufficient to work in this area. Third, students should realize that such a bachelor's degree prepares a person to begin working as a technician performing largely support functions such as preparing reagents for analysis. A master's degree or certification program is most likely required to specialize in a subfield of forensic science and to perform more of the scientific analysis of evidence, while a doctoral degree in an associated field is preferred for advancement to administrative positions such as laboratory director.

SOCIAL CONTEXT AND FUTURE PROSPECTS

Single nucleotide polymorphisms (SNPs), although not used extensively in forensic applications, potentially contain valuable information that can be of use to crime scene investigators. For example, the presence of particular SNPs may indicate a perpetrator's race, while others could indicate hair color. One disadvantage of SNPs, besides the relative difficulty of identifying them, is that many more of them are needed to provide a unique identification (compared to the number of short tandem repeats needed for PCR). Because most SNPs are biallelic, they contain one base or another but generally not all four possible bases, and it is estimated that as many as fifty would have to be analyzed to obtain the same level of confidence as provided by the thirteen STR loci contained in CODIS. This may not prove as difficult as it sounds because it is estimated that there are probably

about 10 million SNP sites scattered throughout the human genome. If accurate, that would mean that SNPs outnumber short tandem repeats to the same degree that short tandem repeats outnumber variable number tandem repeats.

DNA sequencing in some form or another is likely to continue to play an increasing role in DNA analysis. The cost of DNA sequencing probably will drop as it becomes more prevalent and increasingly automated. Although the first human genome sequence was produced at a cost of billions of dollars, scientists have set a goal of reducing the cost of DNA sequencing to about one thousand dollars. At the same time, scientists are developing a number of methods that allow SNPs to be determined without first finding the sequence of the 99.7 percent of DNA bases that do not exist as SNPs. These methods include directed hybridizations, ligations, primer extensions, or nuclease cleavages that specifically involve SNPs while leaving the rest of the DNA alone.

With any increase in the involvement of DNA sequencing in forensics comes the likelihood that debate will intensify concerning privacy issues regarding the use of sequence information. Unlike commonly used methods of PCR analysis, SNP determination will reveal certain details about suspects that could be open to abuse. Ethical issues involving the use and dissemination of DNA data will have to be resolved as the methods of DNA analysis continue to evolve.

James S. Godde, Ph.D.

FURTHER READING

McClintock, J. Thomas. *Forensic DNA Analysis: A Laboratory Manual.* Boca Raton, Fla.: CRC Press, 2008. Examines the various methods of DNA analysis and DNA fingerprinting.

Nakamura, Yusuke. "DNA Variations in Human and Medical Genetics: Twenty-five Years of My Experience." *Journal of Human Genetics* 54 (2009): 1-8. A historic perspective on the progression of DNA analysis techniques with a particular emphasis on human disease characterization.

Pereira, Filipe, Joao Carneiro, and Antonio Amorim. "Identification of Species with DNA-Based Technology: Current Progress and Challenges." *Recent Patents on DNA and Gene Sequence* 2 (2008): 187-200. Contains an excellent table comparing methods of DNA analysis, along with a helpful flowchart.

Also contains clear descriptions of each method, including diagrams.

Roper, Stephan M., and Owatha L. Tatum. "Forensic Aspects of DNA-Based Human Identity Testing." *Journal of Forensic Nursing* 4 (2008): 150-156. A straightforward description of all pertinent methods and applications of DNA analysis, including simple diagrams as well as a glossary of terms.

Rudin, Norah, and Keith Inman. *An Introduction to Forensic DNA Analysis.* 2d ed. Boca Raton, Fla: CRC Press, 2002. Discusses forensic DNA analysis from both the medical and legal standpoints. Examines the advantages and limitations of the various techniques.

Watson, James D., and Andrew Berry. *DNA: The Secret of Life.* New York: Alfred A. Knopf, 2006. This comprehensive introduction to DNA has the famous biologist Watson as one of its authors.

WEB SITES

Association of Forensic DNA Analysts and Administrators
http://www.afdaa.org

The DNA Initiative
Advancing Criminal Justice Through DNA Technology
http://www.dna.gov

Federal Bureau of Investigation
CODIS
http://www.fbi.gov/hq/lab/html/codis1.htm

International Society for Forensic Genetics
http://www.isfg.org

See also: Criminology; DNA Sequencing; Forensic Science; Genetic Engineering; Human Genetic Engineering.

DNA SEQUENCING

FIELDS OF STUDY

Biology; genetics; population genetics; genomics; forensics; bioinformatics; microbiology; biotechnology; biological systematics; chemistry; biochemistry; medical genetics; clinical biochemical genetics; clinical molecular genetics; clinical cytogenetics; genetic counseling; molecular oncology; pharmaceuticals; pharmacogenomics; anthropology; archaeology; genealogy; agriculture; bioethics; engineering; computer science; mathematics; bioremediation.

SUMMARY

DNA (deoxyribonucleic acid) sequencing is a technique used to determine the order of the nitrogenous bases (adenine, guanine, cytosine, and thymine) that make up a gene, DNA molecule, or entire genome. Genome sequencing has been completed for many organisms, including animals, plants, and humans. Applications of this technology can advance the understanding, diagnosing, and treatment of disease; enable personalized health care; and produce innovative techniques that can be used in forensics, agriculture, and archaeology.

KEY TERMS AND CONCEPTS

- **Deoxynucleoside Triphosphate (dNTP):** Nucleotide used to synthesize strands of DNA during sequencing.
- **Dideoxynucleoside Triphosphate (ddNTP):** Nucleotide lacking the 3'-hydroxyl function on its deoxyribose sugar required for formation of bonds between nucleotides.
- **DNA Polymerase:** Enzyme functioning as a catalyst in the connection of deoxynucleoside triphosphates.
- **Electrophoresis:** Technique that exposes molecules in a medium to an electric field in order to separate them by some feature.
- **Nitrogenous Base:** Nitrogen-containing base, such as adenine, guanine, cytosine, and thymine (DNA) or uracil (RNA). Represented as A, G, C, T, and U.
- **Nucleotide:** Basic unit of DNA consisting of a five-carbon sugar, a phosphate group, and a nitrogenous base.
- **Oligonucleotide:** Short strand of DNA synthesized and used as a primer for a polymerase chain reaction.
- **Polymerase Chain Reaction (PCR):** Technique in which target sequences of DNA are amplified to large amounts for use in molecular and genetic analyses.

DEFINITION AND BASIC PRINCIPLES

DNA sequencing is a laboratory technique that allows scientists to determine the structure of DNA at its highest level of resolution. DNA is a double-stranded helix made of building blocks called deoxyribonuceotides. Deoxyribonuceotides are nucleotides that contain the deoxyribose sugar as well as a phosphate group and a nitrogenous base. The information in DNA is stored as a code made up of the nitrogenous bases adenine (A), guanine (G), cytosine (C), and thymine (T).

DNA sequencing has a number of applications that may revolutionize medicine, agriculture, anthropology, and archaeology. The prospective uses of these applications include personalized medicine, the decoding of genes, and the identification of mutations causing genetic diseases. Sequencing, coupled with genetic engineering and agriculture, has produced plants with improved nutritional quality, greater resistance against insects, and better ability to withstand poor soil and drought. Sequencing of entire genomes has been completed for several organisms, including extinct species. This has provided an extensive insight into human migration and the evolution of all living organisms.

Despite the prospective and generated benefits of DNA sequencing, the use of this technique has raised ethical, legal, and social questions. Concerns exist regarding genetic determinism and discrimination, the manipulation of an individual's attributes, the loss of genetic privacy, and the modification of food. Programs have been created in response to these concerns.

BACKGROUND AND HISTORY

In 1953, James D. Watson and Francis Crick proposed the double-helical structure of DNA. This discovery has since yielded revolutionary insights into the genetic code and protein synthesis. However, because of certain properties of DNA, it took about fifteen years before the first sequencing experiments were completed.

The first nucleic acid to be sequenced was yeast alanine transfer RNA (tRNA) because of its size and availability for purification. Following this accomplishment, scientists began to purify genomes of bacteriophages and pursue sequencing. However, whole genome sequencing did not become an actuality until after the discovery of restriction enzymes by Hamilton Smith, Werner Arber, and Daniel Nathans in 1970 and the development of more modern methods of DNA sequencing by Frederick Sanger and Alan R. Coulson in 1975. In 1977, the first genome, belonging to the bacteriophage phiX174, was sequenced.

The increasing amount of information created a need for computer programs capable of compilation and analysis of DNA, followed by databases with rapid searching programs (such as Genbank, created in 1982). These developments, as well as several advances in laboratory techniques (such as automated sequencing), led to the completion in 1998 of the first genome for an animal, a nematode called *Caenorhabditis elegans*, and ultimately the complete mapping of the human genome in 2003.

HOW IT WORKS

Plus and Minus Method. Before the development of direct DNA sequencing, DNA had to be converted into RNA, sequenced, and then decoded. In 1975, Sanger and Coulson introduced plus and minus sequencing, the first direct DNA sequencing method. Plus and minus sequencing begins with several asynchronous cycles of DNA synthesis with radioactively labeled deoxynucleoside triphosphates (dNTPs). The asynchronous cycles lead to an array of DNA fragments varying in nucleotide length (1, 2, 3, . . . 100). Products are separated into eight containers and dNTPs added; however, each container receives either one of the dNTPs (the plus reactions) or three of the four dNTPS (minus reactions). This allows for termination of synthesis in a sequence-specific manner. The products are separated via electrophoresis on a polyacrylamide gel. Subsequently, the gel is exposed to X-ray film that results in a series of bands corresponding to the radiolabeled DNA fragments, allowing the sequence to be constructed. Although

this method revolutionized how sequencing was completed, it was inefficient, and therefore other techniques were developed.

Maxam and Gilbert Method. In 1977, Allan Maxam and Walter Gilbert developed a sequencing method that replaced the plus and minus method. This method was similar, as it required a radioactive label, gel electrophoresis for fragment separation, and the use of X-ray autoradiography for product visualization and inference of the sequence. However, the method differed in that it allowed for the direct analysis of purified double-stranded DNA and used another way of creating products ending in a specific nucleotide.

In the Maxam and Gilbert method, the double-stranded DNA is radioactively labeled, cut with restriction enzymes, and denatured. Subsequently, the DNA is treated chemically in four separate reactions, which cut DNA at different nucleotides. The first reaction, called the A+G reaction, cuts the nucleotides adenine (A) and guanine (G). The second reaction, called the G reaction, cuts at G. The third and fourth reactions are similar, but involve cytosine (C) and thymine (T) in the C+T and the C reaction. The products of these four reactions are run through gel electrophoresis in four adjacent wells and analyzed for the sequence. The G reactions determine the placement of G, the A+G reactions determine the location of A, and so forth.

Sanger Sequencing Method. Named after Sanger, this method was developed in December, 1977, by Sanger and Gilbert. Because of its overall efficiency and limited use of chemicals and radioactivity, it has become the most widely used technique. This method takes advantage of dideoxynucleoside triphosphates (ddNTPs), analogues of the dNTPs. The ddNTPs are nucleotides lacking the 3'-hydroxyl function on its deoxyribose sugar required for the formation of phosphodiester bonds between two nucleotides of a developing DNA strand. Therefore, the ddNTPs are used during the synthesis of DNA strands to terminate DNA extension, resulting in products of different lengths.

Originally, this method required four separate reactions. Each aliquot contained a template DNA primed with an oligonucleotide, DNA polymerase to extend the sequence, and dNTPs. Each reaction received one of the chain-terminating ddNTPs labeled radioactively for detection. Later, sequencing could be completed in one reaction by substituting the radioactive label with four unique fluorescent dyes corresponding to each ddNTP. Further progress was made in 1983, when Kary Mullis introduced polymerase chain reactions (PCR), which allowed target sequences of DNA to be amplified in a fraction of the time. After PCR, the strands are separated and analyzed with automated sequencers, resulting in a chromatogram with a series of four-colored peaks representing each of the DNA bases. Computers are used to assemble sequences and analyze them for a variety of characteristics.

APPLICATIONS AND PRODUCTS

Genetic Diagnostics. The increased knowledge of genes and the organization of the genome have had a significant impact on medicine. Any disorder is caused by a combination of the environment and genetics; however, the role of genetics may be large

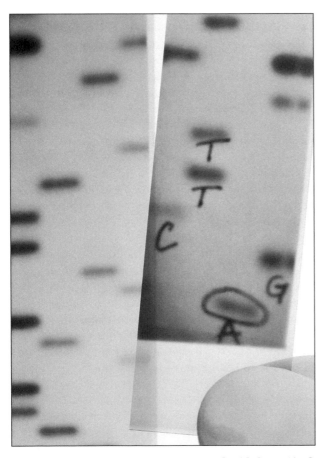

Scientist points to bands representing nucleotide bases (A, C, T, G) in an X-ray image of a gel. (Martin Shields/Photo Researchers, Inc.)

(as in Huntington's disease) or small (as in diabetes). Sequencing has helped elucidate the genetic variation and mutations responsible for predisposing a person to disease, modifying the course of a disease, or causing the disease itself. Understanding the molecular mechanisms of disease allows for the development of tests, diagnoses, treatments, and even cures and preventative options.

Personalized Medicine. Medicine is moving in the direction of using specific treatments based on patients' individual attributes, a development termed as personalized medicine. Although personalized treatments are being developed in many medical fields, the most striking examples are in oncology. For example, physicians are measuring the levels of human epidermal growth factor receptor 2 (HER2) in patients with breast cancer, and if the test is positive, the person is treated with trastuzumab.

Agriculture. Sequencing the genomes of plants and animals has made it possible to create transgenic organisms, which incorporate desired characteristics from other organisms. For example, genes from the bacterium *Bacillus thuringiensis* have been successfully transferred to crops such as rice, cotton, corn, and potatoes, thereby producing plants that are protected from insects. The alteration of genomes has also produced plants that resist drought and disease. Golden Rice, a genetically modified strain of rice, contains high levels of beta-carotene, which is converted to vitamin A in the human body. This rice has the potential to fight vitamin A deficiency in less-developed countries. Interestingly, bananas have also been modified to produce human vaccines against diseases such as hepatitis B.

Comparative/Evolution. The ability to sequence DNA has led to the study of the evolution of all forms of life, including humans. Since sequencing began, genomes have been mapped for innumerable organisms, including chimpanzees, mice, fish, fruit flies, plants, yeasts, bacteria, and viruses. The Human Genome Project, completed in 2003, identified 20,000 to 25,000 genes and about 3 billion bases in humans. DNA sequencing has also been completed on ancient DNA from clinical, museum, archaeological, and paleontological specimens. These data have greatly increased knowledge of genetic variation, thus increasing understanding of human differences, similarities, evolution, and origins.

Microbial Genomics. In 1994, the U.S. Department of Energy (DOE) launched the Microbial Genome

Project. Scientists realized that these organisms, with their ability to withstand extremes of temperature, radiation, acidity, and pressure, provided an excellent resource for the development of applications of renewable energy production, environmental cleanup of toxic waste, management of environmental carbon dioxide, and industrial processing of antibiotics, insecticides, and enzymes. Although this project ended in 2005, the DOE has continued its research in the Genomic Science Project and the Joint Genome Institute's Community Sequencing Program.

Biological Weapons. Although highly controversial, the use of genetic sequencing has led to development of materials for biowarfare. One example is invisible anthrax, which was developed by introducing a gene that altered the immunological properties of the microorganisms *Bacillus anthracis*. Access to the DNA of virulent agents and strains is regulated and restricted, thus preventing the introduction of genes to create novel properties. However, with the advancement of microbiology it is becoming increasingly possible to synthesize agents artificially. In 2002, the poliovirus was synthesized using only the information of the genetic sequence.

IMPACT ON INDUSTRY

Biotechnology. The Biotechnology Industry Organization reports that the development of DNA sequencing and other techniques has enabled the field to boom into a $30 billion a year industry creating more than 1,400 companies in the United States and employing more than 200,000 people. Biotechnology has been responsible for creating vaccines, hundreds of medical diagnostic tests, and treatments for cancer, diabetes, autoimmune disorders, and human immunodeficiency virus (HIV) infection. Additionally, biotechnology has dramatically affected the agricultural industry, one of the world's largest employers, by increasing yields and decreasing the need to apply pesticides.

Government and University Research. One of the largest effects of the development of DNA sequencing and other techniques has been the expansion of research. The goals of the Human Genome Project (1990-2003) were to identify all the genes in human DNA, determine the sequences of the base pairs that make up this DNA, record the information in databases, improve methods of analyzing data, transfer technologies to the private sector, and address all legal, ethical, and social issues stemming from the

project. The cost for the project was about $2.7 billion. This project opened the door for additional research opportunities, including the International HapMap Project, the Genographic Project, and the Human Proteome Project.

The International HapMap Project is aimed at creating a catalog of common genetic variants in human beings. It records the differences and similarities in human DNA, where in the DNA they occur, and distribution patterns within and among human populations. The most common genetic variant, a difference in individual bases, is known as a single nucleotide polymorphism, or SNP. The project is identifying most of the 10,000 SNPs that commonly occur in the human genome. These SNPs are not being linked to specific genetic diseases but rather are being used in genetic association studies to estimate people's risk of developing many common diseases.

The Genographic Project, led by the National Geographic Society, IBM, geneticist Spencer Wells, and the Waitt Family Foundation, is using computer analysis of DNA submitted by people around the world in an effort to shed light on human migratory patterns and origins. The project seeks to collect DNA from 100,000 indigenous and traditional peoples as well as from the general public. Started in 2005, the project was to last five years but was extended to 2011.

The Human Proteome Organisation's Human Proteome Project, initiated in January, 2010, aims to characterize all the genes of the human body and generate a map of the protein-based molecular architecture of the human body to answer the question of how the genetic code can guide the enormous intricacies of human anatomy, physiology, and biochemistry. This research will aid in the diagnosis and treatment of disease. The field of proteomics has enabled many complexities of the regulation and structure of the sequenced DNA to be elucidated.

Direct-to-Consumer Marketing. Genetic testing has traditionally been available only through health care providers such as genetic counselors and doctors, who interpret test results for patients. However, some companies have begun marketing genetic tests directly to consumers. These companies send a DNA testing kit to the customer's home. Typically, the customer swabs the inside of his or her cheek and returns the DNA sample to the company. The company processes the sample and communicates the results through regular mail, telephone, or e-mail.

Fascinating Facts About DNA Sequencing

- A completed sequence of the human genome was revealed in the spring of 2003, fifty years after James D. Watson and Francis Crick described the structure of DNA.

- DNA has been successfully amplified from several sources, including hair, blood, saliva, bone marrow, body tissue, and urine. DNA can be extracted from seeds, plant leaves, vegetables, fruits, bacteria, and any other material containing DNA.

- DNA sequencing is not limited to modern samples. Scientists have amplified DNA from ancient human remains and extinct species, including DNA that was more than 64,800 years old.

- In 1977, Frederick Sanger sequenced the first DNA genome, the bacteriophage phiX174, which had only 5,386 base pairs. Thirty years elapsed before the sequencing of the human genome, which has about 3 billion base pairs.

- There is considerable genetic variation within a species. Any two humans differ by about 3 million, or 1 out of every 1,000 base pairs. Most variation (about 90 percent) is within a population belonging to a specific continent; the additional variation is found among populations.

- The human genome contains coding and noncoding DNA. The coding DNA, the genes that encode for proteins, makes up only 2 percent of the human genome. The remaining 98 percent does not code for proteins; instead, it may function in regulation of expression and maintenance.

- DNA sequencing has shown that genomes vary by size, but the significance of this remains unclear. The human genome contains about 3 billion base pairs, whereas the protozoan *Amoeba proteus* has 290 billion base pairs, which makes it one hundred times larger.

- Analysis of genetic variation has confirmed that humans are remarkably similar to other life-forms. Humans share 98 percent of their genetic material with chimpanzees, 90 percent with mice, 23 percent with yeast, and 12 percent with *Escherichia coli*.

Although direct-to-consumer marketing of genetic testing kits promotes awareness of genetic disease, allows patients to take an active role in their health care, and helps people learn about their ancestral origins, it is highly controversial. The main concern is that guidance from health care providers regarding whether to have a test performed and how to interpret its results is missing. Patients may not fully understand the implications of a positive or negative test result or the predictive ability of these tests.

Regardless of the controversy, companies such as 23andMe, deCODE genetics, Navigenics, and GeneleX have begun offering home-based genetic testing for tracing ancestral origins and determining carrier status of genetic diseases and predisposition to physical and mental illnesses. These kits are priced between $300 and $1,000. In early 2010, Walgreens stated its intent to begin carrying over-the-counter genetic tests produced by Pathway Genomics; however, the company changed its mind after the U.S. Food and Drug Administration began an investigation of the supplier and product. In July, 2010, the Government Accountability Office (GAO) issued a report focusing on the direct-to-consumer kits marketed by four companies. It found the test results to be misleading and contradictory. The report noted "questionable marketing claims, serious quality control and privacy concerns, and questions about the accuracy of information provided to consumers."

CAREERS AND COURSE WORK

Students interested in pursuing careers involving genomic research must take a cross-disciplinary approach. Students should develop a solid background in science, including biology, chemistry, physics, and mathematics, at the undergraduate level. Pursuing higher degrees in the basic sciences or combining studies of journalism, law, computer science, anthropology, archaeology, bioethics, medicine, and pharmaceuticals can result in further specialization. Although higher education, including a master's degree, a medical degree, or doctorate is required for many advanced research positions in academic institutes or industries, opportunities are available for individuals without advanced degrees.

Careers may take several paths, including medicine, public health, pharmaceuticals, agriculture, computer science, engineering, business, law, history, archaeology, and anthropology. Individuals interested in medicine can pursue careers as genetic counselors, medical geneticists, or genetic nurses. Pharmacy students with backgrounds in genetics can pursue innovative research and development of personalized medicine and pharmacogenomics. Scientists with an interest in agriculture may be involved in the genetic modification of food. Other possibilities include programming and maintaining DNA databases, marketing and promoting new technologies, paternity testing, and forensic science. Many of the positions in genetics are likely to be in research into evolution, diagnostic testing, and development of the proteomes and HapMaps.

SOCIAL CONTEXT AND FUTURE PROSPECTS

DNA sequencing has come a long way since its inception in 1975. Enormous accomplishments have been made, creating a future full of many new careers and innovative applications. The knowledge gained from DNA sequencing is a starting point for years of additional research and developments. Armed with the blueprint of life, scientists have begun working to unlock some of biology's most intricate and complex processes, including determining how a human develops from a single cell, how genes regulate the functions of organs and tissues, and what is involved in the preposition of disease.

The completion of the Human Genome Project and the initiation of projects such as the Human Proteome Project and International HapMap Project have demonstrated the commitment of government and society to understanding the nature and role of genetics. However, DNA sequencing, coupled with developments in genetic engineering, have raised profound ethical and social concerns. The heightened ability to determine an individual's genetic profile has raised concerns about confidentiality and privacy, possible stigmatization, negative consequences in the areas of employment and insurance, and the psychological effects of knowing one's predisposition to diseases and conditions. The commercialization of DNA products is another area of concern.

In response, many U.S. governmental agencies such as the National Institutes of Health have created bioethics programs. For example, the Ethical, Legal and Social Implications (ELSI) Research Program, part of the National Human Genome Research Institute, supports research on the ethical, legal, and social implications of genetics research. Other programs include a bioethics component.

Amber M. Mathiesen, M.S.

FURTHER READING

Brown, T. A. *DNA Sequencing*. New York: IRL Press at Oxford University Press, 1995. A basic book examining the technique of DNA sequencing.

Hummel, Susanne. *Ancient DNA Typing: Methods, Strategies, and Applications*. New York: Springer, 2003. A manual for the analysis of ancient and degraded DNA. Includes information on extraction, techniques, and applications, including identification of objects, kinship, and population genetics.

Hutchison, Clyde A., III. "DNA Sequencing: Bench to Bedside and Beyond." *Nucleic Acids Research* 35, no. 18 (August, 2007): 6227-6237. Reviews the history and development of DNA sequencing from the discovery of the structure of DNA to modern times.

Janitz, Michal. *Next-Generation Genome Sequencing: Towards Personalized Medicine*. Weinheim, Germany: Wiley-VCH, 2008. Provides the reader with a comprehensive overview of next generation sequencing techniques, highlighting their implications for research, human health, and society's perception of genetics.

Jones, Martin. "Archaeology and the Genetic Revolution." In *A Companion to Archaeology*, edited by John Bintliff, Timothy Earle, and Christopher S. Peebles. Malden, Mass.: Blackwell, 2004. Reviews the expanding field of archaeogenetics by discussing the history of how DNA entered the field, the study of ancient DNA, human evolutionary studies, existing practices, and future prospects.

Lynch, April, and Vickie Venne. *The Genome Book: A Must-Have Guide to Your DNA for Maximum Health*. North Branch, Minn.: Sunrise River Press, 2009. Provides an explanation of the growing medical benefits provided from decoding the human genome. Discusses several health topics, including cancer, behavior, and heart conditions.

Meyers, Robert A., ed. *Genomics and Genetics: From Molecular Details to Analysis and Techniques*. Weinheim, Germany: Wiley-VCH, 2007. Covers the basics of genomics and genetics and discusses techniques such as DNA sequencing.

WEB SITES

The Genographic Project
https://genographic.nationalgeographic.com/genographic/index.html

The Human Genome Project
http://www.ornl.gov/sci/techresources/Human_Genome/home.shtml

The International HapMap Project
http://hapmap.ncbi.nlm.nih.gov

Microbial Genomics at the U.S. Department of Energy
Microbial Research Programs: Past and Present
http://microbialgenomics.energy.gov/research-programs.shtml

National Human Genome Research Institute
http://www.genome.gov

See also: Animal Breeding and Husbandry; Anthropology; Archaeology; DNA Analysis; Forensic Science; Genetically Modified Food Production; Genetically Modified Organisms; Genetic Engineering; Human Genetic Engineering; Pathology; Plant Breeding and Propagation; Proteomics and Protein Engineering.

DRUG TESTING

FIELDS OF STUDY

Biochemistry; biology; molecular biology; microbiology; chemistry; medical technology; pharmaceutical technology; pharmacology.

SUMMARY

Drug testing is done to ensure the safety of the general public, to maintain standards at schools and places of employment, and to make sure that athletes do not gain unfair advantage through the use of performance-enhancing drugs. The goal of these tests is detect whether a person has used drugs such as alcohol, marijuana, cocaine, amphetamines, barbiturates, benzodiazepines, lysergic acid diethylamide (LSD), opiates, phencyclidine (PCP), synthetic hormones, and steroids. Commonly used drug tests analyze a person's breath, urine, saliva, sweat, blood, or hair.

KEY TERMS AND CONCEPTS

- **Antibody:** Glycoprotein that binds to and immobilizes a substance that the cell recognizes as foreign.
- **Antigen:** Substance that triggers an immune response.
- **Barbiturate:** Class of drugs that depress the central nervous system, thereby inducing sleep and sedation.
- **Benzodiazepine:** Class of drugs that are structurally characterized by a seven-member ring containing two nitrogen atoms and that act as tranquilizers.
- **Beta Blocker:** Class of drugs that compete with beta-adrenergic receptor-stimulating agents.
- **Binding Assay:** Experimental method for selecting one molecule out of a number of possibilities by specific binding.
- **Creatine:** Substance that is a natural product of the kidney and is sometimes used by body builders as a performance-enhancing drug to quickly increase muscle mass.
- **Diuretic:** Any substance that increases the formation of urine.
- **Hormone:** Substance produced by endocrine glands and delivered by the bloodstream to target cells, producing a desired effect.
- **Masking Agent:** Any substance that can prevent the detection of a drug in a person's system.
- **Monoclonal Antibody:** Antibody produced from the progeny of a single cell and specific for a single antigen.
- **Tetrahydrocannabinol (THC):** The active ingredient in marijuana.

DEFINITION AND BASIC PRINCIPLES

Drug testing in the workplace and schools has become commonplace. A variety of tests are used to detect elevated levels of the most common drugs that can impair job performance or are illegal to use. The importance of drug testing has continued to increase since the Controlled Substances Act of 1970 placed all regulated drugs into five classifications based on their medicinal value, their potential to harm people, and their likelihood of being abused or causing addiction. Schedule I drugs have no known medical value and are most likely to be abused, while Schedule V drugs have little potential for abuse. These scheduled drugs are called controlled substances because their use, manufacture, sale, and distribution are subject to control by the federal government.

There are two general types of drug testing. Federally regulated drug testing, according to the National Institute on Drug Abuse (NIDA), requires testing for cannabinoids (THC, marijuana, hashish), cocaine, amphetamines, opiates (morphine, heroin, and codeine), and PCP. Nonfederally regulated drug testing is often used to test athletes in various sports for the use of creatine, hormones, steroids, and other performance-enhancing drugs. Additional tests are used to detect barbiturates and alcohol.

Urinalysis is typically used as a preliminary test because it is less expensive and more convenient than the other tests. Saliva tests and breathalyzers are commonly used. Blood tests, although less frequently employed because they are generally more expensive and invasive, are more dependable, as are hair strand tests. A preliminary positive test using urinalysis must be confirmed by diagnostic tests completed in an analytical laboratory setting, which can take several days to complete. These diagnostic tests include the analytical instruments of gas chromatography (GC), mass spectrometry (MS), ion scanning, high-pressure liquid chromatography (HPLC), immunoassay (IA), and inductively coupled plasma spectrometry (ICP-MS).

BACKGROUND AND HISTORY

The detection of ingested drugs in various body fluids first sparked the interest of the ancient alchemists. In 1936, Rolla N. Harger of Indiana University patented the Drunkometer, a breath test to measure a person's level of alcohol intoxication. In 1954, Robert F. Borkenstein of Indiana University invented the breathalyzer, which had the benefit of greater portability, to measure blood-alcohol content. However, it was not until the widespread use of recreational drugs in the 1960's that the National Institute of Drug Abuse was established to monitor drug use. With its creation, federal funding became available to researchers to develop drug testing methods, which led to rapid advances. In 1973, physician Robert L. DuPont was appointed director of the National Institute of Drug Abuse. As director, DuPont implemented the use of the urine test and further developed immunoassays to test for several controlled substances.

In 1981, an airplane crashed on the USS *Nimitz*, and the investigation into the incident revealed drug use to be a contributing factor. As a result, the United States Navy began random drug testing of all active-duty personnel in 1982. In the 1980's, the U.S. Department of Transportation began to test all of its employees. In September, 1986, President Ronald Reagan signed Executive Order 12564, making drug testing mandatory for federal employees and all employees in safety-sensitive positions, such as employees in the nuclear power industry. The National Institute of Drug Abuse extended this mandatory testing to include truck drivers working in the petroleum industry. This testing has come to be regulated by the Substance Abuse and Mental Health Services Administration (SAMHSA), which is part of the U.S. Department of Health and Human Services.

HOW IT WORKS

Because of the commercial availability of so many masking agents, the most effective drug testing occurs when the subject has had no previous notification. Thus, random drug testing is very effective and has become common in the workplace, schools, and for athletes. The National Collegiate Athletic

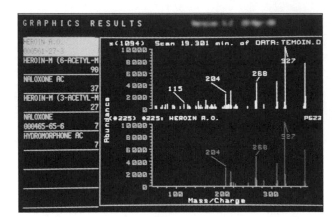

Screenshot showing the results of a mass spectrometry test on a suspect's blood sample. The molecular weights of various chemicals are shown at right, while the names of the drugs found in the sample are at left. (Patrick Landmann/ Photo Researchers, Inc.)

Association and the National Football League provide only one to two days notice before drug testing, and the United States Olympic Committee has a no-notice policy for drug testing.

The first commonly available drug testing method was the breathalyzer, followed by urinalysis. The usage of saliva tests continues to increase, while sweat tests remain the least-used testing method. Blood tests require additional medical staff and are also the most invasive testing method; therefore, although they are very accurate, they are not as commonly used as urinalysis. Hair tests are very accurate but do not detect drug use in the last four to five days. In terms of validity for legal purposes, any of these preliminary, or screening, tests must be confirmed by an analytical technique, most often gas chromatography/mass spectrometry, performed by trained personnel within a diagnostic laboratory.

Breath. Borkenstein received a patent in 1958 for the breathalyzer, which determines an individual's blood-alcohol content (BAC) from a breath sample. The ethanol in the breath of an individual reacts with the dichromate ion, which has a yellow-orange color, in the presence of acid to form the green chromate ion. This color change from pale orange to green can easily be observed. All fifty states and the District of Columbia have laws that forbid a person to drive with a BAC of 0.08 percent or greater, a level at which the individual is judged to be legally impaired.

Urine. Urinalysis became a common method of detecting drugs in the 1980's and has continued to be widely used. The urine sample is collected and sealed to ensure that it remains tamper-free. It is generally subjected to an immunoassay test first because this test is very fast.

Saliva. Testing of oral fluids is becoming increasingly common because of its convenience for random testing, and it is more resistant to adulteration than urine samples. Saliva testing can detect cocaine, amphetamine, methamphetamine, marijuana, bezodiazepines, PCP, opiates, and alcohol if the substance was ingested between six hours and three days before the test was administered.

Sweat. Although traditionally not considered to be as useful as the other methods because of the dilute sample obtained, patches that can be worn on the skin and collect samples over several hours are increasingly popular. This method of drug testing is preferred by government agencies such as parole departments and child protective services in which urine testing is not the method of choice.

Blood. Because blood tests are the most invasive and expensive method, requiring additional medical personnel, they are not as widely used as the other tests as a screening method. However, blood testing is very accurate and reliable, so it is often used to confirm a positive result from another type of drug test.

Hair. Hair samples from any part of the body can be used and are extremely resistant to any type of tampering or adulteration. Special fatty esters are permanently formed in the hair as a result of alcohol and drug metabolism, and therefore, this method is very reliable.

Gas Chromatography/Mass Spectrometry (GC/ MS). This tandem analytical instrumentation must be done by trained personnel within a laboratory setting and therefore is not as convenient as the other testing methods. However, it is much more accurate and is used to confirm more rapid, preliminary tests. The gas chromatograph is able to separate molecules based on their attractive interactions with the material that packs a column. Molecules take varying amounts of time to travel, or elute, from the column, resulting in different retention times, or amounts of time retained on the packing material of the column. These separated molecules are then ionized in the mass spectrometer, which is able to produce molecular weight information.

APPLICATIONS AND PRODUCTS

Drug Test Dips. Test strips known as drug test dips, dip strips, drug test cards, or drug panels use a single immunoassay panel to test for several common drugs at once. Specific reactions between antibodies and antigens allow marijuana, cocaine, amphetamine, opiates, and methamphetamine to be detected in urine samples. These assay strips are so easy to use that the staff of many schools, sports clubs, and offices can use them. However, they must be used only as a preliminary test. Positive results should be confirmed using gas chromatography/mass spectrometry conducted by an independent diagnostic laboratory.

The test strip is removed from its protective pouch and allowed to equilibrate to room temperature. Meanwhile, a urine sample is obtained in a small cup and also allowed to equilibrate to room temperature. The test strip is dipped vertically, with the arrow on the test strip pointing down into the sample, and remains immersed in the urine sample for ten to fifteen seconds. Then the test strip is removed from the urine sample and placed on a flat, nonabsorbent surface. After five minutes, the test strip is checked for the appearance of any horizontal lines. The appearance of a colored line in the control region of the dip strip and a faded color line in the test region of the dip strip indicates a negative test and that the concentration of a drug is too low to be detected. If only one line appears in the control region, with no line visible in the test region, then the test is considered to be positive for the presence of drugs. The test is considered to be invalid if only one line appears in the test region or no line appears at all. An invalid test is usually the result of either not following the procedure correctly or not using a large enough urine sample.

The test strips for marijuana use a monoclonal antibody to detect levels of THC, the active ingredient in marijuana, in excess of 50 nanograms per milliliter (ng/ml), the level recommended by SAMHSA. Methamphetamine can be detected in urine samples for three to five days after usage by using a test strip equipped with a monoclonal antibody. A positive test indicates a level in excess of 1,000 ng/ml. A test strip detects the major metabolite of cocaine, benzoylecgonine, for up to twenty-four to forty-eight hours after use. Morphine in excess of 2,000 ng/ml can also be detected by using a test strip containing a specific antibody. Morphine is the primary metabolite product of heroin and codeine.

Kits. Drug tests used for fast, preliminary screening include easy-to-use kits that can test samples of urine, saliva, breath, hair, or sweat. Of these tests, the hair test for drugs is considered to be the most accurate but is still considered to be a preliminary test. To confirm preliminary results or to obtain results for legal purposes, a sample of saliva or urine must be sent to a laboratory when a more reliable test such as GC/MS must be performed. It can take three to seven days to obtain the results. Urine drug test kits are less expensive than other tests, provide instantaneous results, and are easy to store. However, because of variations in metabolism rate, there can be a three-day to one-month detection window, making these tests easier to adulterate than the saliva, hair, sweat, or blood tests.

Adulteration of Specimens. Adulteration generally refers to intentional tampering with a urine sample, and certain substances can be added to urine to create a false-negative test result. Adulteration of urine samples is a common problem because four types of masking products—dilution substances, synthetic urine, cleaning substances, and adulterants—are readily available. More than four hundred readily available commercial products can mask urine samples. Dilution substances, including diuretics, lower the concentration of drug in a sample. An individual can either ingest one of these substances before submitting a urine sample or add the substance directly to the urine sample. Synthetic and dehydrated human urine can be bought and submitted for testing. An individual can also purchase a cleaning substance such as an herbal supplement for $30 to $70 and ingest the substance before submitting a sample. The herbal supplement reacts chemically with the drug to essentially nullify its active ingredient. Adulterants are chemicals that can actually react with the drugs, but these are actually added to the sample rather than ingested by the individual.

Methods to Detect Adulteration. Several methods can be used to detect the use of some type of adulterant. If the specific gravity of urine is outside the normal range of 1.003 and 1.030, then the sample may have been diluted. Another indication of adulteration via dilution is to test the level of creatine. If the level is too low (less than 5 milligrams, or mg, per deciliter) then dilution took place. Oxidants, such as pyridinium chlorochromate (PCC), bleach, or hydrogen peroxide, can react chemically with the drug to essentially nullify it. One commonly sold PCC adulterant is called Urine Luck. Tests can also detect the presence of any type of additional oxidant. If the pH (acidity-alkalinity) value of the urine is outside the normal range of 4.0 to 9.0, then an adulterant was added. Two common adulterants are sold under the names of Whizzies or Klear, and these react chemically with drugs in the urine by oxidizing the active ingredient in marijuana. Another chemical reaction occurs when an adulterant called Clear Choice or Urine Aid prevents enzyme activity in the test, which results in the presence of glutaraldehyde, which causes a false negative. Among the states that have passed laws to prevent the sale of masking agents are Florida, South Carolina, North Carolina, New Jersey, Maryland, Virginia, Kentucky, Oklahoma, Nebraska, Illinois, Pennsylvania, and Arkansas.

IMPACT ON INDUSTRY

Drug testing was initially developed for military use and then mandated for federal employees.

Fascinating Facts About Drug Testing

- Each year in the United States, drug and alcohol use costs more than $100 billion in lost productivity.

- A single urine testing kit can detect up to twelve drugs.

- The drug detection window for testing urine samples is wider than for other methods for drug testing because it depends on metabolic rate, age, amount and frequency of use, urine pH, drug tolerance, body mass, and overall health. Therefore, urine testing is considered less accurate than other testing methods.

- Sweat patches for drug testing are about the size of a playing card and have a tamper-proof feature. They can be worn for up to seven hours to collect samples.

- The annual cost of conducting more than 100,000 drug tests to detect performance-enhancing drugs in athletes in the United States is about $30 million.

- There are more than four hundred commercial products readily available to mask urine samples.

- An aerosol detecting agent called DrugAlert can be sprayed on a paper towel that has been used to wipe any household surface to detect a variety of drugs on that surface.

However, the widespread use of recreational drugs beginning in the 1960's caused private sector employers and then academic institutions to routinely administer drug tests. As a result, a huge market has developed not only for rapid, on-site testing methods but also for at-home testing methods, which many parents use to monitor their children. This demand produced explosive growth by companies that manufacture on-site or at-home drug testing kits, and the kits are sold by mass merchandisers and in drugstores such as Target, Walmart, Walgreens, and CVS. In addition, many of these inexpensive kits can be purchased on the Internet.

CAREERS AND COURSE WORK

An interest and aptitude for biology, chemistry, and quantitative classes are important prerequisites to pursuing a career in drug testing. Additional required characteristics include the ability to solve problems, pay close attention to detail, work under pressure, have good manual dexterity, and have normal color vision. Depending on the requirements of the state of residence, a person needs a certificate or license or an associate's degree in biology, chemistry, or medical technology to be employed as a clinical laboratory technician, medical technician, or clinical laboratory technician. Specific information regarding certification can be obtained from the board of registry of the American Association of Bioanalysts and the National Accrediting Agency for Clinical Laboratory Sciences. Typical job duties involve drug sample collection and storage and operation of automated analyzers, often wearing protective gloves and safety glasses. In May, 2008, the median annual salary for a technician was about $35,000.

A bachelor's degree is required to work as a technologist or scientist and earn a higher salary of about $50,000. In addition to a higher salary, a bachelor's degree allows a person to take on more responsibility and possibly advance into supervisory and managerial positions. The ideal bachelor's degree is medical technology, which requires courses in chemistry, biology, microbiology, statistics, computers, and mathematics. To pursue research or become a laboratory director, a master's or doctoral degree is necessary. Employment in the drug testing field is projected to grow at the rate of 14 percent through 2016, which is faster than the average, according to the U.S. Department of Labor. Typical employers include

forensic science laboratories, research and development laboratories, and quality assurance laboratories in industry, government, schools, or hospitals.

SOCIAL CONTEXT AND FUTURE PROSPECTS

Mandatory testing is regulated by SAMHSA, part of the U.S. Department of Health and Human Services. This mandatory testing does not yet test for semisynthetic opioids, such as oxycodone, oxymorphone, and hydrocodone, which are often used to relieve pain but have the potential to be abused. However, many employers, athletic organizations, and schools test for these drugs, and ongoing research is directed toward increasing the convenience and reliability of methods of detecting these drugs. Because so many masking agents are readily available, random testing without prior notification is the most effective method, although it is not without controversy. Schools are increasingly performing random drug testing, often leading to protests that the tests are an invasion of privacy and a violation of Fourth Amendment rights.

Organizations such as the International Olympic Committee, National Collegiate Athletic Association, National Basketball Association, and National Football League monitor athletes for the use of more than one hundred anabolic-androgenic steroids. Efforts are being made to eliminate the use of performance-enhancing drugs in all sports. The International Olympic Committee led a collective initiative in creating the World Anti-Doping Agency (WADA) in Switzerland in 1999. WADA created a code in an attempt to standardize regulation and procedures in all sporting countries and keeps a list of prohibited substances. Banned substances include anabolic steroids, hormones, masking agents, stimulants, narcotics, cannabinoids, glucocorticosteroids, and for some sports, alcohol and beta-blockers during competition. Also forbidden are methods of enhancing oxygen transfer (such as blood doping) and gene doping. The UNESCO International Convention Against Doping in Sport, which came into force in 2007, is a global treaty designed to help governments align their policies with the WADA code.

Jeanne L. Kuhler, B.S., M.S., Ph.D.

FURTHER READING

Jenkins, Amanda J., and Bruce A. Goldberger, eds. *On-Site Drug Testing*. Totowa, N.J.: Humana Press, 2002. Discusses on-site methods of testing for drugs in hospital, criminal, workplace, and school settings. Looks at many specific tests, discussing their efficacy and their underlying principles.

Karch, Stephen B., ed. *Workplace Drug Testing*. Boca Raton, Fla.: CRC Press, 2008. Examines regulations and mandatory guidelines for federal workplace drug testing and describes techniques. Provides sample protocols from the nuclear power and transportation industries.

Liska, Ken. *Drugs and the Human Body with Implications for Society*. Upper Saddle River, N.J.: Pearson/ Prentice Hall, 2004. Simply describes the various classes of drugs and drug testing methods.

Mur, Cindy, ed. *Drug Testing*. Farmington Hills, Mich.:Greenhaven Press/Thomson Gale, 2006. A collection of essays on drug testing in schools and the workplace, discussing efficacy and ethical issues such as privacy.

Pascal, Kintz. *Analytical and Practical Aspects of Drug Testing in Hair*. Boca Raton, Fla.: CRC Press, 2006. Looks at advances in the use of strands of hair for drug testing in the workplace and in forensic crime laboratories and techniques for detecting specific drugs.

Thieme, Detlef, and Peter Hemmersbach. *Doping in Sports*. Berlin: Springer, 2010. Examines sports doping from its beginning, covering the use of anabolic steroids, erthyropoietin, human growth hormone, and gene doping in humans and the doping of race horses. Effects of the drugs, detection methods, and regulations are also discussed.

WEB SITES
Drug and Alcohol Testing Industry Association
http://www.datia.org

Substance Abuse and Mental Health Services Administration
http://www.samhsa.gov

Substance Abuse Program Administrators Association
http://www.sapaa.com

U.S. Department of Labor
Drug-Free Workplace Adviser
http://www.dol.gov/elaws/drugfree.htm

World Anti-Doping Agency
http://www.wada-ama.org

See also: Chemical Engineering; Criminology; Forensic Science; Pharmacology; Toxicology.

E

EARTHQUAKE ENGINEERING

FIELDS OF STUDY

Civil engineering; earthquake engineering; engineering seismology; geology; geotechnical engineering.

SUMMARY

Earthquake engineering is a branch of civil engineering that deals with designing and constructing buildings, bridges, highways, railways, and dams to be more resistant to damage by earthquakes. It also includes retrofitting existing structures to make them more earthquake resistant.

KEY TERMS AND CONCEPTS

- **Asperity:** Surface roughness that projects outward from the surface.
- **Epicenter:** Point on the Earth's surface directly above the hypocenter.
- **Hypocenter:** Point beneath the Earth's surface where an earthquake originates.
- **Love Wave:** Wave formed by the combination of secondary waves and primary waves on the surface; causes the ground to oscillate from side to side perpendicular to the propagation direction of the wave and is the most destructive.
- **P Wave:** First wave from an earthquake to reach a seismograph; travels through the body of the Earth, including through a liquid; also called a primary wave.
- **Rammed Earth:** Mixture of damp clay, sand, and a binder such as crushed limestone that is poured into a form and then rammed down by thrusting with wooden posts; after it dries the forms are removed.
- **Rayleigh Wave:** Wave formed by the combination of secondary waves and primary waves on the surface; causes the ground to oscillate in a rolling motion parallel with the direction of the wave.

- **S Wave:** Wave that reaches a seismic station after a primary wave; travels through the body of the Earth, but not through a liquid; also called secondary wave.

DEFINITION AND BASIC PRINCIPLES

Worldwide, each year there are about eighteen major earthquakes (magnitude 7.0 to 7.9) strong enough to cause considerable damage and one great earthquake (magnitude 8.0 or greater) strong enough to destroy a city. The outermost layer of the Earth is the rocky crust where humans live. The continental crust of the Earth is 30 to 50 kilometers thick, while the oceanic crust is 5 to 15 kilometers thick. The crust is broken up into about two dozen plates that fit together like pieces of a jigsaw puzzle, with the larger plates being hundreds to thousands of kilometers across. As these plates move about on the underlying mantle at rates of a few to several centimeters per year, they rub against neighboring plates. Asperities (irregularities) from one plate lock with those of an adjacent plate and halt the motion. While the plate boundary is held motionless, the rest of the plate continues in motion in response to the forces on it, and this action builds up stress in the boundary rocks.

Finally, when the stress on the boundary rocks is too great the asperities are sheared off as the boundary rock surges ahead several centimeters to several meters. It is this sudden lurching of the rock that produces earthquake waves. The point of initial rupture produces the most waves and is called the hypocenter, while the point on the surface directly above the hypocenter is called the epicenter. The epicenter is usually the site of the worst damage on the surface. Earthquake engineers can design structures to reduce the damage and the number of deaths, but the limited resources available means that not everything that might be done is done. The philosophy generally adopted is that while a strong earthquake may damage most structures they should remain

The ATLSS Engineering Research Center's structural testing lab at Lehigh University in Bethlehem, Pennsylvania is home of one of the largest structural testing facilities in the United States. Scientists at the lab test new materials that will help buildings survive earthquakes with little or no structural damage. (AP Photo)

standing at least long enough for the people in them to evacuate safely. Essential structures such as hospitals should not only remain standing but should be still usable after the quake.

BACKGROUND AND HISTORY

Earthquakes have plagued mankind throughout history. The Antioch (now in Turkey) earthquake in 526 killed an estimated 250,000 people. A thousand years later in 1556, an estimated 830,000 died in Shaanxi, China. In ancient times, earthquakes were ascribed to various fanciful causes such as air rushing out of deep caverns. Chinese mathematician Zhang Heng (78-139) is credited with the invention of a Chinese earthquake detector consisting of a large, nearly spherical vessel with eight dragon heads projecting outward from its circumference. A brass ball is loosely held in each dragon's mouth. A pendulum is suspended inside the vessel so that if an earthquake sets it swinging, it will strike a dragon causing the ball to fall from its mouth and into the waiting mouth of a toad figure. The sound of the ball striking the metal toad alerts the operator that an earthquake has occurred, and whichever toad has the ball indicates the direction to the epicenter.

The scientific study of earthquakes did not blossom until the twentieth century. In 1935, American seismologist Charles Richter with German seismologist Beno Gutenberg developed the Richter scale for measuring the intensity of an earthquake based on the amplitude of the swinging motion of the needle on a seismometer. The Richter scale was superseded in 1979 by the moment magnitude scale based on the energy released by the quake. This scale was developed by Canadian seismologist Thomas C. Hanks and Japanese American seismologist and Kyoto Prize winner Hiroo Kanamori and is the same as the Richter scale for quakes of magnitude 3 through 7.

Following the 1880 Yokohama earthquake, the Seismological Society of Japan was founded to see what might be done to reduce the consequences of earthquakes. It was the world's first such society and marks the beginning of earthquake engineering. The Japanese were forced into such a leadership role by being an industrial society sitting on plate boundaries and therefore subject to frequent earthquakes. In 1893, Japanese seismologist Fusakichi Omori and British geologist John Milne studied the behavior of brick columns on a shake table (to simulate earthquakes). Toshikata Sano, a professor of structural engineering at the Imperial University of Tokyo, published "Earthquake Resistance of Buildings" in 1914, and by the 1930's, several nations had adopted seismic building codes. Knowledge from earthquake engineering was beginning to be put into practice.

The early twenty-first century has seen several devastating quakes. On January, 12, 2010, a magnitude-7 earthquake struck Haiti, killing an estimated 316,000 people, injuring 300,000, and leaving more than one million people homeless. On March 11, 2011, a magnitude-8.9 quake occurred off the east coast of

Honshu, Japan, causing a massive tsunami that destroyed entire villages and also affected places as far away as Australia and the West Coast of the United States. This was the strongest quake in Japanese history. In addition to an estimated 10,000 deaths and almost 8,000 people reported missing, Japanese officials had to deal with subsequent leaks at three nuclear reactors in the affected region.

HOW IT WORKS

Earthquakes may be classified by their depth. Shallow-focus quakes have a hypocenter less than 70 kilometers deep and are the most destructive. Mid-focus quakes originate between 70 kilometers and 300 kilometers deep, and deep-focus quakes originate deeper than 300 kilometers and are the least destructive.

P Waves and S Waves. Underground quakes emit two kinds of waves, P waves (primary waves) and S waves (secondary waves). P waves are longitudinal or compression waves just like sound waves. The rock atoms vibrate along the direction in which the wave moves. P waves travel about 5,000 meters per second in granite. S waves are transverse waves, so atoms vibrate up and down perpendicular to the direction of wave travel. The speed of an S wave is about 60 percent that of a P wave. The difference in arrival times of P waves and S waves at a seismic station provides an estimate of the distance to the hypocenter. Therefore during a site evaluation, earthquake engineers must locate any nearby faults and the location of past hypocenters. Then they must try to determine the most likely hypocenters of future earthquakes, which they hope will be deep and far away.

Love Waves and Rayleigh Waves. S waves and P waves interact at the Earth's surface to produce two new types of surface waves: Love waves and Rayleigh waves. These are the waves that destroy buildings and knock people off their feet. Love waves are named for British mathematician Augustus Edward Hough Love, who developed the theory about the waves in his book *Some Problems About Geodynamics* (1911). They cause the ground to oscillate from side to side perpendicular to the propagation direction of the wave. Love waves are the greatest source of destruction outside the epicenter. Rayleigh waves are named for British physicist and Nobel Prize winner John William Strutt (Lord Rayleigh). They cause the ground to oscillate in a rolling motion parallel with the direction of the wave. The greater the wave amplitude, the more violent the shaking. A careful examination of the rocks and soil underlying a site should give information on propagation, damping, and direction of likely earthquake waves.

Core Sampling. To complete the site investigation, core samples may need to be taken to look for damp and insufficiently compacted soil. The shaking of an earthquake can turn damp sand into jelly, a process called liquefaction. Sand grains are surrounded by liquid and cannot cling together. Buildings sink into liquified soil. If it looks like liquefaction may be a problem, support piers must go from the building's foundation down into bedrock. If that is not possible, densification and grout injection may stabilize the soil. Liquefaction caused a segment of roadway to drop 2.4 meters in the 1964 Alaska earthquake. It caused great destruction in the Marina District of the 1989 Loma Prieta earthquake, and contributed to the destruction of Christchurch, New Zealand, in the February 22, 2011, magnitude-6.3 earthquake.

If the site involves a slope, or if the site is a railroad embankment, an earthquake might cause the slope to collapse. In this case the angle of repose of the slope could be reduced. If that is not possible, a retaining wall may be required, or a geotextile (made from polyester or polypropylene) covered with sand could be used to stabilize the slope.

APPLICATIONS AND PRODUCTS

Geologists and geophysicists have learned enough about earthquakes to be able to identify ways in which the damage they cause can be minimized by appropriate human behavior. In general, there are two main approaches to mitigating the effects of earthquakes: Build the structure to withstand a quake, and make the structure invisible to the earthquake waves, but if that cannot be done, dampen the waves as quickly as possible.

Designing for Earthquake Safety. Earthquake engineers can test design ideas by carefully modeling a structure on a computer, inputting the location and strength of the various materials that will be used to construct the building. They can then use the computer to predict the results of various stresses. They can also build a scaled-down model, or even a full-scale model, on a shake table, a platform that can be shaken to simulate an earthquake. When using a scale model, care must be taken so that it is not only geometrically similar to the full-size structure but that other factors such as the velocity of waves moving through the structure are also scaled down. The ultimate test is to build the structure and see how it fares.

Retrofitting Old Structures. The single act that has the potential to save the most lives is to fortify adobe houses against earthquakes. About 50 percent of the population in developing countries live in houses made of adobe or rammed earth, since dirt is cheap and is the ubiquitous building material. Adobe bricks are made by mixing water into 50 percent sand, 35 percent clay, and 15 percent binder-either straw or manure (said to repel insects). The mixture is poured and patted into a mold and then the mold is turned upside down so that the new brick will fall onto the ground to dry in the sun. The bricks may be assembled into a wall using a wet mixture of sand and clay as mortar. Mortar joints should be no more than 20 millimeters thick to avoid cracking. Rammed earth uses a similar mixture of sand and clay but uses lime, cement, or asphalt as a stabilizer. The wet mixture is poured into a form and then tamped down, or rammed, by workers with thick poles. After it sets, the forms are removed.

In an earthquake, both rammed earth and adobe crack and shatter. Walls tumble down, and roofs that had rested on the walls collapse onto people. The magnitude-7 quake that struck Haiti in 2011 flattened a large part of its capital, Port-Au-Prince. It was so devastating because its hypocenter was shallow (only 13 kilometers deep) and only 25 kilometers away, and the poorly constructed adobe houses fell on people and buried them.

Earthquake engineers have figured out relatively inexpensive ways to strengthen adobe houses. Laying bamboo lengthwise in the mortar strengthens the wall as does drilling holes in the bricks and inserting vertical bamboo sticks so that they tie rows of blocks together. A continuous wooden or cement ring should go all around the top of the walls to tie the walls together and to provide a way to fasten ceiling and roof joists to the walls. If an existing house is being retrofitted, vertical and horizontal strips of wire mesh should be nailed onto the walls both inside and outside at corners and around windows and doors. In practice, the mesh strips used range from seven centimeters to sixty centimeters in width. The strips are attached by driving nails through metal bottle caps and into the adobe.

Bridges, Dams, and Isolation Bearings. Structures can be protected by strengthening them to withstand an earthquake or by isolating them from the ground so that earthquake energy does not enter the structure. Dams are built to the first plan, and bridges follow the second plan. To ensure that a concrete dam survives an earthquake, extra reinforcing steel bars (rebars) would be used. The dam must either rest directly on bedrock or on massive pillars that extend downward to bedrock. If there are known fault lines in the area so that it is known in which direction land will move, it may be possible to construct a slip joint. A slip joint works like sliding closet doors, where one door can slide in front of the other. The Clyde Dam in Central Otago, New Zealand, has a slip joint that will allow the land on either side to slip up to two meters horizontally and one meter vertically. Finally, the dam should be built several meters higher than originally planned so that if an earthquake causes the impounded water to slosh back and forth it will not overtop the dam.

Bridges have long used bearings in the form of several hardened steel cylinders between a flat bridge plate and a flat supporting plate. These bearings allow the bridge to move as it expands or contracts with temperature. The same method can be used to allow motion during an earthquake without damaging the bridge, but now the bridge is isolated from the pier since the pier can move back and forth while the bridge remains stationary.

Waves that shake the foundation of a building send some of that vibrational energy up into the building. Placing a bridge or a building on lead and rubber bearings lessens the energy transmitted to the bridge or building. A typical bearing consists of a large block of alternating steel and rubber layers surrounding a vertical lead cylinder. It is quite rigid in the vertical direction but allows considerable motion in the horizontal direction. Since the rubber heats up as it is deformed, it converts the horizontal motion into heat and thereby damps this motion. The Museum of New Zealand and the New Zealand Parliament buildings stand on lead-rubber bearings and are thereby partially isolated from ground motion. Ironically, Christchurch, New Zealand, was considering reinforcing a large number of buildings prior to the February, 2011, quake, but had not proceeded very far because of the cost. That cost was a pittance compared with the rebuilding cost.

Mass Dampers. A building of a certain height, mass, and stiffness will tend to oscillate or resonate at a given frequency. Small oscillations can quickly build up into large oscillations, just as repeatedly giving a small push at the right frequency to a child in a playground swing will cause the swing to move in a large arc. If the upper floors of a tall building sway too much, people get motion sickness, and the structure gets weakened and may eventually fall. Mass dampers are usually huge concrete blocks mounted on tracks on an upper floor. When sensors detect lateral motion of the building's upper floors, motors drive the block in the opposite direction to the building's motion. This pushes the building in the opposite direction from its motion and causes that motion to die out. A tuned mass damper oscillates at the natural frequency of the building. This technique also works with motion caused by earthquake waves.

Taipei 101 in Taiwan is the world's second-tallest building. It is 101 stories (508 meters) tall, and its mass wind damper is a 660-metric ton metal sphere suspended like a pendulum between the eighty-seventh and ninety-second floors. If the building sways, the pendulum is driven in the opposite direction. This passively tuned mass wind damper reduces the building's lateral motion by more than half. Two six-metric-ton pendula are positioned in the tower to control its motion.

X-Braces and Pneumatic Dampers. The vertical and horizontal beams of a building's steel framework form rectangles. Consider a vertical rectangle in the wall at the bottom of the building. The bottom of the rectangle is fastened to the foundation and will move with a seismic wave, but the top of the rectangle is fastened to the building above it so that inertia will tend to hold it fixed. As the foundation moves laterally, the rectangle will be deformed into a parallelogram. Two diagonal beams making an "X" in the rectangle will keep it from deforming very much. The beautiful seventy-one story Pearl River Tower in Guangzhou, China, uses massive X-braces to keep the tower from swaying in the wind or an earthquake. These beams can clearly be seen in construction photographs of the tower. The tower uses integral wind turbines and solar cells to be largely energy self-sufficient. Rather than using the X-braces to stiffen the tower, the centers of the diagonal beams could have been clamped together with break-lining material between them, then they would damp the horizontal

motion. Diagonally mounted massive hydraulic pistons can also be used to strengthen a structure and simultaneously to damp out the earthquake energy in a building.

Pyramids. Large amplitude horizontal motion can also be avoided if earthquake motion is not concentrated at the building's natural frequency, but is spread over many frequencies. A building can be designed not to amplify waves of certain frequencies and to deflect some waves and absorb others. The speed of a wave traveling up a building depends on the amount of stress present, the amount of mass per meter of height, and the frequency of the wave. A bullwhip made of woven leather thongs makes a good analogue. Near the handle, the whip is as thick as the handle, but it tapers to a single thin thong at the far end. When the handle is given a quick backward and forward jerk, a wave speeds down the length of the whip. If the momentum (mass times velocity) of a whip segment is to remain constant, as the mass of a

---

Fascinating Facts About Earthquake Engineering

- San Francisco is moving toward Los Angeles at about 5 centimeters per year. They should be across from each other in about 12 million years.
- There are about 500,000 earthquakes each year, but only about 100,000 are strong enough to be felt by people.
- Only about 100 earthquakes each year are strong enough to do any damage.
- The largest recorded earthquake was a magnitude 9.5 in Chile on May 22, 1960.
- On March 28, 1964, a woman in Anchorage, Alaska, was trying to remove a stuck lid from a jar of fruit. At the instant she tapped the lid against the corner of the kitchen counter, a 9.2 magnitude earthquake struck. For a few moments she feared she had caused the quake.
- Without a tuned mass damper, the top floors of very tall buildings can sway back-and-forth 30 centimeters or more in a strong wind, let alone an earthquake.
- The 2004 Indian Ocean 9.0-magnitude earthquake released energy equivalent to 9,560,000 megatons of TNT, about 1,000 times the world supply of strategic nuclear warheads.

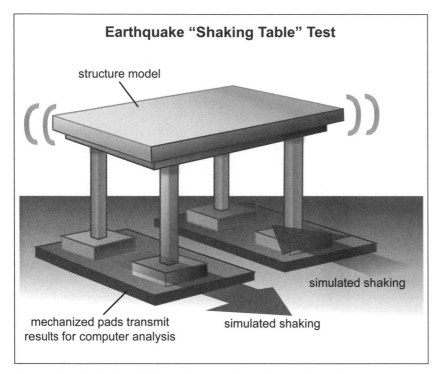

Earthquake "Shaking Table" Test

structure model

mechanized pads transmit results for computer analysis

simulated shaking

simulated shaking

An earthquake shake table is used to test the resistance of certain components or structures to seismic activity.

segment decreases (because of the taper), the speed must increase. The distinctive whip crack occurs as the whip end exceeds the speed of sound. In a similar fashion, the speed of a wave traveling up a pyramid-shaped building changes, and as the speed changes, the frequency changes. If a pyramidal building is properly designed, earthquake waves will attenuate as they try to pass upward. This is the idea behind the design of the forty-eight story Transamerica Pyramid building in San Francisco. It is not essential that the shape of the building be a pyramid since changing the mass density or the tension in the steel structure can have a similar effect.

IMPACT ON INDUSTRY

Government and University Research. Most earthquake engineering work is done by consulting companies, while research is carried out by universities and research institutes. The United States Geological Survey (USGS) is the federal agency tasked with recording and reporting earthquake activity nationwide. Data is provided by the Advanced National Seismic System (ANSS), a nationwide array

of seismic stations. USGS maintains several active research projects. The Borehole Geophysics and Rock Mechanics project drills deep into fault zones to measure heat flow, stress, fluid pressure, and the mechanical properties of the rocks. The Earthquake Geology and Paleoseismology project seeks out and analyzes the rocks pertaining to historic and prehistoric earthquakes.

Earthquake Engineering Research Institute (EERI) of Oakland, California, carries out various research projects. One project involves surveying concrete buildings that failed during an earthquake in an effort to discover the top ten reasons for the failure of these buildings. The Pacific Earthquake Engineering Research Center (PEER) at the University of California, Berkeley, has a 6-meter-by-6 meter shaking table. It can move horizontally and vertically and can rotate about three different axes. It can carry structures weighing up to 45 tons and subject them to horizontal accelerations of 1.5 times gravity. Recent projects include the seismic-qualification testing of three types of 245-kilovolt disconnect switches, testing a friction pendulum system (for damping vibrations), testing a two-story wood-frame house, and testing a reinforced concrete frame.

The John A. Blume Earthquake Engineering Center at Stanford University pursues the advancement of research, education, and practice of earthquake engineering. Scientists there did the research into the earthquake risk for the 2-mile-long Stanford Linear Accelerator, the Diablo Canyon Nuclear Power Plant, and many other sites. The founder, John Blume, is quoted as reminding a reporter that the center designed "earthquake-resistant" buildings not "earthquake-proof" buildings, and added, "Don't say 'proof' unless you're talking about whiskey."

Feeling that more coordinated efforts were needed, the Japanese formed the Japan Association for Earthquake Engineering (JAEE) in January, 2001. The association was to be involved with the evaluation of seismic ground motion and active faults, resistance measures before an earthquake, education on earthquake disaster reduction, and sponsoring meetings and seminars where new techniques could be shared and analyzed. After an earthquake, they hope to aid in damage assessment, emergency rescue and medical care, and in evaluating what building techniques worked and what did not work.

Major Corporations. ABS Consulting, with headquarters in Houston, Texas, is a worldwide risk-management company. It has 1,400 employees and uses earthquake engineers when it evaluates the earthquake risk for a site.

Air Worldwide provides risk analysis and catastrophe modeling software and consulting services. It has offices in Boston, San Francisco, and several major cities in other countries. It hires civil engineers to perform seismic design studies and to prepare plans for structural engineering.

ARUP, an engineering consulting company headquartered in London, employs 10,000 people worldwide. To reduce the lateral movement of the upper stories of buildings, it employs its damped outrigger system. It uses large hydraulic cylinders to tie the central pillar of the building to the outer walls. The alternative is to make the building stiffer with more concrete and steel (which costs several million dollars) and then add a tuned mass damper (which ties up a great deal of space). The company used a few dozen of these dampers in the beautiful twin towers of the St. Francis Shangri-La Place in Manila, Philippines.

International Seismic Application Technology (ISAT), with headquarters in La Mirada, California, uses earthquake engineers to do site studies and to design seismic-restraint systems for plumbing, air-conditioning ducts, and electrical systems in buildings.

Miyamoto International is a global earthquake and structural engineering firm that specializes in designing earthquake engineering solutions. It has offices throughout California, and in Portland-Vancouver and Tokyo, and specializes in viscous and friction cross-bracing dampers.

The Halcrow Group based in London does seismic-hazard analysis, design, and remediation. It also does earthquake site response analysis and liquefaction assessment and remediation. It has done site evaluations for nuclear power plants, dams, and intermediate level nuclear waste storage. It is a large company with 8,000 employees and many interests.

CAREERS AND COURSE WORK

Earthquake engineering is a subset of geotechnical engineering, which itself is a branch of civil engineering. Perhaps the most direct route would be to attend a university such as the University of California, Los Angeles, which offers both graduate and undergraduate degrees in engineering, and get an undergraduate degree in civil engineering. An undergraduate should take principles of soil mechanics, design of foundations and earth structures, advanced geotechnical design, fundamentals of earthquake engineering, soil mechanics laboratory, and engineering geomatics. Graduate courses should include advanced soil mechanics, advanced foundation engineering, soil dynamics, earth retaining structures, advanced cyclic and monotonic soil behavior, geotechnical earthquake engineering, geoenvironmental engineering, numerical methods in geotechnical engineering, and advanced soil mechanics laboratory.

Other schools will have their own version of the program. For example, Stanford University offers a master's degree in structural engineering and geomechanics. The program requires a bachelor's degree in civil engineering including courses in mechanics of materials, geotechnical engineering, structural analysis, design of steel structures, design of reinforced concrete structures, and programming methodology. The University of Southern California; the University of California, San Diego; the University of California, Berkeley; and the University of Alaska at Anchorage all have earthquake engineering programs. On the east coast, the Multidisciplinary Center for Earthquake Engineering Research at the State University of New York at Buffalo and the Center for Earthquake Engineering Simulation at Rensselaer Polytechnic Institute in Troy, New York, are centers for earthquake engineering.

SOCIAL CONTEXT AND FUTURE PROSPECTS

Although earthquake engineering has made a lot of progress, some areas of society have been surprisingly slow to implement proven measures. Many of the buildings that collapsed in the New Zealand earthquake in February, 2011, would have remained standing had they been reinforced to the recommended standard. They were not reinforced because of cost, but that cost was a small fraction of what it will now cost to rebuild. The January, 2010, Haitian earthquake was so deadly because there are no national building standards. The December, 2003, earthquake in Bam, Iran, was so devastating because building codes were not followed. In particular, enough money was budgeted to build the new hospitals to standards that would have kept them standing. The hospitals collapsed into piles of rubble while corrupt officials (according to expatriates) enriched themselves. Building codes will do no good until they are enforced, and they will not be enforced without honest officials.

On a more positive note, an exciting, recent proposal is to make a building invisible to earthquake waves. Earthquake surface waves cause the damage. The speed of such waves depends upon the density and rigidity of the rock and soil they traverse. Consider a wave coming toward a building almost along a radius, and suppose that the building's foundation is surrounded by a doughnut-shaped zone in which the speed of an earthquake wave is increased above that of the surrounding terrain. The incoming wave will necessarily bend away from the radius. One or more properly constructed doughnuts, or rings, should steer the earthquake waves around the building. No doubt there will be problems in implementing this method, but it seems promising. It may even be possible to surround a town with such rings and thereby protect the whole town.

Charles W. Rogers, B.A., M.S., Ph.D.

FURTHER READING

Bozorgnia, Yousef, and Vitelmo V. Bertero, eds. *Earthquake Engineering: From Engineering Seismology to Performance-Based Engineering.* Boca Raton, Fla.: CRC Press, 2004. Provides a good overview of the problems encountered in earthquake engineering and ways to solve them. Requires a good science and math background.

Building Seismic Safety Council for the Federal Emergency Management Agency of the Department of Homeland Security. *Homebuilder's Guide to Earthquake-Resistant Design and Construction.* Washington, D.C.: National Institute of Building Sciences, 2006. A gold mine for the non-engineer or the prospective engineer. Introduces the terms and techniques of earthquake-resistant structures in an understandable fashion.

Kumar, Kamalesh. *Basic Geotechnical Earthquake Engineering.* New Delhi, India: New Age International, 2008. Emphasizes site properties and preparation, when to expect liquefaction and what to do about it. Easily read by the science-savvy layperson.

Stein, Ross S. "Earthquake Conversation." *Scientific American* 288 (January, 2003): 72-79. Active faults are responsive to even a small increase in stress that they acquire when there is a quake in a nearby fault. This may make earthquake prediction more accurate.

Yanev, Peter I., and Andrew C. T. Thompson. *Peace of Mind in Earthquake Country: How to Save Your Home, Business, and Life.* 3d ed. San Francisco: Chronicle Books, 2008. An excellent introductory treatment of earthquakes, how they damage structures, and what may be done beforehand to reduce damage. Discusses building sites and possible problems such as liquefaction.

WEB SITES

ArchitectJaved.com
Earthquake Resistant Structures
http://articles.architectjaved.com

Earthquake Engineering Research Institute
http://www.eeri.org/site

Seismological Society of America
http://www.seismosoc.org

See also: Bridge Design and Barodynamics; Civil Engineering.

EARTHQUAKE PREDICTION

FIELDS OF STUDY

Geology; oceanography; geophysics; seismology; structural engineering; materials science; soil science.

SUMMARY

Earthquakes are among the most potentially devastating catastrophes human beings can face, and ways of predicting their occurrence in advance are urgently needed. Short-range predictions are needed so that people can evacuate dangerous locations or shutdown critical services, such as nuclear power plants and transportation systems. Long-range predictions are needed to help communities upgrade their building codes and identify at-risk areas where construction or other land uses should not be permitted.

KEY TERMS AND CONCEPTS

- **Elastic Rebound:** Sudden release of strain built up by fault creep over a long period of time.
- **Fault:** Crack in the Earth's crust along which differential movement has occurred.
- **Foreshock:** Preliminary vibration before a major earthquake.
- **Locked Fault:** Fault along which anticipated movement is overdue.
- **Moment Magnitude:** Scale for measuring earthquake size based on the amount of energy released.
- **Primary Wave:** First group of vibrations to arrive from an earthquake.
- **Richter Scale:** Scale for measuring earthquake size based on the amount of shaking.
- **Secondary Wave:** Second group of vibrations to arrive from an earthquake.

DEFINITION AND BASIC PRINCIPLES

Earthquake prediction is the act of determining in advance that an earthquake is likely to take place. The prediction must address four factors: the time period during which the earthquake is expected to occur, the area that the earthquake is expected to affect, the type of shaking that will occur, and the likelihood that the earthquake will actually take place. Short-range predictions estimate that an earthquake will occur within the next few hours, or at most, the next few days. These are based on premonitions of an impending earthquake, such as changes in groundwater levels in wells, an increased number of foreshocks, or unusual animal behavior. Long-range predictions, which estimate that an earthquake may take place within a certain number of years, are generally based on the past history of the area, augmented by Global Positioning System (GPS) data or the digging of trenches in critical areas. Short-range warnings are valuable if the earthquake occurs, but if the anticipated earthquake does not materialize, they have disrupted people's lives unnecessarily. Long-range predictions are useful primarily because they often result in reviews of building codes and reconsideration of the suitability of some land areas for construction and development.

BACKGROUND AND HISTORY

The scientific understanding of earthquakes and the development of methods for predicting them date only to the time of the great San Francisco earthquake in 1906. Before that time, the relationship between faulting and earthquakes was not clearly understood. H. F. Reid, a professor at The Johns Hopkins University, was appointed to California's state-funded commission to study the great earthquake. Based on his examination of the terrain surrounding the city, he was able to show that points on opposite sides of the San Andreas fault had moved differentially as much as 6 meters before breakage. Reid coined the term "elastic rebound" to describe what then happened. Strain had built up in the rocks by fault creep, as the two sides of the fault drifted in opposite directions. When breakage finally occurred, the rocks returned to their unstrained positions just as a rubber band returns to its original shape after the band snaps, and this snapping back of the rocks was what had caused the earthquake. Reid led the way to an understanding of how the breakage of faults causes earthquakes, thereby enabling scientists to turn their attention to ways of predicting them in advance.

HOW IT WORKS

Short-Range Predictions. A variety of methods has been used to predict earthquakes on a short-term basis, and several of them are related to changes in underground water. Levels in wells may change, turbidity may appear, or water temperatures may rise. Chemical changes have also been noted. Kobe, Japan, experienced a devastating earthquake in 1995. Subsequent to the earthquake, an analysis of the water produced at the city's bottling plant revealed that the chemistry of the water had been steadily changing for two weeks before the earthquake. The water at the bottling plant is being studied daily for warnings of the next earthquake.

Mexico City uses a different method of predicting earthquakes. Fault lines are located along the Pacific coast, 300 kilometers away from Mexico City. A network of seismographs is used to determine that an earthquake has taken place, and that information is transmitted by satellite radio to the city, warning residents 50 to 80 seconds before the damaging waves arrive. Fault lines lie near and under the Los Angeles and Tokyo metropolitan areas, so there is less time to prepare for an earthquake, but seismographs around the city can detect the less destructive primary waves 10 to 15 seconds before the destructive secondary waves arrive. This gives computerized control systems enough time to shut off the power to specified industries and close down the mass transit systems.

Changes in the behavior of animals before an earthquake has been noted in some instances, most notably before the 1906 San Francisco earthquake. The earthquake occurred around five o'clock in the morning, and afterward residents reported that dogs at the pound had howled incessantly from midnight on and that horses had stamped and neighed in their livery stables during that same period. Presumably the animals heard high-pitched sounds, not audible to humans, caused by dilation of the rock preliminary to the final breaking.

Dilatancy, the opening of small cracks in the rocks of a fault zone before breakage, may cause other signals that an earthquake is about to occur. These signals include the tremors known as foreshocks, which often precede major earthquakes, as well as changes in the bedrock's magnetism, electrical resistance, or ability to transmit seismic waves. Radon gas may also be released into wells, and in some instances, changes in the land surface will also be noted. These changes can be monitored using sensitive instruments known as tiltmeters. Because earthquakes often accompany volcanic eruptions, any volcano overlooking a major city—such as Vesuvius beside Naples, Italy, or Mount Rainier next to Seattle, Washington—is monitored daily by seismographs, watching for the slightest change in vibrations that might indicate that an eruption is imminent. Other warning signs from a volcano include steam clouds coming from the crater, ash falling in the surrounding area, tilting of the volcano's flanks as magma swells into the crater, and changes in water features around the mountain, such as springs and wells.

Long-Range Predictions. The historical record of an area's previous earthquakes is useful in making long-range predictions of future shocks and is helpful in the preparation of seismic risk maps. A seismic risk map of the United States would show the West Coast, Alaska, and Hawaii as the highest risk zones. A repeated pattern of earthquakes is also helpful in making predictions. Parkfield, California, has experienced a magnitude 6 or 8 earthquake about every twenty-two years since 1857. Parkfield experienced an earthquake on September 28, 2004, so the town might expect its next earthquake in 2026. Not all segments of a fault break at the same time, and the inactive segments are known as seismic gaps. People living in areas where the seismic risk is high need to remember that an earthquake is possible at any time.

Careful fieldwork by geologists is another way to predict future shocks. The study of a fault trace across the landscape may reveal offset features that provide clues as to how often and when earthquake movements have taken place. Trenches may be dug for an examination of the soil layers, and a pilot hole is being drilled at Parkfield for a study of the San Andreas fault at depth.

Seismologist have begun using GPS measurements to analyze the differential creep on two sides of a fault. Using such measurements, in April, 2008, scientists predicted that the Enriquillo fault, near Haiti's capital city of Port-au-Prince, was capable of a magnitude 7.2 earthquake. Just twenty months later, in January, 2010, a devastating 7.0 earthquake did occur.

APPLICATIONS AND PRODUCTS

Construction in earthquake-prone areas requires special precautions, and attention must be given to

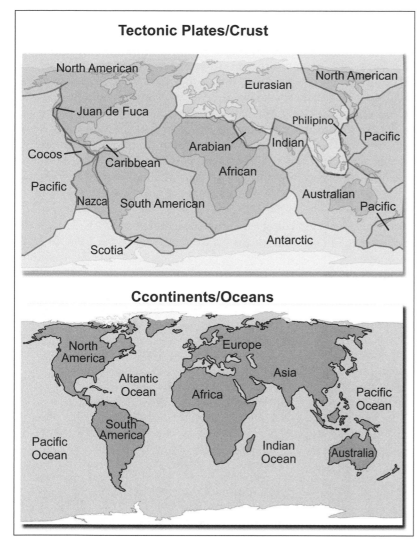

Tectonic Plates/Crust

North American
Eurasian
North American
Juan de Fuca
Philipino
Pacific
Cocos
Arabian
Indian
Caribbean
African
Pacific
Nazca
South American
Australian
Pacific
Scotia
Antarctic

Ccontinents/Oceans

North America
Europe
Altantic Ocean
Asia
Africa
Pacific Ocean
Pacific Ocean
South America
Indian Ocean
Australia

Earthquakes occur along the boundaries of tectonic plates, which span both continents and oceans.

the type of structure planned. Some structures, such as nuclear power plants and large dams, should never be built in areas with active faults. Engineers and developers must determine the type of earthquakes affecting the area, how their vibrations might affect the intended structure, and how long the shaking could last. Knowledge of the area's geology is critical because soil and bedrock characteristics can influence how an earthquake will affect structures. For earthquakes with magnitudes up to 5, damage should be slight; for earthquakes of magnitudes 5 to 7, damage should be easily repaired; and for magnitude 7 or higher earthquakes, the structure must not collapse although significant repairs, or even demolition, may be required.

Nature of the Substratum. Structures built on solid bedrock invariably suffer less damage in an earthquake than structures built on filled land or water-saturated sediments. Filled land or water-saturated sediments lose their cohesiveness because their loose structure amplifies the vibrations from the earthquake. They can begin behaving like a liquid, in a process known as liquefaction, causing severe damage to the structures on the surface. Liquefaction occurred during a 7.5 magnitude earthquake in Niigata, Japan, on June 16, 1964, and during a magnitude 9.2 earthquake in Anchorage, Alaska, on March 18, 1964. In Niigata, part of which was built on sandy soil, hundreds of buildings, including multistory concrete-reinforced apartment buildings, tilted and fell over. In Anchorage, clay beds beneath the Turnagain Heights subdivision liquefied, and more than 200 acres covered with homes slid toward the sea. The damage to homes, streets, sidewalks, water and sewer lines, and electrical systems was massive.

Proper Building Materials. Adobe and mud-walled structures are the weakest buildings and invariably collapse during severe earthquakes. Stucco, unreinforced brick, and concrete-block structures also suffer severe damage; their walls prove inflexible to shaking once the mortar that binds them is broken. Surprisingly, wooden structures survive earthquakes quite well. Wood is lighter weight than metal or concrete, and if the frame of the building is properly secured to the ground, the nails and screws in the wood provide enough flexibility to resist shaking. Wood is preferable for smaller private homes in earthquake-prone areas, although concrete can be used for larger structures, provided it is heavily reinforced with steel bars.

Construction of Larger Buildings. The most important consideration in earthquake-proof construction is that the foundation, floors, walls, and roof of a building be tied together to withstand both horizontal and vertical stresses. Open lower floors are hazardous and should be reinforced with added pillars or diagonal bracing. Tall buildings and skyscrapers need to sway to prevent cracking, and the use of diagonal steel beams will permit this. Such buildings may even be placed on a layered steel or rubber structure that acts as a shock absorber to decrease such swaying. A good example of this is the pyramid-shaped Transamerica building in San Francisco. In addition to its tapering shape, which reduces the amount of mass in the higher floors, the skyscraper sits on a two-story-high cushion of massive diagonal trusses, designed to help the building absorb earthquake vibrations.

Retrofitting Older Structures. Many older structures are being reinforced to provide protection during earthquakes using a process known as retrofitting. The object is to make the existing structure more resistant to stresses, ground motion, and possible soil failure. High-strength steel, fiber-reinforced polymers, or fiber-reinforced concrete can be used to build new structural walls, and columns can be jacketed with steel. Stronger columns can be added and diagonal bracing placed on all floors. Foundation work might include jacking up the building to place springs or other flexible materials between it and the ground.

Quake Preparedness. People in earthquake-prone zones need to be ready to live for long periods without electricity and other services following a severe shock. Battery-powered radios, flashlights, and first-aid kits are essential, and homeowners need to know the location of the fuse box and shut-off valves for water and gas. Heavy appliances must be securely anchored to the floor, and shelves, bookcases, and cupboards firmly attached to walls. Heavy objects should not be placed on top shelves. Families need to have a plan for reuniting in the event that members become separated during an earthquake. During the earthquake itself, people who are outside should remain outside, away from possible falling objects, and people who are inside should seek shelter under a heavy piece of furniture. Those driving cars should come to a stop, away from overhead objects.

IMPACT ON INDUSTRY

Most research related to earthquake prediction is centered in the United States, Russia, Japan, and China because these countries have experienced serious earthquakes in the past. Research includes laboratory and field studies of rock behavior before, during, and after earthquakes, as well as the monitoring of activity along major faults. Both government agencies and universities conduct this research.

Government Agencies. The United States Geological Survey (USGS) plays a major role in earthquake studies and earthquake prediction for the United States. In addition to providing general information regarding earthquakes, it makes available data on the latest earthquakes in the United States and around

---

Fascinating Facts About Earthquake Prediction

- According to an ancient piece of Chinese folklore, before an earthquake, hibernating snakes come out of their holes, sometimes freezing to death in the cold.
- During the 1906 San Francisco earthquake, a cow fell into a small, shallow crack that opened up in the ground and was buried with only her tail protruding.
- A series of strong earthquakes occurred near New Madrid, Missouri, in 1811 and 1812. They were so severe that they made church bells ring in Boston and caused the Mississippi River to flow in the opposite direction during the shaking.
- During the massive 8.8 magnitude earthquake in Chile on February 27, 2010, the city of Concepción moved 3 meters to the west.
- The famed tenor, Enrico Caruso, was in San Francisco for a performance at the time of the 1906 earthquake. Terrified, he ordered his trunks packed, and at great expense, he managed to flee the city, carrying a framed photograph of President Theodore Roosevelt autographed to himself as his identification. He vowed never to sing in the city again and never did.
- The city of Los Angeles is moving north toward San Francisco along the San Andreas fault at the rate of about 1 centimeter a year. If movement continues at this rate, in millions of years, the two cities will meet.

the world. For selected areas, such as California and Nevada where the potential for earthquakes is high, the USGS provides an index map, updated constantly, showing where the latest earthquake activity has been observed. The agency also makes short-range predictions of future earthquake activity, such as the probability of strong shaking in critical areas within the next twenty-four hours, and estimates of the probability of aftershocks in areas where earthquakes have recently occurred. The USGS also makes long-range predictions of future earthquake activity. It forecasts a 62 percent probability of a magnitude 6.7 or greater earthquake, capable of causing widespread damage, striking the San Francisco Bay Area before 2032.

The Alaska Earthquake Information Center (AEIC) provides services similar to those provided by the USGS, which serves primarily the lower forty-eight states. The center is important for Alaska residents because the state is just as prone to earthquakes as California, having had the world's second largest earthquake in 1964. The AEIC receives daily reports of earthquakes from a network of more than four hundred seismograph stations and makes these available on its Web site. The center makes predictions of future earthquakes, although shocks of magnitude 8 and lower are so common that no effort is made to predict them. For earthquakes of magnitude 8 and greater, predictions are made when possible, giving the expected location, magnitude, and time.

A third government agency involved in earthquake monitoring and prediction is the National Oceanographic and Atmospheric Administration (NOAA), which operates tsunami warning centers in Hawaii and Palmer, Alaska. Because of the great loss of life during the Indian Ocean tsunami of 2004, the Hawaii warning center has expanded its coverage to the Indian Ocean and the Caribbean until centers for those two areas can be created. The Alaska center serves the coastal areas of Canada and the United States, but not Hawaii.

Academic Research. The California Institute of Technology (Caltech) is a leading institution involved in earthquake research and prediction on the West Coast of the United States. Charles Richter and Beno Gutenberg developed the Richter scale for measuring the magnitude of earthquakes at Caltech in the 1930's. Geophysicists in the division of geological and planetary sciences supervise Caltech's

seismological laboratory and also help maintain the Southern California Seismic Network and the Southern California Earthquake Data Center.

On the East Coast of the United States, Columbia University's Lamont-Doherty Earth Observatory also has a long tradition of earthquake research. Besides performing observational studies in the field, scientists in the seismology, geology, and tectonophysics division operate the Lamont-Doherty Cooperative Seismograph Network and also contribute to research for the Center for Hazards and Risk at Columbia's Earth Institute.

CAREERS AND COURSE WORK

A basic course in general geology is the logical starting point for a student interested in pursuing a career related to earthquake prediction. Most colleges and universities include such a course among their undergraduate offerings. This course provides the student with basic factual information related to faulting, which is the cause of most earthquakes, and also provides an introduction to the seismograph, which is the instrument scientists use to measure the strength of earthquakes and determine where they have taken place. Graduate study is probably a must for students who wish to obtain research positions in the field of earthquake prediction. Useful courses that the student can take in graduate school include structural geology, which deals with the types of faulting and their causes, and geophysics, which details the types of earthquake waves and how to interpret them using the seismograph. One course in a related field that is useful in earthquake prediction is engineering and materials science. This course provides important information regarding building codes for earthquake-prone areas and appropriate land-use planning for them. Soil science is another useful course because information about the past history of faults is often obtained by digging trenches through them. For students seeking employment opportunities in fields related to earthquake prediction, government agencies and the research departments of colleges and universities would be the logical place to start.

SOCIAL CONTEXT AND FUTURE PROSPECTS

The importance of predicting earthquakes in advance of their occurrence cannot be stressed too greatly. Many of the world's best known cities lie near, or directly on top of, one or more major through-going faults, and the destruction resulting from movement on such faults can be catastrophic. San Francisco and Oakland, California, for example, lie in a narrow sliver of crust, bordered on one side by the San Andreas fault and on the other by the equally dangerous Hayward fault. Destructive earthquakes are likely to take place in Mexico City, Los Angeles, and Tokyo, which have had major earthquakes in the past and need adequate warning systems. Cities that have not had severe shocks in the recent past but are thought to be at high risk are Seattle and Istanbul. Seattle has not had a major earthquake in several hundred years, but geologists doing fieldwork around the city have found evidence that a massive earthquake took place there in the 1700's. Seattle residents need to prepare for a repeat of such an earthquake. Istanbul has not had a major earthquake in many years, but geologists point out that it is at the west end of a fault similar to the San Andreas. Fourteen earthquakes with magnitudes of 6 or higher have taken place along this fault since 1936, with each earthquake moving closer to the city. Despite all the advances that have been made in long-range earthquake prediction, scientists can still give people only a vague idea of when a large earthquake will hit their area.

Donald W. Lovejoy, B.S., M.S., Ph.D.

FURTHER READING

Berke, Philip R., and Timothy Beatley. *Planning for Earthquakes: Risk, Politics, and Policy.* Baltimore: The Johns Hopkins University Press, 1992. Excellent recommendations for land-use planning and construction techniques in earthquake-prone areas. Contains many helpful photographs illustrating the types of hazardous situations discussed.

Bolt, Bruce A. *Earthquakes.* 5th ed. New York: W. H. Freeman, 2006. Written for the lay reader. The chapter "Events that Precede Earthquakes" is a thorough summary of ways to predict earthquakes. Useful diagrams and photographs are included.

Chester, Ray. *Furnace of Creation Cradle of Destruction: A Journey to the Birthplace of Earthquakes, Volcanoes,* and Tsunamis. New York: AMACM, 2008. Explains plate tectonics and faulting, and how they are capable of causing earthquakes, volcanoes, and tsunamis. The section on "Mitigating Against Earthquakes" is of particular interest.

Hough, Susan Elizabeth. *Predicting the Unpredictable: The Tumultuous Science of Earthquake Prediction.* Princeton, N.J.: Princeton University Press, 2010. Examines the science of earthquake prediction, including early attempts, the Chinese prediction efforts, spectacular failures, and social and political issues.

_____. *Richter Scale: Measure of an Earthquake, Measure of a Man.* Princeton, N.J.: Princeton University Press, 2007. The life of the man who invented the Richter scale. Contains a chapter on the problems relating to predicting California earthquakes.

Winchester, Simon. *A Crack in the Edge of the World: America and the Great California Earthquake of 1906.* New York: HarperCollins, 2006. A highly readable account of the most famous earthquake in the United States. Excellent eyewitness accounts and information regarding the behavior of animals just before the earthquake.

Yeats, Robert S., Kerry Sieh, and Clarence R. Allen. *The Geology of Earthquakes.* New York: Oxford University Press, 1997. A comprehensive analysis of the types of earthquakes, with a chapter on assessing the risk of future movement on a fault. Useful diagrams and photographs.

WEB SITES

Alaska Earthquake Information Center
http://www.aeic.alaska.edu

National Oceanic and Atmospheric Administration
Earthquakes
http://www.noaawatch.gov/themes/quake.php

U.S. Geological Survey
http://earthquakes.usgs.gov

See also: Earthquake Engineering; Engineering Seismology; Land-Use Management; Risk Analysis and Management.

ECOLOGICAL ENGINEERING

FIELDS OF STUDY

Chemistry; physics; calculus; biology; agriculture; physiology; genetics; statistics; electronics; GIS systems; geography; geology; hydrology; marine science; biotechnology; ethics; public policy.

SUMMARY

Ecological engineers design sustainable systems using ecological principles to create, restore, and conserve natural systems. The multidisciplinary synergy of the engineering sciences and biological knowledge makes it possible to design and maintain strong, self-evolving ecosystems to support a variety of life forms in a wide range of habitats. Human communities are planned so that they contribute to the balanced flow of energy and products to the surrounding environment. Plant and animal communities are created or restored, often in keeping with regulatory guidelines that mandate the restoration of particular locales to their original, prehuman conditions. Scales of applicability range from microscopic exchanges to the processes of much larger ecosystems.

KEY TERMS AND CONCEPTS

- **Abiotic Environment:** Nonliving and chemical factors that influence the organisms in a biotic environment.
- **Agroecosystem:** Multilevel assemblage of ecological processes, including microbial interactions within soils; the relationships of plants, crops, and herd animals; and the study of farming landscapes and communities.
- **Biotic Environment:** Living organisms within a particular environment and their relationships with one another to form a particular ecosystem.
- **Drilosphere:** All soils affected by earthworm activities, including external structures (middens, burrows, diapause chambers, surface and below-ground casts) and the earthworm's gut.
- **Horizontal Gene Transfer:** Any process that allows an organism to incorporate genetic material from an organism other than a parent or outside its own species; also known as lateral gene transfer.
- **Keystone and Foundation Species:** Species that define the structure and function of an ecosystem; for example, many forests are endangered as pathogens and insects destroy particular foundation tree species on which the forest ecosystem depends for coherence.
- **Laser Scanning:** Laser-based instrumentation used to measure complete environmental systems, including fixed beam, rotation beam, and distance measurers.
- **Light Detection and Ranging (Lidar):** Type of laser scanning used as airborne laser systems to take measurements of landscapes for the collection of detailed spatial data sets.
- **Macroecology:** Study of large-scale patterns and processes and how they relate to local assemblage structures; principles of evolution, ecology, and historical contingency are referenced to explain species richness, range size, abundance, and body size of particular life species.
- **Metal Toxicity:** Inhibition of microbial activity by certain metals—including arsenic, barium, cadmium, chromium, lead, mercury, and nickel—that interact with cellular proteins in complex ways.
- **Microclimate:** Small but distinct climate zone within a larger climate area; slight differences in landscape climates (frequently due to shade trees or mountains and valleys) allow for variegated and unusual species to occupy a common terrain.
- **Micrometeorology:** Study of the atmosphere just above the ground.
- **Mobile Genetic Element:** Small piece of DNA that can replicate and insert copies at random locations within cells.
- **Rhizosphere:** Area of soil that surrounds the roots of plants.
- **Soil Organic Matter (SOM) Dynamics:** Response of soil to its environment; the decomposition of organic residue by microorganisms and its alteration in agroecosystems are essential indicators of soil quality.
- **Soil Properties:** Characteristics that define the soil of a locale, including measurements of color, texture, and acidity; soil properties influence nutrient and water uptake and the disposition of organic materials.

- **Terrestrial Carbon Sequestration:** Photosynthetic process of removing carbon from the atmosphere and storing it in biomass.
- **Urban Ecology:** Impact of urban development on native ecosystems.

DEFINITION AND BASIC PRINCIPLES

The field of ecological engineering represents the synthesis of a variety of knowledge sets coordinated to address the challenges of holistic ecosystem design. Ecological engineering seeks to redress the fragmentation of scientific thought and activity, particularly in reference to the natural world and its processes. The term was first coined by Howard T. Odum, a prolific writer and researcher noted for founding the Center for Wetlands at the University of Florida. He was the son of Howard W. Odum, a noted sociologist and director of the Institute for Research in the Social Sciences at the University of North Carolina. The young Odum and his older brother, Eugene, helped establish the intellectual and scientific foundations of ecology in the twentieth century. The Odum brothers coauthored *Fundamentals of Ecology*, a successful textbook published in 1953 in which they state that mathematics and statistics were essential to the development of a unified system of ecosystem dynamics.

During the nineteenth and twentieth centuries, ecologists made substantial contributions to the knowledge of undisturbed ecosystems. Therefore, it is only within the past century that human relationships have been taken into account in the study of the evolution of natural systems. In 2004, the Ecological Society of America published a report to its governing board outlining a set of responses to the growing need for including sustainable ecological precepts in all phases of land use and development. Plans for a sustainable future are based on four basic principles:

- The rise of human-dominated ecosystems is inevitable.
- Ecosystems of the future will include a variety of conserved, created, and restored landscapes.
- Ecological science must become a core body of knowledge influencing global economics and policies.
- New and unprecedented regional and global partnerships among private and public entities are necessary to promote sustainable life systems.

Practitioners are careful to point out that ecological engineering is not synonymous with environmental engineering or biotechnology. Ecological engineers rely on the self-organizing capacities of ecosystem dynamics to create, restore, or conserve particular communities and habitats. Ecological-systems designers have a wealth of research materials available for consultation and comparison. These include efforts in habitat reconstruction, stream and wetland restoration, and a variety of wastewater, reclamation, and conservation projects. The scientific practices of observation, measurement, and documentation of the complex procedures and results of an ecosystem project are essential. The results of many restoration projects provide valuable information for understanding the processes of succession, the formation of biological macrocommunities in particular locales, and their relationships within broader ecosystems. Time is a fundamental component of ecosystem design. Many projects are undertaken with the realization that their fulfillment will exceed the lifetimes of their founders. Biodiversity, adaptation systems, and succession patterns are important dimensions of ecosystem design that require time.

BACKGROUND AND HISTORY

The use of myriad tools and technologies to construct and engineer life systems is an ancient human practice and a key factor in the evolution of the human species. Many tools and technologies are artifacts of cultures rooted in particular biological communities. Some sustainable traditions have been practiced for centuries and have made important contributions to the design and maintenance of contemporary ecosystems.

In the nineteenth century, the Industrial Revolution accelerated the rate of technological innovation in profound ways. Awareness of the aesthetic value of the natural landscape deepened as whole cultures and biological species disappeared in the wake of rapid urbanization, progressive conservation, and the colonization of the wilderness. Advances in urbanization; the consumption of fossil fuels; the mass fabrication of commodities; the mechanization of agriculture; the transformation of wetlands and deserts into vast farmlands; the unregulated extraction of minerals, metals, and inorganic materials for industrial expansion; increased demands for hydroelectric and nuclear power; a rapidly developing transportation and communications network; widespread deforestation; the unprotected destruction of wildlife; the release of radioactive isotopes, airborne pesticides, and vehicle emissions; and the creation of synthetic chemicals and materials in the wake of World War I and World War II ravaged European and American landscapes. Accelerated industrial production in the postwar period lifted economies while creating vulgar panoramas of urban blight: air and waterways darkened with toxic industrial, agricultural, and residential wastes; scarred mountains and forests; and a host of health and sanitation issues directly related to the toxic environment of the modern world.

The natural world is the subject of many of the supreme artistic and literary expressions of human culture. It is the basis of unique knowledge systems, the value of which is of particular importance to local communities. Since antiquity, classical treatises on the subject of natural history have included detailed studies that have formed the foundation for continuing work in the fields of medicine, chemistry, physics, biology, botany, mineralogy, geography, astronomy, and geology. Natural historians made remarkable contributions to the study of natural systems in the eighteenth and nineteenth centuries. The concept of evolution is an ever-present thread in the dialogue of natural philosophers. That tendril of thought took root in the discourse that followed the publication of *The Origin of Species by Means of Natural Selection* by Charles Darwin in 1859. As of 2011, evolution is understood as a complex process of genetic variation. That reality drives early twenty-first century research in ecology.

The environmental movement of the twentieth century created the grassroots momentum for a global reassessment of the controversial roles of science and technology in designing structures and processes that are functional, aesthetically pleasing, and sustainable. Environmental justice was an important topic of sociopolitical inquiry. Following the enactment of the National Environmental Policy Act in 1970, tens of thousands of regulations were enforced. Under the Toxic Substances Control Act of 1976, more than 80,000 toxic substances were registered by the Environmental Protection Agency (EPA). Similar standards have been adopted worldwide. Extensive research in pollution control provided new sources of expertise needed to remediate critical environmental conditions. Advances in computational technologies greatly facilitated the creation of detailed data sets, microassays, simulations, and measurement technologies. Extensive land surveys helped to substantiate the terms and application of soil and marine biotechnologies. Thermophiles and acidophils, xenobiotics, endocrine disrupters, bioaccumulation, biofilters, bioscrubbers, activated sludge, wastewater stabilization ponds; biosubstitutions, terrestrial phyto systems, and genetic manipulation: These and other terms form a common vocabulary and are found in all substantive research and textbooks addressing topics of environmental engineering and biotechnology.

Fascinating Facts About Ecological Engineering

- Earthworms are among the most celebrated of all ecosystem engineers. They are noted for their amazing capacity to consume litter over large environments of forest and savannah, often consuming an entire seasonal leaf fall. Leaving behind tons of castings, earthworms have profound impacts on local soils and the microorganisms that inhabit them.

- Hydroponically grown sunflowers are used to absorb radioactive metals. Their hyper-accumulation capacities were an ideal solution for soil bioremediation at the Chernobyl nuclear site in Ukraine.

- Salt-tolerant plants, known as halophytes, can be used to restore damaged soils. Once toxic salt levels have been reduced, the plants die and other species of plants and grasses can be planted or restored to previously unusable lands. These technologies are particularly valuable in areas subject to urine accumulation.

- Macrophytes are large aquatic plants found in standing waters common in wetland habitats. Examples include aquatic angiosperms (flowering plants), pteridophytes (ferns), and bryophytes (mosses and liverworts). They are important sources of nutrient uptake. Harvesting the plants and removing decaying organic debris reduces the re-pollution of wetland resources.

HOW IT WORKS

Ecological engineering is know-how applied to the design of a variety of environments. The mechanics of ecological self-maintenance are the foundations of healthy ecosystems design. Practitioners adopt a wide variety of natural techniques and materials to create, manage, or remediate a range of ecosystems.

Technologies rely on the chemical processes of sunlight, soil, temperature, and water in the design of living landscapes that assist in ecosystem management. For instance, constructed wetlands are designed so that natural biofilms and open spaces are formed to break down pollutants into usable bioproducts. Plant evaporation and transpiration, composting, and biogas production are other natural processes with a wide range of applications worldwide. The maintenance of designed ecosystems is similar to that of horticulture. This includes the routine maintenance of beds and screens and sampling and documentation of fluids and solids. The treatment of wastewater effluents in natural systems includes the removal of large solids, the aerobic and anaerobic treatment of organic materials, and regular sampling of safely reusable organic by-products.

APPLICATIONS AND PRODUCTS

The ecological engineering literature embraces ecological technologies compatible with cultures and practices around the globe. Biological processes are the basis for these technologies. The complexity of biological relationships is the essential reality that all ecological engineers strive to replicate and support.

Phytoengineering. Phytoengineering technologies rely on plants as primary ecosystem providers. Designs and processes are coordinated for wastewater treatment, environmental remediation, wetlands remediation, and sustainable processes for industry, agriculture, and urban communities. Traditional biotechnologies favored the removal or destruction of contaminated soils. Phytoremediation is a successful in situ alternative. Plants work with contaminated soils to decompose toxins, accumulate chemical wastes, and create nutrient-rich composting materials.

Natural Wastewater Treatment Systems. Wastewater and storm-water effluents can be treated and reused in natural systems of plants, algae, and other living organisms. Sequenced constructed ecosystems mimic natural processes of land and water habitats. Applications include greenhouses, rain gardens, aquatic systems, and wetlands remediation.

Wetlands Remediation. For some localities, wetlands are an essential component of wastewater treatment. In the Northern European countries of Denmark, Norway, Finland, and Sweden, full-scale filter-bed systems are used to recapture phosphorus from septic effluents. In the European Union, groundwater is treated as a living ecosystem. Microorganisms and subterranean fauna are valuable indicators of the health of a particular source. The mapping of aquifers and other groundwater structures is an essential prerequisite for the accurate assessment and classification of these habitats. Similar assessments of groundwater resources and dependent wetlands, legislated under the Water Framework Directive, set in motion a systematic testing of all groundwater-dependent water resources in Europe. In April, 2009, the International Commission on Groundwater of the International Association of Hydrological Sciences and the UNESCO Division of Water Sciences convened a special session to evaluate the results of this testing program. Sweden, Denmark, Norway, Finland, the Netherlands, England, Wales, Scotland, and Austria participated in the consortium to assess risks of damage to related wetlands. Groundwater salinity is of particular concern to residents of Ravenna, Italy, where subsidence has introduced saltwater into groundwaters that feed pine forests. The forests are dying as a consequence of these shifts in water quality.

Green Roof Technologies. Roofs constructed of living plants and grasses are recognized for their ability to absorb sunlight and serve as effective solar wastewater evaporation systems. Sod roofs were ubiquitous features of Northern European landscapes, where birch bark and turf grasses were plentiful. Many plant, flower, bird, and insect species thrived on established terraces. As of 2011, some species are found only in green roof habitats. Nearly 10 percent of all German homes have green roofs, and 50 percent of all new construction in Germany must include vegetative covering. Green roofs are an important component of that compliance. In Switzerland flat green roof vegetation is required for renovated structures. Similar innovations are occurring in the United States.

Ecological Sanitation. Composting toilets provide a safe and healthy option for sanitation systems worldwide. It relies on "dry" biochemical processes that do not require water. Aerobic decomposition of human feces is a viable technology particularly in rural and desert communities where water is unavailable for flushing toilets.

Agroecology. Self-sustaining agricultural systems engineered to accommodate climate change are of great interest to nations throughout the world. Agroecology addresses these systemic concerns with the intent of balancing crops and animal stocks to maximize their relationships to their environment. This includes impacts on soil quality, effective pest control, discharges into water supplies, and the release of noxious gases and particulate matter into the atmosphere. For example, in southern Mexico, researchers are working with the Lacandon Maya to recover indigenous agroforestry systems of soil fertility and rainforest conservation. These centuries-old practices included the collection of plant species and management of the succession processes of the forest. In Europe and the Americas, similar efforts are under way to restore vast tracts of grasslands and their species and populations.

Light Detection and Ranging (Lidar) Systems. Laser-generated light beams are unique in that they are emitted in a single direction with a constant wavelength and amplitude and total phase correspondence. These characteristics allow for very detailed mappings of surfaces, topical variations, the height of surface objects, and the physical characteristics of different terrains. Low-level and high-level airborne laser scanning provide data of varying resolution. Digital terrain models (DTMs) provide valuable information about forest roads and vegetation, changes in beach topography, channel flows of river floodplains, as well as monitor flood events.

Wireless Sensor Networks (WSNs). Computation and the use of wireless sensors have greatly enhanced scientific knowledge about soil microenvironments. Light, humidity, and temperature in a particular locale can be measured using high-resolution grids of autonomous sensors organized to provide minute-by-minute descriptions of the soil's condition.

Atlantic Rainforest Sensor Net Research. In 2009, Microsoft Research collaborated with researchers from The Johns Hopkins University, the Universidade de São Paolo, the Fundação de Amparo à Pesquisa do Estatdo de São Paulo (FAPESP), and the Instituto Nacional de Pesquisas Espaciais (INPE, the Brazilian National Institute for Space Research) to collect data about the microecology of the Serra do Mar rainforest located in the southeast corner of Brazil. To collect the data, a grid of hundreds of wireless sensors collected 18 million data points over a four-week period. It is hoped that similar data sets will help researchers better understand the relationship of the rainforest to its environment.

Ecosystem Bioinformatics (Ecoinformatics). Undergraduate courses in ecological or ecosystem bioinformatics teach the analysis of gene sequencing and expression within microbial communities. Increasing knowledge in molecular biology and the capacities of computer data systems make it possible to construct algorithms to calculate the mechanics of gene expression within biological systems.

IMPACT ON INDUSTRY

There is an urgent need for of eco-friendly partnerships between private industry and public welfare institutions. Effective communications systems and critical-thinking skills are needed for the broad dissemination of ecological principles across traditional disciplines. The formation of public values that take into consideration the effect of human activity on other life systems should be part of the essential core curriculum at all levels of education. Such partnerships go beyond the local community to include international consortiums concerned with the full range of effects of commodities exchange, the genetic transfer of plants and animals, effective sanitation and waste-disposal practices, and the disciplined use and distribution of chemical substances.

Corporate leadership is an important driver for sustainable global development. Some leading chemical producers are developing novel production processes that reduce hazardous materials and effluents. These include innovative manufacturing systems, solventless processes, and new separation processes. The investment in "green chemistry" technologies is predicted to reap healthy savings, reducing the costs of hazardous-materials containment and on-site contaminant remediation.

CAREERS AND COURSE WORK

In the fall semester of 2007, Oregon State University offered the nation's first accredited ecological engineering degree program. Since then, other universities have begun offering similar accredited programs of study. These include Ohio State University; Purdue University; University of California, Berkeley; University of Maryland; Texas A&M University; and Washington State University.

The bachelor's degree in ecological engineering is remarkable for its breadth and multidisciplinary program of study. Typical requirements include core engineering courses in fluid mechanics, thermodynamics, circuits, vector calculus, differential equations, and mechanics. It will also include course work in ecology, organic chemistry, biology, microbiology, hydrology, biosystems modeling, statistical methods, and environmental technology and design. Graduate and doctoral programs in ecological engineering focus on issues, systems, and technologies related to ecological design. These include water-resource management and pollution control, biotechnology, environmental sensing, systems modeling, informatics, and soil and wetland treatment programs. Ecological engineering graduates pursue careers as environmental engineering project managers and risk advisers, researchers, statisticians, and land use and public policy consultants. Specific design projects include riparian restoration, ecological monitoring with sensor arrays, and the development of sustainable urban, industrial, and agricultural systems. Professionals play important roles in both the private and public sector. As environmental consultants they work in multidisciplinary teams to design and engineer systems for the remediation of wastewaters and the recycling of soil nutrients and contaminants. They work in federal agencies to plan systems for natural resource protection and utilization. Bioremediation, water conservation, and the protection of endangered species are important dimensions of healthy ecosystem design.

SOCIAL CONTEXT AND FUTURE PROSPECTS

There is a universal recognition that environmental protection is not enough to address the continuing degradation of the natural environment. The pace of chemical production and distribution continues to accelerate to meet global demand for industrial and agricultural products. Natural resources continue to be depleted to meet the needs of expanding communities worldwide. Ecological engineering represents a paradigm shift that favors ecological principles as the foundation of global economic and cultural prosperity. Sustainable design is an essential feature of that paradigm and it is the basis of the practice of ecological engineering. Traditional engineering practices and technologies are adapted for use in a variety of self-sustaining eco-

system designs. Many systems accommodate human lifestyle patterns while protecting biodiversity. Other systems are designed to restore or remediate particular biological systems such as a forest, a river, a farmland or pasture, or an industrial site. Sustainable cities can be found in locations around the globe.

Victoria M. Breting-García, B.A., M.A.

FURTHER READING

Allen, T. F. H., M. Giampietro, and A. M. Little. "Distinguishing Ecological Engineering from Environmental Engineering." *Ecological Engineering* 20, no. 5 (2003): 389-407. Helps refine the definition of ecological engineering as a field the practices of which are distinct from those of the environmental engineer.

Cuddington, Kim, et al., eds. *Ecosystem Engineers: Plants to Protists.* Burlington, Mass.: Academic Press, 2007. Presents studies of nonhuman ecosystem engineers; provides a good background for understanding how ecosystems change over time.

Daily, Gretchen C., and Paul R. Ehrlich. "Managing Earth's Ecosystems: An Interdisciplinary Challenge." *Ecosystems* 2, no. 4 (July/August, 1999): 277-280. Provides a substantive framework for understanding the emerging links in science and technology that support ecosystem engineering efforts.

Kingsland, Sharon E. *The Evolution of American Ecology, 1890-2000.* Baltimore: The Johns Hopkins University Press, 2005. A good chronology for understanding the progression of study and observation that strengthened the field of ecology in the twentieth century.

Palmer, Margaret A., et al. "Ecological Science and Sustainability for the Twenty-first Century." *Frontiers in Ecology and the Environment* 3, no. 1 (February, 2005): 4-11. Presents a good conceptual understanding of the core assumptions that motivate the professional concerns of ecological engineers.

Spellman, Frank R., and Nancy E. Whiting. *Environmental Science and Technology: Concepts and Applications.* 2d ed. Lanham, Md.: Government Institutes, 2006. An excellent introduction to the definitions, concepts, and practices of environmental science and technology. These are the foundations for understanding the emergence of ecological engineering as a distinct profession located at the juncture of the biological and applied sciences.

Taylor, Walter P. "What Is Ecology and What Good Is It?" *Ecology* 17, no. 3 (July, 1936): 333-346. Provides a valuable understanding of the history of ecological thought in the twentieth century.

Wargo, John. *Green Intelligence: Creating Environments That Protect Human Health.* New Haven, Conn.: Yale University Press, 2009. A biting update on research in environmental protection, emphasizing the hazards of synthetic chemicals.

Worster, Donald. *Nature's Economy: A History of Ecological Ideas.* 2d ed. New York: Cambridge University Press, 1994. This collection of classic essays in the field of ecology provides a strong foundation for understanding contemporary ecosystem technologies as a part of a flow of ideas and convictions that have flowered since the eighteenth century. The seminal contributions of Carolus Linnaeus, Gilbert White, Charles Darwin, Henry David Thoreau, Rachel Carson, Aldo Leopold, Eugene Odum, and others are presented in the light of twentieth-century research in environmental history.

WEB SITES

Ecological Engineering Group
Ecological Engineers and Designers
http://www.ecological-engineering.com

Ecological Society of America
http://www.esa.org

Michigan State University
Green Roof Research Program
http://www.hrt.msu.edu/greenroof

Oregon State University
Biological and Ecological Engineering (BEE)
http://bee.oregonstate.edu

See also: Agricultural Science; Calculus; Forestry; Hydrology and Hydrogeology.

ECONOMIC GEOLOGY

FIELDS OF STUDY

Geology; chemistry; physics; structural geology; mineralogy; petrology; geochemistry; geochronology; geologic occurrence; mining engineering; geophysics; stratigraphy.

SUMMARY

Economic geology is the study of the origin and distribution of mineral deposits of metals, other useful materials (such as building stone and salt), and fossil fuels (petroleum, natural gas, and coal). The economic geologist explores for these materials, predicts the mineral reserves of known deposits, and assesses the economics for the extraction of the ore. For instance, for an iron ore deposit, the geologist needs to assess the kinds and amount of iron minerals present, the depth of the deposit, the location of the deposit, and the ownership of the land so that estimates for the possibility of mining the iron ore profitably may be made.

KEY TERMS AND CONCEPTS

- **Basalt:** Dark, fine-grained igneous rock composed mostly of plagioclase (a calcium-sodium silicate mineral) and pyroxene (a magnesium iron silicate mineral).
- **Clay Minerals:** Very small minerals that result from the weathering of other silicate minerals.
- **Granite:** Igneous rock composed of coarse crystals (greater than about 2 millimeters) of plagioclase (calcium-sodium silicate mineral), alkali feldspar (potassium-sodium silicate mineral), and quartz (all silicate with no other plus ions), and a few dark silicate minerals.
- **Hydrothermal:** Ore deposit in which the minerals were deposited by "water" vapor.
- **Magma:** Molten silicate rock below the surface with suspended minerals that have crystallized out of the molten material.
- **Mineral:** Naturally occurring element or compound that has a definite ordered arrangement of atoms.
- **Ores:** Minerals that may be economically and legally extracted from the earth.
- **Sandstone:** Sedimentary rock containing mostly quartz, feldspar, or rock fragments with grain sizes mostly in the range of 0.0625 to 2 millimeters.
- **Shale:** Very fine-grained sedimentary rock often consisting mostly of clay minerals and quartz.
- **Subduction Zone:** Boundary where an oceanic plate (oceanic crust and part of the upper mantle) slowly slips below another oceanic or continental plate, creating earthquakes and magma.

DEFINITION AND BASIC PRINCIPLES

Economic geologists explore for nonrenewable resources that formed so slowly, often over millions to even billions of years, that the processes of formation are much slower than the speed at which the resources can be extracted. For example, much of the oil and gas were formed by the slow, deep burial of marine, organic plankton that were alive about 66 million to 144 million years ago. These organisms need to be gradually buried to a certain depth long enough so that they gradually change to petroleum. If they are buried too deeply, the petroleum may break down to form natural gas (methane) or the mineral graphite (pure carbon); if they are not buried deep enough, then useful petroleum or natural gas will not form.

Nonrenewable resources include the abundant metals (such as iron and aluminum), scarce metals (gold and copper), materials used for energy (fossil fuels and uranium minerals), building materials (limestone, crushed stone, sand, and gravel), and other miscellaneous minerals (halite or natural salt). Running water, wind power, and solar power are not included among materials used for energy because they are renewable sources.

Economic geologists use their knowledge of geology to interpret where certain economic deposits might form. For instance, copper and some other associated elements may occur in what are called copper porphyry deposits. These deposits formed as hot "water" vapor carried dissolved copper out of granite magma and upward along rock fractures in which the copper precipitated to form a variety of copper minerals. The porphyry copper deposits occur only in association with subduction zones. Therefore, the economic geologist knows to search for them only where ancient or existing subduction zones occur, such as along the west coasts of North and South America.

BACKGROUND AND HISTORY

The first use of natural resources was to obtain water, salt, and other natural materials to make tools and weapons. Larger cities were located where there was a source of water such as a river. Salt was used as a flavoring and to preserve food. A major discovery around 9000 B.C.E. was that clay minerals could be heated to make pottery so that food and water could be stored and transported much more easily.

The first metals used were those such as gold and copper, which sometimes were found as native metals (not combined with other elements). These native elements could be shaped into useful materials by hammering or cutting. By 4000 to 3000 B.C.E., minerals of copper, zinc, lead, and tin were heated by burning charcoal in a very hot flame so that these elements could be separated from the ore. Much later, this process was applied to iron ore. The purification of iron had to wait until the development of blast furnaces, in which oxygen was blown through molten iron to purify it.

German physician Georgius Agricola wrote about the ways that ore minerals might form in *De re metallica* (1556; English translation, 1912). He divided the origins of ore deposits into those formed in streams and those formed in place. However, Latin was the language of scholars, not those actually working with ore deposits, so his writings were largely ignored.

Up to the nineteenth century, most ores were discovered accidentally because no one understood how ores and the rocks that contained them were geologically formed. Indeed, many people believed that mysterious celestial powers formed the ore minerals. During the period from the latter part of the eighteenth century to the twentieth century, the origin of rocks and ores gradually became understood so that people had a much better understanding of how to search for ores.

HOW IT WORKS

Selection of the Potential Ore Region. An economic geologist may first examine a geologic map, an aerial photograph, or a satellite photograph of a region, usually where other economic deposits have been found, to see if any geographical features provide clues as to where other deposits might be located.

Petroleum and natural gas, for instance, usually occur in sedimentary rocks where certain geologic structures may be seen on photographs or maps. There must be shales that contain organic matter that can potentially be converted to oil or gas if buried to the right depth below the surface. Then there must be an overlying sedimentary rock, such as a permeable sandstone, sandwiched between two impermeable shales, and the sandstone must have spaces between mineral grains through which the oil may move so that it can flow to where it can be trapped. A variety of traps can be used, but one kind uses an impermeable rock to stop the oil from migrating so that it can collect. Once geologists find these circumstances, they can drill below the ground along the trap to see if any petroleum is present. Narrowing the range of possibilities to find oil is important because drilling a single well may cost over $1 million.

The method for exploration differs depending on the type of deposit. For example, copper porphyry deposits often occur as veins in igneous rocks that formed along subduction zones; therefore, the exploration takes place along subduction zones. Some of the large iron ore deposits called banded iron formations found in the Lake Superior region formed only in shallow oceans from 1.8 billion to 2.6 billion years ago. Therefore, geologists search for new iron deposits only in other sedimentary rocks of similar ages.

Geophysical Methods. Remote methods of discovering an economic mineral deposit use instruments that can detect variations in magnetism, electrical conductivity, gravity, and radioactivity in rocks at or close to the surface of the Earth. For example, some iron-rich minerals such as magnetite (Fe_3O_4) and ilmenite ($FeTiO_3$) will produce magnetic attractions that suggest that a region may be hydrothermally mineralized. The variations in magnetism may be detected fairly quickly by flying a magnetic detector over a region in a regular pattern. A Geiger counter may be used to quickly assess if there are any radioactive elements such as uranium or thorium present at the Earth's surface. Also instruments that detect variations in electrical conductivity can be used to determine if certain sulfide minerals might be present.

Petroleum geologists may explode small charges along the surface of the Earth so that sound waves travel through the ground for some distance. The sound waves, which can be detected at a receiver station, travel at varying speeds through different kinds of rocks and geologic structures. A computer program can then be used to construct a cross-section of the characteristics of the rocks below the surface. Sedimentary rocks were deposited in horizontal layers, but in some areas, they may be folded into arches (anticlines), warped downward (synclines), or fractured along faults that displace the sedimentary rock layers. The top portion of an anticline or the side of some fault may provide a possible trap where oil and natural gas can accumulate. Geologists can drill a well into these traps to see if oil or gas is present.

Geochemical Methods. Ore minerals may be weathered out of ore deposits and move into streams, where they slowly move downstream. To search for such deposits, the geologist may collect a series of sediments, often at stream junctions, and look for ore minerals in the sediments or analyze the sediments to see if any metals are present in abnormally high concentrations. This technique was used to discover a number of metal ores, such as the gold and copper porphyry ores in Papua, New Guinea.

Once a suspected ore deposit has been found in a certain region, soil samples may be collected in a systematic grid over the area. The soil samples are analyzed for the elements most likely to be present in the potential ore deposit. Areas that have much higher concentrations of these elements may indicate the presence of ore directly below the surface. Geologists may drill directly into the areas with high concentrations of these elements to obtain subsurface samples to confirm the kinds of minerals and the elemental concentrations. This method, for example, found zinc deposits near Queensland, Australia.

The results of such a drilling program are used to evaluate whether mining the ore prospect might be worthwhile. The decision of whether to mine the prospect depends on many factors, such as the size and location of the deposit, the concentration and

Anticlinal and synclinal folds exposed in an open pit copper mine on the Zambian Copperbelt, southern Africa. (Fletcher & Baylis/Photo Researchers, Inc.)

types of elements present, the finances of the company, the price obtained for the ore, the ownership of the land, and whether open-pit or underground mining can be used. For example, a gold deposit in northern Canada may not be profitable because of the costs of transporting equipment and miners to the area and of transporting the ore to a production facility. If gold prices drastically increase, then a previously uneconomical mine might be able to show some profit. Also large companies sometimes abandon smaller deposits and try to sell them to smaller companies that might still profitably mine the ore.

APPLICATIONS AND PRODUCTS

The mineral resources discovered by economic geologists are used by industries to produce the abundant metals, the scarce metals, fossil fuels, and other natural materials. Economic geologists are not directly concerned with the use of these materials, but they must be aware of the demand for these materials so that they know the amount of money that a company may receive for them.

Abundant Metals and Their Uses. The abundant metals are those found in the highest concentrations within the Earth. Iron minerals are used to make steel, which is used to make automobiles, buildings, roads, bridges, major appliances, and construction materials such as nails. The major iron ore minerals, hematite and magnetite (both iron oxides), are converted to steel by heating them to a high temperature with coke and limestone to form molten steel, which contains up to 2 percent carbon to harden the iron.

Aluminum is another abundant metal that is obtained from the ore bauxite, a mix of the minerals boehmite (aluminum oxyhydroxide) and gibbsite (aluminum hydroxide). The production of aluminum is very expensive because the bauxite must be heated with the compound cryolite to make a molten solution in which the aluminum is concentrated by an electric current. Aluminum is used in products such as cans, foil, windows, vehicles, and household items and mixed with other metals in alloys.

Titanium occurs in the minerals rutile (titanium oxide) and ilmenite (iron titanium oxide). Like aluminum, titanium is expensive to produce from its ore. Most titanium used is in the form of titanium oxide, which produces an intense white color in paint and paper. Titanium is also alloyed with aluminum,

vanadium, and iron and used in ships, airplanes, and missiles. Magnesium oxide is often alloyed with aluminum to provide corrosion-resistant cans and materials used in vehicles. Magnesium metal is expensive to produce because like aluminum, electricity is used in its production.

Silicon is obtained by melting the mineral quartz (silicon oxide) with iron and coke in an electric furnace. Silicon is often alloyed with iron, aluminum, and copper because it improves the strength of these alloys and guards against corrosion. Silicon is also used in transistors in electronic devices.

Scarce Metals and Uses. Scarce metals occur in the Earth in much lower concentrations than the abundant metals. Some minerals, however, may locally concentrate these metals, making them potentially economic to mine. Some scarce metals may be added to steel to give it certain characteristics. For instance, chromium, molybdemun, tungsten, or vanadium may be added to make steel harder, especially at higher temperature. Stainless steel contains more than 11 percent chromium combined with nickel to help keep steel from rusting. Chromium and nickel may improve the strength of steel. Vanadium added to steel decreases the weight of steel, its strength, ductility, and ease of welding.

In addition, chromium is also used in chromium compounds to produce paint and ink pigments with green, yellow, and orange colors. Molybdemun is used in catalysts, pigments, and lubricants. Tungsten is often combined with carbon to produce a very hard compound, tungsten carbide, which is nearly as hard as diamond.

Copper, lead, zinc, tin, mercury, and cadmium are often grouped together simply because they are not used to alloy with iron. Copper metal conducts electricity well and can be shaped into wires, so it has been used mainly in electric lines and electric motors. Lead and mercury are used much less than in the past because of their toxicity. Lead is still used in batteries for vehicles, and mercury is used in some batteries, electrical switches, and for some chemical compounds. Zinc or tin are used in protective coatings for steel to keep it from rusting. Zinc oxide is added to paint to produce a white color and to a variety of lotions to prevent sunburn. Many so-called tin cans contain tin plated on other metals.

The precious metals are gold, silver, and the platinum metals. These metals are often used in jewelry because of their beauty. Gold is also used for money, and in industrial materials such as electronic connectors and dental fillings. Silver is also used in some electrical equipment. Platinum is used as a catalyst, for example, in catalytic converters in cars.

A variety of other metals, such as the rare earth elements, are used for industrial purposes. Rare earth elements have been used as catalysts and to color glass and ceramics and to provide some colors in television screens.

Chemical Minerals and Fertilizers. A number of nonmetallic minerals such as halite, baking soda, and sylvite are used in food. Halite and sylvite are salts used for flavoring. Halite is also used to soften water and to melt ice on roads. Borox is a boron compound that is used in some detergents and cosmetics.

A variety of potassium, nitrogen, and phosphorous minerals are used to fertilize crops. Commonly used nitrate compounds are sodium nitrate and potassium nitrate. Ammonium phosphate and apatite are examples of phosphate minerals.

Gypsum (calcium sulphate) is the main mineral in wallboard. Sulfur compounds derived from petroleum are used to make many industrial compounds such as sulphuric acid.

Building Materials. Building materials include building stones, crushed rocks, sand, gravel, cement, plaster, bricks, and glass. Common building stones are granite, limestone, and sandstone. Granite, for example, can be polished to form an attractive surface for building facings and countertops.

Huge quantities of crushed rocks, sand, and gravel are used for roads, building foundations, and concrete. In the United States, more than 100 billion tons of these materials are used yearly.

Gemstones. The most important gems are diamonds, sapphires, rubies, and emeralds. Gems must be harder than most other minerals, and they must be beautiful. Diamonds are composed of the element carbon, and rubies and sapphires are gem-quality varieties of the mineral corundum. Emeralds are gem-quality varieties of the mineral beryl.

Fossil Fuels. Petroleum, natural gas, and coal are the main fossil fuels. Much of the coal is burned to provide electrical power. Much of the natural gas is burned to provide heat in homes and industry. Petroleum products such as gasoline are used in providing power for automobiles and trucks.

IMPACT ON INDUSTRY

Most of the most easily found minerals and fossil fuels have already been discovered, so economic geologists must continue to devise new ways to search for deposits. The demand for most metals, fossil fuels, and industrial minerals is likely to increase in the future.

Fascinating Facts About Economic Geology

- The desire to find gold and silver motivated the Spanish and Portuguese to explore Central America and South America from the fifteenth to seventeenth centuries.
- In the 1400's, Native Americans dug pits to mine the crude oil at Oil Creek near Titusville, Pennsylvania, where Edwin L. Drake would later drill for oil.
- In spring, 1859, Edwin L. Drake, the general agent of Seneca Oil Company, began drilling for oil in Titusville, Pennsylvania. On August 27, he hit oil, and the U.S. oil industry was born.
- The East Texas Oil Museum commemorates the discovery of oil in the 1930's near Kilgore. On October 3, 1930, a well drilled by wildcatter Columbus Marion Joiner, then seventy years old, produced a gusher. Two other wells were drilled, on the Crim family farm and the J. K. Lathrop lease in Gregg County.
- The commercial mining of coal began in Kanawha County, West Virginia, in 1817. West Virginia would be the site of many mining disasters and union unrest that produced the Matewan Massacre (May 18, 1920) and the Battle of Blair Mountain (summer, 1921).
- The California Gold Rush (1848-1855) drew prospectors who panned for gold in the stream sediment of the American River and significantly raised the population of California.
- Gold Maps Online used a U.S. Bureau of Land Management database listing abandoned and active gold mining claims to create gold maps in Google Earth, which it sells to modern prospectors.
- Oil companies often lease surface and mineral rights from landowners, who are given cash bonuses and paid rental fees. When production begins, the landowner receives a percentage royalty of the gas or oil extracted.

Industry and Business. Private companies control the search for metals, fossil fuels, and industrial minerals. The worldwide supply and demand of many of these commodities controls the prices. Many metals are easy to transport, and many are in high demand, so the prices for them continue to increase. Large groups that control the supply may dictate the prices of some commodities such as diamonds and oil. The Organization of Petroleum Exporting Countries (OPEC) is a cartel that controls the amount of oil exported from it members and hence its price. The DeBeers Group controls much of the supply of diamonds and therefore their price.

Governments. Governments do not usually explore for minerals or fuels, but they may, for instance, periodically stockpile them, thus driving up the prices. For instance, the United States has a stockpile of petroleum, to which it periodically adds. Sometimes it sells part of the petroleum to counteract rapid increases in petroleum prices if OPEC cuts back on the supply.

Governments will usually tax the income from private industry. They may also pass laws to increase the safety in mines and production facilities for a given commodity such as coal.

Universities. Geology and engineering departments in universities provide instruction in economic geology. Academic researchers in geology typically function on how to better explore for mineral resources, and those in engineering concentrate on how to mine minerals and improve on extraction methods.

CAREERS AND COURSE WORK

A person interested in exploring for metals, fossil fuels, and other industrial minerals should major in geology in college with a concentration in economic geology and exploration geology. Courses such as mineralogy, petrology, economic geology, and structural geology are required to obtain a bachelor's degree. A variety of mathematics, chemistry, and physics courses should also be taken. Further study for master's and doctoral degrees would enable the student to carry out research on a specific problem in economic geology. Those interested in solving geochemical aspects of exploration should take many supporting courses in chemistry; those interested in exploring using geophysical techniques should take supporting courses in mathematics and physics. Advanced degrees enable an individual to be given a lot of responsibility if employed by a company searching for natural resources.

Those interested in developing mining operations once a potential site has been discovered can major in mining engineering with course work on how to develop mines and supporting courses in geology, physics, and chemistry. Those interested in solving environmental problems associated with an economic site should take a variety of courses in geology, chemistry, water chemistry, and hydrology.

SOCIAL CONTEXT AND FUTURE PROSPECTS

The increased demand for most metals, oil, natural gas, coal, and other raw materials such as sand and gravel is likely to continue indefinitely. Construction materials such as sand, gravel, and limestone, which are used in large quantities, need to be sourced close to where they will be used because of the high cost of transporting them. Most metals are used in smaller quantities, and they can be shipped economically from sites around the world to where they will be used. In industrial societies such as the United States, the amount of money spent on industrial minerals is much greater than that spent on the metals.

The use of iron, manganese, aluminum, copper, zinc, gold, graphite, nickel, silver, sulphur, vanadium, and zinc have increased substantially from 1950 to the twenty-first century. The use of some metals, such as lead, increased from the 1950's to the 1980's but subsequently declined because of associated environmental problems.

Worldwide, the distribution of many metals and fossil fuels is very uneven, which means that many countries must import these resources. For example, the United States uses much more petroleum than it can produce so it must obtain it from countries such as Saudi Arabia and Venezuela, which are rich in petroleum. The United States must also import a great deal of aluminum, platinum, manganese, tantalum, and tungsten to meet its demand. The United States, however, has an excess of salt, lead, copper, and iron.

Robert L. Cullers, B.S., M.S., Ph.D.

FURTHER READING

Craig, James R., David J. Vaughan, and Brian Skinner. *Earth Resources and the Environment.* Upper Saddle River, N.J.: Prentice Hall, 2010. Gives a basic overview of natural resources with terms that are well defined.

Evans, Anthony M. *An Introduction to Economic Geology and Its Environmental Impact.* 1997. Reprint. London: Blackwell Science, 2005. Has sections on the basics of economic geology and the types and distribution of ores through time.

Guilbert, John M., and Charles Frederick Park. *The Geology of Ore Deposits.* 1986. Reprint. Long Grove, Ill.: Waveland Press, 2007. Based on *Ore Deposits* (3d ed., 1975), by Park and Roy A. MacDiarmid. Gives some history of the development of the theory of finding ore deposits as well as more modern theories for their formation.

Klein, Cornelis, and Barbara Dutrow. *Manual of Mineral Science.* Hoboken, N.J.: John Wiley & Sons, 2008. Gives the methods to identify minerals and the properties of minerals.

Robb, H. G. *Introduction to Ore Forming Processes.* Oxford: Blackwell Science, 2005. Gives a detailed summary of how the major ores are formed.

Singer, D. A., and W. D. Menzie. *Quantitative Mineral Resource Assessments: An Integrated Approach.* New York: Oxford University Press, 2010. Draws on quantitative analyses including deposit density models and frequency distributions to enable a better determination of the likelihood of a mineral deposit and its size.

Wellmer, Friedrich-Wilhelm, Manfred Dalheimer, and Markus Wagner. *Economic Evaluations in Exploration.* New York: Springer, 2008. Gives the reader a feeling for how ore deposits are evaluated in the earlier stages of development.

WEB SITES

American Association of Petroleum Geologists
http://www.aapg.org

Society of Economic Geologists
http://www.segweb.org

U.S. Geological Survey
http://www.usgs.gov

See also: Geoinformatics; Marine Mining; Mineralogy; Plane Surveying.

EGG PRODUCTION

FIELDS OF STUDY

Poultry/animal science; reproduction; food technology; biology; physiology; business management.

SUMMARY

The egg production field includes farm production of shell eggs for direct consumption and further processing of eggs for use in products of the food industry. Egg production includes the development of highly productive strains of laying hens, advances in technology in the production and processing of eggs, and business models that permit the efficient production and marketing of eggs.

KEY TERMS AND CONCEPTS

- **Candling:** Inspecting the internal quality and embryonic development of eggs by shining a bright light through them.
- **Chalaza:** Stringlike attachment that anchors the yolk to the center of an egg.
- **In-Line Production:** Using a single location for production and packaging of eggs.
- **Line:** Group of related chickens that have similar production characteristics.
- **Off-Line Production:** Using different locations for the production and processing of eggs.
- **Pullet:** Immature female chicken destined for egg production.
- **Salmonella:** Genus of bacteria that can contaminate eggs, causing serious illness to humans who consume the eggs.
- **Vertical Integration:** Ownership by a single firm of multiple companies in order to cover all stages of egg production, from the raw materials through distribution, including feed mills, hens, buildings, egg-processing facilities, and transportation vehicles.

DEFINITION AND BASIC PRINCIPLES

Egg production in the United States has undergone a remarkable transformation. Before the twentieth century, hens ran loose around the farmyard, largely fending for themselves. Around the late 1800's, farm flocks came into being, and egg production became a serious part of the farm enterprise. Hens were given their own housing and provided with feeders, waterers, roosts, and nests, as well as a fenced-in yard. The farm flock system allowed for applying important management principles, such as proper feeding, breeding, and egg collection. The next advance took place around the 1960's with the emergence of farms that specialized in egg production. The farmer-manager could then focus entirely on egg production and use the latest in management and feeding techniques and production stock. Later in the twentieth century, egg producers became vertically integrated, with all aspects of production and marketing under the control of the same firm. The farmer-producer became just one part of the entire system.

Egg production involves genetic research to develop strains of highly productive hens; proper management of growing pullets to maximize their potential as laying hens; the use of advanced technology in buildings, equipment, feeding, and lighting for maximal egg production at minimal cost; and the development of new egg products for the consumer. It can also involve support services such as feed mills and transportation. Modern intensive production practices involving millions of birds have come under criticism as factory farming and have raised questions of animal welfare that must be addressed by the producer.

BACKGROUND AND HISTORY

Chickens were probably domesticated from red junglefowl in Southeast Asia. Genetic studies suggest multiple sites of domestication, including China and India. Archaeological studies indicate that chickens were present in the Americas before the time of the Spanish conquistadores.

The modern egg industry is a result of a series of technological advances. In the 1870's, incubators began to be used commercially to hatch chickens, rapidly increasing the number of commercial hatcheries. Poultry breeders applied scientific principles to develop improved breeds and strains of chickens for egg production. Land-grant colleges engaged in research in poultry nutrition and feeding. This led to improved management practices and more efficient

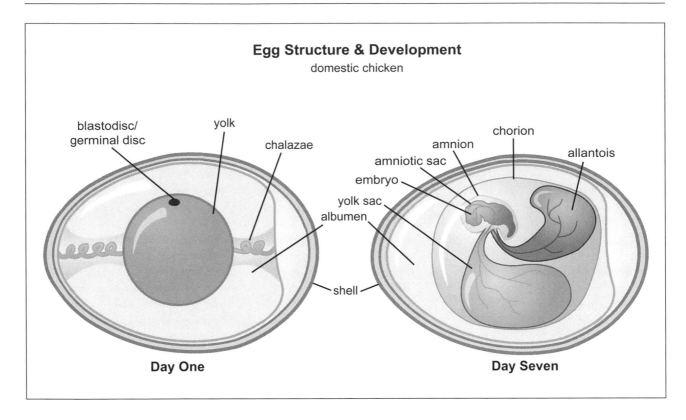

Egg Structure & Development
domestic chicken

blastodisc/germinal disc · yolk · chalazae

amnion · chorion · amniotic sac · allantois · embryo · yolk sac · albumen · shell

Day One   **Day Seven**

production of eggs. Better understanding and treatment of diseases, together with improved sanitation and ventilation, allowed for the creation of confinement systems.

Improved distribution systems and the development of new egg products led to greatly increased consumption of eggs, reaching a maximum of 402 eggs per capita in 1945. Health concerns about the cholesterol content of eggs and changes in lifestyle led to declines in consumption to a low of 230 eggs per capita in 1991. However, after the publication of scientific studies that stated that consuming eggs does not raise blood cholesterol, consumption of eggs began to increase, reaching 248 eggs per capita in 2008.

HOW IT WORKS

Egg production begins with the selection and development of breeding stock. Many breeds of chickens have developed over time, but for commercial purposes, the laying hen (layer) must be highly productive and efficient in converting feed into eggs. These criteria are met by the white leghorn breed, which is light in body weight, is highly active, and

produces a white egg. A very few breeding companies dominate the supply of egg production chicks, and they have their own specialized lines or strains of breeders. The white leghorn has been overwhelmingly adopted by the egg industry, but other breeds are used in markets that prefer a brown egg. Traditionally, this has involved using heavy breeds, such as the New Hampshire or Rhode Island red. The development of specialized lines and crossbreeds has resulted in brown-egg layers that are almost as efficient in feed conversion as the white leghorn. In many countries, including European nations, brown eggs are preferred over white eggs. The breeders must be kept in floor management systems to facilitate the breeders' mating.

From Egg to Layer. Fertilized eggs are transported to commercial incubators for incubation and hatching. After a few days of incubation, the eggs are candled to test for fertility and for viable embryo development. An infertile egg is clear, and a developing embryo shows blood-vessel development. Typically, the eggs are moved to a separate hatching incubator for the final three days of incubation. After hatching, the chicks are vaccinated and sexed, as only the female chicks are useful for egg production. Debeaking (removal of part of the beak) is performed at this time, or after the chicks are seven to ten days old.

The pullets are raised in confinement either on the floor or in cages; outside range rearing is seldom used by commercial breeders. A lighting program is essential for proper development of the pullets. One-day-old chicks receive twenty-three hours of light per day, and for the rest of the growing period, they receive a minimum of ten hours of light per day. They are transferred to laying houses at around sixteen weeks of age. Hens usually begin to lay eggs when they are five months old and continue to lay for about twelve more months.

Egg Production. Several types of management systems are commonly used by egg producers: cages, floor systems, or free-range systems. Cages are used for more than 98 percent of production operations for a variety of reasons. They allow increased population density in the poultry houses, and they are more labor efficient, as feeding, watering, egg collection, and manure removal can all be mechanized. Floor or noncage systems keep hens on litter floors inside buildings that hold feeders, waterers, roosts, and nests. This was the most common management system before the adoption of cage systems. Free-range systems allow hens access to an outdoor yard when weather permits.

The term "organic eggs" refers not so much to a management system but to the feed the hens receive. The feed must be totally vegetarian, the grains used must be pesticide-free, and the hens must not receive hormones or antibiotics.

Because most laying hens produce eggs in windowless houses, artificial lighting is provided. In fact, in all systems, lighting is essential to stimulate the pituitary gland to secrete hormones that help initiate and sustain egg production. Various lighting programs have been developed, but a typical program increases lighting from ten hours at twenty-four weeks of age to seventeen hours at thirty-two weeks and maintains this lighting period until the end of the laying cycle. The length of the lighting period should never be decreased during the laying cycle. The number of eggs produced per hen during a laying cycle can range from 180 to 200 eggs in tropical climates to 250 to 300 eggs in more temperate climates.

In cage systems, after the eggs are laid, they are transported via a conveyor belt to an egg-processing facility, where they are washed, graded for size, and either packed in flats to be shipped to a retail store or broken for further processing.

APPLICATIONS AND PRODUCTS

Breeding Stock. The Institut de Sélection Animale (ISA) holds a dominant position in the egg production industry as it supplies breeding stock for 50 percent of the world's egg production industry. The company began as Hendrix in the Netherlands, where it still has its headquarters. ISA expanded by purchasing many well-known and respected laying-hen breeding companies, including Babcock, J. J. Warren, Kimber, Shaver, Dekalb, Hisex, and Bovans. Many of these companies began as family-owned businesses in the early part of the twentieth century. Many strains of white and brown egg layers under the names of the original companies are sold as day-old chicks. The chicks destined as breeders must have a good egg-production capability, but good fertility is essential.

Laying Stock. Laying stock is also sold by ISA and other breeders as day-old chicks. ISA has strains of white and brown egg layers that are companions to its breeding stock. High egg production and excellent feed efficiency are essential characteristics for these strains.

Ducks for Egg Production. Ducks have never been popular for egg production in the United States and, like quail eggs, are only a niche market. However, ducks are commonly used in Asia for egg production. The Khaki Campbell breed is best known for egg production, and Metzer Farms sells a hybrid duck that produces eggs at a rate similar to the best chicken egg strains. Duck eggs are larger, have a more deeply pigmented yolk, and have firmer albumen than chicken eggs. Compared with chicken eggs, duck eggs have a higher cholesterol content, tend to pick up off-flavors more readily, and are more susceptible to contamination.

Shell Eggs. Eggs are most commonly marketed in the form in which they are laid, still in their shell. There is no difference in nutritional value between white and brown eggs, and although white eggs have a slightly thicker shell than brown eggs, brown eggshells have a stronger structure, so there is no difference in tendency to break. As the laying cycle nears its end, eggs tend to get bigger with thinner shells, leading to a greater tendency for breakage. When eggs are laid, they are coated with a protective layer

Fascinating Facts About Egg Production

- Eggs provide a unique source of balanced nutrients, including protein, essential fatty acids, vitamins, and minerals. The protein is of such high value that it is used as a standard to measure the quality of other food proteins.

- World consumption of eggs is increasing at about 8 percent per year because of higher living standards and the introduction of efficient production methods.

- Consolidation of egg farms has resulted in around three hundred producers supplying most of the nation's eggs. These producers are primarily located in the five top egg production states: Iowa, Ohio, Indiana, Pennsylvania, and California.

- Blood spots in egg yolk do not mean that the egg is fertilized. They are caused by a broken blood vessel on the surface of the yolk as the egg is forming.

- If a carton of eggs bears a U.S. Department of Agriculture grade, it must also have a Julian date, which is the date of packing. A sell-by date, if it appears, can be no more than thirty days after the date of packing, and a use-by date can be no more than forty-five days after packing.

- Pasteurized eggs have been exposed to heat to destroy bacteria. These are the best choice for recipes that call for partially cooked or raw eggs.

called a cuticle. This cuticle is often removed during washing. The shell contains many pores, which nature intended for gaseous exchange for the developing embryo, but which also provide an entry point for bacteria.

The yolk consists of 32 to 36 percent lipids and around 16 percent protein. The lipids include triglycerides (fats), phospholipids, and cholesterol. Triglycerides contain various types of fatty acids. The fatty acid content of yolk can vary according to the diet fed to the hens. A popular modern egg product contains a high content of omega-3 fatty acids, typically 350 milligrams compared with a normal content of 60 milligrams. The eggs also have a lower content of saturated fat, as well as a somewhat lower content of cholesterol. The hens are fed flaxseed to produce these eggs. These eggs have purported health benefits and command a higher price.

Eggs are graded by weight and quality. Egg-processing machinery separates eggs by weight, which can range from jumbo to peewee. Eggs can be grade AA, A, or B in quality. Quality in eggs is determined by candling or breaking them out and measuring albumin height. Grade AA eggs are freshly laid, have a thick, cloudy albumin, and a small air cell. Most eggs in supermarkets are grade A because some time has passed since their laying. Grade A eggs have a larger air cell, and the albumin is clear but thinner. The yolk is more defined in candling but free of defects. Both AA and A eggs can be sold as shell eggs, while grade B eggs are used for further processing. Grade B eggs have poorer quality albumin and minor discoloration or minor blood or meat spots.

Liquid Egg Products. Grade B eggs or other eggs not needed for the shell egg market go to an egg-breaking plant. After breaking, the liquid products obtained include whole egg, egg white, and egg yolk. These products are destined for the food industry and are unlikely to be found in retail stores.

Dried Egg Products. The incentive for developing the technology for drying eggs in the United States began in the 1930's with the availability of large quantities of eggs from China at a very low cost. The industry got a boost during World War II when the military needed dried eggs. Dried eggs have several advantages over shell eggs or liquid eggs: They can be stored at low cost, take less space to store, are not susceptible to spoilage caused by bacteria, are easier to handle in a sanitary manner, and have lower transportation costs.

Dried eggs are used extensively in many products, including bakery foods and mixes, mayonnaise and salad dressings, ice cream, pastas, and convenience foods. Most dried egg products are obtained by spray drying, but before drying, the sugars are removed from the eggs by fermentation or enzymatic treatments. These processes are necessary to avoid reactions of glucose with proteins or phospholipids in the eggs that can result in poor baking qualities or off-flavors. The dried egg products are derived from egg white, egg yolk, whole egg, or blends of whole egg or yolk with carbohydrates such as sucrose or syrups.

Specialty dried egg products include a scrambled-egg mix that has good storage capability and low-cholesterol egg products. Most low-cholesterol egg products contain egg white, with nonfat milk, vegetable oil, and pigments substituting for yolk. The final composition is similar to that of a whole egg.

IMPACT ON INDUSTRY

Worldwide egg production has increased considerably because of the establishment of intensive production systems in developing countries. A study in Southeast Asia has shown an annual growth rate in egg consumption of around 8 percent, with the greatest increase seen in the lower middle class. There is a direct association between increased per capita income and increased egg consumption. However, traditional extensive systems still account for 50 percent of production output and can provide a significant contribution to family nutrition.

Government Research. The U.S. Department of Agriculture (USDA) conducts poultry research at the Beltsville Agricultural Research Center in Maryland. Additional research is conducted at the Southeast Poultry Research Laboratory in Athens, Georgia. The USDA is responsible for determining the quality and grading standards for eggs, provides weekly summaries on national and international markets for eggs, and supplies educational and informational materials.

The most common research topics are feed ingredients, diseases, and management practices such as force molting, which reflect necessities in the egg production industry.

Although the Agricultural Research Service (ARS) does fewer poultry breeding studies than it once did, it developed a gentler line of laying hens that do not need debeaking to control cannibalism. Another interesting study related to breeding selection programs found that when breeders select for productivity traits, they should consider how these traits interact with behavioral traits. Application of such a breeding selection program resulted in dramatic improvements in productivity, livability, and welfare.

University Research and Extension. University research on egg production and egg products largely takes place in land-grant colleges and their associated research stations. Although some of the research is in basic science, much of it is applied research that can be used in a production setting. The results of this research are disseminated to producers or consumers through university extension programs, articles in the scientific and popular press, and electronic media.

Ongoing studies of cage versus alternative management systems indicate no inherent differences in mortality rates. Osteoporosis and other bone disorders may be greater in cage systems, although bone fractures may be greater in floor systems. Researchers have not arrived at definitive conclusions about animal welfare.

Industry Research. The development of specialized breeds and strains of laying hens has become the role of large poultry-breeding companies. The USDA and land-grant universities are less active in this field than previously and are inclined to focus on specialized studies. A few large feed companies still conduct nutrition research on laying hens, but quite often, they enter into contract with universities to conduct research. Poultry equipment manufacturers have played an important role in developing the modern, highly efficient egg industry. In some large facilities, all egg collection, grading, and packaging can be automated so that humans do not handle the eggs.

CAREERS AND COURSE WORK

Course work in poultry science is basic for students interested in pursuing a career in egg production. Because most poultry science departments have been absorbed into animal science departments at land-grant universities, the student should carefully search the curriculum, staff, and programs in these departments to determine if they can provide an adequate focus on egg production and technology.

Suitable undergraduate course work in the field of concentration could include poultry or laying hen production, poultry nutrition, poultry diseases, poultry anatomy and physiology, and reproduction (including breeding, embryonic development, and hatching). Students interested in the egg products industry should consider course work in food science. Supporting course work could include farm management, biology, and business courses such as economics and accounting.

Career opportunities for students with a bachelor's degree include poultry farm manager (as owner or employee), feed mill or hatchery manager, and salesperson for feeds, breeding or production stock, or equipment.

Students interested in a research career in poultry breeding or nutrition will need to complete graduate work leading to a doctorate degree. Advanced courses in animal breeding, statistics, endocrinology, genome analysis, genetics, animal breeding strategies, statistical methods, and biochemistry could be taken. Professional poultry scientists can obtain employment in academia or industry.

SOCIAL CONTEXT AND FUTURE PROSPECTS

The modern cage system of egg production is a marvel of efficiency and low cost. However, the nature of the system has been brought to the attention of animal welfare activists. The hens are kept in very crowded conditions (typically 67 square inches per hen) and are not able to perform their natural or instinctive behaviors, such as sleeping on roosts, laying eggs in nests, and taking a dust bath. Animal activists say that this is not humane. However, egg producers reply that hens kept in cage systems are healthier than those raised in other systems, noting that their productivity is higher. Animal science departments have been aware of these criticisms and have developed a new field of farm animal welfare. Animal welfare can be studied scientifically in a manner that is objective, reliable, and reproducible. However, the demand for answers to animal welfare issues may be outpacing the results of scientific studies. This has resulted in legislation banning the use of cages for egg production in Europe and the passing of Proposition 2 in California. The California legislation will probably phase out cage use in the state, which producers say will increase production costs 40 to 70 percent and drive egg producers out of the state because they will no longer be competitive.

Egg consumption fell because eggs have a high level of cholesterol, but consistent research has shown that egg consumption will not increase blood cholesterol in healthy people. Persons with heart disease may want to consult their physician as their bodies may handle cholesterol differently. The image of eggs suffered, and egg producers must convince the public of the egg's nutritive value if egg consumption is to reach or approach its 1945 peak.

A problem with eggs is possible salmonella contamination. If the shells are contaminated with salmonella, proper washing can eliminate this hazard, but if hens become infected with salmonella during the growing period, the eggs are internally contaminated. In August, 2010, more than 500 million eggs produced by Wright County Egg and Hillandale Farms of Iowa were recalled because of possible salmonella contamination. Programs are being developed to certify hens in large flocks as being salmonella-free.

David Olle, B.S., M.S.

FURTHER READING

Bell, Donald D., William Daniel Weaver, and Mack O. North. *Commercial Chicken Meat and Egg Production.* 5th ed. Norwell, Mass.: Kluwer Academic, 2002. An essential guide for those interested in the poultry industry. This edition emphasizes managerial aspects.

Clancy, Kate. *Greener Eggs and Ham: The Benefits of Pasture-Raised Swine, Poultry, and Egg Production.* Cambridge, Mass.: Union of Concerned Scientists, 2006. The Union of Concerned Scientists looks at egg production, poultry, and pigs and presents an alternative to the intensive production methods in predominant use.

National Agricultural Statistics Service. *U.S. Broiler and Egg Production Cycles.* Washington, D.C.: USDA National Agricultural Statistics Service, 2005. A governmental document providing information on egg production cycles and chickens for those in the poultry industry.

Stedelman, William, and Owen Cotterill. *Egg Science and Technology.* 4th ed. New York: Haworth Press, 1995. Long recognized as the most comprehensive handbook on the egg-processing industry.

WEB SITES

American Egg Board
http://www.aeb.org

American Poultry Association
http://www.amerpoultryassn.com

Institut de Sélection Animale
http://www.isapoultry.com

United Egg Producers
http://www.unitedegg.org

United Egg Producers Certified
http://www.uepcertified.com

U.S. Poultry and Egg Association
http://www.poultryegg.org

See also: Agricultural Science; Animal Breeding and Husbandry; Food Science.

ELECTRICAL ENGINEERING

FIELDS OF STUDY

Physics; quantum physics; thermodynamics; chemistry; calculus; multivariable calculus; linear algebra; differential equations; statistics; electricity; electronics; computer science; computer programming; computer engineering; digital signal processing; materials science; magnetism; integrated circuit design engineering; biology; mechanical engineering; robotics; optics.

SUMMARY

Electrical engineering is a broad field ranging from the most elemental electrical devices to high-level electronic systems design. An electrical engineer is expected to have fundamental understanding of electricity and electrical devices as well as be a versatile computer programmer. All of the electronic devices that permeate modern living originate with an electrical engineer. Items such as garage-door openers and smart phones are based on the application of electrical theory. Even the computer tools, fabrication facilities, and math to describe it all is the purview of the electrical engineer. Within the field there are many specializations. Some focus on high-power analogue devices, while others focus on integrated circuit design or computer systems.

KEY TERMS AND CONCEPTS

- **Alternating Current (AC):** Current that alternates its potential difference, changes its rate of flow, and switches direction periodically.
- **Analogue:** Representation of signals as a continuous set of numbers such as reals.
- **Binary:** Counting system where there are only two digits, 0 and 1, which are best suited for numerical representations in digital applications.
- **Capacitance:** Measure of potential electrical charge in a device.
- **Charge:** Electrical property carried by all atomic particles (protons, neutrons, and electrons).
- **Current:** Flow of electrical charge from one region to another.

- **Digital:** Representation of signals as a discrete number, such as an integer.
- **Digital Signal Processing (DSP):** Mathematics that describes the processing of digital signals.
- **Direct Current (DC):** Current that flows in one direction only and does not change its potential difference.
- **Inductance:** Measure of a device's ability to store magnetic flux.
- **Integrated Circuit (IC):** Microscopic device where many transistors have been etched into the surface and then connected with wire.
- **Resistance:** Measure of how easily current can flow through a material.
- **Transistor:** Three-terminal device where one terminal controls the rate of flow between the other two.
- **Voltage:** Measure of electrical potential energy between two regions.

DEFINITION AND BASIC PRINCIPLES

Electrical engineering is the application of multiple disciplines converging to create simple or complex electrical systems. An electrical system can be as simple as a lightbulb, power supply, or switch and as complicated as the Internet, including all its hardware and software subcomponents. The spectrum and scale of electrical engineering is extremely diverse. At the atomic scale, electrical engineers can be found studying the electrical properties of electrons through materials. For example, silicon is an extremely important semiconductive material found in all integrated circuit (IC) devices, and knowing how to manipulate it is extremely important to those who work in microelectronics.

While electrical engineers need a fundamental background in basic electricity, many (if not most) electrical engineers do not deal directly with wires and devices, at least on a daily basis. An important subdiscipline in electrical engineering includes IC design engineering: A team of engineers are tasked with using computer software to design IC circuit schematics. These schematics are then passed through a series of verification steps (also done by electrical engineers) before being assembled. Because computers are ubiquitous, and the reliance on good computer programs to perform complicated operations is so important, electrical engineers are adept computer programmers as well. The steps would be the same in any of the subdisciplines of the field.

BACKGROUND AND HISTORY

Electrical engineering has its roots in the pioneering work of early experimenters in electricity in the eighteenth and nineteenth centuries, who lent their names to much of the nomenclature, such as French physicist André-Marie Ampère and Italian physicist Alessandro Volta. The title electrical engineer began appearing in the late nineteenth century, although to become an electrical engineer did not entail any special education or training, just ambition. After American inventor Thomas Edison's direct current (DC) lost the standards war to Croatian-born inventor Nicola Tesla's alternating current (AC), it was only a matter of time before AC power became standard in every household.

Vacuum tubes were used in electrical devices such as radios in the early twentieth century. The first computers were built using warehouses full of vacuum tubes. They required multiple technicians and programmers to operate because when one tube burst, computation could not begin until it had been identified and replaced.

The transistor was invented in 1947 by John Bardeen, Walter Brattain, and William Shockley, employees of Bell Laboratories. By soldering together boards of transistors, electrical engineers created the first modern computers in the 1960's. By the 1970's, integrated circuits were shrinking the size of computers and the purely electrical focus of the field.

As of 2011, electrical engineers dominate IC design and systems engineering, which include mainframes, personal computers, and cloud computing. There is, of course, still a demand for high-energy electrical devices, such as airplanes, tanks, and power plants, but because electricity has so many diverse uses, the field will continue to diversify as well.

HOW IT WORKS

In a typical scenario, an electrical engineer, or a team of electrical engineers, will be tasked with designing an electrical device or system. It could be a computer, the component inside a computer, such as a central processing unit (CPU), a national power grid, an office intranet, a power supply for a jet, or an automobile ignition system. In each case, however, the electrical engineer's grasp on the fundamentals of the field are crucial.

Electricity. For any electrical application to work, it needs electricity. Once a device or system has been identified for assembly, the electrical engineer must know how it uses electricity. A computer will use low voltages for sensitive IC devices and higher ones for fans and disks. Inside the IC, electricity will be used as the edges of clock cycles that determine what its logical values are. A power grid will generate the electricity itself at a power plant, then transmit it at high voltage over a grid of transmission lines.

Electric Power. When it is determined how the device or application will use electricity, the source of that power must also be understood. Will it be a standard AC power outlet? Or a DC battery? To power a computer, the voltage must be stepped down to a lower voltage and converted to DC. To power a jet, the spinning turbines (which run on jet fuel) generate electricity, which can then be converted to DC and will then power the onboard electrical systems. In some cases, it's possible to design for what happens in the absence of power, such as the battery backup on an alarm clock or an office's backup generator. An interesting case is the hybrid motor of certain cars such as the Toyota Prius. It has both an electromechanical motor and an electric one. Switching the drivetrain seamlessly between the two is quite a feat of electrical and mechanical engineering.

Circuits. If the application under consideration has circuit components, then its circuitry must be designed and tested. To test the design, mock-ups are often built onto breadboards (plastic rows of contacts that allow wiring up a circuit to be done easily and quickly). An oscilloscope and voltmeter can be used to measure the signal and its strength at various nodes. Once the design is verified, if necessary the schematic can be sent to a fabricator and mass manufactured onto a circuit board.

Digital Logic. Often, an electrical engineer will not need to build the circuits themselves. Using computer design tools and tailored programming languages, an electrical engineer can create a system using logic blocks, then synthesize the design into a circuit. This is the method used for designing and fabricating application-specific integrated circuits (ASICs) and field-programmable gate arrays (FPGAs).

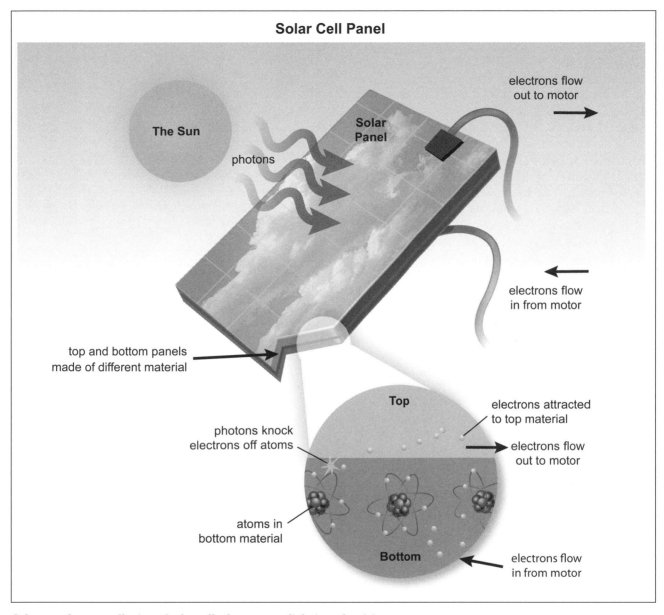

Solar Cell Panel

The Sun

photons

Solar Panel

electrons flow
out to motor

electrons flow
in from motor

top and bottom panels
made of different material

Top

electrons attracted
to top material

electrons flow
out to motor

photons knock
electrons off atoms

atoms in
bottom material

Bottom

electrons flow
in from motor

Solar panels are a collection of solar cells that convert light into electricity.

Digital Signal Processing (DSP). Since digital devices require digital signals, it is up to the electrical engineer to ensure that the correct signal is coming in and going out of the digital circuit block. If the incoming signal is analogue, it must be converted to digital via an analogue-to-digital converter, or if the circuit block can only process so much data at a time, the circuit block must be able to time slice the data into manageable chunks. A good example is an MP3 player: The data must be read from the disk while it is moving, converted to sound at a frequency humans can hear, played back at a normal rate, then converted to an analogue sound signal in the headphones. Each one of those steps involves DSP.

Computer Programming. Many of the steps above can be abstracted out to a computer programming language. For example, in a logical programming language such as Verilog, an electrical engineer can write lines of code that represent the logic. Another program can then convert it into the schematics of an IC block. A popular programming language called SPICE can simulate how a circuit will behave, saving the designer time by verifying the circuit works as expected before it is ever assembled.

APPLICATIONS AND PRODUCTS

The products of electrical engineering are an integral part of our everyday life. Everything from cell phones and computers to stereos and electric lighting encompass the purview of the field.

For example, a cell phone has at every layer the mark of electrical engineering. An electrical engineer designed the hardware that runs the device. That hardware must be able to interface with the established communication channels designated for use. Thus, a firm knowledge of DSP and radio waves went into its design. The base stations with which the cell phone communicates were designed by electrical engineers. The network that allows them to work in concert is the latest incarnation of a century of study in electromagnetism. The digital logic that allows multiple phone conversations to occur at the same time on the same frequency was crafted by electrical engineers. The whole mobile experience integrates seamlessly into the existing landline grid. Even the preexisting technology (low voltage wire to every home) is an electrical engineering accomplishment—not to mention the power cable that charges it from a standard AC outlet.

One finds the handiwork of electrical engineers in such mundane devices as thermostats to the ubiquitous Internet, where everything from the network cards to the keyboards, screens, and software are crafted by electrical engineers. Electrical engineers are historically involved with electromagnetic devices as well, such as the electrical starter of a car or the turbines of a hydroelectric plant. Many devices that aid artists, such as sound recording and electronic musical instruments, are also the inspiration of electrical engineers.

Below is a sampling of the myriad electrical devices that are designed by electrical engineers.

Computers. Computer hardware and often computer software are designed by electrical engineers. The CPU and other ICs of the computer are the

---

Fascinating Facts About Electrical Engineering

- The first transistor, invented in 1947 at the famous Bell Laboratories, fit in the palm of a hand.
- As of 2011, the smallest transistor is the size of a molecule, about a nanometer across, or 1/100,000 the thickness of a hair.
- In the 1890's Thomas Edison partnered with Harold Brown, a self-described electrical engineer, to shock dogs with either AC or DC current to see which was more fatal. Not surprisingly, Edison, a huge backer of DC, found AC to be deadly and DC not so.
- A Tesla coil, which is a high-voltage device, can shoot an arc of electricity in the air. The larger the coil, the bigger the arc. A small coil might spark only a few centimeters, but a large one can spark for a meter or more.
- The Great Northeast Blackout, which occurred on November 9, 1965, was the largest in U.S. history. It affected 30 million people, including 800,000 riders trapped in New York City's subways. It was caused by the failure of a single transmission line relay.
- In the lower 48 states, there are about 300,000 kilometers of interconnected transmission lines, operated by about 500 different companies.
- The term "debug" comes from the days of large vacuum-tube computers when bugs could quite literally get into the circuitry and cause shorts. To debug was to remove all the bugs.

---

product of hundreds of electrical engineers working together to create ever-faster and more miniature devices. Many products can rightfully be considered computers, though they are not often thought of as such. Smart phones, video-game consoles, and even the controllers in modern automobiles are computers, as they all employ a microprocessor. Additionally, the peripherals that are required to interface with a computer have to be designed to work with the computer as well, such as printers, copiers, scanners, and specialty industrial and medical equipment.

Test Equipment. Although these devices are seldom seen by the general public, they are essential to keeping all the other electrical devices in the world working. For example, an oscilloscope can help an electrical engineer test and debug a failing circuit because it can show how various nodes are behaving relative to each other over time. A carpenter might use a wall scanner to find electrical wire, pipes, and studs enclosed behind a wall. A multimeter, which measures voltage, resistance, and current, is handy not just for electrical engineers but also for electricians and hobbyists.

Sound Amplifiers. Car stereos, home theaters, and electric guitars all have one thing in common: They all contain an amplifier. In the past, these have been purely analogue devices, but since the late twentieth century, digital amplifiers have supplanted their analogue brethren due to their ease of operation and size. Audiophiles, however, claim that analogue amplifies sound better.

Power Supplies. These can come in many sizes, both physically and in terms of power. Most people encounter a power supply as a black box plugged into an AC outlet with a cord that powers electrical devices such as a laptop, radio, or television. Inside each is a specially designed power inverter that converts AC power to the required volts and amperes of DC power.

Batteries. Thomas Edison is credited with creating the first portable battery, a rechargeable box that required only water once a week. Batteries are an electrochemical reaction, that is the realm of chemistry, and demonstrate how far afield electrical engineering can seem to go while remaining firmly grounded in its fundamentals. Battery technology is entering a new renaissance as the charge life is extending and the size is shrinking. Edison first marketed his "A" battery for use in electric cars before they went out of fashion. Electric cars that run on batteries may be making a comeback, and their cousin, the hybrid, runs on both batteries and combustion.

The Power Grid. This is one of the oldest accomplishments of electrical engineering. A massive nationwide interdependent network of transmission lines delivers power to every corner of the country. The power is generated at hydroelectric plants, coal plants, nuclear plants, and wind and solar plants. The whole thing works such that if any one section fails, the others can pick up the slack. Wind and solar pose particular challenges to the field, as wind and sunshine do not flow at a constant rate, but the power grid must deliver the same current and voltage at all times of day.

Electric Trains and Buses. Many major cities have some kind of public transportation that involves either an electrified rail, or bus wires, or both. These subways, light rails, and trolleys are an important part of municipal infrastructure, built on many of the same principles as the power grid, except that it is localized.

Automobiles. There are many electronic parts in a car. The first to emerge historically is the electric starter, obviating the hand crank. Once there was a battery in the car to power the starter, engineers came up with all sorts of other uses for it: headlamps, windshield wipers, interior lighting, a radio (and later tape and CD players), and the dubious car alarm, to name a few. The most important electrical component of modern automobiles is the computer-controlled fuel injector. This allows for the right amount of oxygen and fuel to be present in the engine for maximum fuel efficiency (or for maximum horsepower). The recent success of hybrids, and the potentially emerging market of all electric vehicles, means that there is still more electrical innovation to be had inside a century-old technology.

Medical Devices. Though specifically the domain of biomedical engineering, many, if not most, medical devices are designed by electrical engineers who have entered this subdiscipline. Computed axial tomography (CAT) scanners, X rays, ultrasound, and magnetic resonance imaging (MRI) machines all rely on electromagnetic and nuclear physics applied in an electrical setting (and controlled by electronics). These devices can be used to look into things other than human bodies as well. Researchers demonstrated that an MRI could determine if a block of cheese had properly aged.

Telecommunications. This used to be an international grid of telephone wires and cables connecting as many corners of the globe where wire could be strung. As of 2011, even the most remote outposts can communicate voice, data, and video thanks to advances in radio technology. The major innovation in this field has been the ability for multiple connections to ride the same signal. The original cell phone technology picked a tiny frequency for each of its users, thus limiting the number of total users to a fixed division in that band. Mobile communication has multiple users on the same frequency, which opens up the band to more users.

Broadcast Television and Radio. These technologies are older but still relevant to the electrical engineer. Radio is as vibrant as ever, and ham radio is even experiencing a mini renaissance. While there may not be much room for innovation, electrical engineers must understand them to maintain them, as well as understand their derivative technologies.

Lighting. Light-emitting diodes (LEDs), are low-power alternatives to incandescent bulbs (the lightbulb that Thomas Edison invented). They are just transistors, but as they have grown smaller and more colors have been added to their spectrum, they have found their way into interior lighting, computer monitors, flashlights, indicator displays, and control panels.

IMPACT ON INDUSTRY

New electrical devices are being introduced every day in a quantity too numerous to document. In 2006, consumer electronics alone generated $169 trillion in revenue. Nonetheless, electrical engineering has a strong public research and development component. Increasingly, businesses are partnering with universities to better capitalize on independent research. As of 2011, most IC design is done in the United States, Japan, and Europe, and most of the manufacturing is outsourced to Taiwan and Singapore. As wind and solar power become more popular, so does the global need for electrical engineers. Spain, Portugal, and Germany lead the European Union in solar panel use and production. In the United States, sunny states such as California give financial incentives to homes and businesses that incorporate solar.

University Research. University research is funded by the United States government in the forms of grants from organizations such as the Defense Advanced Research Projects Agency (DARPA) and the National Aeronautics and Space Administration (NASA). The National Science Foundation (NSF) indirectly supports research through fellowships. The rest of the funding comes from industry. Research can be directed at any of the subdisciplines of the field: different material for transistors or new mathematics for better DSP or unique circuit configuration that optimizes a function. Often, the research is directed at combining disciplines, such as circuits that perform faster DSP. DARPA is interested in security and weapons, such as better digital encryption and spy satellites. Solar power is also a popular area of research. There is a race to increase the performance of photovoltaic devices so that solar power can compete with gas, coal, and petroleum in terms of price per kilowatt hour. Universities that are heavily dependent on industry funding tend to research in areas that are of concern to their donors.

For example, Intel, the largest manufacturer of microprocessors, is a sponsor of various University of California electrical engineering departments and in 2011 announced their new three-dimensional transistor. The technology is based on original research first described by the University of California, Berkeley, in 2000 and funded by DARPA.

The Internet was originated from a Cold War era DARPA project. The United States wanted a communications network that would survive a first-strike nuclear assault. When the Cold War ended, the technology found widespread use in the civilian sphere. The Internet enabled universities to share research data and libraries a decade before it became a household commodity.

Business Sector. More than half of the electrical engineers employed in the United States are working in electronics. Consumer electronics include companies such as Apple, Sony, and Toshiba that make DVD players, video-game consoles, MP3 players, laptops, and computers. But the majority of the engineering takes place at the constituent component level. Chip manufacturers such as Intel and Advanced Micro Devices (AMD) are household names, but there are countless other companies producing all the other kinds of microchips that find their way into refrigerators, cars, network storage devices, cameras, and home lighting. The FPGA market alone was $2.75 billion, though few end consumers will ever know that they are in everything from photocopiers to cell phone base stations.

In the non-chip sector, there are behemoths such as General Electric, which make everything from lightbulbs to household appliances to jet engines. There are about 500 electric power companies in the contiguous forty-eight states of the United States. Because they are all connected to each other and the grid is aging, smart engineering is required to bring new sources online such as solar and wind. In 2009, the Federal Communications Commission (FCC) issued the National Broadband Plan, the goal of which is to bring broadband Internet access to every United States citizen. Telecommunications companies such as AT&T and cable providers are competing fiercely to deliver ever-faster speeds at lower prices to fulfill this mandate.

CAREERS AND COURSE WORK

Electrical engineering requires a diverse breadth of background course work—math, physics, computer science, and electrical theory—and a desire to specialize while at the same time being flexible to work with other electrical engineers in their own areas of expertise. A bachelor of science degree in electrical engineering usually entails specialization after the general course work is completed. Specializations include circuit design, communications and networks, power systems, and computer science. A master's degree is generally not required for an electrical engineer to work in the industry, though it would be required to enter academia or to gain a deeper understanding in the specialization. An electrical engineer wishing to work as an electrical systems contractor will probably require professional engineer (PE) certification, which is issued by the state after one has several years of work experience and has passed the certification exam.

Careers in the field of electrical engineering are as diverse as its applications. Manufacturing uses electrical engineers to design and program industrial equipment. Telecommunications employs electrical engineers because of their understanding of DSP. More than half of all electrical engineers work in the microchip sector, which uses legions of electrical engineers to design, test, and fabricate ICs on a continually shrinking scale. Though these companies seem dissimilar—medical devices, smart phones, computers (any device that uses an IC)—they have their own staffs of electrical engineers that design, test, fabricate, and retest the devices.

Electrical engineers are being seen more and more in the role of computer scientist. The course work has been converging since the twentieth century. University electrical engineering and computer science departments may share lecturers between the two disciplines. Companies may use electrical engineers to solve a computer-programming problem in the hopes that the electrical engineer can debug both the hardware and software.

SOCIAL CONTEXT AND FUTURE PROSPECTS

Electrical engineering may be the most under-recognized driving force behind modern living. Everything from the electrical revolution to the rise of the personal computer to the Internet and social networking has been initiated by electrical engineers. This field first brought electricity into our homes and then ushered in the age of transistors. Much of the new technology being developed is consumed as software and requires computer programmers. But the power grid, hardware, and Internet that powers it were designed by electrical engineers and maintained by electrical engineers.

As the field continues to diversify and the uses for electricity expands, the need for electrical engineers will expand, as will the demands placed on the knowledge base required to enter the field. Electrical engineers have been working in the biological sciences, a field rarely explored by the electrical engineer. The neurons that comprise the human brain are an electrical system, and it makes sense for both fields to embrace the knowledge acquired in the other.

Other disciplines rely on electrical engineering as the foundation. Robotics, for example, merge mechanical and electrical engineering. As robots move out of manufacturing plants and into our offices and homes, engineers with a strong understanding of the underlying physics are essential. Another related field, biomedical engineering, combines medicine and electrical engineering to produce lifesaving devices such as pacemakers, defibrillators, and CAT scanners. As the population ages, the need for more advanced medical treatments and early detection devices becomes paramount. Green power initiatives will require electrical engineers with strong mechanical engineering and chemistry knowledge. If recent and past history are our guides, the next scientific revolution will likely come from electrical engineering.

Vincent Jorgensen, B.S.

FURTHER READING

Adhami, Reza, Peter M. Meenen III, and Dennis Hite. *Fundamental Concepts in Electrical and Computer Engineering with Practical Design Problems*. 2d ed. Boca Raton, Fla.: Universal Publishers, 2005. A well-illustrated guide to the kind of math required to analyze electrical circuits, followed by sections on circuits, digital logic, and DSP.

Davis, L. J. *Fleet Fire: Thomas Edison and the Pioneers of the Electric Revolution.* New York: Arcade, 2003. The stunning story of the pioneer electrical engineers, many self-taught, who ushered in the electric revolution.

Gibilisco, Stan. *Electricity Demystified.* New York: McGraw-Hill, 2005. A primer on electrical circuits and magnetism.

Mayergoyz, I. D., and W. Lawson. *Basic Electric Circuit Theory: A One-Semester Text.* San Diego: Academic Press, 1997. Introductory textbook to the fundamental concepts in electrical engineering. Includes examples and problems.

McNichol, Tom. *AC/DC: The Savage Tale of the First Standards War.* San Francisco: Jossy-Bass, 2006. The riveting story of the personalities in the AC/DC battle of the late nineteenth century, focusing on Thomas Edison and Nicola Tesla.

Shurkin, Joel N. *Broken Genius: The Rise and Fall of William Shockley, Creator of the Electronic Age.* New York: Macmillian, 2006. Biography of the Nobel Prize-winning electrical engineer and father of the Silicon Valley, who had the foresight to capitalize on invention of the transistor but ultimately went down in infamy and ruin.

WEB SITES
Association for Computing Machinery
http://www.acm.org

Computer History Museum
http://www.computerhistory.org

Institute of Electrical and Electronics Engineers
http://www.ieee.org

National Society of Professional Engineers
http://www.nspe.org/index.html

See also: Bionics and Biomedical Engineering; Computer Engineering; Computer Networks; Computer Science; Mechanical Engineering; Optics; Robotics.

ELECTRICAL MEASUREMENT

FIELDS OF STUDY

Electronics; electronics technology; instrumentation; industrial machine maintenance; electrical engineering; metrology; avionics; physics; electrochemistry; robotics; electric power transmission and distribution services.

SUMMARY

Electrical measurement has three primary aspects: the definition of units to describe the electrical properties being measured; the modeling, design, and construction of instrumentation by which those units may be applied in the measurement process; and the use of measurement data to analyze the functioning of electric circuits. The measurement of any electrical property depends on the flow of electric current through a circuit. A circuit can exist under any conditions that permit the movement of electric charge, normally as electrons, from one point to another. In a controlled or constructed circuit, electrons move only in specific paths, and their movement serves useful functions.

KEY TERMS AND CONCEPTS

- **Ampere:** Measure of the rate at which electrons are passing through an electric circuit or some component of that circuit.
- **Analogue:** Electric current that is continuous and continuously variable in nature
- **Capacitance:** Measure of the ability of a nonconducting discontinuous structure to store electric charge in an active electric circuit.
- **Digital:** Electric current that flows in packets or bits, with each bit being of a specific magnitude and duration.
- **Electromagnetism:** Property of generating a magnetic field by the flow of electricity through a conductor.
- **Hertz:** Unit of measurement defined as exactly one complete cycle of any process over a time span of exactly one second.
- **Inductance:** Measure of the work performed by an electric current in generating a magnetic field as it passes through an induction coil.

- **Left-Hand Rule:** Rule of thumb in determining the direction of the north pole of the magnetic field in a helically wound electromagnetic coil.
- **Phase Difference:** Measure of the extent to which a cyclic waveform such as an alternating voltage follows or precedes another.
- **Resistance:** Measure of the ability of electric current to flow through a circuit or some component part of a circuit or device; the inverse of conductance.

DEFINITION AND BASIC PRINCIPLES

Electrical measurement refers to the quantification of electrical properties. As with all forms of measurement, these procedures provide values relative to defined standards. The basic electrical measurements are voltage, resistance, current, capacitance, and waveform analysis. Other electrical quantities such as inductance and power are generally not measured directly but are determined from the mathematical relationships that exist among actual measured properties of an electric circuit.

An electric circuit exists whenever physical conditions permit the movement of electrons from one location to another. It is important to note that the formation of a viable electric circuit can be entirely accidental and unexpected. For example, a bolt of lightning follows a viable electric circuit in the same way that electricity powering the lights in one's home follows a viable electric circuit.

Electrons flow in an electric circuit because of differences in electrical potential between one end of the circuit and the other. The flow of electrons in the circuit is called the current. When the current flows continuously in just one direction, it is called direct current, or DC. In direct current flow, the potential difference between ends of the circuit remains the same, in that one end is always relatively positive and the other is relatively negative. A second type of electric current is called alternating current, or AC. In alternating current, the potential difference between ends of the circuit alternates signs, switching back and forth from negative to positive and from positive to negative. The electrons in alternating current do not flow from one end of the circuit to the other but instead oscillate back and forth between the ends at a specific frequency.

The movement of electrons through an electric circuit is subject to friction at the atomic level and to other effects that make it more or less difficult for the electrons to move about. These effects combine to restrict the flow of electrons in an overall effect called the resistance of the circuit. The current, or the rate at which electrons can flow through a circuit, is directly proportional to the potential difference (or applied voltage) and inversely proportional to the resistance. This basic relationship is the foundation of all electrical measurements and is known as Ohm's law.

Another basic and equally important principle is Kirchoff's current law, which states that the electric current entering any point in a circuit must always be equal to the current leaving that point in the circuit. In the light of the definition of electric current as the movement of electrons from point to point through a conductive pathway, this law seems concrete and obvious. It is interesting to note, however, that it was devised in 1845, well before the identification of electrons as discrete particles and the discovery of their role in electric current.

Electrical measurement, like all measurement, is a comparative process. The unit of potential difference—called the volt—defines the electrical force required to move a current of one ampere through a resistance of one ohm. Devices that measure voltage are calibrated against this standard definition. This definition similarly defines the ohm but not the ampere. The ampere is defined in terms of electron flow, such that a current of one ampere represents the movement of one coulomb of charge (equivalent to 6.24×10^{18} electrons) past a point in a period of one second.

The capacitance of a device in an electric circuit is defined as the amount of charge stored in the device relative to the voltage applied across the device. The inductance of a device in an electric circuit is more difficult to define but may be thought of as the amount of current stored in the device relative to the voltage applied across the device. In both cases, the current flow is restricted as an accumulation of charge within the device but through different methods. Whereas a capacitor restricts the current flow by presenting a physical barrier to the movement of electrons, an inductor restricts current flow by effectively trapping a certain amount of flowing current within the device.

BACKGROUND AND HISTORY

Wild electricity—lightning and other natural phenomena that result from differences in the oxidation potentials of different materials—has been observed and known for ages. Artificially produced electricity may have been known thousands of years ago, although this has not been proven conclusively. For example, artifacts recovered from some ancient Parthian tombs near Baghdad, Iraq, bear intriguing similarities in construction to those of more modern electrochemical cells or batteries. Reconstructions of the ancient device have produced electric current at about 0.87 volts, and other observations indicate that the devices may have been used to electroplate metal objects with gold or silver.

The modern battery, or voltaic pile, began with the work of Alessandro Volta in 1800. During the nineteenth century, a number of other scientists investigated electricity and electrical properties. Many of the internationally accepted units of electrical measurement were named in their honor.

HOW IT WORKS

Ohm's Law. The basis of electrical measurement is found in Ohm's law, derived by Georg Simon Ohm in the nineteenth century. According to Ohm's law, the current flowing in an electric circuit is directly proportional to the applied voltage and inversely proportional to the resistance of the circuit. In other words, the greater the voltage applied to the circuit, the more current will flow. Conversely, the greater

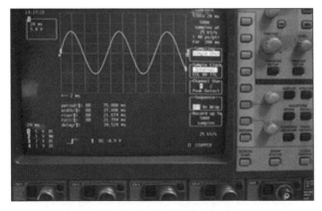

An oscilloscope is a very versatile electronic tool to measure electrical signals. (GIPhotoStock/Photo Researchers, Inc.)

the resistance of the circuit, the less current will flow. This relationship can be stated mathematically as $E = I \times R$ (where E = voltage, I = current, and R = resistance), in which voltage is represented as the product of current and resistance.

Given this relationship, it is a relatively simple matter to design a device that uses two specific properties to determine the third. By constructing a device that employs set values of voltage and resistance, one can measure current. Similarly, by constructing a device that describes a system in which current and resistance are constant, one can measure voltage, and by devising a system in which current and voltage are regulated, one can measure resistance.

If the three primary properties of a circuit are known, then all other properties can be determined by arithmetic calculations. The capacitance of a circuit or circuit component, for example, is the amount of charge stored in the device at a given applied voltage, and the amount of charge is proportional to the current in the device. Similarly, the inductance in a circuit or circuit component depends on the current passing through the device at a given voltage.

Units of Measurement. All electrical properties must have an associated defined standard unit to be measurable. To that end, current is measured in amperes, named after Louis Ampere. Potential difference, sometimes called electromotive force, is measured in volts, named after Volta. Resistance is measured in ohms, named after Ohm. Power is measured in watts, named after James Watt. Capacitance is measured in farads, named after Michael Faraday. Inductance is measured in henrys, named after Joseph Henry. Conductance, the reciprocal of resistance, is measured in siemens, named after Ernst W. von Siemens. Frequencies are measured in hertz, or cycles per second, named after Gerhard Hertz.

Basic Electricity Concepts. Electricity can be produced in a continuous stream known as direct current (DC), in which electrons flow continuously from a negative source to a positive sink. The potential difference between the source and the sink is the applied voltage of the circuit, and it does not change. Electricity can also be produced in a varying manner called alternating current (AC), in which electron flow oscillates back and forth within the circuit. The applied voltage in such a system varies periodically between positive and negative values that are equal

---

Fascinating Facts About Electrical Measurement

- Devices nearly 2,000 years old that were recovered from tombs near Baghdad, Iraq, were almost certainly batteries used to gold plate other metal objects. The technology of the battery was lost and not rediscovered until 1800.

- Henry Cavendish discovered what would be known as Ohm's law about fifty years earlier than Georg Simon Ohm did. However, as Cavendish did not publish his observations, the law has been attributed to Ohm.

- The most commonly and easily measured electrical quantity is voltage, or potential difference.

- Essentially all electrical measurements are determined by the relationship between voltage, resistance, and current (Ohm's law).

- Kirchoff's current law was stated before electrons were discovered and identified, and electricity was still regarded as some kind of mysterious fluid that certain materials contained.

- A bolt of lightning results when electrons flow through a regular electric circuit formed by the presence of charged particles (ions) in the atmosphere between clouds and the surface of the ground.

- An average lightning bolt transfers only about 5 coulombs of charge, but the transfer takes place so quickly that the electric current is about 30,000 amperes.

- Power, as measured in watts, is used to train elite cyclists. Over the course of an hour, professional cyclist Lance Armstrong can consistently generate 350 watts of electricity; the average in-shape cyclist, only about 100 watts.

- A static shock that a person can hear, feel, and see is about 250 volts of electricity.

- Conductance is the reciprocal of resistance; the unit of conductance was originally called the mho, as the reciprocal of the ohm.

---

in magnitude. It is important to understand that circuits designed to operate with one type of applied voltage do not function when the other type of voltage is applied. In other words, a circuit designed to perform certain functions when supplied with a constant voltage and direct current will not perform those functions when supplied with a varying voltage and alternating current. The fundamental concept of Ohm's law applies equally to both cases, but other characteristics such as phase and frequency differences and voltage waveform make the relationships more complex in alternating current applications. Electrical measurement devices are designed to accommodate these characteristics and are capable of extremely fine differentiation and precision.

APPLICATIONS AND PRODUCTS

The easiest electrical properties to measure accurately are voltage and resistance. Thus, the primary application tool of electrical measurement is the common volt-ohm meter (VOM), either as an analogue device or as its digital counterpart, the digital volt-ohm meter (DVOM).

Basic Analogue Measuring Devices. Two systems are required for any measuring device. One system is the structure by which the unknown value of a property is measured, and the other is the method of indicating that value to the user of the device. This latter feature was satisfied through the application of electromagnetic induction in the moving coil to produce a D'Arsonval movement. The strength of a magnetic field produced by current flowing through a coil is directly proportional to the magnitude of that current. In a basic analogue measuring device, the small coil is allowed to pivot freely within a permanent magnetic field. The amount by which the coil pivots is determined by the amount of current flowing through it; a needle attached to the coil indicates the appropriate value being measured on a calibrated scale. Analogue meters used to measure most electrical properties employ the moving-coil system to indicate the value of the property being measured.

Basic Digital Measuring Devices. The advent of digital electronics brought about a revolution in electrical measurement devices. Digital meters use a number of different systems to produce the desired information. The basic operation of digital devices is controlled by a central clock cycle in such a way that the value of the inputs are effectively measured thousands of separate times per second rather than continuously. This is known as sampling. Because the flow of electricity is a continuous or analogue process, sampling converts the analogue input value to a digital data stream. The data values can then be manipulated to be displayed directly as a numerical readout, eliminating the guesswork factor involved in reading a needle scale indicator. Another advantage of digital measurement devices is their inherent sensitivity. Through the use of an operational amplifier, or op-amp integrated circuits (IC), input signals can be amplified by factors of hundreds of thousands. This allows extremely small electrical values to be measured with great accuracy.

Other Measuring Devices. One of the most valuable devices in the arsenal of electrical measurement is the oscilloscope, which is available in both analogue and digital models. Like a typical meter, the oscilloscope measures electrical inputs, but it also has the capability to display the reading as a dynamic trace on a display screen. The screen is typically a cathode-ray tube (CRT) display, but later versions use liquid crystal display (LCD) screens and even can be used with desktop and laptop computers.

Another highly useful device for electrical measurement is the simple logic probe used with digital circuitry. In digital electronics, the application of voltages, and therefore the flow of current, is not continuous but appears in discrete bits as either an applied voltage or no applied voltage. These states are known as logic high and logic low, corresponding to on and off. The manipulation of these bits of data is governed by strict Boolean logic (system for logical operations on which digital electronics is based). Accordingly, when a digital device is operating properly, certain pins (or leads) of an integrated-circuit connection must be in the logic high state while others are in the logic low state. The logic probe is used to indicate which state any particular pin is in, generally by whether an indicator light in the device is on or off. Unlike a meter, the logic probe does not provide quantitative data or measurements, but it is no less invaluable as a diagnostic tool for troubleshooting digital circuitry.

Ancillary Devices. A more specific analytical device is the logic analyzer. Designed to read the input or pin signals of specific central processing unit (CPU) chips in operation, the logic analyzer can provide a running record of the actual programming used in the function of a digital electronic circuit. Another device that is used in electrical measurement, more typically as an electrical source than as a measuring device, is the waveform generator. This device is used to provide a specific shape of input voltage for an electric circuit to verify or test the function of the circuit.

Indirect Applications. Because of the high sensitivity that can be achieved in electrical measurement, particularly in the application of digital electronic devices, the measurement of certain electrical properties is widely used in analytical devices. The most significant electrical property employed in this way is resistance measurement. Often, this is measured as its converse property—conductivity. Gas-phase detectors on analytical devices such as gas chromatographs, high-performance liquid chromatographs, and combination devices with mass spectrometers are designed to measure the resistance, or conductivity, of the output stream from the device. For example, in gas chromatography, a carrier gas is passed through a heated column packed with a chromatography medium. When a mixture of compounds is injected into the gas stream, the various components become separated as they pass through the column. As the materials exit the column, they pass through a detector that measures changes in conductivity that occur with each different material. The changes are recorded either as a strip chart or as a collection of data.

The sensitive measurement of electrical resistance is made possible by the use of specific electric circuits, most notably the Wheatstone bridge circuit. In a Wheatstone bridge circuit, four resistance values are used, connected in a specific order. Two sets of resistance values are connected parallel to each other, with each set containing two resistance values connected in series with each other. One of the resistance values is the unknown quantity to be measured, while the other three are precisely known values. The voltage between the two midpoints of the series circuits of the bridge changes very precisely with any change in one of the resistance values. In addition, the output voltage signal can be amplified by several orders of magnitude, making even very small changes in output voltage meaningful. The role of digital electronics and op-amps cannot be overstated in this area.

IMPACT ON INDUSTRY

Electrical measurement has had a tremendous impact on the manner in which industry is carried out, particularly in regard to automation and robotic control, and in process control. The electrical measurement of process variables in manufacturing and processing can be applied to control circuitry that automatically adjusts and maintains the correct and proper functioning of the particular system. This both renders the process more precise and eliminates problems that are caused by human error, as well as reduces the number of personnel required for hands-on operational checking and process maintenance.

Electrical measurement is central to essentially all modern methods of testing and analysis and can be attributed to several factors. The technology is superbly adaptable to such functions and is fundamental to the operations of detection and electronic control. It is also extremely sensitive and precise, with the capability of measuring extremely small values of voltage, resistance, and current with precision. A thriving industry has subsequently developed to provide the various control mechanisms needed for the operation of modern methods of production and analysis. The procedures of electrical measurement also have a central role in medical analysis and other high-tech sciences such as physics and chemistry. In these fields, the devices used to carry out everything from the routine analysis of large groups of urine samples in a biomedical analytical laboratory to single-run experiments in the newest, largest subatomic particle colliders depend on electrical measurement for their functioning.

Nanoscience and Nanotechnology. Scientists continuously discover new materials and develop them for applications that capitalize on their specific electrical properties. They also modify existing measurement devices and design completely new ones. This often requires the redesign of electric circuitry, particularly the miniaturization of the devices. In the field of nanotechnology, for example, control circuitry for such small devices may consist of no more than a network of metallic tracings on a surface that measures correspondingly small electrical values.

Basic Electric Service. The mainstay of electrical measurement in industry is the service industry that maintains the operation of the continental electric grid and the power supplied to individual locations. Distribution of electric power calls for close monitoring of the network by which electric power is carried to individual residences, factories, and other locations. It also requires the installation and operation of new sources of electric power. This requires a skilled workforce that is capable of using basic electrical measurement practices. Accordingly, a significant component of training programs provided by colleges, universities, and vocational schools focuses on the practical aspects of the physical delivery of electric power, where measurement is a foundation skill. Simple economics demands that the amount of electricity being produced and the amount being consumed must be known. The effective distribution of that electricity, both generally and within specific installations, is vitally important to its continued functioning.

CAREERS AND COURSE WORK

Studies related to the use of electrical devices provide a good, basic knowledge of electrical measurement. A sound basis in physics and mathematics will also be required. College and university level course work will depend largely on the chosen area of specialization. Options at this level range from basic electrical service technician to fundamental physics. At a minimum, students will pursue studies in mathematics, physical sciences, industrial technologies (chemical, electrical, and mechanical), and business at the undergraduate level or as trade students. More advanced studies usually consist of specialized courses in a chosen field. The study of applied mathematics will be particularly appropriate in advanced studies, as this branch of mathematics provides the mathematical basis for phase relationships, quantum boundaries, and electron behavior that are the central electrical measurement features of advanced practices.

In addition, as technologies of electrical measurement and application and regulations governing the distribution of electric energy change, those working in the field can expect to be required to upgrade their working knowledge on an almost continual basis to keep abreast of changes.

SOCIAL CONTEXT AND FUTURE PROSPECTS

Economics drives the production of electricity for consumption, and in many ways, the developed world is very much an electrical world. A considerable amount of research and planning has been devoted to the concept of the smart grid, in which the grid itself would be capable of controlling the distribution of electricity according to demand. Effective electrical measurement is absolutely necessary for this concept to become a working reality.

At a very basic level, the service industry for the maintenance of electrically powered devices will continue to provide employment for many, particularly as the green movement tends to shift society toward recycling and recovering materials rather than merely disposing of them. Accordingly, repair and refurbishment of existing units and the maintenance of residential wiring systems would focus more heavily on effective troubleshooting methods to determine the nature of flaws and correct them before any untoward incident should occur.

The increased focus on alternative energy sources, particularly on the development of solar energy and fuel cells, will also place higher demands on the effectiveness of electrical measurement technology. This will be required to ensure that the maximum amount of usable electric energy is being produced and that the electric energy being produced is also being used effectively.

Richard M. J. Renneboog, M.Sc.

FURTHER READING

Clark, Latimer. *An Elementary Treatise on Electrical Measurement for the Use of Telegraph Inspectors and Operators.* London: E. and F. N. Spon, 1868. An interesting historical artifact that predates the identification of electrons and their role in electric current and measurement.

Herman, Stephen L. *Delmar's Standard Textbook of Electricity.* 5th ed. Clifton Park, N.Y.: Delmar Cengage Learning, 2009. A good basic presentation of the fundamental principles of electricity and electrical measurement.

Herman, Stephen L., and Orla E. Loper. *Direct Current Fundamentals.* 7th ed. Clifton Park, N.Y.: Thomson Delmar Learning, 2007. An introductory level textbook offering basic information regarding direct current electricity and electrical measurement.

Lenk, John D. *Handbook of Controls and Instrumentation.* Englewood Cliffs, N.J.: Prentice Hall, 1980. Describes general electrical measurement applications in principle and practice, as used for system and process control purposes.

Malvino, Albert Paul. *Malvino Electronic Principles.* 6th ed. New York: Glencoe/McGraw-Hill Books, 1999. Presents a detailed analysis of the electronic principles and semiconductor devices behind digital electronic applications in electrical measurement.

Strobel, Howard A., and William R. Heineman. *Chemical Instrumentation: A Systematic Approach.* 3d ed. New York: John Wiley & Sons, 1989. Describes the fundamentals of design and the operation of instrumentation used in chemistry and chemical applications, all of which ultimately use and depend on electrical measurement.

Tumanski, Slawomir. *Principles of Electrical Measurement.* New York: Taylor and Francis, 2006. Provides a thorough, descriptive introduction to the principles of electrical measurement and many of the devices that are used to manipulate and measure electric signals.

WEB SITES

U.S. Department of Energy
Electric Power
http://www.energy.gov/energysources/electricpower.htm

U.S. Energy Information Administration
Electricity
http://www.eia.doe.gov/fuelelectric.html

See also: Electrical Engineering; Electrochemistry; Electrometallurgy; Electronics and Electronic Engineering; Fossil Fuel Power Plants; Integrated-Circuit Design; International System of Units; Measurement and Units; Nanotechnology; Solar Energy; Wind Power Technologies.

ELECTRIC AUTOMOBILE TECHNOLOGY

FIELDS OF STUDY

Automotive engineering; chemical engineering; clean-energy technologies; electrical engineering; informatics; information technologies; materials engineering; mathematics; physics; resource management.

SUMMARY

Electric vehicles have been around even longer than internal combustion engine cars. With health issues resulting from the modern use of internal combustion engines, the automotive industry is intensifying its efforts to produce novel machines that run on electricity. Many cars come with drivetrains that can accept electric propulsion, offering quieter, healthier transportation options. Although consumers still seem to shy away from completely electric vehicles, hybrid vehicles that use both internal combustion engines and electric power are in use in many cities around the world.

KEY TERMS AND CONCEPTS

- **Aromatic Compound:** Carbon-based chemical such as benzene or toluene; sometimes harmful.
- **Battery:** Combination of one or more electrochemical cells used to convert stored chemical energy into an electric current.
- **Combustion:** Process by which a substance (fuel) reacts with oxygen to form carbon dioxide, water, heat, and light.
- **Current:** Flow of electricity through a material.
- **Drivetrain:** Mechanical system that transmits power or torque from one place (such as the motor) to another (such as the wheels).
- **Electrochemical Cell:** Device that can derive electrical energy from chemical reactions or facilitate chemical reactions by applying electricity.
- **Fuel:** Chemical compound that can provide energy by its conversion to water or carbon dioxide, either by combustion or by electrochemical conversion.
- **Fuel Cell:** Device that combines oxygen from the air with a fuel such as hydrogen to form water, heat, and electricity by splitting off protons from the fuel and allowing only those protons to pass through a dense membrane.
- **Hybrid Vehicle:** Car using a combination of several essentially different drive mechanisms, usually an internal combustion engine in combination with an electric motor, which in turn can derive its power from a battery or a fuel cell.
- **Lithium-Ion Battery:** Modern battery that involves lithium in storing and converting energy, has a high energy density, but degrades fast at elevated temperatures.
- **Membrane:** Dense dividing wall between two compartments.
- **Polarization:** Magnetization of a material with a defined polarity.
- **Proton Exchange Membrane Fuel Cell:** Polymer-based electrochemical device that converts various fuels and air into water, heat, and electric energy at low operating temperatures (70 to 140 degrees Celsius).
- **Solid Oxide Fuel Cell:** Ceramic electrochemical device that converts various fuels and air into water, heat, and electric energy at high operating temperatures (600 to 1,000 degrees Celsius).
- **Torque:** Tendency of a force to rotate an object around a defined axis or pivot; also known as moment force.
- **Voltage:** Electromotive force of electricity; also defined as the work required to move a charged object through an electric field.

DEFINITION AND BASIC PRINCIPLES

Electric vehicles are driven by an electric motor. The electricity for this motor can come from different sources. In vehicle technology, electrical power is usually provided by batteries or fuel cells. The main advantages of these devices are that they are silent, operate with a high efficiency, and do not have tailpipe emissions harmful to humans and the environment. In hybrid vehicles, two or more motors coexist in the vehicle. When large quantities of power are required rapidly, the power is provided by combusting fuels in the internal combustion engine; when driving is steady, or the car is idling at a traffic light, the car is entirely driven by the electric motor, thereby cutting emissions while providing the consumer with the normal range typically associated with traditional cars that would rely entirely on internal combustion engines. Electric vehicles make it possible for drivers to avoid having to recharge at a station. Recharging can occur at home, at work, and in parking structures, quietly, cleanly, and without involving potentially carcinogenic petroleum products.

BACKGROUND AND HISTORY

Electric vehicles have been around since the early 1890's. Early electric vehicles had many advantages over their combustion-engine competitors. They had no smell, emitted no vibration, and were quiet. They also did not need gear changes, a mechanical problem that made early combustion-engine cars cumbersome. In addition, the torque exhibited by an electric engine is superior to the torque generated by an equivalent internal combustion engine. However, battery technology was not yet sufficiently developed, and consequently, as a result of low charge, storage capacity in the batteries, and rapid developments in internal combustion engine vehicle technology, electric vehicles declined on the international markets in the early 1900's.

At the heart of electric vehicles is the electric motor, a relatively simple device that converts electric energy into motion by using magnets. This technology is typically credited to the English chemist and physicist Michael Faraday, who discovered electromagnetic induction in 1831. These motors require some electrical power, which is typically provided by

Power control unit of Toyota FCHV (fuel cell hybrid vehicle) car. The unit regulates fuel cell output and battery charging and discharging depending on the driving conditions. (GI-PhotoStock/Photo Researchers, Inc.)

batteries, as it was done in the early cars, or by fuel cells. Batteries were described as early as 1748 by one of the founding fathers of the United States, Benjamin Franklin. These devices can convert chemically stored energy into a flow of electrons by converting the chemicals present in the battery into different chemicals. Depending on the materials used, some of these reactions are reversible, which means that by applying electricity, the initial chemical can be recreated and the battery reused. The development of fuel cells is typically credited to the English lawyer and physicist Sir William Grove in 1839, who discovered that flowing hydrogen and air over the surfaces of platinum rods in sulfuric acid creates a current of electricity and leads to the formation of water. The devices necessary to develop an electric car had been around for many decades before they were first assembled into a vehicle.

A major circumstance that led to the commercial success of combustion-engine vehicles over electric vehicles was the discovery and mining of cheap and easily available oil. Marginal improvements in battery technology compared with internal combustion engine technology occurred during the twentieth century. As a result of stricter emissions standards near the end of the twentieth century, global battery research began to reemerge and has significantly accelerated in the 2010's. Some early results of this research are on the market.

During the 1990's, oil was still very cheap, and consumers, especially in North America, demanded heavier and larger cars with stronger motors. During this decade, General Motors (GM) developed an electric vehicle called the EV1, which gained significant, though brief, international positive attention before it was taken off the market shortly after its introduction. All produced new cars were destroyed, and the electric-vehicle program was shut down. The development of electric vehicles was then left to other companies.

Barely twenty years later, following a significant negative impact on the car manufacturing companies in North America from the 2009 financial crisis and their earlier abandonment of research and development of electric car technology, North American companies tried to catch up with the electric vehicle technology of the global vehicle manufacturing industry. During the hiatus, global competitors surpassed North American companies by creating modern, fast, useful electrical vehicles such as the Nissan Cube from Japan and the BMW ActiveE models from Germany. GM, at least, has made a full turnaround after receiving government financial incentives to develop battery-run vehicles and has actively pursued a new electric concept called Chevrolet Volt. Similar to hybrid cars, the Volt has a standard battery, but because early twenty-first century batteries are not yet meeting desired performance levels, the Volt also has a small engine to extend its range. Installing two different power sources in a vehicle, one electric and one combustion based, makes sense in order to develop a product that has lower emissions but the same range as combustion-engine-based vehicles.

HOW IT WORKS

Power Source. Gasoline, which is mainly a chemical called octane, is a geologic product of animals and plants that lived many millions of years ago. They stored the energy of the Sun either directly from photosynthesis or through digestion of plant matter. The solar energy that is chemically stored in gasoline is released during combustion.

The storage of energy in batteries occurs through different chemicals, depending on the type of battery. For example, typical car-starter lead-acid batteries use the metal lead and the ceramic lead oxide to store energy. During discharge, both these materials convert into yet another chemical called lead sulfate. When a charge is applied to the battery, the original lead and lead oxide are re-created from the lead sulfate. Over time, some of the lead and lead oxide are lost in the battery, as they separate from the main material. This can be seen as a black dust swimming in the sulfuric acid of a long-used battery and indicates that the battery can no longer be recharged to its full initial storage capacity. This happens to all types of batteries. Modern lithium-ion batteries used in anything from vehicles to mobile phones use lithium-cobalt/nickel/manganese oxide, and lithium graphite. These batteries use lithium ions to transport the charges around, while allowing the liberated electrons to be used in an electric motor. Other batteries use zinc to store energy—for example, in small button cells. Toxic materials such as mercury and cadmium have for some years been used in specific types of batteries, but have mostly been phased out because of the potential leaching of these materials into groundwater after the batteries' disposal.

Fuel cells do not use a solid material to store their charge. Instead, low-temperature proton exchange membrane fuel cells use gases such as hydrogen and liquid ethanol (the same form of alcohol found in vodka) or methanol as fuels. These materials are pumped over the surface of the fuel cells, and in the presence of noble-metal catalysts, the protons in these fuels are broken away from the fuel molecule and transported through the electrolyte membrane to form water and heat in the presence of air. The liberated electrons can, just as in the case of batteries, be used to drive an electric motor. Other types of fuel cells, such as molten carbonate fuel cells and solid oxide fuel cells, can use fuels such as carbon in the form of coal, soot, or old rubber tires and operate at 800 degrees Celsius with a very high efficiency.

Converting Electricity into Motion. Most electric motors use a rotatable magnet the polarity of which can be reversed inside a permanent magnet. Once electricity is available to an electric motor, electrons, traveling through an electric wire and coiled around a shaft that can be magnetized, generate an electrical field that polarizes the shaft. As a result, the shaft is aligned within the external permanent magnet, since reverse polarities in magnets are attracted to each other. If the polarity of the rotatable shaft is now reversed by changing the electron flow, the magnet reverses polarity and rotates 180 degrees. If the switching of the magnetic polarity is precisely timed, constant motion will be created. Changes in rotational speed can be achieved by changing the frequency of the change in polarization. The rotation generated by an electric motor can then be used like the rotation generated by an internal combustion engine by transferring it to the wheels of the vehicle.

Research and Development. Many components of electric vehicles can be improved by research and development. In the electric motor, special magnetic glasses can be used that magnetize rapidly with few losses to heat, and the magnet rotation can occur in a vacuum and by using low-friction bearings. Materials research of batteries has resulted in higher storage capacities, lower overall mass, faster recharge cycles, and low degradation over time. However, significant further improvements can still be expected from this type of research as the fundamental understanding of the processes occurring in batteries become better understood.

Novel fuel cells are being developed with the goal of making them cheaper by using non-precious-metal catalysts that degrade slowly with time and are reliable throughout the lifetime of the electric motor. The U.S. Department of Energy has set specific lifetime and performance targets to which all these devices have to adhere to be useful on the commercial market. In fact, electric car prototypes are already available and comparable to vehicles using internal combustion engines. As of 2011, the main factor preventing deep mass-market penetration is cost, but that is continuously addressed by research and development of novel batteries and fuel cells that are lighter, use less expensive precious-metal catalysts, last longer, and are more reliable than previous devices. Since the 2010's, the development of these devices has significantly accelerated, especially because of international funding that is being poured into clean-energy technologies.

There are, however, disadvantages to all energy-conversion technologies. Internal combustion engines require large amounts of metals, including iron, chromium, nickel, manganese, and other alloying elements. They also require very high temperatures in forges during production. Additionally, the petroleum-based fuels contain carcinogenic chemicals, and the exhausts are potentially dangerous to humans and the environment, even when catalytic converters are used; to function well, these devices require large amounts of expensive and rare noble metals such as palladium and platinum. The highest concentrations of oil deposits have been found in politically volatile regions, and oil developments in those regions have been shown to increase local poverty and to cause severe local environmental problems.

Batteries require large quantities of rare-earth elements such as lanthanum. Most of these elements are almost exclusively mined in China, which holds a monopoly on the pricing and availability of these elements. Some batteries use toxic materials such as lead, mercury, or cadmium, although the use of these elements is being phased out in Europe. Lithium-ion batteries can rapidly and explosively discharge when short-circuited and are also considered a health risk. Electricity is required to recharge batteries, and it is often produced off-site in reactors whose emissions and other waste can be detrimental to human health and the environment.

Fuel cells require catalysts that are mostly made from expensive noble metals. Severe price fluctuations make it difficult to identify a stable or predictable cost for these devices. The fuels used in fuel cells, mostly hydrogen and methanol or ethanol, have to be produced, stored, and distributed. As of 2011, the majority of the hydrogen used is derived via a water-gas shift reaction, where oxygen is stripped off the water molecules and binds with carbon molecules from methane gas, producing hydrogen with carbon dioxide as a by-product; the process requires large quantities of natural gas. Methanol or ethanol can be derived from plant matter, but if it is derived from plants originally intended as food, food prices may increase, and arable land once used for food production then produces fuels instead.

Nevertheless, while the advantages and disadvantages of cleaner energy technologies such as fuel cells and batteries must be weighed against their ecological and economic impacts, it is important to remember that they are significantly cleaner than current internal combustion engine technologies.

APPLICATIONS AND PRODUCTS

Batteries. Battery technology still needs to be developed to be lighter without reducing the available charge. This means that the energy density of the battery (both by mass and by volume) needs to increase in order to improve a vehicle's range. Furthermore, faster recharge cycles have to be developed that will not negatively impact the degradation of the batteries. Overnight recharge cycles are possible, and good for home use, but a quick recharge during a shopping trip should allow the car to regain a significant proportion of its original charge. Repeated recharge cycles at different charge levels as well as long-time operation with large temperature fluctuations should not detrimentally affect the microstructure of the batteries, so the power density of the batteries will remain intact. Furthermore, operation in very cold environments, in which the charge carriers inside the battery are less mobile, should be realized for a good market penetration. The introduction of the Chevrolet Volt into the North American market in March, 2011, resulted in a disappointment, as customers appeared unwilling to pay a premium for battery-operated cars. Stricter policies enforcing the conversion of a more significant proportion of cars into electric vehicles are necessary to change the market, especially in North America.

Personal Vehicles. GM's EV1 was an attempt to market electric vehicles in North America in the 1990's. It was fast and lightweight and had all the amenities required by consumers but was discontinued by the manufacturer because it was not commercially viable. The 2011 edition of GM's Chevrolet Volt was almost indistinguishable from other GM station wagons, but the cost for the battery-powered car proved too high for a market that was used to very cheap vehicles with internal combustion engines. Electric vehicle technology is arguably much more advanced in Asia. Asian vehicle manufacturers were up to ten years ahead of the rest of the world in producing hybrid-electric vehicles, and they are set up to be ahead in the manufacturing of completely electric vehicles as well. For example, battery-only vehicles such as the Toyota iQ and the Nissan Leaf can drive up to 100 miles on a single battery charge with performance similar to that of an internal combustion engine car. European manufacturers, such as Renault, have teamed up with Nissan so as to not be left behind in the electric vehicle business and offered their first electric vehicle lineup to the European markets in 2011. Volkswagen has produced several studies including a concept called SpaceUp, but Volkswagen CEO Martin Winterkorn said in 2011 that battery technology was not mature enough for vehicles, and that Volkswagen would not produce any mass-market devices before 2013. Car manufacturer Fisker developed two plug-in hybrid electric vehicles, Nina and Karma, and planned to sell 1,000 units in the United States in 2011. As of 2011, Tesla Motors had two electric vehicles ready for the North American market: Model S and Roadster. Think City also manufactures small electric vehicles. However, while demand outside North America is large, demand in the United States is low. With major international governmental tax incentives in place, all vehicle manufacturers are developing at least some studies of electric vehicles for auto shows. Whether these models will actually go on to be developed for commercial markets remains to be seen.

Utility Vehicles and Trucks. To develop a green image, some municipalities considered switching their fleets to electric vehicles based on fuel cells or batteries. Ford has created a model called Transit Connect, which is an electrified version of its Ford

---

Fascinating Facts About Electric Automobile Technology

- The average power capacity of rechargeable batteries has tripled since the 1970's.
- Fuel cells produce electricity with clean water as the only exhaust. During space missions, this water has been used for drinking.
- The clean and quiet Brammo Enertia electric motorcycle can go 100 miles at a velocity of 100 miles per hour before requiring a recharge.
- Hitting the accelerator in a car with an electric motor results in a much faster acceleration than in a comparable gasoline-run car, since the torque of the electric motor far exceeds that of the internal combustion engine.
- Modern electric vehicles from major carmakers are indistinguishable from their standard line of models, as this reduces potential customer barriers to purchasing electric or electric-hybrid vehicles and lowers manufacturing cost.
- Non-rechargeable batteries that are not properly disposed of may leach toxic chemicals into the ground over time.
- It takes significantly more energy to create a new car than it does to keep an old car in good shape and run it—especially if the new car uses an internal combustion engine.

Transporter. Navistar has developed an electric truck, the eStar, and expects to sell a maximum of 1,000 units each year until 2015 in the United States. Smith Electric Vehicles has developed several models, such as the Newton, for the expected demand in electric-utility vehicles. Additionally, there are many small companies producing small utility trucks, such as the Italian manufacturer Alkè. The products of these companies are small, practical multi-purpose vehicles for cities and municipalities.

Bicycles and Scooters. Small electric motor-assisted bicycles and electric scooters have been in use since the early twentieth century. Other small electric vehicles include wheelchairs, skateboards, golf carts, lawnmowers, and other equipment that typically does not require much power. For customers looking for faster vehicles, Oregon-based Brammo produces electric motorcycles that have a range of 100 miles at a speed of 100 miles per hour. These electric vehicles have comparable performance to any internal combustion motorcycle but lack any tailpipe or noise emissions.

Mass Transit. In North America, many cities had electric public transit similar to the San Francisco cable cars until they were sold to car manufacturers who decommissioned them. As a result, most public transit systems rely heavily on diesel engine buses. Some of the public transit companies have considered testing fuel-cell- or battery-powered electric buses, but all these efforts have remained very small, with a handful of buses running at any time throughout Europe and North America. For example, during the 2010 Olympics, the Canadian Hydrogen and Fuel Cell Association ran fuel-cell electric vehicle busses between the cities of Vancouver and Whistler. United Kingdom-based manufacturer Optare has produced battery-powered buses since 2009. Cities, such as Seattle, that use electric overhead lines to power trolley buses, trams, and trains have had much higher impact in terms of actual transported passengers. All these systems constitute electric vehicles, but all of them are dependent on having electric wires in place before they can operate. On the other hand, once the wires are in place, the public transit systems can operate silently and cleanly, using electricity provided through an electric grid instead of a battery or a fuel cell.

Forklifts. In spaces with little ventilation, the exhaust of internal combustion engines can be harmful and potentially toxic to humans, which is why warehouse forklifts are typically powered by electric engines. Traditionally, these engines are powered by batteries, but the recharging time of several hours often requires the purchase of at least twice as many batteries as forklifts—or twice as many forklifts as drivers—to be able to work around the clock. Using fuel cells as a power source, forklifts such as the ones produced by Vancouver-based company Cellex require only a short time at a hydrogen refueling station before being ready for use. Such a short downtime of fuel-cell powered electric forklifts compared with battery-powered forklifts allows warehouses to operate with less machinery, cutting back on the initial capital cost of operation.

IMPACT ON INDUSTRY

Government Research. Globally, most governments have introduced specific targets for the electrification of vehicles to reduce pollution. In Germany, for example, the federal government plans to have at least one million electric vehicles on the streets by 2020. This is an ambitious target that is backed by significant funding for government, university, and private-sector research. However, in 2009-2010, the German government offered car wreckage flat-rate premium to all car owners, many of whom traded their used but still very usable cars for newer cars with larger motors and increased emissions. Although the funding of this incentive program significantly benefited automakers worldwide, incentives to improve the performance of electric motors and the market penetration of electric vehicles were not included. As a consequence, no net reduction in vehicle-emission pollution was achieved, and the number of electric cars actually on the road remained negligible. About 1,500 electric vehicles were registered in Germany in 2010 out of a total of about 41 million cars. With no incentives in place as of 2011, the number of registered electric vehicles has not significantly changed either. Asian countries appear to invest more into an electric-vehicle infrastructure, with China set to place a half million electric vehicles on its roads in 2012, although it remains doubtful whether this target can be met completely. The U.S. Department of Energy's February, 2011, Status Report stated the ambitious target of having one million electric vehicles on the road by 2015. This is based on 2010 statistics in which 97 percent of cars sold had conventional internal combustion engines and 3 percent had hybrid electric engines. While one million cars sounds like a lot, in the actual car market, this amount will not represent any significant step in reducing pollution from internal combustion engines. Significantly higher numbers of electric vehicle usage are required to make cleaner inner-city air a reality.

University Research. Universities have always been involved in battery research, but only since the 2010's has there been a significant increase in attention to clean-energy storage technologies from super capacitors to flywheels, hydrogen, and battery technologies. Although university projects suffered some detrimental impact from the 2009 world economic crisis, the research, especially in batteries, was minimally affected and has since seen significant increases in funding, scope, and technology-focused applications. One example is the Institute for Electrochemical Energy Storage and Conversion at the University of Ulm in Germany, which has significantly expanded its activities in battery research. Similar expansions in battery-technology research can be seen around the globe.

Industry and Business. One of the main business sectors for electronic vehicles includes forklifts and other indoor equipment for confined or explosive environments, such as mines. Other sectors that have made profits over the past years include electric scooters and small electric bicycles. And while there is significant media attention on electric cars, both fuel cell and battery based, as well as hybrid-electric vehicles, their contribution to the global automobile market is negligibly small and requires significant governmental incentives to become a larger part of the global automotive economy.

Major Corporations. Although North America was a global leader in electric vehicles and battery research, the car industry mostly dismissed electric vehicle developments in the 1990's, as North American consumers demanded larger, heavier, and more powerful vehicles with internal combustion engines. As of 2011, North American car manufacturers lag behind the rest of the world in developing mature vehicles that will be accepted in the market. Both Ford, with its Focus electric vehicle, and GM, with its Chevrolet Volt, have produced early studies of electric vehicles that they intend to develop for the North American market. They compete with all major international car manufacturers that are introducing the electric vehicles worldwide, as well as domestic corporations that focus on electric vehicles, such as the California-based luxury-car manufacturer Tesla Motors and Missouri-based electric truck and commercial vehicle manufacturer Smith Electric Vehicles. Internationally, there is a large number of electric car manufacturing businesses starting up, reviving a business that had become the monopoly of very few corporations. Because of more stringent emissions targets and significant tax incentives, North American car manufacturers have returned to some electric car developments in the 2010's. Whether these are just greening initiatives and convenient tax-reduction programs or whether these cars can become a major business in the North American market remains to be seen.

Most car manufacturers have created spin-off companies that develop customer-specific batteries for modern vehicles. For example, the European car manufacturer Daimler and the international chemical corporation Evonik and its subsidiary Li-Tec formed a new company called Deutsche Accumotive, with the sole aim of producing better batteries for the international vehicle market. This comes as a result of the dominance of lithium-ion batteries in the low-end markets by Chinese manufacturers and high-end markets by Japan and South Korea. Most batteries are used in power tools and small handheld devices such as mobile phones and laptop computers. The clean technology consulting firm Pike Research estimates that the global annual market for lithium-ion batteries for vehicles is $8 billion.

CAREERS AND COURSE WORK

As gasoline prices increase, it becomes more important to have lighter vehicles that require less material during manufacturing, as these have to be mined and transported around the world and machined using energy coming primarily from fossil fuels. Additionally, vehicles should become more efficient, to reduce the operating cost for vehicle owners. All these issues are addressed by selecting and designing better, novel materials. Those interested in a career in electric vehicle manufacturing or design would do well studying materials, mechanical, chemical, mining, or environmental engineering for designing novel cars, highly efficient motors, better batteries, and cheaper, more durable fuel cells. The mathematical modeling of the electrochemistry involved in electric motors is also very important to understand how to improve electric devices, and studies in chemistry and physics may lead to improvements in the efficiency of vehicles.

After earning a bachelor's degree in one of the above-mentioned areas, an internship would be ideal. After an internship, one's career path can be extremely varied. In the research sector, for example, working on catalysts for batteries and fuel cells in a chemical company could include developing new materials that involve inexpensive, nontoxic, durable noble metals that are at least as efficient as traditional catalysts. This is only one example of many potential careers in the global electric vehicle market.

SOCIAL CONTEXT AND FUTURE PROSPECTS

Energy consumption per capita is increasing continuously. The majority of power production uses the combustion of fossil fuels with additional contributions from hydroelectric and nuclear energy conversion. These energy-conversion methods create varying kinds of pollution and dangers to the environment such as habitat destruction, toxic-waste production, or radiation, as seen in nuclear reactors hit by earthquakes, equipment malfunction, or operator errors. The increasing demand for a finite quantity of fossil fuels has the potential to increase the cost of these resources significantly. Another undesirable consequence of the thermochemical conversion of fossil fuels by combustion is environmental contamination. The reaction products from combustion can be harmful to humans on a local scale and have been cited as contributing to global climate change. The remaining ash of coal combustion contains heavy metals and radioactive isotopes that can be severely damaging to health, as seen in the 2008 Kingston Fossil Plant coal ash slurry spill in Tennessee.

Furthermore, fossil fuel resources are unevenly distributed over the globe, leading to geopolitical unrest as a result of the competition for resource access. As a consequence of upheavals in the Middle East and North Africa in 2011, oil and food prices have soared, and may continue doing so.

Clearly, the energy demands of society need to be satisfied in a more appropriate, sustainable, and efficient way. Cleaner devices for energy conversion are batteries and fuel cells. They operate more efficiently, produce less pollution, are modular, and are less likely to fail mechanically since they have fewer moving parts than energy conversion based on combustion.

The advantages of electric vehicles are clear: a world in which all or most vehicles are quiet, with no truck engine brakes to rattle windows from a mile away and no lawnmowers disturbing the quiet or fresh air of a neighborhood; a society with no harmful local emissions from any of the machines being used, allowing people to walk by a leaf blower without having to hold their breath and to live next to major roads without risking chronic diseases from continually breathing in harmful emissions. All this could already be humanity's present-day reality if people were willing to change their habits and simply use electric motors instead of combustion engines.

Lars Rose, M.Sc., Ph.D.

FURTHER READING

Cancilla, Riccardo, and Monte Gargano, eds. *Global Environmental Policies: Impact, Management and Effects.* Hauppauge, N.Y.: Nova Science Publishers, 2010. Features multifaceted chapters on various international aspects of environmental policies, laying the groundwork for a change in consumer attitude toward new clean energy and infrastructure.

Hoel, Michael, and Snorre Kverndokk. "Depletion of Fossil Fuels and the Impact of Global Warming." *Resource and Energy Economics* 18, no. 2 (June, 1996): 115-136. Gives an explanation of the calculation of global oil depletion.

Husain, Iqbal. *Electric and Hybrid Vehicles: Design Fundamentals.* 2d ed. Boca Raton, Fla.: CRC Press, 2011. A very technical but comprehensive book that provides an overview of modern electric and electric-hybrid vehicle technologies.

Root, Michael. *The TAB Battery Book: An In-Depth Guide to Construction, Design, and Use.* New York: McGraw-Hill, 2011. Provides a good, readable background for all different types of batteries and the challenges in research and development.

Taylor, Peter J., and Frederick H. Buttel. "How Do We Know We Have Global Environmental Problems? Science and the Globalization of Environmental Discourse." *Geoforum* 23, no. 3 (1992): 405-416. Outlines the science behind global environmental issues.

U.S. Army Center for Health Promotion and Preventive Medicine. "Engine Emissions—Health and Medical Effects." http://phc.amedd.army.mil/PHC%20Resource%20Library/FS65-039-1205.pdf. Outlines the detrimental acute and chronic effects of engine exhaust intake via the lungs and the skin.

U.S. Department of Energy. "One Million Electric Vehicles By 2015." http://www.energy.gov/media/1_Million_Electric_Vehicle_Report_Final.pdf. Details the U.S. plan to achieve a sale rate of at least 1.7 percent electric vehicles every year until 2015, also indicating preferred companies involved for the North American electric vehicle market.

WEB SITES

American Society for Engineering Education
http://www.asee.org

Electric Auto Association
http://www.electricauto.org

Electric Drive Transportation Organization
http://www.electricdrive.org

European Association for Battery, Hybrid and Fuel Cell Electric Vehicles
http://www.avere.org/www/index.php

See also: Chemical Engineering; Electrical Engineering; Hybrid Vehicle Technologies.

ELECTROCHEMISTRY

FIELDS OF STUDY

Physical chemistry; thermodynamics; organic chemistry; inorganic chemistry; quantitative analysis; chemical kinetics; analytical chemistry; metallurgy; chemical engineering; electrical engineering; industrial chemistry; electrochemical cells; fuel cells; electrochemistry of nanomaterials; advanced mathematics; physics; electroplating; nanotechnology; quantum chemistry; electrophoresis; biochemistry; molecular biology.

SUMMARY

Electrochemists study the chemical changes produced by electricity, but they are also concerned with the generation of electric currents due to the transformations of chemical substances. Whereas traditional electrochemists investigated such phenomena as electrolysis, modern electrochemists have broadened and deepened their interdisciplinary field to include theories of ionic solutions and solvation. This theoretical knowledge has led to such practical applications as efficient batteries and fuel cells, the production and protection of metals, and the electrochemical engineering of nanomaterials and devices that have great importance in electronics, optics, and ceramics.

KEY TERMS AND CONCEPTS

- **Anode:** Positive terminal (or electrode) of an electrochemical cell to which negatively charged ions travel with the passage of an electric current.
- **Battery:** Electrochemical device that converts chemical energy into electrical energy.
- **Cathode:** Negative electrode of an electrochemical cell in which positively charged ions migrate under the influence of an electric current.
- **Electrolysis:** Process by which an electric current causes chemical changes in water, solutions, or molten electrolytes.
- **Electrolyte:** Substance that generates ions when molten or dissolved in a solvent.
- **Electrophoresis:** Movement of charged particles through a conducting medium due to an applied electric field.
- **Electroplating:** Depositing a thin layer of metal on an object immersed in a solution by passing an electric current through it.
- **Faraday's Law:** Magnitude of the chemical effect of an electrical current is directly proportional to the amount of current passing through the system.
- **Fuel Cell:** Electrochemical device for converting a fuel with an oxidant into direct-current electricity.
- **Ion:** Atom or group of atoms carrying a charge, either positive (cation) or negative (anion).
- **Nanomaterials:** Chemical substances or particles the masses of which are measured in terms of billionths of a gram.
- **pH:** Numerical value, extending from 0 to 14, representing the acidity or alkalinity of an aqueous solution (the number decreases with increasing acidity and increases with increasing alkalinity).

DEFINITION AND BASIC PRINCIPLES

As its name implies, electrochemistry concerns all systems involving electrical energy and chemical processes. More specifically, this field includes the study of chemical reactions caused by electrical forces as well as the study of how chemical processes give rise to electrical energy. Some electrochemists investigate the electrical properties of certain chemical substances, for instance, these substances' ability to serve as insulators or conductors. Because the atomic structure of matter is fundamentally electrical, electrochemistry is intimately involved in all fields of chemistry from physical and inorganic through organic and biochemistry to such new disciplines as nanochemistry. No matter what systems they study, chemists in some way deal with the appearance or disappearance of electrical energy into the surroundings. On the other hand, electrochemists concentrate on those systems consisting of electrical conductors, which can be metallic, electrolytic, or gaseous.

Because of its close connection with various branches of chemistry, electrochemistry has applications that are multifarious. Early applications centered on electrochemical cells that generated a steady current. New metals such as potassium, sodium, calcium, and strontium were discovered by electrolysis of their molten salts. Commercial production of such metals as magnesium, aluminum, and zinc were

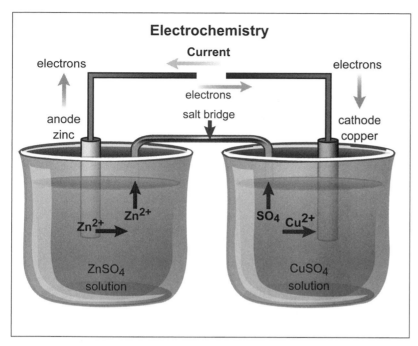

Connecting copper and zinc creates an electrochemical cell.

mainly accomplished by the electrolysis of solutions or melts. An understanding of the electrical nature of chemical bonding led chemists to create many new dyes, drugs, plastics, and artificial plastics. Electroplating has served both aesthetic and practical purposes, and it has certainly decreased the corrosion of several widely used metals.

Electrochemistry played a significant part in the research and development of such modern substances as silicones, fluorinated hydrocarbons, synthetic rubbers, and plastics. Even though semiconductors such as germanium and silicon do not conduct electricity as well as copper, an understanding of electrochemical principles has been important in the invention of various solid-state devices that have revolutionized the electronics industries, from radio and television to computers. Electrochemistry, when it has been applied in the life sciences, has resulted in an expanded knowledge of biological molecules. For example, American physical chemist Linus Pauling used electrophoretic techniques to discover the role of a defective hemoglobin molecule in sickle-cell anemia. A grasp of electrochemical phenomena occurring in the human heart and brain has led to diagnostic

and palliative technologies that have improved the quality and length of human lives. Much research and development are being devoted to increasingly sophisticated electrochemical devices for implantation in the human body, and some even predict, such as American inventor Ray Kurzweil, that these "nanobots" will help extend human life indefinitely.

BACKGROUND AND HISTORY

Most historians of science trace the origins of electrochemistry to the late eighteenth and early nineteenth centuries, when Italian physician Luigi Galvani studied animal electricity and Italian physicist Alessandro Volta invented the first battery. Volta's device consisted of a pile of dissimilar metals such as zinc and silver separated by a moist conductor. This "Voltaic pile" produced a continuous current, and applications followed quickly. Researchers showed that a Voltaic pile could decompose water into hydrogen and oxygen by a process later called electrolysis. English chemist Sir Humphry Davy used the electrolysis of melted inorganic compounds to discover several new elements. The Swedish chemist Jöns Jacob Berzelius used these electrochemical studies to formulate a new theory of chemical combination. In his dualistic theory atoms are held together in compounds by opposite charges, but his theory declined in favor when it was unable to explain organic compounds, or even such a simple molecule as diatomic hydrogen.

Though primarily a physicist, Michael Faraday made basic discoveries in electrochemistry, and, with the advice of others, he developed the terminology of this new science. For example, he introduced the terms "anode" and "anion," "cathode" and "cation," "electrode," "electrolyte," as well as "electrolysis." In the 1830's his invention of a device to measure the quantity of electric current resulted in his discovery of a fundamental law of electrochemistry—that the quantity of electric current that leads to the formation of a certain amount of a particular chemical substance also leads to chemically equivalent amounts of other substances. Even though Faraday's discovery of the relationship between quantity of electricity and electrochemical equivalents was extraordinarily significant, it was not properly appreciated until much later. Particularly helpful was the work of the Swedish chemist Svante August Arrhenius, whose ionic theory, proposed toward the end of the nineteenth century, contained the surprising new idea that anions and cations are present in dilute solutions of electrolytes.

In the twentieth century, Dutch-American physical chemist Peter Debye, together with German chemist Erich Hückel, corrected and extended the Arrhenius theory by taking into account that, in concentrated solutions, cations have a surrounding shell of anions, and vice versa, causing these ions' movements to be retarded in an electric field. Norwegian-American chemist Lars Onsager further refined this theory by taking into account Brownian motion, the movement of these ionic atmospheres due to heat. Other scientists used electrochemical ideas to understand the nature of acids and bases, the interface between dissimilar chemicals in electrochemical cells, and the complexities of oxidation-reduction reactions, whether they occur in electrochemical cells or in living things.

How It Works

Primary and Secondary Cells. The basic device of electrochemistry is the cell, generally consisting of a container with electrodes and an electrolyte, designed to convert chemical energy into electrical energy. A primary cell, also known as a galvanic or Voltaic cell, is one that generates electrical current via an irreversible chemical reaction. This means that a discharged primary cell cannot be recharged from an external source. By taking measurements at different temperatures, chemists use primary cells to calculate the heat of reactions, which have both theoretical and practical applications. Such cells can also be used to determine the acidity and alkalinity of solutions. Every primary cell has two metallic electrodes, at which electrochemical reactions occur. In one of these reactions the electrode gives up electrons, and at the other electrode electrons are absorbed.

In a secondary cell, also known as a rechargeable or storage cell, electrical current is created by chemical reactions that are reversible. This means that a discharged secondary cell may be recharged by circulating through the cell in a quantity of electricity

equal to what had been withdrawn. This process can be repeated as often as desired. The manufacture of secondary cells has grown into an immense industry, with such commercially successful products as lead-acid cells and alkaline cells with either nickel-iron or nickel-cadmium electrodes.

Electrolyte Processes. Electrolysis, one of the first electrochemical processes to be discovered, has increased in importance in the twentieth and twenty-first centuries. Chemists investigating electrolysis soon discovered that chemical reactions take place at the two electrodes, but the liquid solution between them remains unchanged. An early explanation was that with the passage of electric current, ions in the solution alternated decompositions and recombinations of the electrolyte. This theory had to be later revised in the light of evidence that chemical components had different motilities in solution.

For more than two hundred years, the electrolysis of water has been used to generate hydrogen gas. In an electrolytic cell with a pair of platinum electrodes, to the water of which a small amount of sulfuric acid has been added (to reduce the high voltage needed), electrolysis begins with the application of an external electromotive force, with bubbles of oxygen gas appearing at the anode (due to an oxidation reaction) and hydrogen gas at the cathode (due to a reduction reaction). If sodium chloride is added to the water, the electrochemical reaction is different, with sodium metal and chlorine gas appearing at the appropriate electrodes. In both these electrolyses the amounts of hydrogen and sodium produced are in accordance with Faraday's law, the mass of the products being proportional to the current applied to the cell.

Redox Reactions. For many electrochemists the paramount concern of their discipline is the reduction and oxidation (redox) reaction that occurs in electrochemical cells, batteries, and many other devices and applications. Reduction takes place when an element or radical (an ionic group) gains electrons, such as when a double positive copper ion in solution gains two electrons to form metallic copper. Oxidation takes place when an element or radical loses electrons, such as when a zinc electrode loses two electrons to form a doubly positive zinc ion in solution. In electrochemical research and applications the sites of oxidation and reduction are spatially separated. The electrons produced by chemical processes can be forced to flow through a wire, and this electric current can be used in various applications.

Electrodes. Electrochemists employ a variety of electrodes, which can consist of inorganic or organic materials. Polarography, a subdiscipline of electrochemistry dealing with the measurement of current and voltage, uses a dropping mercury electrode, a technique enabling analysts to determine such species as trace amounts of metals, dissolved oxygen, and certain drugs. Glass electrodes, whose central feature is a thin glass membrane, have been widely used by chemists, biochemists, and medical researchers. A reversible hydrogen electrode plays a central role in determining the pH of solutions. The quinhydrone electrode, consisting of a platinum electrode immersed in a quinhydrone solution, can also be used to measure pH (it is also known as an indicator electrode because it can indicate the concentration of certain ions in the electrolyte). Also widely used, particularly in industrial pH measurements, is the calomel electrode, consisting of liquid mercury covered by a layer of calomel (mercurous chloride), and immersed in a potassium chloride solution. Electrochemists have also created electrodes with increasing (or decreasing) power as oxidizing or reducing agents. With this quantitative information they are then able to choose a particular electrode material to suit a specific purpose.

APPLICATIONS AND PRODUCTS

Batteries and Fuel Cells. Soon after Volta's invention of the first electric battery, investigators found applications, first as a means of discovering new elements, then as a way to deepen understanding of chemical bonding. By the 1830's, when new batteries were able to serve as reliable sources of electric current, they began to exhibit utility beyond their initial value for experimental and theoretical science. For example, the Daniell cell was widely adopted by the telegraph industry. Also useful in this industry was the newly invented fuel cell that used the reaction of a fuel such as hydrogen and an oxidant such as oxygen to produce direct-current electricity. However, its requirement of expensive platinum electrodes led to its replacement, in the late nineteenth and throughout the twentieth century, with the rechargeable lead-acid battery, which came to be extensively used in the automobile industry. In the late twentieth and early twenty-first centuries many electrochemists

---

Fascinating Facts About Electrochemistry

- Alessandro Volta, often called the father of electrochemistry, invented the world's first battery, but he did not understand how it worked (he believed in a physical-contact theory rather than the correct chemical-reaction theory).
- English physicist Michael Faraday, with only a few years of rudimentary education, later formulated the basic laws of electrochemistry and became the greatest experimental physicist of the nineteenth century.
- Swedish chemist Svante August Arrhenius, whose ionic theory of electrolytic solutions became fundamentally important to the progress of electrochemistry, was nearly drummed out of the profession by his skeptical doctoral examiners, who gave him the lowest possible passing grade.
- The electrochemical discovery that charged atoms or groups of atoms in solution have a distinct electric charge (or some integral multiple of it) led to the idea that electricity itself is atomic (and not a fluid, as many believed), and in 1891, Irish physicist George Johnstone Stoney gave this electrical unit the name, "electron."
- Charles M. Hall in the United States and Paul L. T. Héroult in France independently discovered the modern electrolytic technique of manufacturing aluminum in 1886; furthermore, both men were born in 1863 and both died in 1914.
- Many molecules in plants, animals, and humans have electric charges, some with two charges (called zwitterions), others with many (called polyelectrolytes).
- In the second half of the twentieth century, many academic institutions in the United States taught courses in electroanalytical chemistry and electrochemical engineering, but no American college, university, or technical institute taught any courses in applied electrochemistry.

---

predicted a bright future for fuel cells based on hydrogen and oxygen, especially with the pollution problems associated with widespread fossil-fuel use.

Electrodeposition, Electroplating, and Electrorefining. When an electric current passes through a solution (for instance, silver nitrate), the precipitation of a material (silver) at an electrode (the cathode) is called electrodeposition. A well-known category of electrodeposition is electroplating, when a thin layer of one metal is deposited on another. In galvanization, for example, iron or steel objects are coated with rust-resistant zinc. Electrodeposition techniques have the advantage of being able to coat objects thoroughly, even those with intricate shapes. An allied technique, electrorefining, transforms metals contaminated with impurities to very pure states by anodic dissolution and concomitant redeposition of solutions of their salts. Some industries have used so-called electrowinning techniques to produce salable metals from low-grade ores and mine tailings.

Advances in electrochemical knowledge and techniques have led to evermore sophisticated applications of electrodeposition. For example, knowledge of electrode potentials has made the electrodeposition of alloys possible and commercial. Methods have also been discovered to provide plastics with metal coatings. Similar techniques have been discovered to coat such rubber articles as gloves with a metallic layer. Worn or damaged metal objects can be returned to pristine condition by a process called electroforming. Some commercial metal objects, such as tubes, sheets, and machine parts, have been totally manufactured by electrodeposition (sometimes called electromachining).

Electrometallurgy. A major application of electrochemical principles and techniques occurs in the manufacture of such metals as aluminum and titanium. Plentiful aluminum-containing bauxite ores exist in large deposits in several countries, but it was not until electrochemical techniques were developed in the United States and France at the end of the nineteenth century that the cost of manufacturing this light metal was sufficiently reduced to make it a commercially valuable commodity. This commercial process involved the electrolysis of alumina (aluminum oxide) dissolved in fused cryolite (sodium aluminum fluoride). During the century that followed this process's discovery, many different uses for this lightweight metal ensued, from airplanes to zeppelins.

Corrosion Control and Dielectric Materials. The destruction of a metal or alloy by oxidation is itself an electrochemical process, since the metal loses electrons to the surrounding air or water. A familiar example is the appearance of rust (hydrated ferric oxide) on an iron or steel object. Electrochemical knowledge of the mechanism of corrosion led researchers to ways of preventing or delaying it. Keeping oxidants away from the metallic surface is an obvious means of protection. Substances that interfere with the oxidizing of metals are called inhibitors. Corrosion inhibitors include both inorganic and organic materials, but they are generally categorized by whether the inhibitor obstructs corrosive reactions at the cathode or anode. Cathodic protection is used extensively for such metal objects as underground pipelines or such structures as ship hulls, which have to withstand the corrosive action of seawater. Similarly, dielectric materials with low electrical conductivity, such as insulators, require long-term protection from high and low temperatures as well as from corrosive forces. An understanding of electrochemistry facilitates the construction of such electrical devices as condensers and capacitors that involve dielectric substances.

Electrochemistry, Molecular Biology, and Medicine. Because of the increasing understanding of electrochemistry as it pertains to plant, animal, and human life, and because of concerns raised by the modern environmental movement, several significant applications have been developed, with the promise of many more to come. For example, electrochemical devices have been made for the analysis of proteins and deoxyribonucleic acid (DNA). Researchers have fabricated DNA sensors as well as DNA chips. These DNA sensors can be used to detect DNA damage. Electrochemistry was involved in the creation of implantable pacemakers designed to regulate heart beats, thus saving lives. Research is under way to create an artificial heart powered by electrochemical processes within the human body. Neurologists electrically stimulate regions of the brain to help mitigate or even cure certain psychological problems. Developments in electrochemistry have led to the creation of devices that detect various environmental pollutants in air and water. Photoelectrochemistry played a role in helping to understand the dramatic depletion of the ozone layer in the stratosphere and the role that chlorofluorocarbons (CFCs) played in exacerbating this problem. Because a large hole in the ozone layer allows dangerous solar radiation to damage plants, animals, and humans, many countries have banned the use of CFCs.

Nanomaterials in Electrochemistry. Miniaturization of electronic technologies became evident and important in the computer industry, where advances have been enshrined in Moore's law, which states that transistor density in integrated circuits doubles every eighteen months. Electrodeposition has proved to be a technique well-suited to the preparation of metal nanostructures, with several applications in electronics, semiconductors, optics, and ceramics. In particular, electrochemical methods have contributed to the understanding and applications of quantum dots, nanoparticles that are so small that they follow quantum rather than classical laws. These quantum dots can be as small as a few atoms, and in the form of ultrathin cadmium-sulfide films they have been shown to generate high photocurrents in solar cells. The electrochemical synthesis of such nanostructured products as nanowires, biosensors, and microelectro-analytical devices has led researchers to predict the ultimate commercial success of these highly efficient contrivances.

IMPACT ON INDUSTRY

Because electrochemistry itself an interdisciplinary field, and is a part of so many different scientific disciplines and commercial applications, it is difficult to arrive at an accurate figure for the economic worth and annual profits of the global electrochemical industry. More reliable estimates exist for particular segments of this industry in specific countries. For example, in 2008, the domestic revenues of the United States battery and fuel cell industry were about $4.9 billion, the lion's share of which was due to the battery business (in 2005, the United States fuel cell industry had revenues of about $266 million). During the final decades of the twentieth century, the electrolytic production of aluminum in the United States was about one-fifth of the world's total, but in the twenty-first century competition from such countries as Norway and Brazil, with their extensive and less expensive hydropower, has reduced the American share.

Government and University Research. In the United States during the decades after World War II, the National Science Foundation provided support for many electrochemical investigations, especially those projects that, because of their exploratory nature, had no guarantee of immediate commercial success. An example is the 1990's electrochemical research on semiconducting nanocrystals. The United States' Office of Naval Research has supported projects on nanostructured thin films. It is not only the federal government that has seen fit through various

agencies to support electrochemical research but state governments as well. For example, the New York State Foundation for Science, Technology and Innovation has invested in investigations of how ultrathin films can be self-assembled from polyelectrolytes, nanoparticles, and nanoplatelets with the hope that these laboratory-scale preparations may have industrial-scale applications.

Government agencies and academic researchers have often worked together to fund basic research in electrochemistry. The Department of Energy through its Office of Basic Energy Sciences has supported research on electrochromic and photochromic effects, which has led to such commercial products as switchable mirrors in automobiles. Researchers at the Georgia Institute of Technology have contributed to the improvement of proton exchange membrane fuel cells (PEMFCs), and scientists at the University of Dayton in Ohio have shown the value of carbon nanotubes in fuel cells. Hydrogen fuel cells became the focus of an initiative promoted by President George W. Bush in 2003, which was given direction and financial support in the 2005 Energy Policy Act and the 2006 Advanced Energy Initiative. However, a few years later, the Obama administration chose to de-emphasize hydrogen fuel cells and emphasize other technologies that will create high energy efficiency and less polluting automobiles in a shorter period of time.

Industry and Business. Because of the widespread need for electrochemical cells and batteries, companies manufacturing them have devoted extensive human and financial resources to the research, development, and production of a variety of batteries. Some companies, such as Exide Technologies, have emphasized lead-acid batteries, whereas General Electric, whose corporate interest in batteries goes back to its founder, Thomas Alva Edison, has made fuel cells a significant part of its diversified line of products. Some businesses, such as Alcoa, the world's leading producer of aluminum, were based on the discovery of a highly efficient electrolytic process, which led to a dramatic decrease in the cost of aluminum and, in turn, to its widespread use (it is second only to steel as a construction metal).

CAREERS AND COURSE WORK

Electrochemistry is an immense field with a large variety of specialties, though specialized education generally takes place at the graduate level. Undergraduates usually major in chemistry, chemical engineering, or materials science engineering, the course work of which involves introductory and advanced physics courses, calculus and advanced mathematics courses, and elementary and advanced chemistry courses. Certain laboratory courses, for example, qualitative, quantitative and instrumental analysis, are often required. Because of the growing sophistication of many electrochemical disciplines, those interested in becoming part of these fields will need to pursue graduate degrees. Depending on their specialty, graduate students need to satisfy core courses, such as electrochemical engineering, and a certain number of electives, such as semiconductor devices. Some universities, technical institutes, and engineering schools have programs for students interested in theoretical electrochemistry, electrochemical cells, electrodeposition, nanomaterials, and many others. For doctoral and often for a master's degree, students are required to write a thesis under the supervision of a faculty director.

Career opportunities for electrochemists range from laboratory technicians at small businesses to research professors at prestigious universities. The battery business employs many workers with bachelor of science degrees in electrochemistry to help manufacture, service, and improve a variety of products. Senior electrochemical engineers with advanced degrees may be hired to head research programs to develop new products or to supervise the production of the company's major commercial offerings. Electrochemical engineers are often hired to manage the manufacture of electrochemical components or oversee the electrolytic production of such metals as aluminum and magnesium. Some electrochemists are employed by government agencies, for example, to design and develop fuel cells for the National Aeronautics and Space Administration (NASA), while others may be hired by pharmaceutical companies to develop new drugs and medical devices.

SOCIAL CONTEXT AND FUTURE PROSPECTS

Even though batteries, when compared with other energy sources, are too heavy, too big, too inefficient, and too costly, they will continue to be needed in the twenty-first century, at least until suitable substitutes are found. Although some analysts predict a bright future for fuel cells, others have been discouraged by their slow rate of development. As advanced industrialized societies continue to expand, increasing demand for such metals as beryllium, magnesium, aluminum, titanium, and zirconium will necessarily follow, forcing electrochemists to improve the electrolytic processes for deriving these metals from dwindling sources. If Moore's law holds well into the future, then computer engineers, familiar with electrochemical principles, will find new ways to populate integrated circuits with more and better microdevices.

Some prognosticators foresee significant progress in the borderline field between electrochemistry and organic chemistry (sometimes called electro-organic chemistry). When ordinary chemical methods have proved inadequate to synthesize desired compounds of high purity, electrolytic techniques have been much better than traditional ones in accomplishing this, though these successes have occurred at the laboratory level and the development of industrial processes will most likely take place in the future. Other new fields, such as photoelectrochemistry, will mature in the twenty-first century, leading to important applications. The electrochemistry of nanomaterials is already well underway, both theoretically and practically, and a robust future has been envisioned as electrochemical engineers create new nanophase materials and devices for potential use in a variety of applications, from electronics to optics.

Robert J. Paradowski, M.S., Ph.D.

FURTHER READING

Bagotsky, Vladimir, ed. *Fundamentals of Electrochemistry.* 2d ed. New York: Wiley-Interscience, 2005. Provides a good introduction to this field for those unfamiliar with it, though later chapters contain material of interest to advanced students. Index.

Bard, Allen J., and Larry R. Faulkner. *Electrochemical Methods: Fundamentals and Applications.* 2d ed. Hoboken, N.J.: John Wiley & Sons, 2001. For many years this "gold standard" of electrochemistry textbooks was the most widely used such book in the world. Its advanced mathematics may daunt the beginning student, but its comprehensive treatment of theory and applications rewards the extra effort needed to understand this field's fundamentals; index.

Brock, William H. *The Chemical Tree: A History of Chemistry.* New York: Norton, 2000. This work, previously published as *The Norton History of Chemistry*, was listed as a *New York Times* "Notable Book" when it appeared. The development of electrochemistry forms an important part of the story of chemistry; includes an extensive bibliographical essay, notes, and index.

Ihde, Aaron J. *The Development of Modern Chemistry*. New York: Dover, 1984. Makes available to general readers a well-organized treatment of chemistry from the eighteenth to the twentieth century, of which electrochemical developments form an essential part. Illustrated, with extensive bibliographical essays on all the chapters; author and subject indexes.

MacInnes, Duncan A. *The Principles of Electrochemistry*. New York: Dover, 1961. This paperback reprint brings back into wide circulation a classic text that treats the field as an integrated whole; author and subject indexes.

Schlesinger, Henry. *The Battery: How Portable Power Sparked a Technological Revolution*. New York: HarperCollins, 2010. The author, a journalist specializing in technology, provides an entertaining, popular history of the battery, with lessons for readers familiar only with electronic handheld devices.

Zoski, Cynthia G., ed. *Handbook of Electrochemistry*. Oxford, England: Elsevier, 2007. After an introductory chapter, this book surveys most modern research areas of electrochemistry, such as reference electrodes, fuel cells, corrosion control, and other laboratory techniques and practical applications.

WEB SITES

Electrochemical Science and Technology Information Resource (ESTIR)
http://electrochem.cwru.edu/estir

The Electrochemical Society
http://www.electrochem.org

International Society of Electrochemistry
http://www.ise-online.org

See also: Chemical Engineering; Electrical Engineering; Electrometallurgy; Metallurgy; Nanotechnology.

ELECTROMAGNET TECHNOLOGIES

FIELDS OF STUDY

Mathematics; physics; electronics; materials science.

SUMMARY

Electromagnetic technology is fundamental to the maintenance and progress of modern society. Electromagnetism is one of the essential characteristics of the physical nature of matter, and it is fair to say that everything, including life itself, is dependent upon it. The ability to harness electromagnetism has led to the production of most modern technology.

KEY TERMS AND CONCEPTS

- **Curie Temperature:** The temperature at which a ferromagnetic material is made to lose its magnetic properties.
- **Induction:** The generation of an electric current in a conductor by the action of a moving magnetic field.
- **LC Oscillation:** Alternation between stored electric and magnetic field energies in circuits containing both an inductance (L) and a capacitance (C).
- **Magnetic Damping:** The use of opposing magnetic fields to damp the motion of a magnetic oscillator.
- **Sinusoidal:** Varying continuously between maximum and minimum values of equal magnitude at a specific frequency, in the manner of a sine wave.
- **Toroid:** A doughnut-shaped circular core about which is wrapped a continuous wire coil carrying an electrical current.

DEFINITION AND BASIC PRINCIPLES

Magnetism is a fundamental field effect produced by the movement of an electrical charge, whether that charge is within individual atoms such as iron or is the movement of large quantities of electrons through an electrical conductor. The electrical field effect is intimately related to that of magnetism. The two can exist independently of each other, but when the electrical field is generated by the movement of a charge, the electrical field is always accompanied by a magnetic field. Together they are referred to as an electromagnetic field. The precise mathematical relationship between electric and magnetic fields allows electricity to be used to generate magnetic fields of specific strengths, commonly through the use of conductor coils, in which the flow of electricity follows a circular path. The method is well understood and is the basic operating principle of both electric motors and electric generators.

When used with magnetically susceptible materials, this method transmits magnetic effects that work for a variety of purposes. Bulk material handling can be carried out in this way. On a much smaller scale, the same method permits the manipulation of data bits on magnetic recording tape and hard-disk drives. This fine degree of control is made possible through the combination of digital technology and the relationship between electric and magnetic fields.

BACKGROUND AND HISTORY

The relationship between electric current and magnetic fields was first observed by Danish physicist and chemist Hans Christian Oersted (1777-1851), who noted, in 1820, how a compass placed near an electrified coil of wires responded to those wires. When electricity was made to flow through the coil, these wires changed the direction in which the compass pointed. From this Oersted reasoned that a magnetic field must exist around an electrified coil.

In 1821, English chemist and physicist Michael Faraday (1791-1867) found that he could make an electromagnetic field interact with a permanent magnetic field, inducing motion in one or the other of the magnetized objects. By controlling the electrical current in the electromagnet, the permanent magnet can be made to spin about. This became the operating principle of the electric motor. In 1831, Faraday found that moving a permanent magnetic field through a coil of wire caused an electrical current to flow in the wire, the principle by which electric generators function. In 1824, English physicist and inventor William Sturgeon (1783-1850) discovered that an electromagnet constructed around a core of solid magnetic material produced a much stronger magnetic field than either one alone could produce.

Since these initial discoveries, the study of electromagnetism has refined the details of the mathematical relationship between electricity and magnetism as it is known today. Research continues to refine understandings of the phenomena, enabling its use in new and valuable ways.

HOW IT WORKS

The basic principles of electromagnetism are the same today as they have always been, because electromagnetism is a fundamental property of matter. The movement of an electrical charge through a conducting medium induces a magnetic field around the conductor. On an atomic scale, this occurs as a function of the electronic and nuclear structure of the atoms, in which the movement of an electron in a specific atomic orbital is similar in principle to the movement of an electron through a conducting material. On larger scales, magnetism is induced by the movement of electrons through the material as an electrical current.

Although much is known about the relationship of electricity and magnetism, a definitive understanding has so far escaped rigorous analysis by physicists and mathematicians. The relationship is apparently related to the wave-particle duality described by quantum mechanics, in which electrons are deemed to have the properties of both electromagnetic waves and physical particles. The allowed energy levels of electrons within any quantum shell are determined by two quantum values, one of which is designated the magnetic quantum number. This fundamental relationship is also reflected in the electromagnetic spectrum, a continuum of electromagnetic wave phenomena that includes all forms of light, radio waves, microwaves, X rays, and so on. Whereas the electromagnetic spectrum is well described by the mathematics of wave mechanics, there remains no clear comprehension of what it actually is, and the best theoretical analysis of it is that it is a field effect consisting of both an electric component and a magnetic component. This, however, is more than sufficient to facilitate the physical manipulation and use of electromagnetism in many forms.

Ampere's circuit law states that the magnetic field intensity around a closed circuit is determined by the sum of the currents at each point around that circuit. This defines a strict relationship between electrical current and the magnetic field that is produced around the conductor by that current. Because electrical current is a physical quantity that can be precisely controlled on scales that currently range from single electrons to large currents, the corresponding magnetic fields can be equally precisely controlled.

Electrical current exists in two forms: direct current and alternating current. In direct current, the movement of electrons in a conductor occurs in one direction only at a constant rate that is determined by the potential difference across the circuit and the resistance of the components that make up that circuit. The flow of direct current through a linear conductor generates a similarly constant magnetic field around that conductor.

In alternating current, the movement of electrons alternates direction at a set sinusoidal wave frequency. In North America this frequency is 60 hertz, which means that electrons reverse the direction of their movement in the conductor 120 times each second, effectively oscillating back and forth through the wires. Because of this, the vector direction of the magnetic field around those conductors also reverses at the same rate. This oscillation requires that the phase of the cycle be factored into design and operating principles of electric motors and other machines that use alternating current.

Both forms of electrical current can be used to induce a strong magnetic field in a magnetic material that has been surrounded by electrical conductors. The classic example of this effect is to wrap a large steel nail with insulated copper wire connected to the terminals of a battery so that as electrical current flows through the coil, the nail becomes magnetic. The same principle applies on all scales in which electromagnets are utilized to perform a function.

APPLICATIONS AND PRODUCTS

The applications of electromagnetic technology are as varied and widespread as the nature of electromagnetic phenomena. Every possible variation of electromagnetism provides an opportunity for the development of some useful application.

Electrical Power Generation. The movement of a magnetic field across a conductor induces an electrical current flow in that conductor. This is the operating principle of every electrical generator. A variety of methods are used to convert the mechanical energy of motion into electrical energy, typically in generating stations.

Hydroelectric power uses the force provided by falling water to drive the magnetic rotors of generators. Other plants use the combustion of fuels to generate steam that is then used to drive electrical generators. Nuclear power plants use nuclear fission for the same purpose. Still other power generation projects use renewable resources such as wind and ocean tides to operate electrical generators.

Solar energy can be used in two ways to generate electricity. In regions with a high amount of sunlight, reflectors can be used to focus that energy on a point to produce steam or some other gaseous material under pressure; this material can then drive an electrical generator. Alternatively, and more commonly, semiconductor solar panels are used to capture the electromagnetic property of sunlight and drive electrons through the system to generate electrical current.

Material Handling. Electromagnetism has long been used on a fairly crude scale for the handling and manipulation of materials. A common site in metal recycling yards is a crane using a large electromagnet to pick up and move quantities of magnetically susceptible materials, such as scrap iron and steel, from one location to another. Such electromagnets are powerful enough to lift a ton or more of material at a time. In more refined applications, smaller electromagnets operating on exactly the same principle are often incorporated into automated processes under robotic control to manipulate and move individual metal parts within a production process.

Machine Operational Control. Electromagnetic technology is often used for the control of operating machinery. Operation is achieved normally through the use of solenoids and solenoid-operated relays. A solenoid is an electromagnet coil that acts on a movable core, such that when current flows in the coil, the core responds to the magnetic field by shifting its position as though to leave the coil. This motion can be used to close a switch or to apply pressure according to the magnetic field strength in the coil.

When the solenoid and switch are enclosed together in a discrete unit, the structure is known as a relay and is used to control the function of electrical circuitry and to operate valves in hydraulic or pneumatic systems. In these applications also, the extremely fine control of electrical current facilitates the design and application of a large variety of solenoids. These solenoids range in size from the very small (used in micromachinery) and those used in typical video equipment and CD players, to very large

Hydroelectric generators inside the power station at the Hoover Dam on the Colorado River, Nevada. The power station contains seventeen generators, of which eight are seen. The generators convert the energy in falling water into electricity. (Paul Avis/Photo Researchers, Inc.)

ones (used to stabilize and operate components of large, heavy machinery).

A second use of electromagnetic technology in machine control utilizes the opposition of magnetic fields to effect braking force for such precision machines as high-speed trains. Normal frictional braking in such situations would have serious negative effects on the interacting components, resulting in warping, scarring, material transfer welding, and other damage. Electromagnetic braking forces avoid any actual contact between components and can be adjusted continuously by control of the electrical current being applied.

Magnetic Media. The single most important aspect of electromagnetic technology is also the smallest and most rigidly controlled application of all. Magnetic media have been known for many years, beginning with the magnetic wire and progressing to magnetic recording tape. In these applications, a substrate material in tape form, typically a ribbon of an unreactive plastic film, is given a coating that contains fine granules of a magnetically susceptible material such as iron oxide or chromium dioxide. Exposure of the material to a magnetic field imparts the corresponding directionality and magnitude of that magnetic field to the individual granules in the coating. As the medium moves past an electromagnetic "read-write head" at a constant speed, the variation of the magnetic properties over time is embedded into the medium. When played back through the system, the read-write head senses the variation in magnetic patterns in the medium and translates that into an electronic signal that is converted into the corresponding audio and video information. This is the analogue methodology used in the operation of audio and videotape recording.

With the development of digital technology, electromagnetic control of read-write heads has come to mean the ability to record and retrieve data as single bits. This was realized with the development of hard drive and floppy disk-drive technologies. In these applications, an extremely fine read-write head records a magnetic signal into the magnetic medium of a spinning disk at a strictly defined location. The signal consists of a series of minute magnetic fields whose vector orientations correspond to the 1s and 0s of binary code; one orientation for 1, the opposite orientation for 0. Also recorded are identifying codes that allow the read-write head to locate the position of specific data on the disk when requested. The electrical control of the read-write head is such that it can record and relocate a single digit on the disk. Given that hard disks typically spin at 3,000 rpm (revolutions per minute) and that the recovery of specific data is usually achieved in a matter of milliseconds, one can readily grasp the finesse with which the system is constructed.

Floppy disk technology has never been the equal of hard disk technology, instead providing only a means by which to make data readily portable. Other technologies, particularly USB (universal serial bus) flash drives, have long since replaced the floppy drive as the portable data medium, but the hard disk drive remains the staple data storage device of modern computers. New technology in magnetic media and electromagnetic methods continues to increase the amount of data that can be stored using hard disk technology. In the space of twenty years this has progressed from merely a few hundred megabytes (10^6 bytes) of storage to systems that easily store 1 terabyte (10^{12} bytes) or more of data on a single floppy disk. Progress continues to be made in this field, with current research suggesting that a new read-write head design employing an electromagnetic "spin-valve" methodology will allow hard disk systems to store as much as 10 terabytes of data per square inch of disk space.

IMPACT ON INDUSTRY

It is extremely difficult to imagine modern industry without electromagnetic technologies in place. Electric power generation and electric power utilization, both of which depend upon electromagnetism, are the two primary paths by which electromagnetism affects not only industry but also modern society as a whole. The generation of alternating current electricity, whether by hydroelectric, combustion, nuclear, or other means, energizes an electrical grid that spans much of North America. This, in turn, provides homes, businesses, and industry with power to carry out their daily activities, to the extent that the system is essentially taken for granted and grinds to an immediate halt when some part of the system fails. There is a grave danger to society in this combination of complacency and dependency, and efforts are expended both in recovering from any failure and in preventing failures from occurring.

That modern society is entirely dependent upon a continuous supply of electrical energy is an unavoidable conclusion of even the most cursory of examinations. The ability to use electricity at will is a fundamental requirement of many of the mainstays of modern industry. Aluminum, for example, is the most abundant metal known, and its use is essentially ubiquitous today, yet its ready availability depends on electricity. This metal, as well as magnesium, cannot be refined by heat methods used to refine other metals because of their propensity to undergo oxidation in catastrophic ways. Without electricity, these materials would become unavailable.

Paradoxically, the role of electromagnetism in modern electronic technology, although it has been very large, has not been as essential to the development of that technology as might be expected. It is a historical fact that the development of portable storage and hard disk drives greatly facilitated development and spawned whole new industries in the process. Nevertheless, it is also true that portable storage, however inefficient in comparison, was already available in the form of tape drives and nonmagnetic media, such as punch cards. It is possible that other methods, such as optical media (CDs and DVDs) or electronic media (USB flash drives), could have developed without magnetic media.

One area of electromagnetic technology that would be easy to overlook in the context of industry is

---

Fascinating Facts About Electromagnet Technologies

- It may soon be possible for hard disk drives to store as much as 10 terabytes of data per square inch of disk space.
- The giant electromagnet at work in a junkyard works on the same principle as the tiny read-write heads in computer disk drives.
- What came first: electricity or magnetism?
- An electron moving in an atomic orbital has the same magnetic effect as electric current moving through a wire.
- Before generators were invented to make electricity readily available for the refining of aluminum metal from bauxite ore, the metal was far more valuable than gold, even though it is the most common known metal.
- Electromagnetic "rail gun" tests have accelerated a projectile from 0 to 13,000 miles per hour in 0.2 seconds.
- Non-ionizing electromagnetic fields are being investigated as treatments for malaria and cancer.
- Coupling an electric field and a magnet produces a magnetic field that is stronger than either one alone could produce.
- In 1820, Oerstad first observed the existence of a magnetic field around an electrified coil.
- Because of the electrical activity occurring inside the human body, people generate their own magnetic fields.
- Mass spectrometers use the interaction of an electric charge with a magnetic field to analyze the mass of fragments of molecules.

---

that of analytical devices used for testing and research applications. Because light is an electromagnetic phenomenon, any device that uses light in specific ways for measurement represents a branch of electromagnetic technology. Foremost of these are the spectrophotometers that are routinely used in medical and technical analysis. These play an important role in a broad variety of fields. Forensic analysis utilizes spectrophotometry to identify and compare materials. Medical analytics uses this technology in much the same way to quantify components in body fluids and other substances. Industrial processes often use some form of spectrophotometric process to monitor system functions and obtain feedback data to be used in automated process control.

In basic research environments, electromagnetic technologies are essential components, providing instrumentation such as the mass spectrometer and the photoelectron spectrometer, both of which function strictly on the principles of electric and magnetic fields. Nuclear magnetic resonance and electron spin resonance also are governed by those same principles. In the former case, these principles have been developed into the diagnostic procedure known as magnetic resonance imaging (MRI). Similar methods have application in materials testing, quality control, and nondestructive structural testing.

CAREERS AND COURSEWORK

Because electromagnetic technology is so pervasive in modern society and industry, students are faced with an extremely broad set of career options. It is not an overstatement to say that almost all possible careers rely on or are affected in some way by electromagnetic technology. Thus, those persons who have even a basic awareness of these principles will be somewhat better prepared than those who do not.

Students should expect to take courses in mathematics, physics, and electronics as a foundation for understanding electromagnetic technology. Basic programs will focus on the practical applications of electromagnetism, and some students will find a rewarding career simply working on electric motors and generators. Others may focus on the applications of electromagnetic technology as they apply to basic computer technology. More advanced careers, however, will require students to undertake more advanced studies that will include materials science, digital electronics, controls and feedback theory, servomechanism controls, and hydraulics and pneumatics, as well as advanced levels of physics and applied mathematics. Those who seek to understand the electromagnetic nature of matter will specialize in the study of quantum mechanics and quantum theory and of high-energy physics.

SOCIAL CONTEXT AND FUTURE PROSPECTS

Electromagnetic technologies and modern society are inextricably linked, particularly in the area of communications. Despite modern telecommunications swiftly becoming entirely digital in nature, the transmission of digital telecommunications signals is still carried out through the use of electromagnetic carriers. These can be essentially any frequency, although regulations control what range of the electromagnetic spectrum can be used for what purpose. Cellular telephones, for example, use the microwave region of the spectrum only, while other devices are restricted to operate in only the infrared region or in the visible light region of the electromagnetic spectrum.

One cannot overlook the development of electromagnetic technologies that will come about through the development of new materials and methods. These will apply to many sectors of society, including transportation and the military. In development are high-speed trains that use the repulsion between electrically generated magnetic fields to levitate the vehicle so that there is no physical contact between the machine and the track. Electromagnetic technologies will work to control the acceleration, speed, deceleration, and other motions of the machinery. High-speed transit by such machines has the potential to drastically change the transportation industry.

In future military applications, electromagnetic technologies will play a role that is as important as its present role. Communications and intelligence, as well as analytical reconnaissance, will benefit from the development of new and existing electromagnetic technologies. Weaponry, also, may become an important field of military electromagnetic technology, as experimentation continues toward the development of such things as cloaking or invisibility devices and electromagnetically powered rail guns.

Richard M. Renneboog, M.Sc.

FURTHER READING

Askeland, Donald R. "Magnetic Behaviour in Materials." In *The Science and Engineering of Materials*, edited by Donald Askeland. New York: Chapman & Hall, 1998. Provides an overview of the behavior of magnetic materials when acted upon by electromagnetic fields.

Dugdale, David. *Essentials of Electromagnetism.* New York: American Institute of Physics, 1993. A thorough discussion of electromagnetism through electric and magnetic properties, from first principles to practical applications and materials in electrical circuits and magnetic circuits.

Funaro, Daniele. *Electromagnetism and the Structure of Matter.* Hackensack, N.J.: World Scientific, 2008. An advanced treatise discussing deficiencies of Maxwell's theories of electromagnetism and proposing amended versions.

Han, G. C., et al. "A Differential Dual Spin Valve with High Pinning Stability." *Applied Physics Letters* 96 (2010). Discusses how hard disk drives could include ten terabytes of data per square inch using a new read-write head design.

Sen, P. C. *Principles of Electric Machines and Power Electronics.* 2d ed. New York: John Wiley & Sons, 1997. An advanced textbook providing a thorough treatment of electromagnetic machine principles, including magnetic circuits, transformers, DC machines, and synchronous and asynchronous motors.

See also: Applied Physics; Audio Engineering; Computer Engineering; Electrical Engineering; Electronics and Electronic Engineering; Energy Storage Technologies; Forensic Science; Hydroelectric Power Plants; Magnetic Resonance Imaging; Magnetic Storage; Music Technology; Spectroscopy; Transportation Engineering; Wind Power Technologies.

ELECTROMETALLURGY

FIELDS OF STUDY

Chemistry; physics; engineering; materials science.

SUMMARY

Electrometallurgy includes electrowinning, electroforming, electrorefining, and electroplating. The electrowinning of aluminum from its oxide (alumina) accounts for essentially all the world's supply of this metal. Electrorefining of copper is used to achieve levels of purity needed for its use as an electrical conductor. Electroplating is used to protect base metals from corrosion, to increase hardness and wear resistance, or to create a decorative surface. Electroforming is used to produce small, intricately shaped metal objects.

KEY TERMS AND CONCEPTS

- **Anode:** Electrode where electrons return to the external circuit.
- **Cathode:** Electrode where electrons react with the electrolyte.
- **Electrode:** Interface between an electrolyte and an external circuit; usually a solid rod.
- **Electroforming:** Creation of metal objects of a desired shape by electrolysis.
- **Electrolysis:** Chemical change produced by passage of an electric current through a substance.
- **Electrolyte:** Water solution or molten salt that conducts electricity.
- **Electroplating:** Creation of an adherent metal layer on some substrate by electrolysis.
- **Electrorefining:** Increasing the purity of a metal by electrolysis.
- **Electrowinning:** Obtaining a metal from its compounds by the use of electric current.
- **Equivalent Weight:** Atomic weight of a metal divided by the integer number of electron charges on each of its ions.
- **Faraday:** Least amount of electric charge needed to plate out one equivalent weight of a metal.
- **Pyrometallurgy:** Reduction of metal ores by heat and chemical reducing agents such as carbon.

DEFINITION AND BASIC PRINCIPLES

Electrometallurgy is that part of metallurgy that involves the use of electric current to reduce compounds of metals to free metals. It includes uses of electrolysis such as metal plating, metal refining. and electroforming. Electrolysis in this context does not include the cosmetic use of electrolysis in hair removal.

BACKGROUND AND HISTORY

The field of electrometallurgy started in the late eighteenth century and was the direct result of the development of a source of electric current, the cell invented by Alessandro Volta of Pavia, Italy. Volta placed alternating disks of zinc and copper separated by brine-soaked felt disks into a long vertical stack. In March, 1800, Volta described his "pile" in a letter to the Royal Society in London. His primitive device could produce only a small electric current, but it stimulated scientists all over Europe to construct bigger and better batteries and to explore their uses. Luigi Valentino Brugnatelli, an acquaintance of Volta, used a voltaic pile in 1802 to do electroplating. British surgeon Alfred Smee published *Elements of Electrometallurgy* in 1840, using the term "electrometallurgy" for the first time.

Michael Faraday discovered the basic laws of electrolysis early in the nineteenth century and was the first to use the word "electrolysis." Both Sir Humphry Davy and Robert Bunsen used electrolysis to prepare chemical elements. The development of dynamos based on Faraday's ideas made possible larger currents and opened up industrial uses of electrolysis, most notably the Hall-Héroult process for the production of aluminum around 1886.

The Aluminum Company of America (known as Alcoa since 1999) was founded in 1909 and became a major producer of aluminum by the Hall-Héroult process. Canada-based company Rio Tinto Alcan and Norway-based Norsk Hydro are also major producers of aluminum. Magnesium was first made commercially in Germany by I. G. Farben in the 1890's, and later in the United States by Dow Chemical. In both instances, electrolysis of molten magnesium chloride was used to produce the magnesium.

HOW IT WORKS

Electrolysis involves passage of an electric current through a circuit containing a liquid electrolyte, which may be a water solution or a molten solid. The current is carried through the electrolyte by migration of ions (positively or negatively charged particles). At the electrodes, the ions in the electrolyte undergo chemical reactions. For example, in the electrolysis of molten sodium chloride, positively charged sodium ions migrate to one electrode (the cathode) and negatively charged chloride ions to the other electrode (the anode). As the sodium ions interact with the cathode, they absorb electrons from the external circuit, forming sodium metal. This absorption of electrons is known as reduction of the sodium ions. At the anode, chloride ions lose electrons to the external circuit and are converted to chlorine gas—an oxidation reaction.

The chemical reaction occurring in electrolysis requires a minimum voltage. Sufficient voltage also must be supplied to overcome internal resistance in the electrolyte. Because an electrolyte may contain a variety of ions, alternative reactions are possible. In aqueous electrolytes, positively charged hydrogen ions are always present and can be reduced at the cathode. In a water solution of sodium chloride, hydrogen ions are reduced to the exclusion of sodium ions, and no sodium metal forms. The only metals that can be liberated by electrolysis from aqueous solution are those such as copper, silver, cadmium, and zinc, whose ions are more easily reduced than hydrogen ions.

An electrochemical method of liberating active metals depends on finding a liquid medium more resistant to reduction than water. The use of molten salts as electrolytes often solves this problem. Salts usually have high melting points, and it can be advantageous to choose mixtures of salts with lower melting temperatures, as long as the new cations introduced do not interfere with the desired reduction reaction. Choice of electrode materials is important because it can affect the purity of the metal being liberated. Usually inert electrodes are preferable.

In electroplating, it is necessary to produce a uniform adherent coating on the object that forms the cathode. Successful plating requires careful control of several factors such as temperature, concentration of the metal ion, current density at the cathode, and cleanliness of the surface to be plated. The metal ions in the electrolyte may need to be replenished as they are being depleted. Current density is usually kept low and plating done slowly. Sometimes additives that react with the metal ions or modify the viscosity or surface tension of the medium are used in the electrolyte. Successful plating conditions are often discovered by experiment and may not be understood in a fundamental sense.

Electroforming consists of depositing a metal plate on an object with the purpose of preparing a duplicate of the object in all its detail. Nonconducting objects can be rendered conductive by coating them with a conductive layer. The plated metal forms on a part called a mandrel, which forms a template and from which the new metal part is separated after the operation. The mandrel may be saved and used again or dissolved away with chemicals to separate it from the new object.

In electrorefining, anodes of impure metals are immersed in an electrolyte that contains a salt of the metal to be refined—often a sulfate—and possibly sulfuric acid. When current is passed through, metal dissolves from the anode and is redeposited on the cathode in purer form. Control of the applied voltage is necessary to prevent dissolution of less easily oxidized metal impurities. Material that is not oxidized at the anode falls to the bottom of the electrolysis cell as anode slime. This slime can be further processed for any valuable materials it may contain.

APPLICATIONS AND PRODUCTS

Electrowinning of Aluminum. Aluminum, the most abundant metal in the Earth's crust, did not become readily available commercially until the development of the Hall-Héroult process. This process involves electrolysis of dry aluminum oxide (alumina) dissolved in cryolite (sodium aluminum hexafluoride). Additional calcium fluoride is used to lower the melting point of the cryolite. The process runs at about 960 degrees Celsius and uses carbon electrodes. The alumina for the Hall-Héroult process is obtained from an ore called bauxite, an impure aluminum oxide with varying amounts of compounds such as iron oxide and silica. The preparation of pure alumina follows the Bayer process: The alumina is extracted from the bauxite as a solution in sodium hydroxide (caustic soda), reprecipitated by acidification, filtered, and dried. The electrolysis cell has a carbon coating at the bottom that forms a cathode, while the anodes are graphite rods extending down into the molten salt electrolyte. As aluminum forms, it forms a pool at the bottom of the cell and can be removed periodically

The anodes are consumed as the carbon reacts with the oxygen liberated by the electrolysis. The alumina needs to be replenished from time to time in the melt. The applied voltage is about 4.5 volts, but it rises sharply if the alumina concentration is too low. The electrolysis cells are connected in series, and there may be several hundred cells in an aluminum plant. The consumption of electric power amounts to about 15,000 kilowatt-hours per ton of aluminum. (A family in a home might consume 150 kilowatt-hours of power in a month.) Much of the power goes for heating and melting the electrolyte. The cost of electric power is a significant factor in aluminum manufacture and makes it advantageous to locate plants where power is relatively inexpensive, for example, where hydropower is available. Although some other metals are produced by electrolysis, aluminum is the metal produced in the greatest amount, tens of millions of metric tons per year.

Aluminum is the most commonly used structural metal after iron. As a low-density strong metal, aluminum tends to find uses where weight saving is important, such as in aircraft. When automobile manufacturers try to increase gas mileage, they replace the steel in vehicles with aluminum to save weight. Aluminum containers are commonly used for foods and beverages and aluminum foil for packaging.

Alcoa has developed a second electrolytic aluminum process that involves electrolysis of aluminum chloride. The aluminum chloride is obtained by chlorinating aluminum oxide. The chlorine liberated at the anode can be recycled. Also the electric power requirements of this process are less than for the Hall-Héroult process. This aluminum chloride reduction has not been used as much as the oxide reduction.

Electrowinning of Other Metals. Magnesium, like aluminum, is a low-density metal and is also manufactured by electrolysis. The electrolysis of molten magnesium chloride (in the presence of other metal chlorides to lower the melting point) yields magnesium at the cathode. The scale of magnesium production is not as large as that of aluminum, amounting to several hundred thousand metric tons per year. Magnesium is used in aircraft alloys, flares, and chemical syntheses.

Sodium metal comes from the electrolysis of molten sodium chloride in an apparatus called the Downs cell after its inventor J. C. Downs. Molten sodium forms at the cathode. Sodium metal was formerly very important in the process for making the gasoline additive tetraethyl lead. As leaded fuel is no longer sold in the United States, this use has declined, but sodium continues to be used in organic syntheses, the manufacture of titanium, and as a component (with potassium) in high-temperature heat exchange media for nuclear reactors.

Lithium and calcium are obtained in relatively small quantities by the electrolysis of their chlorides. Lithium is assuming great importance for its use in

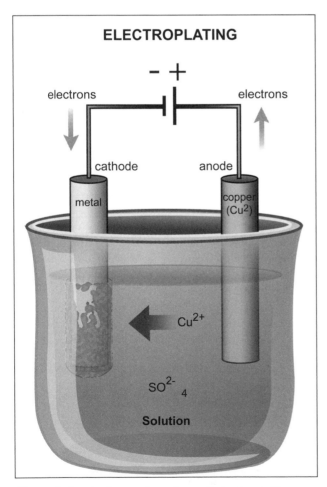

Electroplating is the creation of an adherent metal layer on some substrate by electrolysis.

Fascinating Facts About Electrometallurgy

- The engine block of the airplane built by Orville and Wilbur Wright in 1903 was made of aluminum obtained by electrolysis.
- In the early nineteenth century, aluminum was so expensive that it was used in jewelry.
- Roughly 5 percent of the electricity in the United States is used to make aluminum.
- The price of aluminum fell to its lowest recorded level in 1993; $0.53 per pound.
- The U.S. industry that uses the most aluminum is the beverage industry. Americans use 80 billion aluminum cans per year.
- When Prince Charles of England was invested as Prince of Wales in 1969, the coronet he received was made partly by electroforming.
- The Voyager space vehicle launched in 1977 contained a gold-plated copper disk electroplated with a patch of pure uranium-238 isotope. If this disk were recovered by some advanced alien civilization, the disk's age could be determined from the radioactive decay rate of the isotope.

high-performance batteries for all types of applications but particularly for powering electric automobiles. A lightweight lithium-aluminum alloy is used in the National Aeronautics and Space Administration's Ares rocket and the external tank of the space shuttle.

Electrorefining Applications. Metals are often obtained from their ores by pyrometallurgy (sequences of heating) and the use of reducing agents such as carbon. Metals obtained this way include iron and copper. Iron can be refined electrolytically, but much iron is used for steel production without refining.

Copper, however, is used in applications where purity is important. Copper, when pure, is ductile and an excellent electrical conductor, so it needs to be refined to be used in electrical wiring. Copper anodes (blister copper) are suspended in a water solution containing sulfuric acid and copper sulfate with steel cathodes. Electrolysis results in dissolution of copper from the anode and migration of copper ions to the cathode, where purified metal is deposited. The result is copper of 99.9 percent purity. A similar procedure may be used in recycling copper. Other metals that are electrorefined include aluminum,

zinc, nickel, cobalt, tin, lead, silver, and gold. Materials are added to the electrolyte to make the metal deposit more uniform: glue, metal chlorides, levelers, and brighteners may be included. The details of the conditions for electrorefining vary depending on the metal. The electrorefining done industrially worldwide ranges from 1,000 to 100,000 metric tons of metals per year.

Electroforming Applications. The possibility of reproducing complicated shapes on both large and small scale is an advantage of electroforming. Objects that would be impossible to produce because of their intricate shapes or small sizes are made by electroforming. The metal used may be a single metal or an alloy. The metals used most often are nickel and copper. The manufacture of compact discs for recording sound makes use of electroforming in the reproduction of the bumps and grooves in a studio disk of glass, which is a negative copy, and is then used to make a mold from which plastic discs can be cast. Metal foil can be electroformed by using a rotating mandrel surrounded by a cylindrical anode. The foil is peeled off the mandrel in a continuous sheet. The electroforming of copper foil for electronic applications is the largest application of electroforming. About $2 billion per year is spent on electroforming.

Electroplating Applications. Plating is done to protect metal surfaces from corrosion, to enhance the appearance of a surface, or to modify its properties in other ways such as to increase hardness or reflectivity. Familiar applications include chromium plating of automobile parts, silver plating of tableware and jewelry, and gold plating of medals and computer parts. Plating can also be done on nonmetallic surfaces such as plastic or ceramics. The manufacture of circuit boards requires a number of steps, some of which involve the electroplating of copper, lead, and tin. Many switches and other electrical contacts are plated with gold to prevent corrosion. The metals involved in commercial electroplating are mostly deposited from an aqueous electrolyte. This excludes metals such as aluminum or magnesium, which cannot be liberated in water solution. If a molten salt electrolyte is used, aluminum plating is possible, but it is seldom done.

IMPACT ON INDUSTRY

In 2006, 33.4 million tons of primary aluminum was produced worldwide, with the majority of it being made in China, the United States, Russia, Canada, Brazil, Australia, Norway, and India. Magnesium production in 2008 was 719,000 tons. These two metals account for most of the electrochemical metal production. Because of the high economic value of these industries, the processes involved are the subjects of research at government institutions and universities and at major corporations. Continuing research is done to improve the energy efficiency of processes and to mitigate pollution problems. A continuing problem with the Hall-Héroult process is emission of fluorine-containing compounds that are either toxic or are greenhouse gases. The high temperature of the process leads to corrosion problems in the cells, but attempts to lower the temperature can lead to reduced solubility of alumina in the electrolyte.

Alcoa and Rio Tinto Alcan are two of the largest aluminum producers in the world. Each has tens of thousands of employees spread over forty countries. Alcoa maintains a technical research center near Pittsburgh. In 2007, the company signed an agreement with a consortium of universities in Russia to sponsor research and development. Similar agreements were made in both India and China. Rio Tinto Alcan makes grants to several Canadian universities for research on aluminum production.

The National Research Council of Canada has an aluminum technical center in Saguenay, Quebec, and the E. B. Yeager Center for Electrochemical Science at Case Western University is in Cleveland, Ohio. The center sponsors workshops, lectures, and research in all areas of electrochemistry, including the use of computer-controlled processes to achieve optimum conditions in manufacturers. U.S. government laboratories such as Argonne National Laboratory in Illinois also do research on electrochemistry. The U.S. Department of Energy supports university research at various institutions, including Michigan Technological University, which received a $2 million grant for research on uses for aluminum smelting wastes.

CAREERS AND COURSE WORK

The path to a career in electrometallurgy is through a bachelor's or master's degree in chemistry or chemical engineering, although careers in research require a doctorate. The course work involves a thorough grounding in physical science (chemistry, physics), two years of calculus, and courses in computer science and engineering principles. Most large state universities offer programs in chemistry and chemical engineering. A few universities, including the University of Utah and the University of Nevada at Reno, offer specialized work in electrometallurgy. Rio Tinto Alcan and Alcoa offer summer internships for students and Rio Tinto Alcan also offers scholarships.

Multiday, personal development courses in electroplating are widely available through groups such as the National Society for Surface Finishing.

SOCIAL CONTEXT AND FUTURE PROSPECTS

The electometallurgy industry, like many industries, poses challenges for society. Metals have great value and many uses, which are an essential part of modern life. Unfortunately, electrometallurgy consumes huge amounts of energy and uses tons of unpleasant chemicals. In addition, aluminum plants emit carbon dioxide and fluorine compounds. However, the use of electricity to produce metals probably remains the cleanest and most efficient method.

In the future, the process of electrometallurgy probably will become more efficient and less polluting. Also, new techniques are likely to be developed to permit additional metals to be obtained by electrometallurgy. Titanium metal continues to be made by reduction of its chloride by sodium metal, but the Fray Farthing Chen (FFC) Cambridge process announced in 2000 shows promise as an electrolytic method for obtaining titanium. The process involves reduction of titanium oxide by electrically generated calcium in a molten calcium chloride medium. Titanium is valued for its strength, light weight, high temperature performance, and corrosion resistance. These qualities make it essential in jet engine turbines. Titanium is stable in the human body and can be used for making artificial knees and hips for implantation.

John R. Phillips, B.S., Ph.D.

FURTHER READING

Chen, G. Z., D. J. Fray, and T. W. Farthing. "Direct Electrochemical Reduction of Titanium Dioxide to Titanium in Molten Calcium Chloride." *Nature* 407 (2000): 361-364. First description of the FFC-Cambridge method—a possibly game-changing procedure for reducing metal oxides.

Curtis, Leslie. *Electroforming*. London: A & C Black, 2004. A simple description of electroforming with practical directions for applications.

Geller, Tom. "Common Metal, Uncommon Past." *Chemical Heritage* 25 no. 4 (Winter, 2007): 32-36. Historical details on the discovery and manufacture of aluminum.

Graham, Margaret B. W., and Bettye H. Pruitt. *R&D for Industry: A Century of Technical Innovation at Alcoa.* New York: Cambridge University Press, 1990. An extensive history that is mostly nontechnical. Much discussion of organizational and management matters.

Kanani, Nasser. *Electroplating: Basic Principles, Processes and Practice.* New York: Elsevier, 2005. A thorough discussion of electroplating processes including measurements of thickness and adherence using modern instruments.

Mertyns, Joost. "From the Lecture Room to the Workshop: John Frederic Daniell, the Constant Battery and Electrometallurgy Around 1840." *Annals of Science* 55, no. 3 (July, 1998): 241-261. Early developments in batteries, electroplating, and electroforming are described.

Pletcher, Derek, and Frank C. Walsh. *Industrial Electrochemistry.* 2d ed. New York: Blackie Academic & Professional, 1992. Particularly good discussion of metal plating and electroforming with both theoretical and practical aspects treated. List of definitions and units.

Popov, Konstantin I., Stojan S. Djokić, and Branimir N. Grgur, eds. *Fundamental Aspects of Metallurgy.* New York: Kluwer Academic/Plenum, 2002. Provides background on metallurgy, then the specifics of metal disposition.

WEB SITES
The Electrochemical Society
http://www.electrochem.org

International Society of Electrochemistry
http://www.ise-online.org

See also: Aeronautics and Aviation; Chemical Engineering; Coating Technology; Electrochemistry; Jet Engine Technology; Metallurgy.

ELECTRONIC COMMERCE

FIELDS OF STUDY

Business information systems; computer science; computer programming; database administration; retailing; marketing; Web development; software engineering; Web design.

SUMMARY

Electronic commerce (e-commerce) refers to the buying, selling, and transfer of products and services using the Internet. It offers enormous advantages for supply-chain management and the coordination of distribution channels, in addition to being convenient for consumers. Convenience is a big selling point: In 2007, online retail generated $175 billion and is expected to climb to $335 billion by 2012.

KEY TERMS AND CONCEPTS

- **Bandwidth:** Difference between highest and lowest number of data that a medium can transmit, expressed in cycles per second or bits per second. The amount of data that can be transmitted increases as the bandwidth increases.
- **Browser:** Software program that is used to view Web pages on the Internet.
- **Download:** Copy digital data and transmit it electronically.
- **E-Book:** Electronic book that can exist with or without a printed form of the same content. E-books are downloaded from a Web site to an electronic reader (e-reader).
- **Encryption:** Any procedure used in cryptography to convert plaintext into ciphertext to prevent anyone but the intended recipient from reading the data.
- **Enterprise Resource Planning (ERP):** Software integration of projects, distribution, manufacturing, and employees.
- **Firewall:** Hardware device placed between the private network and Internet connection that prevents unauthorized users from gaining access to data on the network.
- **Hacking:** Process of penetrating the security of other computers by using programming skills.
- **Lead Time:** Time required for a product to be received from a supplier after an order has been placed.
- **Operating System:** Computer program designed to manage the resources (including input and output devices) used by the central processing unit and that functions as an interface between a computer user and the hardware that runs the computer.
- **Overhead Costs:** Daily operating expenses.
- **Server:** Computer that is dedicated to managing resources shared by users (clients).

DEFINITION AND BASIC PRINCIPLES

E-commerce refers to the communications between customers, vendors, and business partners over the Internet. The term e-business incorporates the additional activities carried out within a business using intranets, such as communications related to production management and product development. Many also view e-business as referring to collaborations with partners and e-learning organizations.

In addition to online purchases of goods and services, e-commerce involves bill payments, online banking, e-wallets, smart cards, and digital cash. E-commerce depends on secure connections to the Internet. Many precautions to ensure security are necessary to maintain successful e-commerce, including public key cryptography, the Secure Sockets Layer (SSL), digital signatures, digital certificates, firewalls, and antivirus programs. New technology is constantly being developed to protect against worms, viruses, and other cyber attacks.

BACKGROUND AND HISTORY

In 1969, the Advanced Research Projects Agency (now the Defense Advanced Research Projects Agency) of the U.S. Department of Defense proposed a method to link together the computers at several universities to share computational data via networks. This network became known as ARPANET, which was the precursor of the Internet. As a result, electronic mail (e-mail) was developed, along with protocols for sending information over phone lines in packets. The protocols for the transmission of these packets of data came to be known as Transmission Control

Protocol (TCP) and Internet Protocol (IP). Together these two protocols, known as TCP/IP, are still in use and are responsible for the efficient communication conducted through the network of networks referred to as the Internet.

Oxford University graduate Tim Berners-Lee initially created the World Wide Web in 1989 while working at CERN, the European Council for Nuclear Research, and made it available in 1991. During this time, Cisco Systems was growing to become the first company to produce the broad range of hardware products that allowed ordinary individuals to access the Internet. In 1993, Marc Andreessen and Eric Bina, employees at National Center for Supercomputing Applications (NCSA), created Mosaic, the first Web browser that supported clickable buttons and links and allowed users to view text and images on the same page. New software and programming-language developments rapidly followed, allowing ordinary consumers easy access to the Internet. As a result, companies saw the opportunity to gain customers, resulting in the creation of online businesses, including Amazon in 1994, eBay in 1995, and PayPal and Priceline in 1998.

How It Works

The engineers who developed ARPANET created the use of digital packets for transmitting data via packet switching. The general idea was that it would be faster and cheaper to transmit digital data using small packets that could be sent, or routed, to their destination in the most efficient way possible, even if the original message had to be split up into smaller packets that were then joined back together at their destination. In order to accomplish the packaging of data and transmission via the best routes, the engineers who developed ARPANET also developed Transmission Control Protocol (TCP). The first and most common access method to the Internet was through the wiring that has transmitted telephone calls. However, wireless Internet connections can now be made much faster from many locations— even from cell phones. The economy has become dependent on digital communication, and the companies that sell the most goods and services have a strong Web presence. A great deal of planning goes into the maintenance of an effective Web site, and some of the most important steps in establishing an e-commerce company are discussed next.

E-Commerce Business Establishment. After first developing a practical business plan, the process for establishing an online business that will be able to compete for sales successfully could be overwhelming. One way to start an e-business is by using a turnkey solution, which is essentially a prepackaged type of software specifically for a new business. An alternative is to use the services of an Internet incubator, which is a company that specializes in e-business development. Both eToys and NetZero used Internet incubators to help them get started. An Internet incubator typically obtains ownership of at least 50 percent of the business and may also enlist funding help from venture capitalists to get started. Web-hosting companies sell space on a Web server to customers and maintain enough storage space for the Web site and to provide support services as well. A domain name for the Web site must be chosen and registered. This domain name is to be used in the URL (uniform resource locator) for the Web site. The URL, or Web site's Internet address, consists of three parts: the host name, which is shown by the www for World Wide Web; the domain name, which is usually the name of the company; and lastly, the top-level domain (TLD), which describes the type of organization that owns the domain name, such as .com for a commercial organization, or .gov for a government organization. An initial public offering (IPO) of stock to assist with funding usually follows for enterprises that achieve a certain level of success.

Design of Markets and Mechanisms of Transactions. Initially the business-to-business (B2B) types of transactions were the primary e-commerce activities. These activities quickly expanded to include sales to consumers via electronic retailing (e-tailing), often called business-to-consumer (B2C). Since the late 1990's e-commerce has expanded to include consumer-to-consumer (C2C) Web sites, including eBay, and consumer-to-business (C2B), such as Priceline, where several airlines or hotels will compete for the purchase dollars of consumers. Each of these types of transactions can be completed within the general structure of one of the many different types of e-commerce models to generate revenue.

Automated Negotiation and Peer-to-Peer Distribution Systems. Auction models allow an Internet user to assume the role of a buyer using either the reverse-auction model (where the buyer sets a price and sellers have to compete to beat that price) or the reverse-price model (where the seller sets the minimum price that will be accepted). Auction sites, such as eBay, update listings, feature items, and earn submission and commission fees, but they leave the processes of payment and delivery up to the actual buyers and sellers.

Dynamic-pricing models include the name-your-price companies, such as Priceline.com, that use a shopping bot to collect bids from customers and deliver these bids to the providers of services to see if they are accepted. A shopping bot is a computer program that searches through vast amounts of information, then collects, summarizes, and reports the information. This is one example of intelligent agents, software programs that have been designed to gather information, used by e-businesses. Priceline's immense success is due in part to its use of this technology.

Network Resource Allocation: Electronic Data Interchange (EDI). Portal models present a whole variety of news, weather, sports, and shopping all on one Web page that allows a visitor to see an overview and then choose to obtain more in-depth information. Vertical portals are specific for a single item, while horizontal portals function as search engines with access to a large range of items. Storefront models require a product line to be accessible online via the merchant server so that customers can select items from the database of products and collect them in the order-processing technology called a shopping cart. Businesses use EDI as a standardized protocol for communication to monitor daily inventory, shipments, and payments. Standardized forms for invoices and purchase orders are routinely accessible via the use of extensible markup language (XML). Companies such as Commerce One and TIBCO Software were created with the sole purpose of helping companies to move their businesses to the Web via B2B techniques. The transition of traditional brick-and-mortar stores to click-and-mortar stores has helped to decrease lead time and has caused an increase of just-in-time (JIT) inventory management. JIT inventory management allows e-businesses to save money because the companies do not overbuy goods and create an inventory surplus that they then have to worry about storing and selling. JIT in turn decreases overhead costs.

APPLICATIONS AND PRODUCTS

The development of computer technologies and the Internet has given rise to e-commerce, which, in turn, has spawned a variety of applications and products that facilitate e-business as well as enhance people's lives.

Consumer Products. Several tablets were among the most popular consumer digital purchases in 2010. Apple's iPad, Barnes & Noble's NookColor, Samsung's Galaxy Tab, Sony's Reader, and Amazon's Kindle are all capable of accessing the Internet and downloading magazines and books. The iPad is the most expensive of these tablets, but it is also capable of downloading video and audio files, while the Kindle is the least expensive and functions exclusively as an e-reader.

Both contact and contactless "smart cards," which resemble credit cards, have been developed to store much more information (banking, retail, identification, health care) because of a microprocessor embedded in the card. The E-ZPass, which is used by New York and New Jersey commuters to pay tolls, is an example of a contactless smart card. Smart cards are more secure than credit cards because they are encrypted and password protected.

Security Applications and Products. Companies have been created to help merchants accept credit card payments online, which are called card-not-present (CNP) transactions. These companies, such as CyberCash and PAYware, offer services to facilitate the authentication and authorization processes through the Secure Socket Layer (SSL) using Secure Electronic Transaction (SET) technology to minimize fraud. Additional security features include firewalls, encryption, and antivirus software. Visa and other major credit card companies have introduced e-wallets that allow customers to save their shipment address and payment information securely in an online database so that purchases can be made with one click of the mouse, instead of having to reenter the same information each time. In 1999, the Electronic Commerce Modeling Language (ECML) emerged as the protocol for e-wallet usage by merchants. PayPal can be used to transfer payments between consumers securely by simply creating an account using an e-mail address and a credit card, which is used to pay for goods and services. PayPal is ideally suited for use on an auction site, such as eBay. PayPal is especially secure because credit card information is checked before the transaction actually begins. This allows for payment to take place in real time, minimizing the opportunity for fraud.

Applications Using Wireless Transactions. Mobile business (m-business) made possible by wireless technology will continue to grow in importance. The third generation, called 3G technology, is allowing wireless devices to transmit data more than seven times faster than the 56K modem, and 4G technology began to replace it in 2011. Sprint PCS provides access to the Internet using the Code Division Multiple Access (CDMA) technology. CDMA technology assigns a unique code to each transmission on a specific channel, which allows each transmission to use the entire bandwidth available for that channel, greatly decreasing the time it takes to complete an e-commerce transaction.

IMPACT ON INDUSTRY

The developments in technology that have made the Internet more easily accessible to the average

---

Fascinating Facts About Electronic Commerce

- In 2010, e-commerce sales increased by 15.4 percent over 2009 e-commerce sales.
- E-book purchases in 2010 increased 150 percent over e-book purchases in 2009.
- By the year 2015 it is estimated that there will be more than 29.4 million e-books. Some believe that traditional book stores will eventually cease to exist. Borders bookstore is the largest national bookstore chain, and its financial difficulties (closing stores) could be an indication of this trend.
- The Apple iPhone is essentially a miniature, handheld computer, because it can download application programs ("apps") directly from the Internet, in addition to functioning as a phone. There are apps designed to perform just about any task, from accessing files on an office server to making the iPhone function as a flashlight. As of 2011 there were more than 100,000 apps available on the Apple App Store Web site.
- Almost 8 million Kindles were sold in 2010.
- Online sales for the 2010 holiday shopping season were estimated to total more than $36 billion.

---

consumer have been made primarily by American companies. These companies have continued to grow and expand to reach consumers all around the world. Other nations have grown economically because of the explosion in e-commerce, and much of the e-commerce growth is expected to involve the BRIC nations (Brazil, Russia, India, and China).

Amazon.com was founded in Jeff Bezos's Seattle garage in 1994. Bezos was named *Time* magazine's Person of the Year in 1999, and the company that he started in his garage continues to grow at an amazing rate. In addition to books, products available on Amazon have come to include jewelry, sporting goods, shoes, digital downloads of music, videos, games, software, health and beauty aids, and just about everything else possible, including groceries. The company reported net sales of $7.56 billion for the third quarter of 2010, which was a 39 percent increase over the same period in 2009.

eBay.com has become the largest online auction site in the world. Visitors to the site can buy or sell just about anything, including iPods, laptops, digital cameras, tickets to concerts and sporting events, jewelry, books, antiques, crafts, sporting goods, pet supplies, and clothes. eBay is headquartered in San Jose, California, and revenue for the third quarter of 2010 was $2.2 billion, an increase of 1 percent over the same period of 2009. Its PayPal business, acquired in 2002, has grown by at least an additional 1 million accounts every month, and it can accept twenty-four different types of currencies worldwide.

PayPal was initially founded in 1998 in Palo Alto, California, and like many other original U.S.-based tech companies, it has long since expanded its operations worldwide. PayPal has locations in Berlin, Tel Aviv, Dublin, Luxembourg, and China. Its purpose is to facilitate the processing of online payments for various e-commerce businesses.

Priceline.com, located in Norwalk, Connecticut, developed its name-your-own-price system for its online auction type of business and has grown since it emerged in 1998. Visitors to the Web site can list the price they want to pay for travel-related services and items, including hotels, vacation packages, cruises, airplane tickets, and car rentals. Although it briefly tried to expand on its e-commerce activities to include home loans, long-distance telephone service, and cars, it discontinued these ventures in 2002 and has continued to excel in travel-related services. Its chief e-commerce competitors are Expedia.com, Travelocity.com, Orbitz.com, and Hotwire.com. As of 2011, Priceline.com was the leader. The company had more than three hundred employees, and its third-quarter 2010 revenue was reported to be $1 billion, a 37.1 percent increase over 2009.

FedEx is the leader in air and ground transportation and won recognition in *Fortune* magazine's "World's Most Admired Companies" in 2006, 2007, 2008, 2009, and 2010 and in *Business Week*'s "50 Best Performers" in 2006. In 2010, its second-quarter revenue was $9.63 billion, an increase of 12 percent from the same period in 2009. FedEx continues its commitment to innovation with its FedEx Institute of Technology at the University of Memphis, in the company's hometown. The institute focuses on nanotechnology, artificial intelligence, biotechnology, and multimedia arts.

While an undergraduate at Yale University in 1965, Fred W. Smith wrote a term paper describing the implementation of an airfreight system that could transport computer parts and medications in a timely fashion. Although at the time his idea was not viewed as feasible by Yale faculty, Smith later bought an interest in Arkansas Aviation Sales, which eventually grew to become FedEx. E-commerce has continued to grow because of the rapid and efficient transportation of goods made possible by FedEx and the United Parcel Service (UPS).

CAREERS AND COURSE WORK

The job titles, career paths, and salaries vary a great deal within the field of e-commerce. Some of the typical job titles include Web site developer, Web designer, database administrator, Web master, and Web site manager. Jobs related to Web content require skills in Web development tools and software languages, including hypertext markup language (HTML), extensible markup language (XML), Java, JavaScript, Visual Basic, Visual Basic Script, and Active Server Pages (ASP). Database administrators focus less on these Web tools and languages and more on database-related tools, such as structured query language (SQL), Microsoft Access, and Oracle. Knowledge of computer networks and operating systems is also very helpful.

SOCIAL CONTEXT AND FUTURE PROSPECTS

Due in part to the easy accessibility of goods and services via the Internet, online businesses have become increasingly competitive, with more made-to-order goods being produced by companies such as Dell and corresponding decreases in the costs associated with the maintenance of a large inventory. Since so much more data are exchanged digitally via stock trades, mortgages, purchases of consumer goods, payment of bills, and banking transactions, it is conceivable that eventually digital cash and smart cards could replace traditional cash. Because consumers enjoy the comparison shopping among goods offered by companies all over the world, as well as the twenty-four-hour-a-day, seven-days-a-week convenience of online shopping, sales at the traditional brick-and-mortar stores will no doubt continue to decline, or at least migrate toward those goods that consumers prefer not to purchase online (such as those they wish to consider physically and those they wish to obtain immediately).

Hacking, identity theft, and other types of cyber theft have become problems, which have only increased the need for better security tools, such as digital certificates and digital signatures. Increased security needs will continue to fuel the ever-expanding Internet security industry.

Jeanne L. Kuhler, B.S., M.S., Ph.D.

FURTHER READING

Byrne, Joseph. *I-Net+ Certification Study System.* Foster City, Calif.: IDG Books Worldwide, 2000. This review guide provides technical information regarding hardware for networks, as well as software, important for e-commerce.

Castro, Elizabeth. *HTML, XHTML, and CSS, Sixth Edition: Visual Quick Start Guide.* Berkeley, Calif.: Peachpit Press, 2007. An introductory text describing how a novice can set up an individual Web site and includes plenty of helpful screen shots.

Deitel, Harvey M., Paul J. Deitel, and Kate Steinbuhler. *E-Business and E-Commerce for Managers.* Upper Saddle River, N.J.: Prentice Hall, 2001. This introductory textbook describes e-commerce in fairly nontechnical terms from the business perspective.

Longino, Carlo. "Your Wireless Future." *Business 2.0.* May 22, 2006. http://money.cnn.com/2006/05/18/technology/business2_wirelessfuture_intro/. Discusses the future of wireless technology in all sectors—business, entertainment, and communications.

Turban, Efraim, and Linda Volonino. *Information Technology for Management: Improving Performance in the Digital Economy*. 7th ed. Hoboken, N.J.: John Wiley & Sons, 2010. This introductory textbook provides both technical and nontechnical information related to e-commerce.

Umar, Amjad. "IT Infrastructure to Enable Next Generation Enterprises." *Information Systems Frontiers* 7, no. 3 (July, 2005): 217-256. Describes advances in network protocols and design.

WEB SITES
E-Commerce Times
http://www.ecommercetimes.com

National Retail Foundation's Digital Division
http://www.shop.org

See also: Computer Graphics; Computer Networks; Computer Science; Internet and Web Engineering; Software Engineering; Typography; Wireless Technologies and Communication.

ELECTRONIC MATERIALS PRODUCTION

FIELDS OF STUDY

Mathematics; physics; chemistry; crystallography; quantum theory; thermodynamics

SUMMARY

While the term "electronic materials" commonly refers to the silicon-based materials from which computer chips and integrated circuits are constructed, it technically includes any and all materials upon which the function of electronic devices depends. This includes the plain glass and plastics used to house the devices to the exotic alloys and compounds that make it possible for the devices to function. Production of many of these materials requires not only rigorous methods and specific techniques but also requires the use of high-precision analytical methods to ensure the structure and quality of the devices.

KEY TERMS AND CONCEPTS

- **Biasing:** The application of a voltage to a semiconductor structure (transistor) to induce a directional current flow in the structure.
- **Czochralski Method:** A method of pulling material from a molten mass to produce a single large crystal.
- **Denuded Zone:** Depth and area of a silicon wafer that contains no oxygen precipitates or interstitial oxygen.
- **Epi Reactor:** A thermally programmable chamber in which epitaxial growth of silicon chips is carried out.
- **Gettering:** A method of lowering the potential for precipitation from solution of metal contaminants in silicon, achieved by controlling the locations at which precipitation can occur.
- **Polysilicon (Metallurgical Grade Silicon):** A form of silicon that is 99 percent pure, produced by the reaction of silicon dioxide (SiO_2) and carbon (C) to produce silicon (Si) and carbon monoxide (CO) at a temperature of 2000 degrees Celsius.

DEFINITION AND BASIC PRINCIPLES

Electronic materials are those materials used in the construction of electronic devices. The major electronic material today is the silicon wafer, from which computer chips and integrated circuits (ICs) are made. Silicon is one of a class of elements known as semiconductors. These are materials that do not conduct electrical currents appreciably unless acted upon, or "biased," by an external voltage. Another such element is germanium.

The construction of silicon chips requires materials of high purity and consistent internal structure. This, in turn, requires precisely controlled methods in the production of both the materials and the structures for which they are used. Large crystals of ultrapure silicon are grown from molten silicon under strictly controlled environmental conditions. Thin wafers are sliced from the crystals and then polished to achieve the desired thickness and mirror-smooth surface necessary for their purpose. Each wafer is then subjected to a series of up to five hundred, and sometimes more, separate operations by which extremely thin layers of different materials are added in precise patterns to form millions of transistor structures. Modern CPU (central processing unit) chips have between 10^7 and 10^9 separate transistors per square centimeter etched on their surfaces in this way.

One of the materials added by the thin-layer deposition process is silicon, to fill in spaces between other materials in the structures. These layers must be added epitaxially, in a way that maintains the base crystal structure of the silicon wafer.

Other materials used in electronic devices are also formed under strictly controlled environmental conditions. Computers could not function without some of these materials, especially indium tin oxide (ITO) for what are called transparent contacts and indium nitride for light-emitting diodes in a full spectrum range of colors.

BACKGROUND AND HISTORY

The production of modern electronic materials began with the invention of the semiconductor bridge transistor in 1947. This invention, in turn, was made possible by the development of quantum theory and the vacuum tube technology with which electronic devices functioned until that time.

The invention of the transistor began the development of electronic devices based on the semiconducting character of the element silicon. Under the influence of an applied voltage, silicon can be induced to conduct an electrical current. This feature allows silicon-based transistors to function somewhat like an on-off switch according to the nature of the applied voltage.

In 1960, the construction of the functional laser by American physicist and Nobel laureate Arthur Schawlow began the next phase in the development of semiconductor electronics, as the assembly of transistors on silicon substrates was still a tedious endeavor that greatly limited the size of transistor structures that could be constructed. As lasers became more powerful and more easily controlled, they were applied to the task of surface etching, an advance that has produced ever smaller transistor structures. This development has required ever more refined methods of producing silicon crystals from which thin wafers can be cut for the production of silicon semiconductor chips, the primary effort of electronic materials production (though by no means the most important).

HOW IT WORKS

Melting and Crystallization. Chemists have long known how to grow large crystals of specific materials from melts. In this process, a material is heated past its melting point to become liquid. Then, as the molten material is allowed to cool slowly under controlled conditions, the material will solidify in a crystalline form with a highly regular atomic distribution.

Now, molten silicon is produced from a material called polysilicon, which has been stacked in a closed oven. Specific quantities of doping materials such as arsenic, phosphorus, boron, and antimony are added to the mixture, according to the conducting properties desired for the silicon chips that will be produced. The polysilicon melt is rotated in one direction (clockwise); then, a seed crystal of silicon, rotating in the opposite direction (counterclockwise), is introduced. The melt is carefully cooled to a specific temperature as the seed crystal structure is drawn out of the molten mass at a rate that determines the diameter of the resulting crystal.

To maintain the integrity of the single crystal that results, the shape is allowed to taper off into the form of a cone, and the crystal is then allowed to cool completely before further processing. The care with which this procedure is carried out produces a single

crystal of the silicon alloy as a uniform cylinder, whose ends vary in diameter first as the desired extraction rate was achieved and then due to the formation of the terminal cone shape.

Wafers. In the next stage of production, the non-uniform ends of the crystal are removed using an inner diameter saw. The remaining cylinder of crystal is called an ingot, and is then examined by X ray to determine the consistency and integrity of the crystal structure. The ingot then will normally be cut into smaller sections for processing and quality control.

To produce the rough wafers that will become the substrates for chips, the ingot pieces are mounted on a solid base and fed into a large wire saw. The wire saw uses a single long moving wire to form a thick network of cutting edges. A continuous stream of slurry containing an extremely fine abrasive provides the cutting capability of the wire saw, allowing the production of many rough wafers at one time. The rough wafers are then thoroughly cleaned to remove any residue from the cutting stage.

Another procedure rounds and smooths the edges of each wafer, enhancing its structural strength and resistance to chipping. Each wafer is also laser-etched with identifying data. They then go on to a flat lapping procedure that removes most of the machining marks left by the wire saw, and then to a chemical etching process that eliminates the marking that the lapping process has left. Both the lapping process and the chemical etching stage are used to reduce the thickness of the wafers.

Polishing. Following lapping and rigorous cleaning, the wafers move into an automated chemical-mechanical polishing process that gives each wafer an extremely smooth mirror-like and flat surface. They are then again subjected to a series of rigorous chemical cleaning baths, and are then either packaged for sale to end users or moved directly into the epitaxial enhancement process.

Epitaxial Enhancement. Epitaxial enhancement is used to deposit a layer of ultrapure silicon on the surface of the wafer. This provides a layer with different properties from those of the underlying wafer material, an essential feature for the proper functioning of the MOS (metal-oxide-semiconductor) transistors that are used in modern chips. In this process, polished wafers are placed into a programmable oven and spun in an atmosphere of trichlorosilane gas. Decomposition of the trichlorosilane deposits silicon atoms on the surface of the wafers. While this produces an identifiable layer of silicon with different properties, it also maintains the crystal structure of the silicon in the wafer. The epitaxial layer contains no imperfections that may exist in the wafer and that could lead to failure of the chips in use.

From this point on, the wafers are submitted to hundreds more individual processes. These processes build up the transistor structures that form the functional chips of a variety of integrated circuit devices and components that operate on the principles of digital logic.

APPLICATIONS AND PRODUCTS

Microelectronics. The largest single use of silicon chips is in the microelectronics industry. Every digital device functions through the intermediacy of a silicon chip of some kind. This is as true of the control pad on a household washing machine as it is of the most sophisticated and complex CPU in an ultramodern computer.

Digital devices are controlled through the operation of digital logic circuits constructed of transistors built onto the surface of a silicon chip. The chips can be exceedingly small. In the case of integrated circuit chips, commonly called ICs, only a few transistors may be required to achieve the desired function.

The simplest of these ICs is called an inverter, and a standard inverter IC provides six separate inverter circuits in a dual inline package (DIP) that looks like a small rectangular block of black plastic about 1 centimeter wide, 2 centimeters long, and 0.5 centimeters thick, with fourteen legs, seven on each side. The actual silicon chip contained within the body of the plastic block is approximately 5 millimeters square and no more than 0.5 millimeters thick. Thousands of such chips are cut from a single silicon wafer that has been processed specifically for that application.

Inverters require only a single input lead and a single output lead, and so facilitate six functionalities on the DIP described. However, other devices typically use two input leads to supply one output lead. In those devices, the same DIP structure provides only four functionalities. The transistor structures are correspondingly more complex, but the actual chip size is about the same. Package sizes increase according to the complexity of the actual chip and the number of leads that it requires for its

Older image of a semiconductor IC chip and wafer. (Robert A. Isaacs/Photo Researchers, Inc.)

function and application to physical considerations such as dissipation of the heat that the device will generate in operation.

In the case of a modern laptop or desktop computer, the CPU chip package may have two hundred leads on a square package that is approximately 4 centimeters on a side and less than 0.5 centimeters in thickness. The actual chip inside the package is a very thin sheet of silicon about 1 square centimeter in size, but covered with several million transistor structures that have been built up through photo-etching and chemical vapor deposition methods, as described above. Examination of any service listing of silicon chip ICs produced by any particular manufacturer will quickly reveal that a vast number of different ICs and functionalities are available.

Solar Technology. There are several other current uses for silicon wafer technology, and new uses are yet to be realized. Large quantities of electronic-grade silicon wafers are used in the production of functional solar cells, an area of application that is experiencing high growth, as nonrenewable energy resources become more and more expensive. Utilizing the photoelectron effect first described by Albert Einstein in 1905, solar cells convert light energy into an electrical current. Three types are made, utilizing both thick (> 300 micrometers [μm]) and thin (a few μm) layers of silicon. Thick-layer solar cells are constructed from single crystal silicon and from large-grain polycrystalline silicon, while thin-layer solar cells are constructed by using vapor deposition to deposit a layer of silicon onto a glass substrate.

Microelectronic and Mechanical Systems. Silicon chips are also used in the construction of microelectronic and mechanical systems (MEMS). Exceedingly tiny mechanical devices such as gears and single-pixel mirrors can be constructed using the technology developed for the production of silicon chips. Devices produced in this way are by nature highly sensitive and dependable in their operation, and so the majority of MEMS development is for the production of specialized sensors, such as the accelerometers used to initiate the deployment of airbag restraint systems in automobiles. A variety of other products are also available using MEMS technology, including biosensors, the micronozzles of inkjet printer cartridges, microfluidic test devices, microlenses and arrays of microlenses, and microscopic versions of tunable capacitors and resonators.

Other Applications. Other uses of silicon chip technology, some of which is in development, include mirrors for X-ray beams; mirrors and prisms for application in infrared spectroscopy, as silicon is entirely transparent to infrared radiation; and the material called porous silicon, which is made electrochemically from single-crystal silicon and has itself presented an exceptionally varied field of opportunity in materials science.

As mentioned, there are also many other materials that fall into the category of electronic materials. Some, such as copper, gold, and other pure elements, are produced in normal ways and then subjected to methods such as zone refining and vapor deposition techniques to achieve high purity and thin layers in the construction of electronic devices. Many exotic elements and metallic alloys, as well as specialized plastics, have been developed for use in electronic devices. Organic compounds known as liquid crystals, requiring no extraordinary synthetic measures, are normally semisolid materials that have properties of both a liquid and a solid. They are extensively used as the visual medium of thin liquid crystal display (LCD) screens, such as would be found in wristwatches, clocks, calculators, laptop and tablet computers, almost all desktop monitors, and flat-screen televisions.

Another example is the group of compounds made up of indium nitride, gallium nitride, and aluminum nitride. These are used to produce light-emitting diodes (LEDs) that provide light across the full visible spectrum. The ability to grow these LEDs on the same chip now offers a technology that could completely replace existing CRT (cathode ray tube) and LCD technologies for visual displays.

IMPACT ON INDUSTRY

Electronic materials production is an entire industry unto itself. While the products of this industry are widely used throughout society, they are not used in the form in which they are produced. Rather, the products of the electronic materials industry become input supplies for further manufacturing processes. Silicon chips, for example, produced by any individual manufacturer, are used for in-house manufacturing or are marketed to other manufacturers, who, in turn, use the chips to produce their own particular products, such as ICs, solar cells, and microdevices.

This intramural or business-to-business market aspect of the electronic materials production industry, with its novel research and development efforts and especially given the extent to which society now relies on information transfer and storage, makes ascribing an overall economic value to the industry impossible. One has only to consider the number of computing devices produced and sold each year around the world to get a sense of the potential value of the electronic materials production industry.

Ancillary industries provide other materials used by the electronic materials production industry, many of which must themselves be classified as electronic materials. An electric materials company, for example, may provide polishing and surfacing materials, photovoltaic materials, specialty glasses, electronic packaging materials, and many others.

Given both the extremely small size and sensitivity of the structures created on the surface of silicon chips and the number of steps required to produce those structures, quality control procedures are stringent. These steps may be treated as part of a multi-step synthetic procedure, with each step producing a yield (as the percentage of structures that meet functional requirements). In silicon-chip production, it is important to understand that only the chips that are produced as functional units at the end of the process are marketable. If a process requires two hundred individual construction steps, even a 99 percent success rate for each step translates into a final yield of functional chips of only 0.99^{200}, or 13.4 percent. The majority of chip structures fail during construction, either through damage or through a step failure. It is therefore imperative that each step in the construction of silicon chips be precisely carried out.

To that end, procedures and quality control methods have been developed that are applicable in other situations too. Clean room technology that is essential for maximizing usable chip production is equally valuable in biological research and medical treatment facilities, applied physics laboratories, space exploration, aeronautics repair and maintenance facilities, and any other situations in which steps to protect either the environment or personnel from contamination must be taken.

CAREERS AND COURSEWORK

Electronic materials production is a specialist field that requires interested students to take specialist training in many subject areas. For many such careers, a university degree in solid state physics or electronic engineering is required. For those who will specialize in the more general field of materials science, these subject areas will be included in the overall curriculum. Silicon technology and semiconductors are also primary subject areas. The fields of study listed here are considered prerequisites for specialist study in the field of electronic materials production, and students can expect to continue studies in these subjects as new aspects of the field develop.

Researchers are now looking into the development of transistor structures based on graphene. This represents an entirely new field of study and application, and the technologies that develop from it will also set new requirements for study. High-end spectrometric methodologies are essential tools in the study and development of this field, and students can expect to take advanced study and training in the use of techniques such as scanning probe microscopy.

SOCIAL CONTEXT AND FUTURE PROSPECTS

Moore's law has successfully predicted the progression of transistor density that can be inscribed onto a silicon chip. There is a finite limit to that density, however, and the existing technology is very near or at that limit. Electronic materials research continues to improve methods and products in an effort to push the Moore limit.

New technologies must be developed to make the use of transistor logic as effective and as economic as possible. To that end, there exists a great deal of research into the application of new materials. Foremost is the development of graphene-based transistors and quantum dot technology, which will drive

---

Fascinating Facts About Electronic Materials Production

- Large single crystals of silicon are grown from a molten state in a process that literally pulls the molten mass out into a cylindrical shape.

- About 70 percent of silicon chips fail during the manufacturing process, leaving only a small percentage of chips that are usable.

- Silicon is invisible to infrared light, making it exceptionally useful for infrared spectroscopy and as mirrors for X rays.

- Quantum dot and graphene-based transistors will produce computers that are orders of magnitude more powerful than those used today.

- The photoelectric effect operating in silicon allows solar cells to convert light energy into an electrical current.

- Semiconductor transistors were invented in 1947 and integrated circuits in 1970, and the complexity of electronic components has increased by about 40 percent each year.

- Copper and other metals dissolve very quickly in liquid silicon, but precipitate out as the molten material cools, often with catastrophic results for the silicon crystal.

- Porous silicon is produced electrochemically from single crystals of silicon. Among its other properties, porous silicon is highly explosive.

the level of technology into the molecular and atomic scales.

Richard M. Renneboog, M.Sc.

FURTHER READING

Akimov, Yuriy A., and Wee Shing Koh. "Design of Plasmonic Nanoparticles for Efficient Subwavelength Trapping in Thin-Film Solar Cells." *Plasmonics* 6 (2010): 155-161. This paper describes how solar cells may be made thinner and lighter by the addition of aluminum nanoparticles on a surface layer of indium tin oxide to enhance light absorption.

Askeland, Donald R. *The Science and Engineering of Materials.* London: Chapman & Hall, 1998. A recommended resource, this book provides a great deal of fundamental background regarding the physical behavior of a wide variety of materials and processes that are relevant to electronic materials production.

Falster, Robert. "Gettering in Silicon: Fundamentals and Recent Advances." *Semiconductor Fabtech* 13 (2001). This article provides a thorough description of the effects of metal contamination in silicon and the process of gettering to avoid the damage that results from such contamination.

Zhang, Q., et al. "A Two-Wafer Approach for Integration of Optical MEMS and Photonics on Silicon Substrate." *IEEE Photonics Technology Letters* 22 (2010): 269-271. This paper examines how photonic and micro-electromechanical systems on two different silicon chips can be precisely aligned.

Zheng, Y., et al. "Graphene Field Effect Transistors with Ferroelectric Gating." *Physical Review Letters* 105 (2010). This paper discusses the experimental development and successful testing of a graphene-based field-effect transistor system using gold and graphene electrodes with SiO_2 gate structures on a silicon substrate.

WEB SITES

SCP Symposium (June 2005) "Silicon Starting Materials for Sub-65nm Technology Nodes."
http://www.memc.com/assets/file/technology/papers/SCP-Symposium-Seacrist.pdf

University of Kiel "Electronic Materials Course."
http://www.tf.uni-kiel.de/matwis/amat/elmat_en/index.html

See also: Applied Physics; Computer Engineering; Computer Science; Electrochemistry; Electronics and Electronic Engineering; Integrated-Circuit Design; Liquid Crystal Technology; Nanotechnology; Surface and Interface Science; Transistor Technologies; Zone Refining.

ELECTRONICS AND ELECTRONIC ENGINEERING

FIELDS OF STUDY

Mathematics; physics; electronics; electrical engineering; automotive mechanics; analytical technology; chemical engineering; aeronautics; avionics; robotics; computer programming; audio/video technology; metrology; audio engineering; telecommunications; broadcast technology; computer technology; computer engineering; instrumentation

SUMMARY

A workable understanding of the phenomenon of electricity originated with proof that atoms were composed of smaller particles bearing positive and negative electrical charges. The modern field of electronics is essentially the science and technology of devices designed to control the movement of electricity to achieve some useful purpose. Initially, electronic technology consisted of devices that worked with continuously flowing electricity, whether direct or alternating current. Since the development of the transistor in 1947 and the integrated circuit in 1970, electronic technology has become digital, concurrent with the ability to assemble millions of transistor structures on the surface of a single silicon chip.

KEY TERMS AND CONCEPTS

- **Cathode Rays:** Descriptive term for energetic beams emitted from electrically stimulated materials inside of a vacuum tube, identified by J. J. Thomson in 1897 as streams of electrons.
- **Channel Rays:** Descriptive term for energetic beams having the opposite electrical charge of cathode rays, emitted from electrically stimulated materials inside a vacuum tube, also identified by J. J. Thomson in 1897.
- **Gate:** A transistor structure that performs a specific function on input electrical signals to produce specific output signals.
- **Operational Amplifier (Op-Amp):** An integrated circuit device that produces almost perfect signal reproduction with high gains of amplification and precise, stable voltages and currents.

- **Sampling:** Measurement of a specific parameter such as voltage, pressure, current, and loudness at a set frequency determined by a clock cycle such as 1 MHz.
- **Semiconductor:** An element that conducts electricity effectively only when subjected to an applied voltage.
- **Zener Voltage:** The voltage at which a Zener diode is designed to operate at maximum efficiency to produce a constant voltage, also called the breakdown voltage.

DEFINITION AND BASIC PRINCIPLES

The term "electronics" has acquired different meanings in different contexts. Fundamentally, "electronics" refers to the behavior of matter as affected by the properties and movement of electrons. More generally, electronics has come to mean the technology that has been developed to function according to electronic principles, especially pertaining to basic digital devices and the systems that they operate. The term "electronic engineering" refers to the practice of designing and building circuitry and devices that function on electronic principles.

The underlying principle of electronics derives from the basic structure of matter: that matter is composed of atoms composed of smaller particles. The mass of atoms exists in the atomic nucleus, which is a structure composed of electrically neutral particles called neutrons and positively charged particles called protons. Isolated from the nuclear structure by a relatively immense distance is an equal number of negatively charged particles called electrons. Electrons are easily removed from atoms, and when a difference in electrical potential (voltage) exists between two points, electrons can move from the area of higher potential toward that of lower potential. This defines an electrical current.

Devices that control the presence and magnitude of both voltages and currents are used to bring about changes to the intrinsic form of the electrical signals so generated. These devices also produce physical changes in materials that make comprehensible the information carried by the electronic signal.

BACKGROUND AND HISTORY

Archaeologists have found well-preserved Parthian relics that are now believed to have been rudimentary, but functional, batteries. It is believed that these ancient devices were used by the Parthians to plate objects with gold. The knowledge was lost until 1800, when Italian physicist Alessandro Volta reinvented the voltaic pile. Danish physicist and chemist Hans Christian Oersted demonstrated the relationship between electricity and magnetism in 1820, and in 1821, British physicist and chemist Michael Faraday used that relationship to demonstrate the electromagnetic principle on which all electric motors work. In 1831, he demonstrated the reverse relationship, inventing the electrical generator in the process.

Electricity was thought, by American statesman and scientist Benjamin Franklin and many other scientists of the eighteenth and nineteenth centuries, to be some mysterious kind of fluid that might be captured and stored. A workable concept of electricity was not developed until 1897, when J. J. Thomson identified cathode rays as streams of light electrical particles that must have come from within the atoms of their source materials. He arbitrarily ascribed their electrical charge as negative. Thomson also identified channel rays as streams of massive particles from within the atoms of their source materials that are endowed with the opposite electrical charge of the electrons that made up cathode rays. These observations essentially proved that atoms have substructures. They also provided a means of explaining electricity as the movement of charged particles from one location to another.

With the establishment of an electrical grid, based on the advocacy of alternating current by Serbian American engineer and inventor Nikola Tesla (1856-1943) , a vast assortment of analogue electrical devices were soon developed for consumer use, though initially these devices were no more than electric lights and electromechanical applications based on electric motors and generators.

As the quantum theory of atomic structure came to be better understood and electricity better controlled, electronic theory became much more important. Spurred by the success of the electromagnetic telegraph of American inventor Samuel Morse (1791-1872), scientists sought other applications. The first major electronic application of worldwide importance was wireless radio, first demonstrated by Italian inventor Guglielmo Marconi (1874-1937). Radio depended on electronic devices known as vacuum tubes, in which structures capable of controlling

Electronic components soldered into a printed circuit board (PCB). The black squares are microprocessor silicon chips, the central component of most electronics devices. (Arno Massee/Photo Researchers, Inc.)

currents and voltages could operate at high temperatures in an evacuated tube with external contacts. In 1947, American physicist William Shockley and colleagues invented the semiconductor-based transistor, which could be made to function in the same manner as vacuum tube devices, but without the high temperatures, electrical power consumption, and vacuum construction of those analogue devices.

In 1970, the first integrated circuit "chips" were made by constructing very small transistor structures on the surface of a silicon chip. This gave rise to the entire digital technology that powers the modern world.

APPLICATIONS AND PRODUCTS

Electronics are applied in practically every conceivable manner today, based on their utility in converting easily-produced electrical current into mechanical movement, sound, light, and information signals.

Basic Electronic Devices. Transistor-based digital technology has replaced older vacuum tube technology, except in rare instances in which a transistorized device cannot perform the same function. Electronic circuits based on vacuum tubes could carry out essentially the same individual operations as transistors, but they were severely limited by physical size, heat production, energy consumption, and mechanical failure. Nevertheless, vacuum tube technology was the basic technology that produced radio, television, radar, X-ray machines, and a broad variety of other electronic applications.

Electronic devices that did not use vacuum tube technology, but which operated on electronic and electromagnetic principles, were, and still are, numerous. These devices include electromagnets and all electric motors and generators. The control systems for many such devices generally consisted of nothing more than switching circuits and indicator lights. More advanced and highly sensitive devices required control systems that utilized the more refined and correspondingly sensitive capabilities available with vacuum tube technology.

Circuit Boards. The basic principles of electricity, such as Ohm's resistance law and Kirchoff's current law and capacitance and inductance, are key features in the functional design and engineering of analogue electronic systems, especially for vacuum-tube control systems. An important application that facilitated the general use and development of electronic systems of all kinds is printed circuit board technology. A printed circuit board accepts standardized components onto a nonconducting platform made initially of compressed fiber board, which was eventually replaced by a resin-based composite board. A circuit design is photo-etched onto a copper sheet that makes up one face of the circuit board, and all nonetched copper is chemically removed from the surface of the board, leaving the circuit pattern. The leads of circuit components such as resistors, capacitors, and inductors are inserted into the circuit pattern and secured with solder connections.

Mass production requirements developed the flotation soldering process, whereby preassembled circuit boards are floated on a bed of molten solder, which automatically completes all solder connections at once with a high degree of consistency. This has become the most important means of circuit board production since the development of transistor technology, being highly compatible with mechanization and automation and with the physical shapes and dimensions of integrated circuit (IC) chips and other components.

Digital Devices. Semiconductor-based transistors comprise the heart of modern electronics and electronic engineering. Unlike vacuum tubes, transistors do not work on a continuous electrical signal. Instead, they function exceedingly well as simple on-off switches that are easily controlled. This makes them well adapted to functions based on Boolean algebra. All transistor structures consist of a series of "gates" that perform a specific function on the electronic signals that are delivered to them.

Digital devices now represent the most common (and rapidly growing) application of electronics and electronic engineering, including relatively simple consumer electronic devices such as compact fluorescent light bulbs and motion-detecting air fresheners to the most advanced computers and analytical instrumentation. All applications, however, utilize an extensive, but limited, assortment of digital components in the form of IC chips that have been designed to carry out specific actions with electrical or electromagnetic input signals.

Input signals are defined by the presence or absence of a voltage or a current, depending upon the nature of the device. Inverter gates reverse the sense of the input signal, converting an input voltage (high input) into an output signal of no voltage (low output), and vice versa. Other transistor structures (gates) called AND, NAND, OR, NOR and X-OR function to combine input signals in different ways to produce corresponding output signals. More advanced devices (for example, counters and shift registers) use combinations of the different gates to construct various functional circuits that accumulate signal information or that manipulate signal information in various ways.

One of the most useful of digital IC components is the operational amplifier, or Op-Amp. Op-Amps contain transistor-based circuitry that boosts the magnitude of an input signal, either voltage or current, by five orders of magnitude (100,000 times) or more, and are the basis of the exceptional sensitivity of the modern analytical instruments used in all fields of science and technology.

Electrical engineers are involved in all aspects of the design and development of electronic equipment. Engineers act first as the inventors and designers of electronic systems, conceptualizing the specific functions a potential system will be required to carry out. This process moves through the specification of the components required for the system's functionality to the design of new system devices. The design parameters extend to the infrastructure that must support the system in operation. Engineers determine the standards of safety, integrity, and operation that must be met for electronic systems.

Consumer Electronics. For the most part, the term "electronics" is commonly used to refer to the electronic devices developed for retail sale to consumers. These devices include radios, television sets, DVD and CD players, cell phones and messaging devices, cameras and camcorders, laptops, tablets, printers, computers, fax and copy machines, cash registers, and scanners. Millions such devices are sold around the world each day, and numerous other businesses have formed to support their operation.

IMPACT ON INDUSTRY

With electrical and electronic technology now intimately associated with all aspects of society, the impact of electronics and electronic engineering on industry is immeasurable. It would be entirely fair to say that modern industry could not exist without electronics. Automated processes, which are ubiquitous, are not possible without the electronic systems that control them.

The transportation industry, particularly the automotive industry, is perhaps the most extensive user of electronics and electronic engineering. Modern automobiles incorporate an extensive electronic network in their construction to provide the ignition and monitoring systems for the operation of their internal combustion engines and for the many monitoring and control systems for the general safe operation of the vehicle; an electronic network also informs and entertains the driver and passengers. In some cases, electronic systems can completely take control of the vehicle to carry out such specific programmable actions as speed control and parallel parking. Every automated process in the manufacture of automobiles and other vehicles serves to reduce the labor required to carry out the corresponding tasks, while increasing the efficiency and precision of the process steps. Added electronic features also increase the marketability of the vehicles and, hence, manufacturer profits.

Processes that have been automated electronically also have become core components of general manufacturing, especially in the control of production machinery. For example, shapes formed from bent tubing are structural components in a wide variety of applications. While the process of bending the tubing can be carried out by the manual operation of a suitably equipped press, an automated process will produce tube structures that are bent to exact angles

and radii in a consistent manner. Typically, a human operator places a straight tube into the press, which then positions and repositions the tube for bending over its length, according to the program that has been entered into the manufacturing system's electronic controller. Essentially, all continuous manufacturing operations are electronically controlled, providing consistent output.

Electronics and electronic engineering make up the essence of the computer industry; indeed, electronics is an industry worth billions of dollars annually. Electronics affects not only the material side of industry but also the theoretical and actuarial side. Business management, accounting, customer relations, inventory and sales data, and human resources

---

Fascinating Facts About Electronics and Electronic Engineering

- In 1847, George Boole developed his algebra for reasoning that was the foundation for first-order predicate calculus, a logic rich enough to be a language for mathematics.

- In 1950, Alan Turing gave an operational definition of artificial intelligence. He said a machine exhibited artificial intelligence if its operational output was indistinguishable from that of a human.

- In 1956, John McCarthy and Marvin Minsky organized a two-month summer conference on intelligent machines at Dartmouth College. To advertise the conference, McCarthy coined the term "artificial intelligence."

- Digital Equipment Corporation's XCON, short for eXpert CONfigurer, was used in-house in 1980 to configure VAX computers and later became the first commercial expert system.

- In 1989, international chess master David Levy was defeated by a computer program, Deep Thought, developed by IBM. Only ten years earlier, Levy had predicted that no computer program would ever beat a chess master.

- In 2010, the Haystack group at the Computer Science and Artificial Intelligence Laboratory at the Massachusetts Institute of Technology developed Soylent, a word-processing interface that lets users edit, proof, and shorten their documents using Mechanical Turk workers.

management all depend on the rapid information-handling that is possible through electronics.

XML (extensible markup language) methods and applications are being used (and are in development) to interface electronic data collection directly to physical processes. This demands the use of specialized electronic sensing and sampling devices to convert measured parameters into data points within the corresponding applications and databases. XML is an application that promises to facilitate information exchange and to promote research using large-scale databases. The outcome of this effort is expected to enhance productivity and to expand knowledge in ways that will greatly increase the efficiency and effectiveness of many different fields.

An area of electronics that has become of great economic importance in recent years is that of electronic commerce: the exclusive use of electronic communication technology for the conduct of business between suppliers and consumers. Electronic communications encompasses interoffice faxing, e-mail exchanges, and the Web commerce of companies such as eBay, Amazon, Google, and of the New York and other stock exchanges. The commercial value of these undertakings is measured in billions of dollars annually, and it is expected to continue to increase as new applications and markets are developed.

The fundamental feature here is that these enterprises exist because of the electronic technology that enables them to communicate with consumers and with other businesses. The electronics technology and electronics engineering fields have thus generated entirely new and different daughter industries, with the potential to generate many others, all of which will depend on persons who are knowledgeable in the application and maintenance of electronics and electronic systems.

CAREERS AND COURSEWORK

Many careers depend on knowledge of electronics and electronic engineering because almost all machines and devices used in modern society either function electronically or utilize some kind of electronic control system. The automobile industry is a prime example, as it depends on electronic systems at all stages of production and in the normal operation of a vehicle. Students pursuing a career in automotive mechanics can therefore be expected to study electronic principles and applications as a significant part of their training. The same reasoning applies in all other fields that have a physical reliance on electronic technology.

Knowledge of electronics has become so essential that atomic structure and basic electronic principles, for example, have been incorporated into the elementary school curriculum. Courses of study in basic electronics in the secondary school curriculum are geared to provide a more detailed and practicable knowledge to students.

Specialization in electronics and other fields in which electronics play a significant role is the province of a college education. Interested students can expect to take courses in advanced mathematics, physics, chemistry, and electronics technology as part of the curriculum of their specialty programs. Normally, a technical career, or a skilled trade, requires a college-level certification and continuing education. In some cases, recertification on a regular schedule is also required to maintain specialist standing in that trade.

Students who plan to pursue a career in electronic engineering at a more theoretical level will require, at minimum, a bachelor's degree. A master's degree can prepare a student for a career in forensics, law, and other professions in which an intimate or specialized knowledge of the theoretical side of electronics can be advantageous. (The Vocational Information Center provides an extensive list of careers involving electronics at http://www.khake.com/page19.html.)

SOCIAL CONTEXT AND FUTURE PROSPECTS

It is difficult, if not impossible, to imagine modern society without electronic technology. Electronics has enabled the world of instant communication, wherein a person on one side of the world can communicate directly and almost instantaneously with someone on the other side of the world. As a social tool, such facile communication has the potential to bring about understanding between peoples in a way that has until now been imagined only in science fiction.

Consequently, this facility has also resulted in harm. While social networking sites, for example, bring people from widely varied backgrounds together peacefully to a common forum, network hackers and so-called cyber criminals use electronic technology to steal personal data and disrupt financial markets.

Electronics itself is not the problem, for it is only a tool. Electronic technology, though built on a foundation that is unlikely to change in any significant way, will nevertheless be transformed into newer and better applications. New electronic principles will come to the fore. Materials such as graphene and quantum dots, for example, are expected to provide entirely new means of constructing transistor structures at the atomic and molecular levels. Compared with the 50 to 100 nanometer size of current transistor technology, these new levels would represent a difference of several orders of magnitude. Researchers suggest that this sort of refinement in scale could produce magnetic memory devices that can store as much as ten terabits of information in one square centimeter of disk surface. Although the technological advances seem inevitable, realizing such a scale will require a great deal of research and development.

Richard M. Renneboog, M.Sc.

FURTHER READING

Gates, Earl D. *Introduction to Electronics.* 5th ed. Clifton Park, N.Y.: Cengage Learning, 2006. This book presents a serious approach to practical electronic theory beginning with atomic structure and progressing through various basic circuit types to modern digital electronic devices. Also discusses various career opportunities for students of electronics.

Mughal, Ghulam Rasool. "Impact of Semiconductors in Electronics Industry." *PAF-KIET Journal of Engineering and Sciences* 1, no. 2 (July-December, 2007): 91-98. This article provides a learned review of the basic building blocks of semiconductor devices and assesses the effect those devices have had on the electronics industry.

Petruzella, Frank D. *Introduction to Electricity and Electronics 1.* Toronto: McGraw-Hill Ryerson, 1986. A high-school level electronics textbook that provides a beginning-level introduction to electronic principles and practices.

Platt, Charles. *Make: Electronics.* Sebastopol, Calif.: O'Reilly Media, 2009. This book promotes learning about electronics through a hands-on experimental approach, encouraging students to take things apart and see what makes those things work.

Robbins, Allen H., and Wilhelm C. Miller. *Circuit Analysis Theory and Practice.* Albany, N.Y.: Delmar, 1995. This textbook provides a thorough exposition and training in the basic principles of electronics, from fundamental mathematical principles through the various characteristic behaviors of complex circuits and multiphase electrical currents.

Segura, Jaume, and Charles F. Hawkins. *CMOS Electronics: How It Works, How It Fails.* Hoboken, N.J.: John Wiley & Sons, 2004. The introduction to basic electronic principles in this book leads into detailed discussion of MOSFET and CMOS electronics, followed by discussions of common failure modes of CMOS electronic devices.

Singmin, Andrew. *Beginning Digital Electronics Through Projects.* Woburn, Mass.: Butterworth-Heinemann, 2001. This book presents a basic introduction to electrical properties and circuit theory and guides readers through the construction of some simple devices.

Strobel, Howard A., and William R. Heineman. *Chemical Instrumentation: A Systematic Approach.* 3d ed. New York: John Wiley & Sons, 1989. This book provides an exhaustive overview of the application of electronics in the technology of chemical instrumentation, applicable in many other fields as well.

WEB SITES

Institute of Electrical and Electronics Engineers
http://www.ieee.org

See also: Audio Engineering; Automated Processes and Servomechanisms; Communication; Computer Engineering; Electrical Engineering; Electrical Measurement; Electromagnet Technologies; Electronic Commerce; Electronic Materials Production; Electronics and Electronic Engineering; Information Technology; Integrated-Circuit Design; Nanotechnology; Radio; Telecommunications; Television Technologies; Transistor Technologies.

ELECTRON MICROSCOPY

FIELDS OF STUDY

Physics; cell biology; chemistry; morphology; materials science; nanotechnology; medicine; atomic theory; toxicology; virology; nanometrology; forensic engineering; geology; engineering; microanalysis.

SUMMARY

Electron microscopy is the use of electrons, instead of light, to study biological specimens and other nonliving materials under much greater resolution than the conventional light microscope. The electron microscope, first built in the 1930's, saw a number of technological advances, including better lenses and higher voltages for accelerating electrons for increased resolution and imaging. As technology and sample preparation techniques improved, electron microscopy has found multiple applications in the natural sciences, particularly in areas such as medicine, cell biology, morphology, materials sciences, and engineering.

KEY TERMS AND CONCEPTS

- **Contrast:** Distinction between an object and its background, or between two adjacent objects.
- **Field Of View:** What the microscope user sees when he or she looks through its lens.
- **Fixative:** Chemical (usually a solution) that will kill the cells to preserve the tissue for viewing under the microscope. A good fixative will keep the tissue as close to its living state as possible.
- **Light Microscopy:** Earliest form of microscopy, also called optical microscopy, it uses light to illuminate small objects under a series of magnifying glass
- **Magnification:** Amount by which an object can be enlarged.
- **Resolution:** Measure of the sharpness and quality of an image.
- **Resolving Power:** Ability to distinguish the difference between two adjacent objects as distinct.
- **Section:** Slice of a specimen thin enough to be illuminated.
- **Vacuum:** Environment devoid of air.
- **Whole Mount:** Placing an entire specimen or organism under a microscope.

DEFINITION AND BASIC PRINCIPLES

In contrast to traditional microscopes, which use light as an illuminating source, electron microscopes use a beam of electrons to visualize objects. A typical electron microscope is made of a long, hollow cylinder in which the electron beam (the illuminating source) passes through or is reflected off a specimen. At the top of the column is a cathode (an electrode or terminal), usually a heated tungsten filament, that provides electrons. After a high voltage is applied between the cathode and anode, the electrons are accelerated as a thin beam. It is also necessary to create a vacuum within the cylinder by pumping out the air—otherwise, the electrons would collide with gas molecules present in the air and be scattered before the beam reaches the specimen. The beam of electrons is focused using electromagnetic lenses located along the side of the column; the strength of the magnets is controlled by the current that is applied to them by the user. Depending on the current, the object can be magnified from 1,000 to 250,000 times.

The high resolving power of the electron microscope is due to the wave properties of electrons. The resolution of a microscope is limited by the wavelength of the light source in a proportional manner; in other words, the longer the wavelength, the poorer the resolution. The wavelengths of photons (of light) are constant, whereas wavelengths of electrons depend on the speed at which they are traveling, which in turn depends on the voltage that is applied. There are two main types of electron microscopy: transmission electron microscopy (TEM), and scanning electron microscopy (SEM). Both methods provide much greater resolution of the specimen than the light microscope does.

BACKGROUND AND HISTORY

The electron microscope was invented by Max Knoll and Ernst Ruska in 1931. The technology, developed at the Technical University of Berlin, was important because it was the first major improvement on the resolving power of microscopes. Six years later, the first version of the electron microscope was

James Hillier, left, helped supervise the development of electron microscopes in RCA laboratories. (AP Photo)

ready for the market. Ruska received the 1986 Nobel Prize in Physics for his design and creation of the first electron microscope.

When the electron microscope was first conceptualized, the goal was the ability to visualize individual atoms. Although this was not achieved for a few decades, the early electron microscopes were able to use electron beams to view objects. By the late 1930's, electron microscopes, which had theoretical resolutions of 10 nanometers (nm), were being produced. By 1944, the resolution was decreased to 2 nm—a vast improvement on the light microscope, which had a theoretical resolution of 200 nm. Other parts of the instrument were also improved; for example, the voltage accelerations were increased, which resulted in better resolution. Also, better technology in the electron lens decreased the amount of optical aberrations, and the vacuum systems and the electron guns were refined.

Other researchers helped further electron microscopy technology. Ladislaus L. Marton of Brussels assembled the first micrograph of a biological sample, while Manfred von Ardenne of Berlin constructed the first scanning electron microscope in Berlin in 1937. At the University of Toronto, Cecil Hall, Albert Prebus, and James Hillier built a model of the electron microscope that would later be used by the Radio Corporation of America (RCA) to build its own Model B, the first commercially produced electron microscope in North America. The first electron microscope for commercial use was produced by Siemens in Germany.

In the 1940's, electron microscopy was used to study materials and particles such as carbon black (a material that gives strength to car tires), paints, and pigments. Over the next ten years, the technology improved slightly, with lenses being stabler and brighter electron guns being produced.

In the 1960's, electron microscopy moved forward to include commercialization of the first scanning electron microscope and ultrahigh-voltage transmission electron microscopy. In addition, specimen stages that were able to rotate and tilt allowed more angles of the specimen to be viewed. In the 1980's and 1990's, environmental electron microscopes were developed to examine samples under more natural temperature and pressure conditions.

HOW IT WORKS

Transmission Electron Microscopy. In transmission electron microscopy (TEM), electrons are transmitted through the specimen. The electron beam is evenly illuminated across the entire field of view, and an image is generated by the specimen, which will scatter the electrons.

Scanning Electron Microscopy. In scanning electron microscopy (SEM), the principles of using electrons as an illuminating source are the same as in TEM. However, in SEM, the electrons are reflected off the surface of a specimen, making it the better tool for studying the external structure and surface of objects. SEM generally provides magnification in the range of 15 to 150,000 times, and its resolving power is related to the diameter of the electron beam.

Specimen Preparation. A number of methods are used to prepare specimens for electron microscopy. Because electron microscopy is much more powerful than traditional light microscopy, the

Fascinating Facts About Electron Microscopy

- Cellular structure is typically measured using the angstrom (Å). One angstrom is equal to 1×10^{-10} meters. While the practical limit of most transmission electron microscopes is 3-5 Å, a range of 10-15 Å is required to observe most cellular structures.

- Cryosections, or sections of tissue that are rapidly frozen (using a liquid with a very low boiling point such as propane), can be used to study enzymes, which are proteins that catalyze reactions in cells. Cryosections can be a quicker process than methods such as using chemical fixatives and subsequently dehydrating the specimen.

- The use of electron microscopy in materials science began with examining carbon and plastics in the 1940's, to metals and semiconductors and minerals in the 1960's, to essentially all types of materials in the 1980's. The Microscopy Society of America, previously known as the Electron Microscopy Society of America, held meetings beginning in 1942 for microscopists and manufacturers to discuss the latest technology in electron microscopy.

- Ernst Ruska, winner of the 1986 Nobel Prize in Physics, realized that the focal length of waves could be decreased by using an iron cap. The polschuh lens was built using this concept and was eventually used in all magnetic electron microscopes.

- The 1938 Model Toronto Electron Microscope in the Physics Department of the University of Toronto was constructed in six sections, each joined by vacuum-tight seals. It was 6 feet tall (with the electron beam standing vertically). This design still exists in almost all commercially produced electron microscopes.

steps and care required to fix and prepare specimens are significant to viewing success. Any damage sustained during the preservation process will be apparent under the microscope. Chemical fixing or rapid freezing can be used to preserve biological specimens quickly and prevent degradation. Commonly, chemical fixatives such as glutaraldehyde are used to link cellular proteins so that they can be removed. Next, water is removed by a dehydration process using alcohol, and the specimen is sectioned into extremely thin slices (often of less than 0.1 micrometer).

For SEM, the specimens (particularly biological ones) have to be prepared so that the surface morphology is preserved but completely devoid of liquid. Because water consists of a large portion of living cells, it has to be removed in a way that will not be destructive to the structure. The specimens cannot be air dried, because surface tension between air and water will disfigure the surface. Therefore, specimens for SEM are usually fixed, passed through a series of dehydrating alcohols, and dried using a process called critical-point drying. Briefly, the specimens are placed in a liquid such as carbon dioxide, which slowly replaces the water in the sample over a period of time. The carbon dioxide can then be vaporized into gas when the pressure is increased, leaving the surface of the specimen intact. When the specimen is dried using the method described above, it has to be coated with a thin layer of metal (usually gold) to reflect the electrons from the electron beam.

APPLICATIONS AND PRODUCTS

Cellular Biology. When one thinks of microscopy, one often thinks of observing two-dimensional images through a narrow lens. An electron microscope allows scientists to scan a series of two-dimensional, high-resolution images to create a three-dimensional reconstruction of the specimen. The final result is called a tomogram, and it has changed the way that scientists study structures within cells that have not been fixed or dehydrated with chemicals.

Material Science and Engineering. An application of scanning electron microscopy involves using an energy-dispersive X-ray spectroscopy system, which examines the phases of minerals and metals, as well as determines the size, shape, and distribution of particles. Using high-resolution electron microscopy, the molecular arrangements of polymers (such as plastics) can be viewed.

Medicine. If a patient is having a tumor removed, any tissue removed during the surgery can be examined immediately using either light or electron microscopy. The specimen can be rapidly frozen and examined by pathologists rather than undergoing more time-consuming techniques that require the tissue to be fixed and dehydrated.

IMPACT ON INDUSTRY

Major Corporations. Germany's Siemens and Japan's JEOL are two companies that began developing electron microscopy technology and applications early in the field's history, and they remain important manufacturers of these instruments. JEOL produced its first electron microscope in 1948. By 1956, the company had entered the overseas market, selling an electron microscope to the Atomic Energy Commission in France. The Japanese company Hitachi also produces electron microscopes.

Government and University Research. In the United States, the National Institutes of Health and the National Science Foundation award grants to fund research related to electron microscopy research and technology. Many university laboratories deal directly with electron microscopy. They include the biological electron microscopy facilities at the University of Illinois and the University of Hawaii; the Center for Cell Imaging at Yale University; and the Penn Regional Nanotechnology Facility at the University of Pennsylvania.

CAREERS AND COURSE WORK

Electron microscopy is often used to examine biological specimens and other materials on a magnification level that cannot be approached by light microscopy. Microscopists typically have a bachelor's degree in biology and have taken courses in cell and molecular biology concepts, as well as in the principles and techniques of electron microscopy, including specimen preparation and research methods.

Microscopists often work as technicians at a university, in government, or in private engineering firms, where they will analyze samples for various clients. The study of electron microscopy also leads to careers in the pharmaceutical industry and in manufacturing, medical research, and environmental agencies.

SOCIAL CONTEXT AND FUTURE PROSPECTS

The increasing resolution of electron microscopy will enable more scientists and engineers to observe structures on the molecular scale. For example, in 2009, JEOL, in partnership with the Japan Science and Technology Agency, the National Institute of Advanced Industrial Science and Technology in Japan, and the National Institute for Materials Science in Japan, developed a new electron microscope capable of analyzing individual atoms and molecules. This new technology, with which they were able to observe a single atom of calcium, makes use of a new correction mechanism such that the accelerating voltage can be decreased. An important application of this new electron microscopic technique is that a biological or organic specimen can easily be altered by high-voltage electron beams.

Jessica C. Y. Wong, B.Sc., M.Sc.

FURTHER READING

Chandler, Douglas E., and Robert W. Roberson. *Bioimaging: Current Concepts in Light and Electron Microscopy.* Sudbury, Mass.: Jones and Bartlett, 2009. Begins with the history of electron microscopy and examines topics such as specimen preparation, transmission and scanning electron microscopes, and fluorescence microscopy.

Kuo, John, ed. *Electron Microscopy: Methods and Protocols.* 2d ed. Totowa, N.J.: Humana Press, 2007. Presents clear instructions on how to process biological specimens. Protocols are described by experienced experts. Both transmission and scanning electron microscopy are covered.

McIntosh, J. Richard, ed. *Cellular Electron Microscopy.* Boston: Elsevier/Academic Press, 2007. Experts discuss various aspects of electron microscopy, including specimen preparation, analysis of data, imaging from frozen-hydrated samples, and three-dimensional imaging.

Slayter, Elizabeth M., and Henry S. Slayter. *Light and Electron Microscopy.* New York: Cambridge University Press, 2000. Examines the principles behind the electron microscope and how understanding these basics helps researchers interpret results.

WEB SITES

Microscopy Society of America
http://www.microscopy.org

Museum of Science
Scanning Electron Microscope
http://www.mos.org/sln/sem

National Center for Electron Microscopy
http://ncem.lbl.gov/index.html

See also: Electronics and Electronic Engineering; Histology; Microscopy; Optics; Pathology; Scanning Probe Microscopy; Spectroscopy.

ELECTRON SPECTROSCOPY ANALYSIS

FIELDS OF STUDY

Mathematics; physics; chemistry; biology; materials science; biochemistry; biomedical technology; metallurgy

SUMMARY

Electron spectroscopy analysis is a scientific method that uses ionizing radiation, such as ultraviolet, X-ray, and gamma radiation, to eject electrons from atomic and molecular orbitals in a given material. The properties of these electrons are then interpreted to provide information about the system from which they were ejected.

KEY TERMS AND CONCEPTS

- **Auger Electron:** An electron emitted from a valence orbital as a result of an energy cascade initiated by the photoemission of an electron from a core orbital.
- **Balmer Series:** A set of absorption lines in the electromagnetic spectrum of the hydrogen molecule, corresponding to the specific frequencies of the energy differences between molecular orbitals in the H_2 molecule.
- **Bond Strength (or Bond Dissociation Energy):** The amount of energy required to overcome the bond between two atoms and separate them from each other.
- **Electron-Volt:** The energy acquired by any charged particle with a unit charge on passing through a potential difference of one volt, equal to 23,053 calories per mole.
- **Ionization Potential:** The amount of work required to completely remove a specific electron from an atomic or molecular orbital.
- **Paramagnetism:** A measurable increase in the strength of an applied magnetic field caused by alignment of electron orbits in the material.
- **Photoelectron Emission:** The emission of an electron from an orbital caused by the impingement of a photon.

DEFINITION AND BASIC PRINCIPLES

The quantum mechanical theory of matter describes the positions and energies of electrons within atoms and molecules. When ionizing radiation is applied to a sample of a material, electrons are ejected from atomic and molecular orbitals in that material. The measured energies of those ejected electrons provide information that corresponds to the chemical identity and molecular structure of the material.

The analytical methods that employ this technique, such as mass spectrometry, typically study the properties of the molecular ions themselves rather than the electrons that were removed. The two processes are related, however, because the energies observed for one technique are often identical to those observed for the other. This can be understood at a rudimentary level by considering the law of conservation of energy as it must apply to the overall process of rearrangement. Electrons move from one orbital to another after one has been removed from an inner orbital and rearrangement of the electron distribution takes place to "fill in the hole."

Electron spectroscopic methods require that the electron emission process be carried out under high vacuum and with the use of sensitive electronic equipment to capture and measure the emitted electrons and their properties. Each technique utilizes unique methods, but similar devices, to carry out its tasks.

BACKGROUND AND HISTORY

The beginning of the science of electron spectroscopy can only be equated to the experiments of British physicist and Nobel laureate J. J. Thomson in 1897. These experiments first identified electrons and protons as the electrically charged particles of which atoms were composed, according to the atomic model propounded by British chemist and physicist Ernest Rutherford. Thomson's experiments were also the first to measure the ratio of the charge of the electron to the mass of the electron. This feature must be known to utilize the interaction of electrons and electromagnetic fields quantitatively.

In 1905, Albert Einstein identified and explained the photoelectron effect, in which light is observed to provide the energy by which electrons are ejected from within atoms. This work, one of only a handful of scientific papers actually published by Einstein, earned him the Nobel Prize in Physics in 1921.

German physicist Wilhelm Röntgen's discovery of X rays in 1895, and the subsequent development of the means to precisely control their emission, provided an important way to probe the nature of matter. X rays are designated in the electromagnetic spectrum as intermediate between ultraviolet light and gamma rays. High-vacuum technology and, most recently, digital electronic technology, all combine in the construction of devices that permit the precise measurement of minute changes in the properties of electrons in atoms and molecules.

How It Works

Photoelectron Spectroscopy. Two general categories of photoelectron spectroscopy are commonly used. These are ultraviolet photoelectron spectroscopy (UPS) and X-ray photoelectron spectroscopy (XPS). Both methods function in precisely the same manner, and both utilize the same devices. The difference between them is that UPS uses ultraviolet radiation as the ionizing method, while XPS uses X rays to effect ionization.

A typical photoelectron spectrometer consists of a high-vacuum chamber containing a sample target; both are connected to an ionizing radiation emitter and a detection system constructed around a magnetic field. In operation, the vacuum chamber is placed under high vacuum. When the system has been evacuated, the sample is introduced and the emitter irradiates the sample, bringing about the emission of electrons from atomic or molecular orbitals in the material. The emitted electrons are then free to move through the magnetic field, where they impinge upon the detector.

The ability to precisely control the strength of the magnetic field allows an equally precise measurement of the energy of the emitted electrons. This measured energy must correspond to the energy of the electrons within the atomic or molecular orbitals of the target material, according to the mathematics of quantum mechanical theory, and so provides information about the intimate internal structure of the atoms and molecules in the material. The

Electron spectroscopy for chemical analysis at Lehigh University in Pennsylvania. (Joseph Nettis/Photo Researchers, Inc.)

methodology has been developed such that measurements are obtainable using matter in any phase as a solid, liquid, or gas. Each phase requires its own modification of the general technique.

The direct measurement of emitted electron energies through the use of photomultiplying devices is displacing more complex methods based on magnetic field because of the inherent difficulties of providing adequate magnetic shielding to the ever more sensitive components of the devices.

XPS is also known as electron spectroscopy for chemical analysis, or ESCA. The use of this identifier, however, is becoming less common in practice and in the chemical literature.

Auger Electron Spectroscopy (AES). The Auger electron process is a secondary electron emission process that begins with the normal ejection of a core electron by ultraviolet or X-ray radiation. In the

Auger process, electrons from higher energy levels shift to lower levels to fill in the gap left by the emission of the core electron. Excess energy that accrues from the difference in orbital energies as the electrons shift then brings about the secondary emission of an electron from a valence shell. The overall process is in accord with both quantum mechanics and the law of conservation of energy, which requires the total energy of a system before a change occurs to be exactly the same as the total energy of the system after a change occurs.

Unlike UPS and XPS, AES generally utilizes an electron beam to effect core electron emission. Detection of emitted electrons is entirely by direct measurement through photomultiplying devices rather than by any magnetic field methods. The methodology of the technique is otherwise similar to that of UPS and XPS. It is amenable to the study of matter in all phases, except for hydrogen and helium, but is generally valuable for use only with solids as a surface analysis technique. This is true because sample materials must be stable under vacuum at pressures of 10^{-9} Torr. Also, AES is known to be highly sensitive and capable of fast response.

Electron Spin Resonance (ESR). The principles of ESR are based on an entirely different physical property of electrons in their atomic or molecular orbitals. In quantum mechanics, each electron is allowed to exist only in very specific states with very specific energies within an atom or molecule.

One of the allowed states is designated as "spin." In this state, the electron can be thought of as an electrical charge that is physically spinning about an axis, thus generating a magnetic field. Only two orientations are allowed for the magnetic fields generated in this way, and according to the Pauli exclusion principle, pairs of electrons must occupy both states in opposition. This requirement means that ESR can be used only with materials that contain single or unpaired electrons, including ions. When placed in an external magnetic field, the magnetic fields of the single electrons align with the external magnetic field.

Subsequent irradiation with an electromagnetic field fluctuating at microwave frequencies acts to invert the magnetic fields of the electrons. Measurement of the frequencies at which inversion takes place provides specific information about the structure of the particular material being examined. The precise locations of inversion signals depend upon the atomic or molecular structure of the material, as these environments affect the nature of the magnetic field surrounding the electron.

APPLICATIONS AND PRODUCTS

In application, electron spectroscopy is strictly an analytical methodology, and it serves only as a probe of material composition and properties. It does not serve any other purpose, and all applications and products related to electron spectroscopy are the corresponding spectroscopic analyzers and the ancillary products that support their operation.

Spectroscopic analyzers come in a variety of forms and designs, according to the environment in which they will be expected to function, but more with respect to the nature of the use to which they will be put. These range from machines for routine analysis of a limited range of materials and properties at the low end of the scale, to the complex machines used in high-end research that must be capable of extreme sensitivity and finely detailed analysis.

The applications of electron spectroscopic analysis are, in contrast, wide ranging and are applicable in many fields. In its roles in those fields, the methodology has enabled some of the most fundamental technology to be found in modern society.

One application in which electron spectroscopy has proven unequaled in its role is submicroscopic surface analysis. Both XPS and AES are the methods of choice in this application, because each can probe to a depth of about 30 microns below the actual surface of a solid material, enabling analysts to see and understand the physical and chemical changes that occur in that region.

The surface of a solid typically represents the point of contact with another solid, and physical interaction between the two normally effects some kind of change to those surfaces because of friction, impact, or electrochemical interaction. A good example of this is the tribological study of moving parts in internal combustion engines. In normal operation, a piston fitted with sealing rings moves with a reciprocating motion within a closely fitted cylinder. The rings physically interact with the wall of the cylinder with intense friction, even though well lubricated, under the influence of the high heat produced through the combustion of fuel. At the same time, the top of the piston is subjected to intense pressures and heat from the explosive combustion of the fuel. At an engine revolution rate of 2,400 revolutions per minute (rpm), each cylinder in a four-cylinder internal combustion engine goes through its reciprocating motion six hundred times each minute, or ten times each second.

In turbine and jet engines, for example, parts are subjected to such stress and friction at a rate of hundreds and even thousands of times per second. Engine and automobile manufacturers and developers must understand what happens to the materials used in the corresponding parts under the conditions of operation. Both XPS and AES are used to probe the material effects at these surfaces for the development of better formulations and materials, and for understanding the weaknesses and failure modes of existing materials.

ESR, on the other hand, is used entirely for the study of the chemical nature of materials in the liquid or gaseous phase. In this role, researchers and analysts use the methodology to study the reactions and mechanisms involving single-electron chemical species. This includes the class of compounds known as radicals, which are essentially molecules containing their full complement of electrons but not of atoms. The methyl radical, for example, is basically a molecule of methane (CH_4) that has lost one hydrogen atom. The remaining CH_3 portion is electrically neutral, because it has all of the electrons that would normally be present in a neutral molecule of CH_4, but with one of its electrons free to latch on to the first available molecule that comes along.

Radical reactions are understood to be responsible for many effects: aging in living systems, especially humans; atmospheric reactions, especially in the upper atmosphere and ozone layer; the detrimental effects of singlet oxygen; combustion processes; and many others. In biological systems, specially designed molecules are used to tag other nonparamagnetic molecules so that they can be studied by ESR. Such molecules often include a "nitroso" functional group in their structures to provide a paramagnetic radical site that can be monitored by ESR. The production and testing of these specialty chemicals is another area of application.

IMPACT ON INDUSTRY

The impact on industry of electron spectroscopy is not obvious. The methodology plays very much a

Fascinating Facts About Electron Spectroscopy Analysis

- The photoelectron effect, one of the basic principles of electron spectroscopy, was first explained by Albert Einstein in 1905, for which he received the 1921 Nobel Prize in Physics.
- Spinning electrons generate a magnetic field around themselves in the same way that moving electrons through a wire produces a magnetic field around the wire.
- X rays and ultraviolet light can both eject electrons from the inner orbitals of atoms.
- Auger electrons are electrons emitted by the extra energy released when electrons cascade into lower orbitals to replace electrons that have been ejected by X rays or ultraviolet light.
- Electron spectroscopy can measure the chemical and physical properties of materials up to 30 microns below the surface of a solid.
- Electron spin resonance measures the frequencies required to switch the orientation of the magnetic fields of single electrons.
- Scanning Auger microscopy can produce detailed maps of the distribution of specific metal atoms at the surface of an alloy, allowing metallurgists to see how the material is structured.

behind-the-scenes or supportive role that is not apparent from outside any industry that uses electron spectroscopy. Although this is true, the role played by electron spectroscopy in the development and improvement of products and materials has been valuable to those same industries.

Many advances in metallurgy and tribology, the study of friction and its effects, have been made possible through knowledge obtained by electron spectroscopy, especially AES and XPS. Given the untold millions of moving parts—bearings, pistons, slides, shafts, link chains—that are in operation around the world every single minute of every single day, lubrication and lubricating materials represents a huge world-wide industry. With the vast majority of lubricating materials (oils and greases) being derived from nonrenewable resources, the need to enhance the performance of those materials through better understanding of material interactions has been one driving force behind the application of electron spectroscopy in industry.

Other industrial processes require that materials undergo a chemical process called passivation, which is essentially the rendering of the surface of a material inert to chemical reaction through the formation of a thin coating layer of oxide, nitride, or some other suitable chemical form. With its ability to accurately measure the thickness and properties of thin films such as oxide layers on a surface, electron spectroscopy is uniquely appropriate to use in industries that rely on passivation or on the formation of thin layers with specific properties. One such industry is the semiconductor industry, upon which the computer and digital electronics fields have been built. AES and XPS are commonly used to monitor the quality and properties of thin layers of semiconductor materials used to construct computer chips and other integrated circuits.

CAREERS AND COURSEWORK

Electron spectroscopy is used to examine the intimate details of the atomic and molecular structure of matter, making electron spectroscopy an advanced career. Students who look to a career in this field will undertake highly technical foundation courses in mathematics, physics, inorganic chemistry, organic chemistry, surface chemistry, physical chemistry, chemical physics, and electronics. The minimum requirement for a career in this field is an associate's degree in electronics technology or an honors (four-year) bachelor's degree in chemistry, which will allow a student to specialize in the practice as a technician in a research or analytical facility. More advanced positions will require a postgraduate degree (master's or doctorate).

As the field finds more application in materials research and forensic investigation, general opportunities should further develop. These applications will require, however, that those wishing a career involving electron spectroscopy have specialist training. The vast majority of opportunities in the field are to be found in such academic research facilities as surface science laboratories and in materials science research. Forensic analysis also holds a number of opportunities for electron spectroscopists.

SOCIAL CONTEXT AND FUTURE PROSPECTS

Electron spectroscopy analysis is a methodology with an important role behind the scenes. The field is neither well known nor readily recognized. Nevertheless, it is a critical methodology for advancing the understanding of materials and the nature of matter. As such, electron spectroscopy adds to the general wealth of knowledge in ways that permit the development of new materials and processes and to advancing the understanding of how existing materials function.

One development of electron spectroscopy, known as scanning Auger microscopy (SAM) has the potential to become an extremely valuable technique because of its ability to generate detailed maps of the surface structure of materials at the atomic and molecular level. By tuning SAM to focus on specific elements, the precise distribution of those elements in the surface being examined can be identified and mapped, providing detailed knowledge of the granularity, crystallinity, and other structural details of the material. This is especially valuable in such widely varied fields as metallurgy, geology, and advanced composite materials.

XPS and AES have been applied in a variety of different fields and are themselves becoming very important surface analytical methods in those fields. These areas include the aerospace and automotive industries, biomedical technology and pharmaceuticals, semiconductors and electronics, data storage, lighting and photonics, telecommunications, polymer science, and the rapidly growing fields of solar cell and battery technology.

Richard M. Renneboog, M.Sc.

FURTHER READING

Chourasia, A. R., and D. R. Chopra. "Auger Electron Spectroscopy." In *Handbook of Instrumental Techniques for Analytical Chemistry*, edited by Frank Settle. New York: Prentice Hall Professional Reference, 1997. This chapter provides a thorough and systematic description of the principles and practical methods of Auger spectroscopy, including its common applications and limitations.

Kolasinski, Kurt W. *Surface Science: Foundations of Catalysis and Nanoscience.* 2d ed. Chichester, England: John Wiley & Sons, 2008. Provides a complete study of the utility of electron spectroscopy as applied to the study of processes that occur on material surfaces.

Merz, Rolf. "Nano-analysis with Electron Spectroscopic Methods: Principle, Instrumentation, and

Performance of XPS and AES." In *NanoS Guide 2007*. Weinheim, Germany: Wiley-VCH, 2007. A lucid and readable presentation of the principles, capabilities, and limitations of XPS and AES based on actual applications and comparisons with methods such as scanning electron microscopy.

Strobel, Howard A., and William R. Heineman. *Chemical Instrumentation: A Systematic Approach*. 3d ed. New York: John Wiley & Sons, 1989. This book provides a valuable resource for the principles and practices of electron spectroscopy and for many other analytical methods. Geared toward the operation and maintenance of the devices used in those practices.

WEB SITES
Farach, H. A., and C. P. Poole "Overview of Electron Spin Resonance and Its Applications"
http://www.uottawa.ca/publications/interscientia/inter.2/spin.html

Molecular Materials Research Center, California Institute of Technology "Overview of Electron Spin Resonance and Its Applications"
http://mmrc.caltech.edu/SS_XPS/XPS_PPT/XPS_Slides.pdf

See also: Applied Physics; Computer Engineering; Electrical Engineering; Electrochemistry; Electromagnet Technologies; Electrometallurgy; Electronics and Electronic Engineering; Electron Microscopy; Forensic Science; Spectroscopy.

ENDOCRINOLOGY

FIELDS OF STUDY

Bariatrics; diabetes medicine; internal medicine; laboratory medicine; neuroendocrinology; obstetrics and gynecology; pediatric endocrinology; radiology reproductive endocrinology; thyroid medicine.

SUMMARY

Endocrinology is a medical field focused on the diagnosis and treatment of abnormalities of the endocrine system. The endocrine system consists of glands that produce hormones: the adrenal gland, hypothalamus, ovaries, pancreas, parathyroid glands, pituitary gland, testes, and thyroid gland. These hormones control metabolism (utilization of food by the body), reproduction, and growth. Endocrinology is practiced by medical doctors with specialized training in that field. Physicians in other fields (such as obstetricians, gynecologists, and internists) may devote some of their practice to endocrinology. Some endocrinologists specialize in one area of endocrinology (such as neuroendocrinology or pediatric endocrinology). Conditions treated by endocrinologists include diabetes, hypertension (high blood pressure), inadequate growth, infertility, obesity, osteoporosis (weak bones), menopause, metabolic disorders, and thyroid disorders.

KEY TERMS AND CONCEPTS

- **Adrenal Glands:** Glands that are situated above each kidney and produce hormones, which respond to stress, such as cortisol and adrenaline.
- **Hypothalamus:** Portion of the brain that contains a variety of specialized cells; an important function of the hypothalamus is the linkage of the brain to the endocrine system through the pituitary gland.
- **Ovaries:** Paired organs adjacent to the uterus, which release eggs (ova) for reproduction and produce a variety of hormones, primarily estrogen and progesterone.
- **Pancreas:** Organ located in the upper abdomen, which produces hormones that regulate blood sugar levels (insulin, glucagon, and somatostatin) and secretes digestive enzymes, which pass into the small intestine.
- **Parathyroid Glands:** Small glands, usually located within the thyroid gland, which produce parathyroid hormone; this hormone regulates calcium levels in the bloodstream and bones.
- **Pituitary Gland:** Small gland, located at the base of the brain, which is sometimes referred to as the master gland because it produces hormones that stimulate or suppress the secretions of other endocrine glands.
- **Testes:** Male reproductive organs, which produce sperm and male sex hormones, such as testosterone.
- **Thyroid Gland:** Butterfly-shaped organ located in the neck; it controls the metabolic rate, protein production, and the sensitivity of the body to other hormones.

DEFINITION AND BASIC PRINCIPLES

Endocrinology is a medical field dealing with the endocrine system, which is a complex system of organs that secrete hormones into the bloodstream. Hormones are chemicals that are released from cells in one location and affect cells located elsewhere in the body. To respond to these chemical messengers, a cell must possess a receptor to the hormone. Hormones control many bodily functions, including metabolism, reproduction, and growth. Diseases of the endocrine system often involve the abnormal production of a hormone or a cell's resistance to the effects of a hormone. For example, excess thyroid hormones produce a condition known as hyperthyroidism in which metabolism is increased. Patients with hyperthyroidism have increased nervousness, irritability, tremors, and a rapid heart rate. Individuals with hypothyroidism (inadequate level of thyroid hormones) have fatigue, poor muscle tone, constipation, and dry skin. Either thyroid condition may be caused by an abnormal functioning of the thyroid gland. These conditions may also be secondary to an abnormal functioning of the pituitary gland or hypothalamus, both of which regulate the thyroid gland. The hypothalamus produces a hormone, thyrotropin-releasing hormone (TRH), which causes the pituitary gland to release thyroid-stimulating hormone (TSH), which signals the thyroid gland to release thyroid hormones. Patients with diabetes usually have inadequate levels of insulin, which controls glucose (sugar) metabolism. Some cases of diabetes are caused by insulin resistance, a condition in which the cells do not respond well to insulin circulating in the bloodstream.

BACKGROUND AND HISTORY

Endocrinology is derived from the Greek words *endo* (within), *krīnō* (to separate), and *logia*, which supplies the suffix "-ology" (referring to a field of knowledge). Endocrinology originated in 200 B.C.E. in China, when pituitary and sex hormones were isolated from the urine for medicinal purposes. In the Western world, an organ basis for pathology did not develop until the nineteenth century. In 1841, the German physician Friedrich Henle described "ductless glands" that secrete products directly into the bloodstream. In 1902, William Bayliss and Ernest Starling discovered secretin, which they described as a hormone. They defined a hormone as a chemical that is produced in an organ, then travels via the bloodstream to a distant organ and exerts a specific function on it. The field of endocrinology is based on the replacement of inadequate levels of a hormone with purified extracts. In cases in which hormone levels are unusually high, treatment involves lowering the hormonal level through surgical removal of a portion of the gland or destruction of some of the gland's cells by radiation. For example, hyperthyroidism is commonly treated with radioactive iodine, which concentrates in the thyroid gland and destroys some of the cells that produce hormones.

HOW IT WORKS

Endocrinology is a medical specialty, and usually patients are referred from other physicians (such as an internist or family physician) for evaluation. The endocrinologist examines the patient and makes a diagnosis. If the referring physician has made a preliminary diagnosis, the endocrinologist often orders further tests to confirm it. The endocrinologist must be well versed in biochemistry and clinical chemistry to properly interpret these tests.

The endocrinologist frequently relies on the radiologist for diagnosis and treatment of an endocrine disorder. This involves the use of imaging equipment such as ultrasound and scintigraphy. Scintigraphy is a two-dimensional visualization of a radionuclide in the body. A radionuclide is a radioactive substance that is taken up by an endocrine grand. Computed tomography (CT) and magnetic resonance imaging (MRI) are also used to visualize organs such as the thyroid and adrenal glands. After diagnosing an endocrine condition, the radiologist might be called on to treat the condition with the injection of a radionuclide. The treatment levels of radiation are much higher than the diagnostic level and are designed to destroy cells producing excessive amounts of a hormone.

Once a diagnosis is made, a course of treatment must be developed. Initial treatment might include referral to a surgeon for excision of a tumor or an abnormally functioning organ. Radiotherapy is employed to treat certain endocrine conditions such as hyperthyroidism (overactive thyroid) and Cushing's syndrome (overactive adrenal glands). Drug therapy is an option in some cases. Many diseases of the endocrine system are chronic and require lifelong treatment. A classic example is diabetes, which can develop in children and young adults (type 1 diabetes) and in adults (type 2 diabetes). Milder forms of type 2 diabetes can often be treated with medication; however, type 1 diabetes and severer forms of type 2 diabetes require insulin injections. Patients requiring insulin must be educated as to the importance of controlling their blood sugar level with self-administered insulin injections and of frequent monitoring of their blood glucose (sugar) through blood sampling. Diabetics are more prone to many conditions such as cardiovascular disease and loss of vision. Diabetics whose condition is under good control are less likely to develop serious health problems. Obesity greatly increases the risk of developing type 2 diabetes; therefore, a weight-loss program can sometimes return blood glucose levels to normal levels.

Some diseases are congenital (present at birth) or occur at a young age; the pediatrician plays a crucial role in diagnosing an endocrine problem in young patients. Inadequate levels of many hormones can severely affect a child's development, both physically and mentally. Children with endocrine abnormalities are often referred to a pediatric endocrinologist.

Couples with an infertility problem may seek the help of a reproductive endocrinologist. The problem may be caused by either male or female factors. Some of these problems are nonendocrine in origin (for example, Fallopian tubes blocked from an infection); however, many infertile women can greatly increase their chances of becoming pregnant with assisted reproductive technology (ART), which involves the administration of hormones and other medications to stimulate and regulate the ovulation process.

Gynecologists practice endocrinology related to the female reproductive system. A common problem dealt with by gynecologists is a menstrual irregularity. When women approach the menopause, which is caused by a drop in the level of female hormones (estrogen and progesterone), they often develop distressing symptoms, such as hot flashes, dry skin, and depression. Many women do not consult an endocrinologist at that time. Instead, they seek the advice of a gynecologist, family physician, or an internist. That physician will often diagnose and treat the condition; however, the patient may occasionally be referred to an endocrinologist.

APPLICATIONS AND PRODUCTS

Most hormones can be administered orally or by injection, skin patch, vaginal cream, or nasal inhalation. Some hormones cannot be administered orally. Hormones can be derived from animal or human sources. Some can be retrieved from urine, and others are obtained from animal organs harvested at slaughterhouses. In many cases, hormones have been synthesized in the laboratory.

Hormones derived from animal or human sources must be subjected to bioassay.

Bioassay involves administering a hormone sample to an animal and measuring its effect. By this process, the pharmaceutical manufacturer can adjust the hormonal level of that batch to a standardized level. Thus, an individual ingesting a hormone can be assured that the dose is uniform from day to day. Synthesized hormones are easier to standardize; however, they still may require a bioassay. They are less likely to contain impurities, which could produce an adverse effect, including an allergic reaction.

Many products on the market are used for diabetes, thyroid disease, osteoporosis, infertility, and the menopause. A significant market exists for performance-enhancing products, although athletes are banned from using hormones in this manner.

Specialized surgical procedures exist for medical conditions with an endocrine component, such as morbid obesity. Surgery for morbid obesity is known as bariatric surgery. An increasing proportion of surgical procedures are being done with laparoscopy, which has the advantage of a small incision and quicker recovery.

Medical Laboratory. Endocrinology is a field that uses laboratory services to a greater degree than many other specialties. Some medical laboratories contain specialized equipment to accurately measure specific hormonal levels. Inasmuch as this equipment can be quite expensive and may measure only one specific hormone, these services are usually found in specialty laboratories that analyze samples from a large geographic area.

Diabetes. Many products are marketed for diabetic patients. Meters that can reliably measure blood glucose levels with minimal discomfort are a necessity for all diabetics. Syringes for injection are also essential for management of the condition. Also available are insulin pumps that can be programmed to administer the appropriate insulin dosage throughout a twenty-four-hour period. The patient can adjust the rate at any time depending on any variance from the normal routine or an abnormal blood glucose reading.

Thyroid Disease. In addition to laboratory tests to measure thyroid levels, diagnosis of thyroid disease often involves radioactive iodine (iodine 131), which is manufactured in specialized laboratories. A small amount of the substance is given by injection or in tablet form for diagnosis. Diagnosis is facilitated by radiologists using a specialized device, a gamma camera. Treatment of hyperthyroidism involves the use of a much higher dosage of iodine 131, which destroys thyroid cells. Hyperthyroidism can also be treated with antithyroid drugs, such as methimazole and propylthiouracil. These drugs become concentrated in the thyroid gland and block production of thyroid hormones. For the immediate treatment of the symptoms of hyperthyroidism, beta-blockers such as propranolol, atenolol, and metoprolol can be used. These medications lower metabolism; however, they do not alter thyroid hormone levels in the circulation. Surgical removal of a portion of the thyroid gland is sometimes done. It is usually reserved for pregnant patients, children with an adverse reaction to antithyroid medications, and patients with a very large thyroid gland.

Osteoporosis. Osteoporosis is a weakening of the bones that can be triggered by an imbalance of a number of hormones. It commonly occurs in women at the time of the menopause. Osteoporotic bone is susceptible to fracture and disfiguring deformities. Specialized equipment, using specialized X-ray equipment or other imaging modalities, can measure the degree of osteoporosis. These devices are marketed not only for radiology facilities but also for physician offices and other health care facilities. Dual energy X-ray absortiometry (DEXA) scanning is the most common procedure used to measure the amount of calcium and other minerals present in bone. The amount of minerals present is known as the bone mineral density (BMD). The most commonly scanned areas are the hip and spine. If a patient is diagnosed with osteoporosis, hormonal and nonhormonal products are available to lessen the severity of or reverse osteoporosis. For women, estrogen preparations can slow or halt the progression of osteoporosis; however, they cannot reverse it. Bisphosphonates, pharmaceuticals that can reverse osteoporosis, are available.

Infertility. Reproductive endocrinologists are major users of specialized products. Assisted reproduction technology (ART) involves stimulating the ovaries to produce ova (eggs), extracting the ova, fertilizing the ova in the laboratory, culturing the ova (multiple cell stage or embryonic stage), then implanting the ova in the uterus. Medications such as clomiphene are used to stimulate the ovaries to release eggs; hormones, such as human menopausal hormone (HMG) and gonadotropins are also used to stimulate the ovaries. The menstrual cycle is often regulated with hormones such as estrogen and progesterone. Specialized equipment is used to extract the ova from the patient's ovaries. The process involves inserting a needle through the vaginal wall and into an ovarian follicle; it is conducted under the guidance of ultrasound. The extracted ova are placed in a culture medium and fertilized with sperm supplied by the husband or donor. The fertilized ova are then placed in specialized incubators for growth to the desired embryonic stage. The ova are then inserted into the uterine cavity for development. Ova and embryos are placed in vials and frozen with liquid nitrogen. At a future date, the embryos are thawed and inserted into the uterus.

Menopause. A number of products are on the market for replacement of hormones lost after the menopause. They include oral medication, injections, and transdermal skin patches (the hormones

Fascinating Facts About Endocrinology

- Gigantism is a condition marked by excessive growth because of the secretion of a growth hormone by a pituitary tumor. Trijntje Keever, born in 1616 in the Netherlands, suffered from the condition. She was the tallest woman in recorded history. When she died at age seventeen from cancer, her height was 8 feet, 4 inches.

- A deficiency of growth hormone results in dwarfism. General Tom Thumb (the stage name of Charles Sherwood Stratton), born in 1883, probably suffered from the condition. His height at the age of eighteen was 2 feet, 8.5 inches. Stratton became wealthy as a performer in P. T. Barnum's circus.

- Within a year of giving birth, 5 to 10 percent of women develop hypothyroidism secondary to postpartum thyroiditis. Initially, thyroid hormone levels may rise, then either return to normal or drop to hypothyroid levels. Of those women who become hypothyroid, about 20 percent will require lifelong treatment.

- Premarin is an estrogen preparation that is sometimes prescribed to women at the time of the menopause as hormone replacement therapy. The name is derived from "pregnant mare's urine," the source of the hormonal preparation.

- Girls stop growing in height at puberty because estrogen closes the epiphyses (growth plates), located in the arm and leg bones.

- Newborn girls sometimes have vaginal bleeding; this is caused by a drop in the level of estrogen, which the fetus was exposed to while in the uterus.

are absorbed through the skin and pass into the bloodstream). Many nonhormonal products are also available to treat menopausal symptoms.

Performance-Enhancing Hormones. Performance-enhancing hormones have been used by many athletes, both professional and amateur. Although most of these products are banned because they confer an unfair advantage on their user, their use continues, supported by a thriving black market enterprise. Hormones that can enhance athletic performance are anabolic steroids (male hormones), which promote muscle growth and strength, and erythropoietin (EPO), which increases red blood cell production. Stronger muscles allow a baseball player to hit more home runs and increases a cyclist's speed and endurance. Increased red cell production from EPO raises the blood's oxygen-carrying capacity.

The athlete who uses these hormones may not only be disqualified but also experience adverse effects. Anabolic steroids change cholesterol levels. They increase the level of low-density lipoprotein (LDL, or bad cholesterol) and decrease the level of high-density lipoproteins (HDL, or good cholesterol), which can raise a person's risk of developing cardiovascular disease and having a heart attack. Furthermore, these drugs can cause direct damage to the heart and liver. A prominent example of an athlete who was disqualified is cyclist Floyd Landis (although he continues to deny using anything). Landis was the overall leader of the 2006 Tour de France, a grueling multistage bicycle race, but in stage 16, he lost eight minutes to the second-place rider, and most experts believed he could not make up the time. The following day, he made a dramatic comeback and went on to win the event. A mandatory urine test taken after stage 17 revealed high levels of testosterone, and he was stripped of his title.

IMPACT ON INDUSTRY

Endocrinology plays a significant role in laboratory medicine, medical imaging, radiotherapy, assisted reproductive technology, and pharmacology. The medical facilities and industries associated with these areas derive significant revenue from endocrinology. Laboratory procedures for detecting and measuring hormone levels are complex and highly technical, and accurate measurement is imperative for proper diagnosis and treatment. Diagnostic and therapeutic radiology for endocrine conditions requires sophisticated equipment, which is often updated frequently as improved models come on the market. Both laboratory medicine and radiology require a team of skilled physicians, supervised by physicians with specialized training. Assisted reproductive technology is a high-ticket item and often is not covered by insurance; however, for many of those with moderate incomes, the hope of achieving a pregnancy is enough to justify the expense. The field of reproductive endocrinology requires a team of skilled technicians, supervised by one or more physician specialists.

The manufacturing of hormonal products and medications to treat hormonal disorders is a significant segment of the pharmaceutical industry. Patients with chronic conditions, such as diabetes and hypothyroidism, are lifelong consumers. Many women take hormonal medications to deal with menopausal symptoms or dysmenorrhea (painful menstruation), and many use hormonal contraceptives (pills, patches, or vaginal rings). In addition to hormonal products, many pharmaceuticals are designed to treat endocrine problems. These include medications to boost energy, aid in weight loss, induce sleep, and lower blood pressure.

Research. Significant endocrine research is conducted by the government and universities. A branch of the National Institutes of Health, the National Institute of Diabetes and Digestive and Kidney Diseases, funds research in many fields, including diabetes, digestive diseases, genetic metabolic diseases, immunologic diseases, and obesity. The institute also provides health information for the public in these fields. Virtually all developed nations have extensive endocrine research programs. Beyond government and university programs, many practicing endocrinologists devote a significant amount of their time to research in their field.

CAREERS AND COURSE WORK

To become an endocrinologist, one must first graduate from college and then complete a four-year course of medical training. Initial specialty training, typically a three- or four-year residency program, focuses on internal medicine, pediatrics, or gynecology. Subsequently, the physician completes a fellowship (two or more years) in the field of general endocrinology, pediatric endocrinology, reproductive endocrinology, or neuroendocrinology.

Most endocrinologists will locate their practice in a densely populated urban area, and many will join a group of specialists. A large number of endocrinologists practice in the medical university setting. Most endocrinologists are certified in the specialty of internal medicine, pediatrics, and gynecology. They may ultimately be board certified in endocrinology or a subspecialty of endocrinology.

Most endocrinologists become a member of one or more professional organizations, which provide continuing education and forums for physician members and educational material for the general public. In the United States, the main professional organizations for endocrinologists are the American Association of Clinical Endocrinologists, the American Diabetes Association, the American Thyroid Association, the Pediatric Endocrine Society, and the Society for Reproductive Endocrinology and Infertility. In the United Kingdom, the two principal organizations are the British Society for Paediatric Endocrinology and Diabetes and the Society for Endocrinology. The world's largest professional organization for pediatric endocrinology is the European Society for Paediatric Endocrinology. Most developed nations throughout the globe have similar organizations.

SOCIAL CONTEXT AND FUTURE PROSPECTS

Research in endocrinology examines how hormones work in the body and how they relate to diseases and conditions. This research aims at improving the medical treatment of endocrine conditions. Some researchers are focusing on the development of synthetic hormones, which do not have the problems associated with animal-derived products. Others are looking at possible cures, in the form of transplanted or regrown organs or other ways for the body to manufacture hormones.

Organ transplantation has made significant progress since the 1960's. Research continues on transplanting endocrine organs, such as the pancreas for the treatment of diabetes. The major problem faced by transplanted organs is rejection of the organ by the recipient's body. Therefore, research is focusing on developing better medications to combat rejection and also on the creation of artificial endocrine organs. Although an artificial pancreas is still a distant possibility, insulin pumps have been developed to continuously administer insulin to diabetics. These devices are likely to become more sophisticated, possibly with sensors to monitor glucose levels and administer the proper insulin dosage.

A controversial topic, the subject of much debate on medical, political, and religious grounds, is embryonic stem cell research. These cells, derived from early embryos, have the potential to develop into any organ within the human body. For example, in theory, a diabetic could grow a new pancreas, which would produce insulin. Proponents of stem cell research allude to a future in which a patient could grow a new adrenal gland, a paraplegic could walk again, and a child with cystic fibrosis could be cured. Opponents of the research claim that it involves the destruction of an embryo and therefore a human life. Proponents counter that researchers use only surplus embryos—those that are destined for destruction. Treatment with stem cells is still in its infancy, but the ability to grow a replacement endocrine organ would cure many endocrine diseases.

Robin L. Wulffson, M.D., F.A.C.O.G.

FURTHER READING

American Diabetes Association. *American Diabetes Association Complete Guide to Diabetes.* 5th ed. Alexandria, Va.: Author, 2011. Provides information to help diabetics manage their disease. Begins with a discussion of the causes and effects of diabetes. Contains a glossary, an appendix on self-monitoring and injection techniques, and a list of resources and organizations.

Borer, Katarina T. *Exercise Endocrinology.* Champaign, Ill.: Human Kinetics, 2003. Looks at the role of hormones in exercise and athletic performance. Topics include regulation of hydration and fuel use during exercise, gender and performance, biological rhythms, and exercise as a stressor.

Gardner, David, and Dolores Shoback. *Greenspan's Basic and Clinical Endocrinology.* 9th ed. New York: McGraw-Hill Medical, 2011. Examines the molecular biology of endocrine glands and discusses metabolic bone disease, pancreatic hormones and diabetes mellitus, hypoglycemia, obesity, geriatric endocrinology, and many other diseases and disorders.

Hadley, Mac E., and Jon E. Levine. *Endocrinology.* 6th ed. Upper Saddle River, N.J.: Prentice Hall, 2007. Presents explanations of basic concepts and applications. Focuses on how glands and hormones control physiological processes.

Lebovic, Dan I., John D. Gordon, and Robert N. Taylor. *Reproductive Endocrinology and Infertility: Handbook for Clinicians.* Arlington, Va.: Scrub Hill Press, 2005. A ready reference for endocrinologists treating conditions and disorders related to reproduction. Information from textbooks, articles, and endocrinologists was gathered and analyzed to provide evidence-based approaches and strategies.

Potter, Daniel A., and Jennifer S. Hanin. *What to Do When You Can't Get Pregnant: The Complete Guide to All the Technologies for Couples Facing Fertility*

Problems. New York: Marlowe, 2005. A thorough guide for couples with fertility problems.

Skugor, Mario, and Jesse Bryant Wilder. *The Cleveland Clinic Guide to Thyroid Disorders.* New York: Kaplan, 2009. Skugor, an endocrinologist, teamed with writer Wilder to present detailed information on thyroid diseases and treatment options.

Thacker, Holly. *The Cleveland Clinic Guide to Menopause.* New York: Kaplan, 2009. Thacker, a physician at the Center for Specialized Women's Health at the Cleveland Clinic, offers safe treatments for the menopause and explains myths and facts regarding hormonal replacement therapy.

WEB SITES
American Association of Clinical Endocrinologists
http://www.aace.com

American Diabetes Association
http://www.diabetes.org

American Thyroid Association
http://www.thyroid.org

Pediatric Endocrine Society
http://www.lwpes.org/index.cfm

Society for Reproductive Endocrinology and Infertility
http://www.socrei.org

See also: Bionics and Biomedical Engineering; Cell and Tissue Engineering; Geriatrics and Gerontology; Obstetrics and Gynecology; Pediatric Medicine and Surgery; Pharmacology; Reproductive Science and Engineering; Sports Engineering.

ENERGY-EFFICIENT BUILDING

FIELDS OF STUDY

Physics; thermodynamics; architecture; materials science; construction methods; electronics; plumbing; carpentry; building technology; fluid dynamics

SUMMARY

Energy-efficient building practices involve the construction of buildings using as little energy as possible and with minimal environmental impact, beginning with construction techniques and materials and continuing with the ongoing, energy-efficient operation and maintenance of a building.

KEY TERMS AND CONCEPTS

- **Convective Air Current:** A circulatory movement of air caused by the tendency of warm air to rise and cool air to descend.
- **Mass Effect:** A physical characteristic or behavior of a material that is related to the mass of that material present.
- **Thermal Mass:** The thermal energy that can be stored within the material of a body such as the inner section of a Trombe wall or the interior of a water-filled thermal window.
- **Trombe Wall:** A two-part wall structure consisting of a transparent thermal capture outer wall and a dense, massive thermal storage inner wall.

DEFINITION AND BASIC PRINCIPLES

Energy-efficient building is a construction methodology designed and regulated to minimize the energy consumed in building, operating, and maintaining a given structure. The basic principle of minimizing the energy required for the operation of the building is simplicity itself. The practice of energy-efficient building is at the beginning of its development, however, and a great deal of future research will conceptualize and test materials and structures to minimize the cost and maximize the effectiveness of the technology.

The most basic principle of energy-efficient building is the capture of solar thermal or geothermal energy (or both) that can be stored in some mass of material, such as water or concrete. The energy is then released into the building in a controlled manner to maintain environmental conditions within the building.

A second main principle of energy-efficient building is the prevention of thermal transfer through the use of insulation and insulating structures. This works to prevent heat loss in winter and heat gain in summer. The third principle ensures that the daily operation and maintenance of a building is to consume as little energy as possible.

BACKGROUND AND HISTORY

Builders have long known how to construct buildings to make full use of the natural environment for heating, cooling, and ventilation. With the Industrial Revolution of the eighteenth century, and particularly since the establishment of electrical grids, artificial heating, cooling, ventilating, and lighting systems have rapidly developed. They now represent one of the largest general uses of energy throughout the world. As the human population continues to grow, traditional methods of energy production have not been able to meet demand, leading to the necessary development of new materials and more efficient methods of using energy.

Central to this movement are methods that utilize natural heat and light sources to reduce the demand for electrical energy and combustible fuels, and methods to reduce or eliminate energy losses from structures. Initially a fringe philosophy of the 1950's and 1960's, passive solar heating and other methods, for example, have now become a rapidly developing part of mainstream building techniques.

French engineer Felix Trombe (1906-1985) invented what came to be called the Trombe wall structure in the late 1950's. The photoelectric effect that makes solar cells function was first explained by Albert Einstein in 1905, but it was not until the transistor revolutionized electronics in the mid-twentieth century that silicon became a sufficiently viable commodity to produce solar cells on a large scale. The physical principles that drive these different applications now form the foundation of the energy-efficiency movement.

The new Environmental Nature Center building in Newport Beach, California incorporates many sustainable features like photovoltaic roof, composite wood products, renewable materials, water efficient fixtures, natural ventilation, and many others. Visible here is the building's composite siding that is made from wood by-products and recycled plastic but looks like wood. (GIPhotoStock/ Photo Researchers, Inc.)

HOW IT WORKS

The basic concept of energy-efficient building is to construct buildings with as little energy consumption and environmental impact as possible, beginning with construction techniques and materials to the ongoing operation and maintenance of the completed building. Traditional building techniques and materials remain fundamental components of construction practice, although a significant outcome of the move to higher energy efficiency is the development and application of new materials and methods.

Applied physics provides the theoretical framework of energy-efficient building. Convection currents and how they arise have long been known, but it is only relatively recently that this knowledge has been applied to modern building practices. The principle of convection states that warm air, having a lower density than cooler air, rises to float on top of the cooler air, setting up a circulatory movement of air within a room. Passive solar heating makes extensive use of this principle, and building designers consider the fluid dynamics of air movement to maximize solar-heating efficiency. Mechanical devices such as fans to assist convection are often integral to

building design. Structures such as the Trombe wall capture and store heat energy. Openings at the top and bottom of the inner wall facilitate convection, while the massive structure itself radiates heat into the interior of the building.

Foundations represent a significant heat exchange mechanism between the interior and exterior of a building, often producing undesirable moisture-related side effects. Traditional methods rely on sealing the exterior of the foundation wall and either lining the interior of the wall with Styrofoam or creating a dead space with furring strips and paneling. Concrete blocks containing a dead air-space help in insulating foundation walls, but their strength is limited. Poured concrete is a far more effective and structurally versatile construction method, but concrete is a poor thermal barrier. For energy-efficient building, poured concrete has been rethought so that insulated concrete forms are now used in many applications to provide a more efficient thermal barrier.

Above-ground construction, particularly of residential buildings, traditionally utilizes stud-frame construction, in which an insulating material is inserted between the vertical studs of wall structures. The studs themselves serve as thermal conductors connecting the interior and exterior of the building, especially if steel studs are used in place of wooden studs. The use of structural insulated panels, or SIPs, in exterior wall construction provides an uninterrupted layer of insulation throughout the entire wall, with wall strength that meets or exceeds the requirements of a standard stud wall, and an insulating factor between R-45 and R-48 compared with the typical R-20 to R-25 rating for standard stud walls.

Standard glazed windows have very low insulation ratings, generally no more than R-10. A thermal mass window system, however, functions as a solar energy collector by using the mass of a 7.5-centimeter thick layer of water sandwiched between inner and outer glass plates to capture passive solar heat. Such windows are able to maintain a temperature of between

65 degrees and 115 degrees Fahrenheit, radiating heat into a building, even when the exterior temperature is as low as 34 degrees Fahrenheit.

The heating and cooling of buildings traditionally represents high energy consumption, either through the combustion of fuels or through the use of electricity to provide heat or to power cooling systems. Concerns about the long-term availability of fuels and electrical energy demand that more efficient means be utilized. Geothermal heating and cooling systems, coupled with specialized materials that permit facile heat exchange and capture at relatively low temperatures, have thus become more and more important and are rapidly becoming the standard systems for new constructions.

An energy-efficient building quickly ceases to be so if its ongoing operation and maintenance turn to the use of materials and mechanisms that consume more energy. For example, replacing compact fluorescent light bulbs with standard incandescent light bulbs increases energy consumption. Building maintenance may also specify the use of certain cleaning materials based on the energy consumed in their manufacture or transportation requirements.

APPLICATIONS AND PRODUCTS

LEED Certification. LEED (Leadership in Energy and Environmental Design) is an accreditation program of the Green Building Certification Institute, with several designated areas of specialization and three levels of achievement for accreditation. The basic level of accreditation, LEED green associate, demonstrates fundamental knowledge of green construction practices, design principles, and operations. In this context, "green" indicates that the process is effective at reducing energy consumption and at conserving resources. Essentially, it serves to assure others that the completed building is in accord with green construction policies and standards. To qualify for green associate accreditation, a candidate must be able to provide evidence of having been associated with a LEED-registered project, show evidence of employment in a sustainable field, or prove completion of a program of education incorporating green construction principles. The candidate must also successfully complete an examination within specified constraints.

The second tier of LEED certification is accredited professional and the third is accredited professional fellow. Each level represents a significantly higher degree of experience and training in LEED practices. LEED currently provides accreditations in the areas of new construction, existing buildings' operation and maintenance, commercial interiors, core and shell, schools, retail, health care, homes, and neighborhood development. Operations and maintenance provides standards by which services can be measured to maximize operational efficiency and minimize environmental impact and energy consumption. LEED for homes provides a consensus-based, voluntary rating system that is also third-party verified to ensure that homes meet the applicable standards. The new construction protocol ensures that LEED standards and green principles are included in every aspect of building design.

LEED for commercial interiors is geared to ensuring that the interior environments of buildings are healthy and productive workspaces with reduced operating and maintenance costs and reduced environmental impacts. It also provides the direction for tenants and interior designers to make sustainable choices for their own inhabited spaces. LEED certification enables a better understanding of decisions made in project design and construction. Certification better ensures that the work conforms to the spirit of the design and to the design drawings.

Building Materials. Reconsideration of the performance of materials has resulted in numerous modifications to those materials and the structures in which they are used. Energy-efficient modifications are incorporated into the building structure even before the foundation is formed. Foundations themselves are historically responsible for unregulated heat transfer into and out of buildings in the underground levels. Heated basements tend to lose much of their heat to conductive basement walls, while the coolness of those same walls in warm weather contributes to moisture, mildew, and mold problems.

Energy-efficient building design using insulated concrete forms minimizes unregulated heat transfer across basement walls. Primarily made from reinforced Styrofoam or a proprietary wood fiber and concrete mixture called Durisol, insulated forms are stacked like bricks inside a retaining form and then filled with concrete in the usual way. Once the concrete has set, a well-sealed, insulated wall remains.

Structural insulated panels (SIPs) are thick foam panels that have been bonded on both sides to oriented strand board or to plywood. SIP construction replaces the standard stud-wall structure of a building, locking together in such a way that studs are not required for the formation of a straight, strong wall and an air-tight building envelope. The energy requirements of a house built using SIP construction can be as much as 50 percent lower with no further modifications.

Thermal mass (TM) windows are designed to use the heat-storing capacity of a mass of material to absorb solar thermal energy and radiate the stored heat into the interior. Their construction is no more complex than two sheets of thick glass enclosing a mass of water. A TM window system may contain nearly 1,500 kilograms of water and have the ability to capture and store tens of thousands of British thermal units (BTUs) of energy. Other designs incorporate aerogel insulation with water or use phase-changing materials to capture thermal energy.

Trombe walls also use the principle of thermal mass. A Trombe wall consists of a thick solid wall of concrete or other dense material, open at the top and bottom to allow convective air circulation. A second outer wall of glass windows adjacent to the solid wall acts as the solar thermal collector.

Geothermal and In-Floor Heating Systems. The constant underground temperature provides a ready means of heating and cooling through the use of geothermal loop systems. A circulating carrier fluid can be used to transfer heat into the building in winter or to remove excess heat from the building in summer. Though geothermal systems can be used to retrofit an existing building, it is more commonly used in new construction to incorporate an in-floor radiant heating system. The circulating fluid, usually just water, moves through an array of piping built into the floor of the structure to heat the floor itself, which then radiates heat upward to warm the interior of the building. The system also can be used to cool floors in the heat of summer.

Passive Solar Heating. This is the original method of solar heating used to reduce the energy requirements of a building. It is based on the capture of solar thermal energy to directly heat the interior of a building, just as a glass greenhouse captures thermal energy. The method heats air that then circulates by convection, requiring that the building be oriented appropriately when built to maximize the effect.

Fascinating Facts About Energy-Efficient Building

- Thermal mass windows can capture and store as much as 100,000 BTUs of thermal energy on a sunny winter day, which will hold a building's temperature between 60 degrees and 115 degrees Fahrenheit.
- A Trombe wall captures solar energy to heat a massive inner wall that then radiates that heat into the interior of a building through air convection.
- Energy-efficient houses can be built from discarded tires that have been packed full of earth.
- A thermal mass window may hold 1,500 kilograms or more of water.
- Geothermal ground-loop systems for heating and cooling make use of the constant temperature of the earth that exists a short distance below ground level.
- Insulated foundation walls in a building can reduce energy consumption for heating and cooling by as much as 20 percent.
- Structural insulated panel (SIP) construction can reduce a building's energy requirements by 50 percent.
- It is possible to construct modern houses with all the conveniences entirely self-sufficient in energy needs and therefore requiring no additional heating or cooling.

Other Methods of Energy Consumption. Energy-efficient buildings by definition consume less energy than do other buildings. Part of the reduction in energy consumption, defined in standards, includes the use of such low-energy lighting as compact fluorescent light bulbs, the use of insulated flooring, and the use of windows that reflect away excess thermal energy to reduce or eliminate the need for air conditioning.

IMPACT ON INDUSTRY

Energy-efficient building represents a paradigm shift in the ways industry approaches infrastructure and facilities. Historically, industry constructed commercial and residential buildings to be simply functional. In this paradigm, the construction of the building and its ongoing operation and maintenance were not regarded as integral features, but as tasks to be accomplished separately. While the functional stability and integrity of buildings have always been considerations, this construction focused on the use of standard building materials to produce a structure that needed only to withstand the external environment while providing the desired interior capacity.

Integral to design in the new paradigm of energy-efficient building, however, is the ongoing nature, functionality, and maintenance of the building after it has been constructed. Plain poured concrete foundations shaped within bare wooden forms become poured concrete foundations formed with and poured within precision structures of insulated concrete forms. Energy consuming forced-air heating and central air conditioning is replaced, or, at minimum, augmented by passive solar heating, geothermal ground-loop heating and cooling, and Trombe walls. Plain glazed, double-glazed, and even triple-glazed windows are supplanted by water-filled thermal mass windows. To meet energy-efficiency certification standards, the building industry now considers the energy costs associated with products used in the upkeep of energy-efficient buildings.

As with all such shifts in industry practices, associated costs are high at the outset, as the new methods and technologies begin to supplant the older ones. In this period of adjustment, the primary and secondary industries needed to provide materials and education are in short supply and, therefore, carry premium values. In contrast, the secondary industries that support the older established methodologies are functioning to economy of scale and, therefore, represent considerably lower costs. Inevitably, however, as new methods prove their worth, they attain the desirability that leads to established secondary industries, and their associated costs decrease as those new secondary industries move toward economy-of-scale operation to match the increasing demand for their products.

At the same time, new industries are produced that provide specialized expertise with the new methods of energy-efficient building. A concrete-forming company, for example, will add proficiency with insulated concrete forms to the traditional services it provides and so continue as a viable business, else it may find that the demands for its services steadily decrease while those for a newly established competitor that specializes in energy-efficient methods rapidly increase.

One of the fastest growing demands for energy-efficient methodologies in recent years, particularly in areas where seasonal temperatures vary widely from winter to summer, has been for geothermal heating and cooling systems. Because they are amenable to augmentation via heat exchangers by a variety of heating and cooling technologies, such as heat pumps and boiler systems, and because they work well with in-floor systems, geothermal loop systems have become an attractive means of energy-efficient heating and cooling in new structures. They also have proved valuable in reducing the energy consumption of existing heating and cooling systems as a retrofitted augmentation method, particularly for residential purposes.

Retrofitting of existing structures for energy efficiency represents both a significant cost and, over time, a substantial return on investment. While a small retrofitting project may be carried out in a matter of months, large projects may require a number of years. During the retrofit period, funding must be made available, which could negatively affect cash flow. However, as savings accrue from the functioning retrofit, those funds and the overall economic value generated by the building as additional cash flow eventually surpass the initial costs.

CAREERS AND COURSEWORK

Standardized techniques and theoretical principles of energy-efficient building practices are in their infancy as subject matter in academic curricula. Persons interested in pursuing a career in the building trades as specialists in energy-efficient building can expect to take courses and trades-training based on established and traditional building methods. Specialization will be acquired through upgrading programs and on-the-job experience. This training focus can be expected to change as the demand for expertise in energy-efficient building increases and the methodology becomes the accepted norm of building practices.

Certification programs such as LEED are only beginning to become incorporated into trades-training. The certification procedure is offered only outside standard training curricula, although this is often provided through the facilities of a local trade school. Students can therefore expect to take basic courses in mathematics, general physics and chemistry, architectural drafting and design, and the strength of materials as part of the training curriculum for basic trades. Specialized trades such as electrician, plumber, framer, and general construction worker will have more advanced hands-on requirements to ensure knowledge of standard, safe working methods according to appropriate regulations. With such training in hand, or with documented equivalent experience, a person is then able to pursue certification. This hands-on experience is considered an essential prerequisite for a career in energy-efficient building.

SOCIAL CONTEXT AND FUTURE PROSPECTS

Access to the energy required to power an ever-growing society has become an issue of major concern, especially in recent years. Accordingly, developing the means to minimize the use of energy and to maximize the production of energy has also become important.

Electrical energy is the mainstay of modern industry, while both electricity and combustible fuels are the primary means of providing temperature control in buildings. Traditional methods of producing electrical energy have been hydroelectric generation, coal and gas-fired generating plants, and nuclear-powered generating stations. These methods are aging and the technologies are becoming outdated, but significant efforts are being made to augment or replace their capabilities with renewable sources such as solar- and wind-generated electricity. At the same time, efforts are expanding to reduce the energy requirements of buildings and infrastructure. At the forefront of this movement are energy-efficient building design and construction methods.

It is now possible to construct residential buildings that require no additional input of energy to maintain and operate. Such zero-energy housing, however, is a specialty or niche market. The cost of constructing such a building is somewhat higher than the cost of a traditionally built home. This factor alone is sufficient to hold the demand for zero-energy housing to a minimum. That said, as secure access to energy becomes increasingly difficult, the demand and market for zero-energy buildings and construction methods is likely to follow an exponential growth curve.

Richard M. Renneboog, M.Sc.

FURTHER READING

Gulli, Cathy. "The Big Chill." *Maclean's* 123, no. 13 (June, 2010): 63-64. This article explores the energy drain associated with the wide-scale use of air conditioners.

Hordeski, Michael F. *Megatrends for Energy Efficiency and Renewable Energy*. Lilburn, Ga.: Fairmont Press, 2011. This book examines the entwined relationships between energy and industries concerned with building, power, fuels, conservation, and automation.

Jefferey, Yvonne, Liz Barclay, and Michael Grosvenor. *Green Living for Dummies*. Hoboken, N.J.: John Wiley & Sons, 2008. This brief introductory book provides many insights into the philosophy of energy-efficient building.

Johnston, David, and Scott Gibson. *Green from the Ground Up: Sustainable, Healthy, and Energy-efficient Home Construction*. Newtown, Conn.: Taunton Press, 2008. A builder's guide that presents the rationale of green construction and focuses on applying those principles with standard construction practices.

WEB SITES

Energy Efficient Building Technologies
"Affordable Zero Energy Homes"
http://www.eebt.org

Fraser Basin
"Energy Efficiency and Buildings"
http://www.fraserbasin.bc.ca/publications/documents/caee_manual_2007.pdf

International Energy Agency
"Technology Roadmap: Energy-efficient Buildings—Heating and Cooling Equipment"
http://www.iea.org/papers/2011/buildings_roadmap.pdf

United States Department of Energy
"Elements of an Energy-efficient House"
http://www.nrel.gov/docs/fy00osti/27835.pdf

See also: Applied Physics; Architecture and Architectural Engineering; Civil Engineering; Electronics and Electronic Engineering; Engineering; Environmental Engineering; Fluid Dynamics; Solar Energy; Structural Composites; Urban Planning and Engineering.

ENERGY STORAGE TECHNOLOGIES

FIELDS OF STUDY

Biochemistry; chemistry; electrical engineering; electronics; hydraulics; materials science; mechanical engineering; physics; quantum mechanics; superconductivity; thermodynamics.

SUMMARY

Energy storage technologies provide primary power sources for portable devices and vehicles and are employed in electrical grids to act as backups for generators in order to ensure a stable, steady energy supply. Energy storage is particularly needed for grids that rely on renewable energy sources, such as solar and wind power, so that during periods without sunlight or wind when generators are not operating, electricity can still be sent to consumers. Storage technologies fall into three broad categories: mechanical energy (kinetic or potential) and thermal energy systems; electrochemical systems; and electrical storage systems.

KEY TERMS AND CONCEPTS

- **Electrolyte:** Solution, or the molten state of a substance, that conducts electricity through the presence of ions.
- **Energy Density:** Amount of energy stored in a system per unit of volume.
- **Frequency Regulation:** Continuous and instantaneous balancing of supply and demand on an electrical grid.
- **Peak Load:** Period when there is high demand on an electricity grid.
- **Power:** Rate at which energy is transferred or work done, measured in watts.
- **Power Density:** Amount of energy a system transfers (power) per unit of volume.
- **Storage Time:** Discharge time of stored energy at a given power.
- **Watt Hour (Wh):** Energy delivery of one watt for one hour, a basic unit in electrical generation and transfer.

DEFINITION AND BASIC PRINCIPLES

Energy storage is the artificial containment of energy for controlled release. It is different in kind from the various forms of natural energy storage, such as the storage of radiant energy from the sun in wood, oil, or coal, which frequently supplies the energy for the generation of the electrical power that supports modern technological civilization. Energy storage technologies provide the physical means for energy containment and are based on a variety of mechanical, electrochemical, or electromagnetic principles. As such, they are rated for usefulness and efficiency based upon how much energy they can contain (energy density) and how they release that energy (power density).

Energy storage devices can supply either primary power or secondary power. Devices such as batteries, capacitors, and fuel cells, for instance, may provide primary power, usually for portable electronics or vehicles—anything that must be used apart from the steady supply of an electrical power grid. Portable radios, smoke detectors, watches, electric cars, and emergency lighting are examples. As secondary, or supplemental, power sources, energy storage technologies are charged by a power grid and then return the energy back to the grid as needed to manage peak electrical loads, improve power quality, ensure frequency regulation, or make up for failing production, as when a generator must be taken off-line or when absence of sunlight or wind idles a solar- or wind-powered generator.

BACKGROUND AND HISTORY

Archaeologists found evidence of what might have been a battery dating from the third century B.C.E. in the form of a clay pot containing iron and copper rods; the pots could have produced electrical discharges if filled with vinegar, although scholars disagree about this interpretation. It is even less well known when another major energy storage device first saw use, the reservoir or millpond. It consisted of a dam or diverted stream that filled a pond that supplied water on demand to turn a waterwheel.

The first modern artificial energy storage systems were the Leyden jar and the electrochemical battery.

German physicist Ewald Georg von Kleist invented the Leyden jar in 1744. A capacitor, it stored static electricity in a glass jar coated inside and out with metal foil and had a conducting rod stuck through an insulating stopper. In a 1800 letter to England's Royal Society, Italian physicist Alessandro Volta described the first electrochemical battery, an arrangement of stacked disks of tin or zinc paired with disks of copper, brass, or silver, each pair separated by a disk of paper or leather moistened in an electrolyte solution. Whereas the Leyden jar delivered electricity only in a burst, the battery gave a steady current, which proved invaluable to experiments in physics and chemistry.

Other devices for storing energy followed steadily: the fuel cell in 1839, compressed air energy storage in 1870, the rechargeable flow battery in 1884, thermal energy storage in 1886, pumped hydro storage in 1908, flywheel energy storage in 1950, super capacitors in 1966, and superconducting magnetic energy storage in 1986.

How It Works

Mechanical Systems. Pumped water (pumped hydro) is the most common system worldwide for storing energy to serve electrical grids. During low demand on the grid, usually at night, water is pumped from a lower to an upper reservoir. When electrical demand increases and load leveling or supplemental electricity is needed, water in the upper reservoir is released to flow downhill and turn generators. It is also possible to use underground cavities or the open sea for storage. Compressed air energy storage (CAES) also takes advantage of off-peak electricity. Air is compressed, stored underground, and used when needed to turn a gas turbine. The main difficulty with pumped hydro and CAES is finding suitable terrain for reservoirs or underground cavities to store air.

Flywheels store kinetic energy in a rotating mass in the form of a disk or rotor. The flywheel is accelerated during off-peak load by an electrical motor. Later this stored energy can be used to turn an electrical generator and return energy to a grid. Flywheels are categorized as low speed if the rotor spins at fewer than 6,000 revolutions per minute (rpm); these usually have steel rotors and turn on metal ball bearings. High-speed flywheels turn at up to 50,000 rpm and have rotors of composite materials that turn on magnetic bearings. The drawback of flywheels is primarily the degree of standby energy loss. Springs, long used to store mechanical energy in clocks, for instance, are also proposed to store energy for electrical generation.

Thermal energy storage (TES) involves heat stored in saltwater ponds, molten salt, bricks, water reservoirs, pressurized water, or steam by converting electrical energy or by collecting heat directly, as through geothermal or solar heating. Conversely, during low demand in a cooling system, electricity can run refrigerators to make ice, which during high demand can be returned to the system to enhance cooling. Waste electricity during off-peak demand can also be used to produce liquid oxygen and nitrogen that can later be boiled to operate a turbine. In all cases, mechanical and thermal systems are in theory continually renewable as long as machinery and materials remain sound.

Electrochemical Systems. Large or small, batteries combine cells in a sequence. A cell comprises an electron source, the anode, and an electron acceptor, the cathode, that, immersed in an electrolyte, directly converts chemical energy to electricity. Batteries come in two major types: primary, or disposable (alkaline, lithium cells, and zinc-carbon batteries), and secondary, or rechargeable (lead-acid, nickel-cadmium, nickel-iron, nickel-metal hydride, and lithium-ion batteries). The number of charge-discharge cycles that secondary batteries can go through before wearing out, their charging and storage times, and their energy density and power density vary and determine their range of applications.

Flow cells (also redox flow cells, flow batteries) are similar to batteries, except that the electrodes are catalysts for the chemical reaction, which occurs as a microporous membrane allows ions to pass from one electrolyte solution to another. Among flow cells are types that use zinc and bromine, vanadium in two types with different states, or polysulfide and bromine as the pairs of electrolytes. The advantages of flow cells are that they are capable of a large number of cycles, and the electrolytes can be replenished.

Hydrogen fuel cells store energy by employing an electrolyzer to produce hydrogen. It is stored until a fuel cell splits the hydrogen into ions and electrons. The electrons flow through a wire, producing an electric current, while the ions, after passing through an electrolyte between the anode and cathode, are reattached to the electrons reentering the cell from the wire and in the presence of oxygen-form wastewater.

Biological storage is yet another means of storing energy chemically through biological conversion, as is done with adenosine triphosphate (ATP) in the mitochondria of cells. Although interest in artificial biological systems is likely to increase, it is a nascent technology.

Electricity Storage Systems. Superconducting magnetic energy storage (SMES) entails storing energy in a magnetic field produced by passing direct current through a coil. This technique does not work with normal electrical wire because resistance and heat loss dissipates energy from the coil. Superconducting wire, having almost no resistance, is required. The superconductors, however, must be kept at very low temperatures to function. Niobium-based wires, for instance, require a temperature of just more than 9 Kelvin. Ceramic materials that superconduct at much higher temperatures may be used, and room-temperature superconductors are theoretically possible, obviating the need for complex cooling systems for the coil.

Super capacitors are also known as double-layer capacitors, electrochemical capacitors, ultra capacitors, and pseudo capacitors. They store energy by separating electrostatic charges in electrode plates on either side of a liquid or organic electrolyte. One plate attracts positive ions from the electrolyte, the other negative ions, thereby storing electrical energy in two layers. Super capacitors are reversible and capable of 10,000 charge-discharge cycles.

APPLICATIONS AND PRODUCTS

Batteries and Cells. Batteries are ubiquitous in technological society, needed to power virtually all devices that are portable and many vehicles, but also used in utility-scale energy storage. Disposable batteries have long powered household devices, such as the D-cell 1.5 volt batteries in flashlights and toys, but rechargeable batteries are the power source in developing technologies, such as laptop computers, cell phones, and MP3 players. Lithium-ion batteries quickly dominated the portable-devices market because they have high-energy density (300 to 400 kilowatt hours, or kWh, per cubic meter), high efficiency, and long recharging life (about 3,000 cycles). Arrays of batteries are also used for large-scale energy

Fascinating Facts About Energy Storage Technologies

- The first modern battery was described in 1800 by Alessandro Volta, a professor at the University of Pavia in Italy. His stacked zinc-silver cells, moistened with an electrolyte solution, became known as a voltaic pile.

- The world market for batteries in 2010 was estimated at about $15 billion and is expected to grow steadily.

- In 2010, the U.S. Information Administration predicted that total world consumption of marketed energy would rise from 495 quadrillion British Thermal Units (Btu) in 2007 to 739 quadrillion Btu's in 2035, an increase of 49 percent, or an average of 1.4 percent a year. To meet the demand, energy storage systems will be needed to make energy available efficiently.

- Three advantages of flywheels for energy storage are that they are low maintenance, environmentally benign, and last long—up to twenty years.

- Global productions from solar cells reached 21 gigawatts in 2009, enough to supply electricity to 5.5 million houses yearly. Because solar cells do not generate without sunlight, batteries are required to supply a continuous current.

- Pumped hydro is the most commonly employed non-electrochemical energy storage technology and has been used in the United States since 1929.

- In the United States, interest in energy storage technologies soared during the energy crisis of the mid-1970's.

- A laptop computer battery and a full Thermos of hot coffee store about the same amount of energy, but the Thermos does it for less than 1 percent of the battery's cost, which is why thermal energy storage systems are attractive.

- The lithium-ion batteries storing solar-generated power on the Mars rovers Spirit and Opportunity were designed to provide about 900 watt hours of power per day for at least 90 days. They were still working six years later.

storage. Zinc-bromide battery systems, sometimes mounted on trailers for transportation, have capacities of 1 megawatt (MW) for 3 megawatt hours (MWh); units can be linked for further capacity. Sodium-sulfur batteries are widely used in Japan, storing up to 34 MW per 245 MWh for frequency regulation and to receive energy from wind generators. The lead-acid battery, the most developed battery technology, provided the storage core for vehicles for generations. Nickel-cadmium and nickel-metal hydride batteries are competing systems, especially for hybrid and electric vehicles, because they are more efficient and powerful. Home-energy storage from solar photovoltaic cell arrays include lead-acid, lithium, and metal-hydride batteries connected in banks. Nickel-hydrogen, nickel-cadmium, and lithium-ion batteries power space exploration, variously serving on satellites, the Hubble Space Telescope, interplanetary probes, and Opportunity and Spirit, the two rovers that landed on Mars in 2004.

Flow cells are ideal for storage systems in remote locations. Vanadium redox systems, for instance, deliver up to 500 kW for up to ten hours. Zinc-bromine systems have been produced for 50-kWh and 500-kWh systems to reinforce weak distribution networks or prevent power fluctuations. Hydrogen fuel cells can potentially do almost anything a battery can do: provide backup power, perform power leveling, run handheld devices, and supply primary or auxiliary power to cars, trucks, buses, and boats. In many cases they are more efficient than petrochemical fuels. A hydrogen fuel cell in a vehicle that uses an electric motor, for example, can be 40 to 60 percent efficient, compared with the 35 percent peak efficiency of the internal combustion engine.

Mechanical and Thermal Systems. Although flywheels have been tested as power storage in cars and have aerospace applications, mechanical and thermal systems are primarily backups for large-scale power generation or cooling systems. Pumped hydro serves electrical grids worldwide: more than 90 gigawatts (GW) of storage and about 3 percent of global generation capacity in 2010. Systems have efficiencies of 70 to 85 percent. Some flywheels can convert 90 percent of their kinetic energy to electricity that will last for several hours. Telecommunications systems employ flywheels delivering 2 kW for 6 kWh, and megawatts can be stored in linked arrays of flywheels. CAES is used to store energy for electrical grids. The first large-scale commercial system, capable of 290 MW, came online in Huntorf, Germany, in 1978. Later systems varied in output from 110 to 300 MW, and as of 2011, a 2.7 GW system was planned for Norton, Ohio, using an abandoned limestone mine some 670 meters deep.

TES systems have somewhat more variety of use. Passive solar power systems used in houses and commercial buildings soak up energy from the sun that can be used to heat interior water or air. Likewise, systems that use off-peak energy to make ice that later contributes to refrigeration operate in large buildings to save electricity costs. For instance, the Dallas Veterans Affairs Medical Center incorporated a chilled-water TES that reduced electricity consumption nearly 3,000 kW and saved more $200,000 in its first year of operation.

Superconducting Magnetic Energy Storage (SMES). SMES is used by some electrical utilities to improve the system reliability and transfer capacity so that grid operators can compensate for damage from storms, voltage variation, and increasing consumer demand. The first distributed SMES system began operation in Wisconsin in 2000; it employs SMES mounted on 48-foot trailers that can be deployed to support the grid in rural areas. The largest users from the 1990's on, however, were industries making plastics, paper, aluminum, and chemicals.

Super Capacitors. Super capacitors are widely used in consumer electronics of all kinds for power backup and flash charging. They also figure in braking systems, such as energy buffers for elevators, and in motor ignition, especially for the big engines in buses, trucks, tanks, and submarines. However, they also have application in larger systems to compensate for voltage drops and can be used in hybrid systems with rechargeable batteries, such as electric and hybrid vehicles.

IMPACT ON INDUSTRY

Energy storage is a rapidly growing field worldwide. Developed nations, especially those with a heavy industry or high tech base, look to energy storage to help solve problems in maintaining high-quality electrical transmission on their grids, but many nations also use energy storage in remote areas and for conservation. As grids rely more and more on renewable-energy generation—wind, wave, and solar power—energy storage becomes increasingly important to provide current throughout the day. Among leaders in energy storage technologies are the European Union, Japan, the United States, Great Britain, Australia, and Brazil. Several international and regional organizations foster research and development, such as Instituto Para o Desenvolvimento de Energias Alternativas na América Latina (IDEAL); the International Energy Agency (IEA), an autonomous organization based in Paris; and the Electricity Storage Association (ESA) in the United States.

Government and University Research. The Department of Energy (DOE) started its Energy Storage Systems Program at Sandia National Laboratories in New Mexico in the mid-1970's as a response to the first oil crisis. It concentrates on developing technologies in batteries, CAES, and flywheels for use in electrical grids. The DOE also fosters research at university and industry laboratories and holds an annual conference. The Northeastern Center for Chemical Energy Storage, based at the State University of New York at Stony Brook, brings together researchers from American universities and the Lawrence Berkeley National Laboratory, Argonne National Laboratory, and Brookhaven National Laboratory. The Center for Advanced Energy Studies at Idaho National Laboratory in Idaho Falls is another leading research facility.

Industry and Business. Many utilities and industries are expanding both research facilities and development budgets for energy storage systems. ESA and the U.S. Advanced Battery Consortium promote long-range research and development in common with national laboratories and universities, and industry newsletters, such as the *Clean Technology Business Review*, keep potential users apprised of new products. In the car industry alone, for example, battery development to support electric or hybrid cars became a major research focus in the 2000's. The leaders include Chrysler, Daimler, Ford, General Motors, Honda, and Toyota.

Energy Storage Suppliers. The number of established and start-up companies offering energy storage systems grew rapidly beginning in the 1990's. Among them are: A123 Systems, Cobasys, Compact Power, EVionyx, Exide Technologies, NGK Insulators, Optima, Saft (batteries), Deeya Energy, Premium Power Corporation, Prudent Energy, RedFlow, ZBB Energy Corporation (flow cells), GE Energy Industrial Solutions, Saft (capacitors), Active Power, Beacon Power, Piller Power Systems, Urenco Power Technologies (flywheels), CAES Development Company, Dresser-Rand, Energy Storage and Power, Ridge Energy Storage (CAES), American Superconductor, Bruker Energy and Supercon Technologies, and SuperPower (SMES).

CAREERS AND COURSE WORK

The substantial increase in research and development since the 1990's, as well as proliferating venture capital investments, in energy storage technologies means that they will continue to offer career opportunities for scientists, engineers, and businesspeople worldwide through the twenty-first century.

College students interested in pursing a career in energy storage can begin by taking science, mathematics, and engineering courses. At some point selecting a major will prepare one for a particular type of energy storage technology: For example, physics for those interested in SMES and flywheels; electrical engineering, chemistry, or automotive engineering for those interest in batteries, capacitors, and fuel cells and their applications; and mechanical and civil engineering for CAES and pumped hydro. Refining the specialty in graduate school to obtain a master's degree or doctorate would position a person to work at the forefront of a crucial technology. However, someone with a science background would be prepared for a marketing or management position within an energy storage company or utility after obtaining a master of business administration (M.B.A.).

For instance, a physicist who wrote a doctoral dissertation on high-temperature superconducting materials in energy storage systems might find employment at a company such as American Superconductor, a utility, a university or government research laboratory, or a university faculty. An undergraduate degree in mechanical engineering could lead a person, after obtaining an M.B.A., to joining junior management for a corporation like General Electric, working for a utility or construction company, or starting an energy storage company. The prospect of a system based upon biological storage of energy for utilities, consumer devices, or industry was, as of 2011, promising. Because research is not far advanced, experts in biochemistry have the opportunity to be pioneers.

SOCIAL CONTEXT AND FUTURE PROSPECTS

Energy storage helps solve several pressing problems in costs and availability of power. Studies repeatedly argue that electrical grids cannot be expanded sufficiently or fast enough to meet projected consumer demand during the twenty-first century without incorporating energy storage systems. Moreover, in many cases although the initial construction or purchasing and installation costs are high, energy storage systems save money in the long run by obviating investment in expensive new generators, load management systems, and transmission lines. The savings can help keep the price of electricity low, especially for manufacturers, and so reduce the power industry's contribution to inflation. Energy storage will also prove crucial to ensure steady supply from solar-, wind-, or wave-powered energy generation for large grids. This is true on a smaller scale for houses or buildings that rely on autonomous renewable energy systems—most likely solar or wind—for their electricity and heating. As consumers demand better batteries for their portable devices and vehicles, research in batteries and fuel cells will concentrate on reducing recharging time and increasing cycling life, storage time, power density, and energy density.

Energy storage systems will also affect the environment, although not always beneficially. In operation most do not pollute or disrupt an ecosystem and so are environmentally benign. On the other hand, construction of CAES or pumped hydro is disruptive, and manufacture of the other systems frequently entails waste products that are toxic. In balance, however, energy storage systems will prove to have less environmental impact than electricity produced from coal-, nuclear-, or gas-powered or hydroelectric generators.

Roger Smith, Ph.D.

FURTHER READING

Baxter, Richard. *Energy Storage: A Nontechnical Guide.* Tulsa, Okla.: PennWell, 2006. Intended for policy makers and business analysts, this guide argues for a distributed energy supply strategy and in that light surveys several storage technologies, such as pumped-hydrogen, compressed air, batteries, capacitors, and magnetic energy.

Brunet, Yves, ed. *Energy Storage.* Hoboken, N.J.: John Wiley & Sons, 2010. The author discusses hydrogen, super capacitors, flywheels, thermal, and gravitational storage systems as he considers applications in electric power generation, transmission, and distribution systems and in vehicles and buildings.

Dinçer, Ibrahim, and Marc A. Rosen. *Thermal Energy Storage: Systems and Applications.* 2d ed. Hoboken, N.J.: John Wiley & Sons, 2011. Although concentrating on the nature and applications of thermal energy storage, this technical introduction also describes mechanical, chemical, biological, magnetic, and hydrogen systems.

Huggins, Robert A. *Energy Storage.* New York: Springer, 2010. For engineering students, this book presents the fundamentals of energy storage, covering various fuels, phase transitions, heat capacity, and mechanical, hydrogen, and electromagnetic storage systems.

Press, Roman J., et al. *Introduction to Hydrogen Technology.* Hoboken, N.J.: John Wiley & Sons, 2009. Addresses available energy resources, the chemical background of hydrogen reactions and properties, hydrogen technology, and the science and application of fuel cells for an audience who understands basic chemistry and mathematics.

Root, Michael. *The TAB Battery Book: An In-Depth Guide to Construction, Design, and Use.* New York: McGraw-Hill, 2011. Although this books requires some knowledge of chemistry and mathematics to appreciate fully, the introductory chapters about the history and basics of batteries provides a clear overview.

Zito, Ralph. *Energy Storage: A New Approach.* Hoboken, N.J.: Wiley-Scrivener, 2010. Discusses storage of massive energy and explains the basic chemical processes that underlie storage no matter how the energy is generated.

WEB SITES
Electricity Storage Association
http://www.electricitystorage.org

Sandia National Laboratory
Energy Storage Systems Program
http://www.sandia.gov/ess/index.html

World Energy Council
http://www.electricitystorage.org

See also: Electrical Engineering; Mechanical Engineering.

ENGINEERING

FIELDS OF STUDY

Physics; chemistry; computer science; mathematics; calculus; design; systems; processes; materials; circuitry; electronics; environmental science; miniaturization; biology; aeronautics; fluids; gases; technical communication.

SUMMARY

Engineering is the application of scientific and mathematical principles for practical purposes. Engineering is subdivided into many disciplines; all create new products and make existing products or systems work more efficiently, faster, safer, or at less cost. The products of engineering are ubiquitous and range from the familiar, such as microwave ovens and sound systems in movie theaters, to the complex, such as rocket propulsion systems and genetic engineering.

KEY TERMS AND CONCEPTS

- **Analogue:** Technology for recording a wave in its original form.
- **Design:** Series of scientifically rigorous steps engineers use to create a product or system.
- **Digital:** Technology for sampling analogue waves and turning them into numbers; these numbers are turned into voltage.
- **Energy:** Capacity to do work; can be chemical, electrical, heat, kinetic, nuclear, potential, radiant, radiation, or thermal in nature.
- **Feasibility Study:** Process of evaluating a proposed design to determine its production difficulty in terms of personnel, materials, and cost.
- **Force:** Anything that produces or prevents motion; a force can be precisely measured.
- **Matter:** Anything that occupies space and has weight.
- **Power:** Time rate of doing work.
- **Prototype:** Original full-scale working model of a product or system.
- **Quantum Mechanics:** Science of understanding how a particle can act like both a particle and a wave.
- **Specifications:** Exact requirements engineers must comply with to create products or services.
- **Work:** Product of a displacement and the component of the force in the direction of the displacement.

DEFINITION AND BASIC PRINCIPLES

Engineering is a broad field in which practitioners attempt to solve problems. Engineers work within strict parameters set by the physical universe. Engineers first observe and experiment with various phenomena, then express their findings in mathematical and chemical formulas. The generalizations that describe the physical universe are called laws or principles and include gravity, the speed of light, the speed of sound, the basic building or subatomic particles of matter, the chemical construction of compounds, and the thermodynamic relationship that to produce energy requires energy. The fundamental composition of the universe is divided into matter and energy. The potential exists to convert matter into energy and vice versa. The physical universe sets the rules for engineers, whether the project is designing a booster rocket to lift thousands of tons into outer space or creating a probe for surgery on an infant's heart.

Engineering is a rigorous, demanding discipline because all work must be done with regard to the laws of the physical universe. Products and systems must withstand rigorous independent trials. A team in Utah, for example, must be able to replicate the work of a team in the Ukraine. Engineers develop projects using the scientific method, which has four parts: observing, generalizing, theorizing, and testing.

BACKGROUND AND HISTORY

The first prehistoric man to use a branch as a lever might be called an engineer although he never knew about fulcrums. The people who designed and built the pyramids of Giza (2500 B.C.E.) were engineers. The term "engineer" derives from the medieval Latin word *ingeniator*, a person with "ingenium," connoting curiosity and brilliance. Leonardo da Vinci, who used mathematics and scientific principles in everything from his paintings to his designs for military fortifications, was called the Ingegnere Generale (general engineer). Galileo is credited with seeking a systematic explanation for phenomena and adopting a scientific approach to problem solving. In 1600, William Gilbert, considered the first electrical engineer, published *De magnete, magneticisque corporibus et de magno magnete tellure* (*A New Natural Philosophy of the Magnet, Magnetic Bodies, and the Great Terrestrial Magnet*, 1893; better known as *De magnete*) and coined the term "electricity." Until the Industrial Revolution of the eighteenth and nineteenth centuries, engineering was done using trial and error. The British are credited with developing mechanical engineering, including the first steam engine prototype developed by Thomas Savery in 1698 and first practical steam engine developed by James Watt in the 1760's.

Military situations often propel civilian advancements, as illustrated by World War II. The need for advances in flight, transportation, communication, mass production, and distribution fostered growth in the fields of aerospace, telecommunication, computers, automation, artificial intelligence, and robotics. In the twenty-first century, biomedical engineering spurred advances in medicine with developments such as synthetic body parts and genetic testing.

How It Works

Engineering is made up of specialties and subspecialties. Scientific discoveries and new problems constantly create opportunities for additional subspecialties. Nevertheless, all engineers work the same way. When presented with a problem to solve, they research the issue, design and develop a solution, and test and evaluate it.

For example, to create tiles for the underbelly of the space shuttle, engineers begin by researching the conditions under which the tiles must function. They examine the total area covered by the tiles, their individual size and weight, and temperature and frictional variations that affect the stability and longevity of the tiles. They decide how the tiles will be secured and interact with the materials adjacent to them. They also must consider budgets and deadlines.

Collaboration. Engineering is collaborative. For example, if a laboratory requires a better centrifuge, the laboratory needs designers with knowledge in materials, wiring, and metal casting. If the metal used is unusual or scarce, mining engineers need to determine the feasibility of providing the metal. At the assembly factory, an industrial engineer alters the

Working alongside a rapidly deployable temporary floodwall, U.S. Army Corps of Engineers employees survey rising waters near Hannibal, Missouri. (PR Newswire/AP Images)

assembly line to create the centrifuge. Through this collaborative process, the improved centrifuge enables a biomedical engineer to produce a life-saving drug.

Communication. The collaborative nature of engineering means everyone relies on proven scientific knowledge and symbols clearly communicated among engineers and customers. The increasingly complex group activity of engineering and the need to communicate it to a variety of audiences has resulted in the emergence of the field of technical communications, which specializes in the creation of written, spoken, and graphic materials that are clear, unambiguous, and technically accurate.

Design and Development. Design and development are often initially at odds with each other. For example, in an architectural team assigned with creating the tallest building in the world, the design engineer is likely to be very concerned with the aesthetics of the building in a desire to please the client and the city's urban planners. However, the development engineer may not approve the design, no matter how beautiful, because the forces of nature (such as wind shear on a mile-high building) might not allow for facets of the design. The aesthetics of design and the practical concerns of development typically generate a certain level of tension. The ultimate engineering challenge is to develop materials or methods that withstand these forces of nature or otherwise circumvent them, allowing designs, products, and processes that previously were impossible.

Testing. With computers, designs that at one time took days to draw can be created in hours. Similarly, computers allow a prototype (or trial product) to be quickly produced. Advances in computer simulation make it easier to conduct tests. Testing can be done multiple times and under a broad range of harsh conditions. For example, computer simulation is used to test the composite materials that are increasingly used in place of wood in building infrastructures. These composites are useful for a variety of reasons, including fire retardation. If used as beams in a multistory building, they must be able to withstand tremendous bending and heat forces. Testing also examines the materials' compatibility with the ground conditions at the building site, including the potential for earthquakes or other disasters.

Financial Considerations. Financial parameters often vie with human cost, as in biomedical advancements. If a new drug or stent material is rushed into production without proper testing to maximize the profit of the developing company, patients may suffer. Experimenting with new concrete materials without determining the proper drying time might lower the cost of their development, but buildings or bridges could collapse. Dollars and humanity are always in the forefront of any engineering project.

APPLICATIONS AND PRODUCTS

The collaborative nature of engineering requires the cooperation of engineers with various types of knowledge to solve any single problem. Each branch of engineering has specialized knowledge and expertise.

Aerospace. The field of aerospace engineering is divided into aeronautical engineering, which deals with aircraft that remain in the Earth's atmosphere, and astronautical engineering, which deals with spacecraft. Aircraft and spacecraft must endure extreme changes in temperature and atmospheric pressure and withstand massive structural loads. Weight and cost considerations are paramount, as is reliability. Engineers have developed new composite materials to reduce the weight of aircraft and enhance fuel efficiency and have altered spacecraft design to help control the friction generated when spacecraft leave and reenter the Earth's atmosphere. These developments have influenced earthbound transportation from cars to bullet trains.

Architectural. The field of architectural engineering applies the principles of engineering to the design and construction of buildings. Architectural engineers address the electrical, mechanical, and structural aspects of a building's design as well as its appearance and how it fits in its environment.

Fascinating Facts About Engineering

- The Bering Strait is 53 miles of open water between Alaska and Russia. In the 2009 Bering Strait Project Competition, people submitted plans for bridging the strait, including a combination bridge-tunnel with passageways for migrating whales and the capability of circulating the frigid arctic waters of the north to help fend off global warming.

- The Gotthard Base Tunnel network is being built under the Alps from Switzerland to Milan, Italy. At about 35 miles, the two-way tunnel will be the longest in the world and will reduce the time to make the trip by automobile from 3.5 hours to 1 hour. High-speed trains traveling at speeds of 155 miles per hour will make more than two hundred trips through the tunnel per day.

- A typical desktop computer can handle 100 million instructions per second. As of June, 2010, the fastest supercomputer was the Cray Jaguar at Oak Ridge National Laboratory. Its top speed is 1.75 petaflops (1 quadrillion floating point operations) per second.

- The world's smallest microscope weighs about as much as an egg. Instead of using a lens to magnify, it generates holographic images of microscopic particles or cells using a light emitting diode (LED) to illuminate and a digital sensor to capture the image.

- The National Institutes of Health has developed an implant made of silk and metal that when placed in the brain can detect impending seizures and send out electric pulses to halt them. It can also send out electric signals to prostheses used by people with spinal cord injuries.

- Stanford University is developing the computers and technology for a driverless car. The 2010 autonomous race car, dubbed Shelley, is an Audi TTS equipped with a differential Global Positioning System accurate to an inch. It calculates the right times to brake and accelerate while turning.

Areas of concern to architectural engineers include plumbing, lighting, acoustics, energy conservation, and heating, ventilation, and air conditioning (HVAC). Architectural engineers must also make sure that buildings they design meet all regulations regarding accessibility and safety in addition to being fully functional.

Bioengineering. The field of bioengineering involves using the principles of engineering in biology, medicine, environmental studies, and agriculture. Bioengineering is often used to refer to biomedical engineering, which involves the development of artificial limbs and organs, including ceramic knees and hips, pacemakers, stents, artificial eye lenses, skin grafts, cochlear implants, and artificial hands. However, bioengineering also has many other applications, including the creation of genetically modified plants that are resistant to pests, drugs that prevent organ rejection after a transplant operation, and chemical coatings for a stent placed in a heart blood vessel that will make the implantation less stressful for the body. Bioengineers must concern themselves with not only the biological and mechanical functionality of their creations but also financial and social issues such as ethical concerns.

Chemical. Everything in the universe is made up of chemicals. Engineers in the field of chemical engineering develop a wide range of materials, including fertilizers to increase crop production, the building materials for a submarine, and fabric for everything from clothing to tents. They may also be involved in finding, mining, processing, and distributing fuels and other materials. Chemical engineers also work on processes, such as improving water quality or developing less-polluting, readily available, inexpensive fuels.

Civil. Some of the largest engineering projects are in the field of civil engineering, which involves the design, construction, and maintenance of infrastructure such as roads, tunnels, bridges, canals, dams, airports, and sewage and water systems. Examples include the interstate highway system, the Hoover Dam, and the Brooklyn Bridge. Completion of civil engineering projects often results in major shifts in population distribution and changes in how people live. For example, the highway system allowed fresh produce to be shipped to northern states in the wintertime, improving the diets of those who lived there. Originally, the term "civil engineer" was used to distinguish between engineers who worked on public projects and "military engineers" who worked on military projects such as topographical maps and the building of forts. The subspecialties of civil engineering include construction engineering, irrigation engineering, transportation engineering, soils and foundation engineering, geodetic engineering, hydraulic engineering, and coastal and ocean engineering

Computer. The field of computer engineering has two main focuses, the design and development of hardware and of the accompanying software. Computer hardware refers to the circuits and architecture of the computer, and software refers to the computer programs that run the computer. The hardware does only what the software instructs it to do, and the software is limited by the hardware. Computer engineers may research, design, develop, test, and install hardware such as computer chips, circuit boards, systems, modems, keyboards, printers, or computers embedded in various electronic products, such as the tracking devices used to monitor parolees. They may also create, maintain, test, and install software for mainframes, personal computers, electronic devices, and smartphones. Computer programs range from simple to complex and from familiar to unfamiliar. Smartphone applications are extremely numerous, as are applications for personal computers. Software is used to track airplanes and other transportation, to browse the Web, to provide security for financial transactions and corporations, and to direct unmanned missiles to a precisely defined target. Computers can operate from a remote location. For example, anaerobic manure digesters are used to convert cattle manure to biogas that can be converted to energy, a biosolid that can be used as bedding or soil amendment, and a nonodorous liquid stream that can be used as fertilizer. These digesters can be placed on numerous cattle farms in different states and operated and controlled by computers miles away.

Electrical. Electrical engineering studies the uses of electricity and the equipment to generate and distribute electricity to homes and businesses. Without electrical engineering, digital video disc (DVD) players, cell phones, televisions, home appliances, and many life-saving medical devices would not exist. Computers could not turn on. The Global Positioning System (GPS) in cars would be useless, and starting a car would require using a hand crank.

This field of engineering is increasingly is involved in investigating different ways to produce electricity, including alternative fuels such as biomass and solar and wind power.

Environmental. The growth in the population of the world has been accompanied by increases in consumption and the production of waste. Environmental engineering is concerned with the reduction of existing pollution in the air, in the water, and on land, and the prevention of future harm to the environment. Issues addressed include pollution from manufacturing and other sources, the transportation of clean water, and the disposal of nonbiodegradable materials and hazardous and nuclear waste. Because pollution of the air, land, and water crosses national borders, environmental engineers need a broad, global perspective.

Industrial. Managing production and delivery of any product is the expertise of industrial engineers. They observe the people, machines, information, and technology involved in the process from start to finish, looking for any areas that can be improved. Increasingly, they use computer simulations and robotics. Their goals are to increase efficiency, reduce costs, and ensure worker safety. For example, worker safety can be improved through ergonomics and the use of less-stressful, easier-to-manipulate tools. The expertise of industrial engineers can have a major impact on the profitability of companies.

Manufacturing. Manufacturing engineering examines the equipment, tools, machines, and processes involved in manufacturing. It also examines how manufacturing systems are integrated. Its goals are to increase product quality, safety, output, and profitability by making sure that materials and labor are used optimally and waste—whether of time, labor, or materials—is minimized. For example, engineers may improve machinery that folds disposable diapers or that machines the gears for a truck, or they may reconfigure the product's packaging to better protect it or facilitate shipping. Increasingly, robots are used to do hazardous, messy, or highly repetitive work, such as painting or capping bottles.

Mechanical. The field of mechanical engineering is the oldest and largest specialty. Mechanical engineers create the machines that drive technology and industry and design tools used by other engineers. These machines and tools must be built to specifications regarding usage, maintenance, cost, and delivery. Mechanical engineers create both power-generating machinery such as turbines and power-using machinery such as elevators by taking advantage of the compressibility properties of fluids and gases.

Nuclear. Nuclear engineering requires expertise in the production, handling, utilization, and disposal of nuclear materials, which have inherent dangers as well as extensive potential. Nuclear materials are used in medicine for radiation treatments and diagnostic testing. They also function as a source of energy in nuclear power plants. Because of the danger of nuclear materials being used for weapons, nuclear engineering is subject to many governmental regulations designed to improve security.

IMPACT ON INDUSTRY

The U.S. economy and national security are closely linked to engineering. For example, engineering is vital for developing ways to reduce the cost of energy, decrease American reliance on foreign sources of energy, and conserve existing natural resources. Because of its importance, engineering is supported and highly regulated by government agencies, universities, and corporations. However, some experts question whether the United States is educating enough engineers, especially in comparison with China and India. The number of engineering degrees awarded each year in the United States is not believed to be keeping pace with the demand for new engineers.

Government Research. The U.S. government has a vested interest in the commerce, safety, and military preparedness of the nation. It both funds and regulates development through subcontractors, laws, guidelines, and educational initiatives. For example, the Americans with Disabilities Act of 1990 made it mandatory that public buildings be accessible for all American citizens. Its passage spurred innovations in engineering such as kneeling buses, which make it possible for those in wheelchairs to board a bus. The U.S. Department of Defense (DOD) is the largest contractor and the largest provider of funds for engineering research. For example, it issued 80,000 specifications for the creation of a synthetic jet fuel. The Federal Drug Administration (FDA) concentrates its efforts on supplier control and testing of products and materials. Funding for research in engineering is also provided by the National Science Foundation.

The government has also sponsored educational initiatives in science, technology, engineering, and mathematics, an example of which is the America Competes Act of 2007.

Academic Research. Universities, often in collaboration with governmental agencies, conduct research in engineering. Also, universities have entered into partnerships with private industry as investors have sought to capitalize on commercial possibilities presented by research, as in the case of stem cell research in medicine. Universities are also charged with providing rigorous, up-to-date education for engineers. Numerous accrediting agencies, including the Accreditation Board for Engineering and Technology, ensure that graduates from engineering programs have received an adequate and appropriate education. Attending an institution without accreditation is not advisable.

Industry and Business. Engineering has a role in virtually every company in every industry, including nonprofits in the arts, if only because these companies use computers in their offices. Consequently, some fields of engineering are sensitive to swings in the economy. In a financial downturn, no one develops office buildings, so engineers working in architecture and construction are downsized. When towns and cities experience a drop in tax income, projects involving roads, sewers, and environmental cleanup are delayed or canceled, and civil, mechanical, and environmental engineers lose their jobs. However, some economic problems can actually spur developments in engineering. For example, higher energy costs have led engineers to create sod roofs for factories, which keep the building warmer in winter and cooler in summer, and to develop lighter, stronger materials to use in a airplanes.

CAREERS AND COURSE WORK

To pursue a career in engineering, one must obtain a degree from an accredited college in any of the major fields of engineering. A bachelor's degree is sufficient for some positions, but by law, each engineering project must be approved by a licensed professional engineer (P.E.). To gain P.E. registration, an engineer must pass the comprehensive test administered by the National Society of Professional Engineers and work for a specified time period. In addition, each state has its own requirements for being licensed, including an exam specific to the state. An engineer with a bachelor's degree may work as an engineer with or without P.E. registration, obtain a master's degree or doctorate to work in a specialized area of engineering or pursue an academic career, or obtain an M.B.A. in order to work as a manager of engineers and products.

SOCIAL CONTEXT AND FUTURE PROSPECTS

Engineering can both prolong life through biomedical advances such as neonatal machinery and destroy life through unmanned military equipment and nuclear weaponry. An ever-increasing number of people and their concentration in urban areas means that ways must be sought to provide more food safely and to ensure an adequate supply of clean, safe drinking water. These needs will create projects involving genetically engineered crops, urban agriculture, desalination facilities, and the restoration of contaminated rivers and streams. The never-ending quest for energy will remain a fertile area for research and development. Heated political debates about taxing certain fuels and subsidizing others are part of the impetus behind solar, wind, and biomass development and renewed discussions about nuclear power.

The lack of minorities, including women, Hispanics, African Americans, and Native Americans, in engineering is being addressed through education initiatives. Women's enrollment in engineering schools has hovered around 20 percent since about 2000. African Americans make up about 13 percent of the U.S. population yet only about 3,000 blacks earn bachelor's degrees in engineering each year. About 4,500 Hispanics, who represent about 15 percent of the U.S. population, earn bachelor's degrees in engineering each year.

Judith L. Steininger, B.A., M.A.

FURTHER READING

Addis, Bill. *Building: Three Thousand Years of Design Engineering and Construction.* New York: Phaidon Press, 2007. Traces the history of building engineering in the Western world, covering the people, buildings, classic texts, and theories. Heavily illustrated.

Baura, Gail D. *Engineering Ethics: An Industrial Perspective.* Boston: Elsevier Academic Press, 2006. Thirteen case studies examine problems with products, structures, and systems and the role of engineers in each. Chapters cover the Ford Explorer rollovers, the San Francisco-Oakland Bay Bridge earthquake collapse, the Columbia space shuttle explosion, and the 2003 Northeast blackout.

Dieiter, George E., and Linda C. Schmidt. *Engineering Design.* 4th ed. Boston: McGraw-Hill Higher Education, 2009. Looks at design as the central activity of engineering. Provides a broad overview of basic topics and guidance on the design process, including materials selection and design implementation.

Nemerow, Nelson L., et al., eds. *Environmental Engineering: Environmental Health and Safety for Municipal Infrastructure, Land Use and Planning, and Industry.* Hoboken, N.J.: John Wiley & Sons, 2009. Covers environmental issues such as waste disposal for industry, the residential and institutional environment, air pollution, and surveying and mapping for environmental engineering.

Petroski, Henry. *Success Through Failure: The Paradox of Design.* 2006. Reprint. Princeton, N.J.: Princeton University Press, 2008. Examines failure as a motivator for engineering and defines success as "anticipating and obviating failure." Chapters deal with bridges, buildings, and colossal failures.

Yount, Lisa. *Biotechnology and Genetic Engineering.* 3d ed. New York: Facts On File, 2008. Covers the history of genetic engineering and biotechnology, including the important figures. Contains a bibliography and index.

WEB SITES
American Association of Engineering Societies
http://www.aaes.org

American Engineering Association
http://www.aea.org

American Society for Engineering Education
http://www.asee.org

Institute of Electrical and Electronics Engineers (IEEE)
http://www.ieee.org

National Society of Professional Engineers
http://www.nspe.org

See also: Biochemical Engineering; Bioengineering; Biomechanical Engineering; Chemical Engineering; Civil Engineering; Computer Engineering; Electrical Engineering; Electronics and Electronic Engineering; Engineering Mathematics; Environmental Engineering; Genetic Engineering; Mechanical Engineering; Military Sciences and Combat Engineering; Software Engineering; Spacecraft Engineering.

ENGINEERING MATHEMATICS

FIELDS OF STUDY

Algebra; geometry; trigonometry; calculus (including vector calculus); differential equations; statistics; numerical analysis; algorithmic science; computational methods; circuits; statics; dynamics; fluids; materials; thermodynamics; continuum mechanics; stability theory; wave propagation; diffusion; heat and mass transfer; fluid mechanics; atmospheric engineering; solid mechanics.

SUMMARY

Engineering mathematics focuses on the use of mathematics as a tool within the engineering design process. Such use includes the development and application of mathematical models, simulations, computer systems, and software to solve complex engineering problems. Thus, the solution of the engineering problem might be a component, system, or process.

KEY TERMS AND CONCEPTS

- **Algorithmic Science:** Study, implementation, and application of real-number algorithms for solving problems of continuous mathematics that arise in the realms of optimization and numerical analysis.
- **Applied Mathematics:** Branch of mathematics devoted to developing and applying mathematical methods, including math-modeling techniques, to solve scientific, engineering, industrial, and social problems. Areas of focus include ordinary and partial differential equations, statistics, probability, operational analysis, optimization theory, solid mechanics, fluid mechanics, numerical analysis, and scientific computing.
- **Bioinformatics:** Field that uses sophisticated mathematical and computational tools for problem solving in diverse biological disciplines.
- **Computational Science:** Broad field blending applied mathematics, computer science, engineering, and other sciences that uses computational methods in problem solving.

DEFINITION AND BASIC PRINCIPLES

Engineering mathematics entails the development and application of mathematics (such as algorithms, models, computer systems, and software) within the engineering design process. In engineering mathematics course work and professional research, a variety of tools may be used in the collection, analysis, and display of data. Standard tools of measurement include rules, spirit levels, micrometers, calipers, and gauges. Software tools include Maplesoft, Mathematica, MATLAB, and Excel. Engineering mathematics has roots and applications in many areas, including algorithmic science, applied mathematics, computational science, and bioinformatics.

BACKGROUND AND HISTORY

Since engineering is such a broad area, engineering mathematics includes a variety of applications. In general, a link between engineering and mathematics is established when mathematical descriptions of physical systems are formulated.

Links may involve precise mathematical relationships or formulas. For example, Galileo Galilei's pioneering work on the study of the motion of physical objects led to the equations of accelerated motion, $v = at$ and $d = \frac{1}{2} at^2$, in which velocity is v, acceleration a, time t, and distance d. His work paved the way for Newtonian physics.

Other links are established through empirical relationships that have the status of laws. French physicist Charles-Augustin de Coulomb discovered that the force between two electrical charges is proportional to the product of the charges and inversely proportional to the square of the distance between them: $F = kq1q2/d2$ (force is F, constant of variation k, charges $q1$ and $q2$, and distance d). The coulomb, a measure of electrical charge was named for him. Coulomb's work was the first in a sequence of related discoveries by other notable scientists, many of whose findings led to additional laws. The list includes Danish physicist Hans Christian Ørsted, French physicists André-Marie Ampère, Jean-Baptiste Biot, and Félix Savart, British physicist and chemist Michael Faraday, and Russian physicist Heinrich Lenz.

Sometimes mathematical expressions of principles apply almost universally. In physics, for example, the conservation laws indicate that in a closed system certain measurable quantities remain constant: mass, momentum, energy, and mass-energy. Lastly, systems of equations are required to describe physical phenomena of various levels of complexity. Examples include English astronomer and mathematician Sir Isaac Newton's equations of motion, Scottish physicist and mathematician James Clerk Maxwell's equations for electromagnetic fields, and Swiss mathematician and physicist Leonhard Euler's and French engineer Claude-Louis Navier and British mathematician and physicist George Gabriel Stokes's (Navier-Stokes) equations in fluid mechanics.

Further links between engineering and mathematics are discovered through the ongoing development, extension, modification, and generalization of equations and models in broader physical systems. For example, the Euler equations used in fluid mechanics can be connected to the conservation laws of mass and momentum.

How It Works

Many problems in engineering mathematics lead to the construction of models that can be used to describe physical systems. Because of the power of technology, a model may be derived from a system of a few equations that may be linear, quadratic, exponential, or trigonometric—or a system of many equations of even greater complexity. In engineering, such equations include ordinary differential equations, differential algebraic equations, and partial differential equations.

As the system of equations is solved, the mathematical model is formulated. Models are expressed in terms of mathematical symbols and notation that represent objects or systems and the relationships between them. Computer software, such as Maplesoft, Mathematica, MATLAB, and Excel, facilitates the process.

Engineers have available many models of physical systems. The development, extension, and modification of existing models, and the development of new models, are the subject of ongoing research. In this way, engineering mathematics continues to advance.

The ultimate test of a mathematical model is whether it truly reflects the behavior of the physical system under study. Computational experiments can be run to test the model for unexpected characteristics of the system and possibly optimize its design. However, models are approximations, and the accuracy of computed results must be evaluated through some form of error analysis.

The level of complexity of the construction and use of models depends on the engineering application. Further appreciation of the utility of a model may be gained by examining the use of a new model in engineering mathematics that impacts several scientific and technological areas. A mathematical model can now be used to investigate how materials break: One led to a new law of physics that depicts fracturing before it happens, or even as it occurs. In addition to the breakage of materials such as glass and concrete used in construction, the model enables better examination of bone breakage in patients with pathologies such as osteoporosis.

APPLICATIONS AND PRODUCTS

Cell Biology. Advances in research in cell growth and division have proved helpful in disease detection, pharmaceutical research, and tissue engineering. Biologists have extensively explored cell growth and mass and the relationship between them. Using microsensors, bioengineers can now delineate colon cancer cell masses and divisions over given time periods. They have found that such cells grow faster as they grow heavier. With additional cell measurements and mathematical modeling, the scientists examined other properties such as stiffness. They also performed simulations to study the relationship between cell stiffness, contact area, and mass measurement.

Genetics. New genes involved in stem-cell development can be found, quickly and inexpensively, along the same pathway as genes already known. When searching for genes involved in a particular biological process, scientists try to find genes with a symmetrical correlation. However, many biological relationships are asymmetric and can now be found using Boolean logic in data-mining techniques. Engineering and medical researchers can then examine whether such genes become active, such as those in developing cancers. This research is expected to lead to advances in disease diagnosis and cancer therapy.

Energy. A new equation could help to further the use of organic semiconductors. The equation represents the relationship between current and voltage at the junctions of the organic semiconductors. Research in the use of organic semiconductors may lead to advances in solar cells, displays, and lighting. Engineers have been studying organic semiconductors for about 75 years but have only recently begun to discover innovative applications.

IMPACT ON INDUSTRY

Government and University Research. Engineering mathematics has roots and applications in many research areas, including algorithmic science, applied mathematics, computational science, and bioinformatics. Theoretical, empirical, and computational research includes the development of effective and efficient mathematical models and algorithms. Professional research includes diverse medical, scientific, and industrial applications. Even

French physicist Charles Augustin de Coulomb is celebrated for his researches in electricity and magnetism. He was one of the first to measure how electric charge behaved, by observing the attracting and repelling forces they exerted on wire. (SSPL via Getty Images)

research in university-level engineering mathematics includes work that challenges students to integrate and apply course-work knowledge from many disciplines as they examine cracks in solids, mixing in small geometries, the crumpling of paper, lattice packings in curved geometries, materials processes, and predict optimal mechanics for device applications.

Engineering and Technology. Engineering mathematics is among the tools used to create, develop, and maintain products, systems, and processes. Applications are found in many areas, including nanotechnology. For example, radio-frequency designs are approaching higher frequency ranges and higher levels of complexity. The underlying mathematics of such designs is also complex. It requires new modeling techniques, mathematical methods, and simulations with mixed analogue and digital signals. Ordinary differential equations, differential algebraic equations, and partial differential algebraic equations are used in the analyses. The goal is to predict the circuit behavior before costly production begins. In such research, algorithms have been modified and new algorithms created to meet simulation demands.

Other Applied Sciences. There has been a dramatic rise in the power of computation and information technology. With it have come vast amounts of data in various fields of applied science and engineering. The challenge of understanding the data has led to new tools and approaches, such as data mining. Applied mathematics includes the use of mathematical models and control theory that facilitate the study of epidemics, pharmacokinetics, and physiologic systems in the medical industry. In telematics, models are developed and used in the enhancement of wireless mobile communications.

CAREERS AND COURSE WORK

Data Analyst or Data Miner. Data mining involves the discovery of hidden but useful information in large databases. In applications of data mining, career opportunities emerge in medicine, science, and engineering. Data mining involves the use of algorithms to identify and verify previously undiscovered relationships and structure from rigorous data analysis. Course work should include a focus on higher-level mathematics in such areas as topology, combinatorics, and algebraic structures.

Materials Science. Materials science is the research, development, and manufacture of such items

Fascinating Facts About Engineering Mathematics

- In aircraft design, computational simulations have been used extensively in the analysis of lift and drag. Advanced computation and simulation are now essential tools in the design and manufacture of aircraft.

- Auto-engineering researchers have developed a simulation model that can significantly decrease the time and cost of calibrating a new engine. Unlike statistics-based models, the new physics-based model can generate data for transient behavior (acceleration or deceleration between different speeds).

- Although the number of nuclear weapons held by each country in the world is securely guarded information, the rising demands of regulatory oversight require computing technology that exceeds current levels. A knowledge of the mathematical development and use of uncertainty models is critical for this application.

- Researchers have used nontraditional mathematical analyses to identify evolving drug resistance in strains of malaria. Their goal is to enable the medical community to react quickly to inevitable drug resistance and save lives. They also want to increase the life span of drugs used against malaria. The researchers used mathematical methods that involved graphing their data in polar coordinates and in rectangular coordinates. The results of the study are of particular interest to biomedical engineers focusing on genetics research.

- Bioengineers studying Alzheimer's disease found that amyloid beta peptides generate calcium waves. Following these waves in brain-cell networks, the scientists discovered voltage changes that signify intracellular communication. This research involves mathematical modeling of brain networks.

- Medical and engineering researchers have developed a mathematical model reflecting how red blood cells change in size and hemoglobin content during their four-month life span. This model may provide valuable clinical information that can be used to predict who is likely to become anemic.

- Mathematically gifted philosophers Sir Isaac Newton and Gottfried Leibniz invented the calculus that is now part of engineering programs.

- French mathematician Joseph Fourier introduced a mathematical series using sines and cosines to model heat flow.

- Albert Einstein proposed his general theory of relativity, incorporating the Riemannian or elliptical geometry concept that space can be unbounded without being infinite. The theory of relativity, along with later nuclear-energy research, led to development of the atomic and hydrogen bombs and theoretical understanding of thermonuclear fusion as the energy source powering the Sun and stars.

- Among the prominent figures involved in the design and development of electronic computers is American mathematician Norbert Wiener. Wiener is the founder of the science of cybernetics, the mathematical study of the structure of control systems and communications systems in living organisms and machines.

as metallic alloys, liquid crystals, and biological materials. There are many career opportunities in aerospace, electronics, biology, and nanotechnology. Research and development uses mathematical models and computational tools. Course work should include a focus on applied mathematics, including differential equations, linear algebra, numerical analysis, operations research, discrete mathematics, optimization, and probability.

Ecological and Environmental Engineering. The work of professionals in these fields covers many areas that transcend pollution control, public health, and waste management. It might, for example, involve the design, construction, and management of an aquatic ecosystem or the research and development of appropriate sustainable technologies.

Course work should include a focus on higher-level mathematics in such areas as calculus, linear algebra, differential equations, and statistics.

Meteorology and Climatology. These career areas incorporate not only atmospheric, hydrologic, and oceanographic sciences but modeling, forecasting, geoengineering, and geophysics. In general, historical weather data and current data from satellites, radar, and monitoring equipment are combined with other measurements to develop, process, and analyze complex models using high-performance computers. Current research areas include global warming and the impact of atmospheric radiation and industrial pollutants. Mathematics courses in meteorology and atmospheric science programs include calculus, differential equations, linear algebra, statistics, computer science, numerical analysis, and matrix algebra or computer systems.

SOCIAL CONTEXT AND FUTURE PROSPECTS

Within engineering mathematics, an interdisciplinary specialty has emerged, computational engineering. Computational engineering employs mathematical models, numerical methods, science, engineering, and computational processes that connect various fields of engineering science. Computational engineering emerged from the impact of supercomputing on engineering analysis and design. Computational modeling and simulation are vitally important for the development of high-technology products in a globally competitive marketplace. Computational engineers develop and use advanced software for real-world engineering analysis and design problems. The research work of engineering professionals and academics has potential for applications in several engineering disciplines.

June Gastón, B.A., M.S.Ed., M.Ed., Ed.D.

FURTHER READING

Gribbin, John. *The Scientists: A History of Science Told Through the Lives of Its Greatest Inventors.* New York: Random House, 2003. This compelling text on the history of modern science marks the subject's discoveries and milestones through the great thinkers who were integral to its creation, both well known and not as well known.

Merzbach, Uta C., and Carl B. Boyer. *A History of Mathematics.* 3d ed. Hoboken, N.J.: John Wiley & Sons, 2011. An excellent and highly readable history of the subject that chronicles the earliest principles as well as the latest computer-aided proofs.

Schäfer, M. *Computational Engineering, Introduction to Numerical Methods.* New York: Springer, 2006. Schäfer includes applications in fluid mechanics, structural mechanics, and heat transfer for newer fields such as computational engineering and scientific computing, as well as traditional engineering areas.

Shiflet, Angela B., and George W. Shiflet. *Introduction to Computational Science: Modeling and Simulation for the Sciences.* Princeton, N.J.: Princeton University Press, 2006. Two approaches to computational science problems receive focus: system dynamics models and cellular automaton simulations. Other topics include rate of change, errors, simulation techniques, empirical modeling, and an introduction to high-performance computing. Numerous examples, exercises, and projects explore applications.

Stroud, K.A. *Engineering Mathematics.* 6th ed. New York: Industrial Press, 2007. Providing a broad mathematical survey, this innovative volume covers a full range of topics from basic arithmetic and algebra to challenging differential equations, Laplace transforms, and statistics and probability.

Velten, Kai. *Mathematical Modeling and Simulation: Introduction for Scientists and Engineers.* Weinheim, Germany: Wiley-VCH, 2009. Velten explains the principles of mathematical modeling and simulation. After treatment of phenomenological or data-based models, the remainder of the book focuses on mechanistic or process-oriented models and models that require the use of differential equations.

WEB SITES

American Mathematical Society
http://www.ams.org/home/page

National Society of Professional Engineers
http://www.nspe.org

Society for Industrial and Applied Mathematics
http://www.siam.org

See also: Algebra; Applied Mathematics; Bioinformatics; Calculus; Geometry; Numerical Analysis; Trigonometry.

ENGINEERING SEISMOLOGY

FIELDS OF STUDY

Engineering; geology; physics; geophysics; earth science; electrical engineering; volcanology; volcanic seismology; plate tectonics; geodynamics; mineral physics; tectonic geodesy; mantle dynamics; seismic modeling; seismic stratigraphy; statistical seismology; computer science; mathematics; earthquake engineering; paleoseismology; archeoseismology; historical seismology; structural engineering; geography

SUMMARY

Engineering seismology is a scientific field focused on studying the likelihood of future earthquakes and the potential damage such seismic activity can cause to buildings and other structures. Engineering seismology utilizes computer modeling, geological surveys, existing data from historical earthquakes, and other scientific tools and concepts. Engineering seismology is particularly useful for the establishment of building codes and for land-use planning.

KEY TERMS AND CONCEPTS:

- **Duration:** The length of time ground motion occurs during an earthquake.
- **Epicenter:** The surface-level geographic point located directly above an earthquake's hypocenter.
- **Focal Depth:** The depth of an earthquake's hypocenter.
- **Ground Motion:** The ground-level shaking that occurs during an earthquake.
- **Hypocenter:** The point of origin of an earthquake.
- **Love Waves:** Seismic waves that occur in a side-to-side motion.
- **Magnitude:** An earthquake's size and relative strength.
- **Rayleigh Waves:** Seismic waves that occur in a circular, rolling fashion.
- **Richter Scale:** Logarithmic scale used to assign a numerical value to the magnitude of an earthquake.
- **Source Parameters:** A series of earthquake characteristics, including distance, duration, energy, and the types of waves that occur.
- **Stress Drop:** The amount of energy released when locked tectonic plates separate, causing an earthquake.
- **Wave Propagation Path:** The directions in which seismic waves travel in an earthquake.

DEFINITION AND BASIC PRINCIPLES

Engineering seismology (also known as earthquake engineering) is a multidisciplinary field that assesses the effects of earthquakes on buildings, bridges, roads, and other structures. Seismology engineers work in the design and construction of structures that can withstand seismic activity. They also assess the damages and effects of seismic activity on existing structures. Engineering seismologists analyze such factors as quake duration, ground motion, and focal depth in assessing the severity of seismic events and how those events affect fabricated structures. They also consider source parameters, which help seismologists zero in on a seismic event's location and the speed and trajectory at which the quake's resulting waves are traveling.

Earthquake engineers also study theoretical concepts and models related to potential earthquakes and historical seismic events. Such knowledge can help engineers and architects design structures that can withstand as powerful an earthquake as the geographic region has produced (or possibly will produce). Mapping systems and programs and mathematical and computer-based models are essential to engineering seismologists' work. Such techniques are also useful for archeologists and paleontologists, both of whom may use engineering seismology concepts to understand how the earth has evolved over millions of years and how ancient civilizations were affected by seismic events.

BACKGROUND AND HISTORY

Throughout human history, people have struggled to understand the nature of earthquakes and, as a result, have faced the challenges of preparing for these seismic events. Some ancient civilizations attributed earthquakes to giant snakes, turtles, and other creatures living and moving beneath the earth's surface. In the fourth century B.C.E., Aristotle was the first to speculate that earthquakes were not caused by supernatural forces but rather were natural events. However, little scientific study on earthquakes took place for hundreds of years, despite the occurrence of many major seismic events (including the eruption of Mount Vesuvius in Italy in 79 C.E., which was preceded by a series of earthquakes).

In the mid-eighteenth century, however, the British Isles experienced a series of severe earthquakes, which created a tsunami that destroyed Lisbon, Portugal, killing tens of thousands of people. Scientists quickly developed an interest in cataloging and understanding seismic events. In the early nineteenth century, Scottish physicist and glaciologist James D. Forbes invented the inverted pendulum seismometer, which gauged not only the severity of an earthquake but also its duration.

Throughout history, seismology has seen advances that immediately followed significant seismic events. Engineering seismology, which is proactive, represents a departure from reactionary approaches to the study of earthquakes. Today, engineering seismology uses seismometers, computer modeling, and other advanced technology and couples it with historical data for a given site. The resulting information helps civil engineers and architects construct durable buildings, bridges, and other structures and assess the risks to existing structures posed by an area's seismic potential.

How it Works

To understand engineering seismology, one must understand the phenomenon of earthquakes. Earthquakes may be defined as the sudden shaking of the earth's surface as caused by the movement of subterranean rock. These massive rock formations (plates), resting on the earth's superheated core, experience constant movement caused predominantly by gravity. While some plates move above and below one another, others come into contact with one another as they pass. The boundaries formed by these passing plates are known as faults. When passing plates lock together, stored energy builds up gradually. The plates eventually give, causing that energy to be released and sent from the quake's point of origin (the hypocenter) outward to the surface in the form of seismic (or surface) waves. Such waves occur either in a circular, rolling fashion (Rayleigh waves) or in a twisting, side-to-side motion (Love waves).

The field of seismology has developed only over the last few centuries, largely because of major, devastating seismic events. The practice of engineering seismology has grown in demand in recent years, mainly because of the modern world's dependency on major cities, infrastructure (such as bridges, roadways, and rail systems), and energy resources (including nuclear power plants and offshore oil rigs). Earthquake engineers, therefore, have two main areas of focus: studying seismology and developing structures that can withstand the force of an earthquake.

To study seismic activity and earthquakes, engineering seismologists may use surface-based detection systems, such as seismometers, to monitor and catalog tremors. They also employ equipment—including calibrators and accelerometers—that is lowered into deep holes. Such careful monitoring practices help seismologists and engineering seismologists better understand a region's potential for seismic activity.

When earthquakes occur, engineering seismologists quickly attempt to locate the hypocenter and the epicenter (the surface point that lies directly above the hypocenter). They are able to do so by monitoring two types of waves—P and S waves—that move much quicker than surface waves and, therefore, act as precursors to surface waves. These engineers also work to determine the magnitude (a measurement of an earthquake's size) of the event.

Magnitude may be based on a number of key factors (or source parameters), including duration, distance to the epicenter and hypocenter, the size and speed of the surface waves, the amount of energy (known as the stress drop) that is released from the hypocenter, P and S waves, and the directions in which surface waves move (the wave propagation path). Analyzing an earthquake's magnitude provides an accurate profile of the quake and the conditions that caused it.

In addition to developing a profile of a region's past seismic activity, earthquake engineers use such information to ascertain the type of activity a geographic region may experience in the future. For example, scientific evidence suggests that the level of stress drop is a major contributor to the severity of seismic activity that can cause massive destruction in major urban centers. Similarly, studies show that the duration of ground motion (the "shaking" effects of an earthquake) may be more of a factor in the

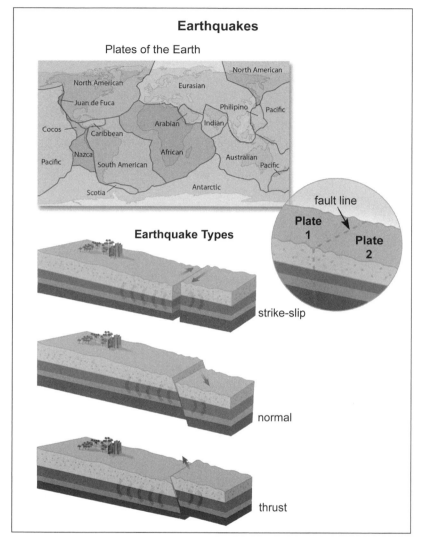

Earthquakes

Plates of the Earth

Earthquake Types

strike-slip

normal

thrust

Earthquakes mostly occur along fault lines, which are cracks in the Earth's crust, formed from the movement of two tectonic plates.

amount of damage to buildings and other structures than stress drop.

The field of engineering seismology is less than one century old, but in the twenty-first century, it plays an important role in urban development and disaster prevention. Earthquake seismologists work with civil engineers and architects to design buildings, roads, bridges, and tunnels that may withstand the type of seismic activity that has occurred in the past.

APPLICATIONS AND PRODUCTS

Engineering seismology applies knowledge of seismic conditions, events, and potential to the design and development of new and existing fabricated structures. Among the methods and applications employed by earthquake seismologists are the following:

Experimentation. Engineering seismologists may construct physical scale models of existing structures or proposed structures. Using data from a region's known seismic history, the engineers attempt to recreate an earthquake by placing these models on so-called shake tables, large mechanical platforms that simulate a wide range of earthquake types. After the "event," engineering seismologists examine the simulation's effects on the model structure, including its foundations, support beams, and walls. This approach enables the engineers and architects to directly examine the pre- and post-simulation structure and determine what sort of modifications may be warranted.

Computer Models. One of the most effective tools utilized by engineering seismologists is computer modeling. Through the application of software, engineering seismologists can input a wide range of source parameters, ground motion velocities, wave types, and other key variables. They also can view how different structural components withstand (or fail to withstand) varying degrees of seismic activities without the expense and construction time of a shake table. Computer modeling has become increasingly useful when attempting to safeguard against earthquake damage to dams, nuclear power plants, and densely developed urban centers. Computer modeling is also used by engineering seismologists to predict the path of destruction that often occurs after an earthquake, destruction such as that caused by fires or flooding.

Seismic Design Software. Earthquake engineers study seismic activity in terms of how it affects structures. To this end, engineers must attempt to predict how earthquakes will strike an area. Seismic design software is used to create a map of a region's seismic activity and how those conditions will potentially cause structural damage. The software enables government officials to establish formalized building codes for buildings, bridges, power plants, and other structures. This software is easily obtained on the Web and through the U.S. Geological Survey (USGA) and other organizations.

Mathematics. Engineering seismology is an interdisciplinary field that relies heavily on an understanding of physics and mathematics. To quantify the severity of earthquakes, to calculate the scope of seismic activity, and, in general, to create a profile of a region's seismic environment, engineering seismologists utilize a number of mathematical formulae. One of the most-recognized of these formulae is the Richter scale, which was developed in 1935 by American seismologist and physicist Charles Richter. The Richter scale uses a logarithm to assign a numerical value (with no theoretical limit) to establish the magnitude of an earthquake. The Richter scale takes into account the amplitude (the degree of change) between seismic waves and the distance between the equipment that detects the quake and the quake's epicenter.

Earthquake engineers use such mathematical data as part of their analyses when working with civil engineers on construction projects. Earthquake engineers also are increasingly called upon by government officials to use this data to assess individual structure and citywide structural deficiencies that resulted in earthquake destruction. Forensic engineering was called into service in 2009, when Australian emergency officials intervened in Padang, Indonesia, after a magnitude 7.6 quake devastated that city. Engineers used mathematical formulae and statistical data to assess system-wide structural deficiencies in Padang rather than analyzing damage on a structure-by-structure basis. In light of the countless variables involved with studying earthquakes and their effects on fabricated structures, the use of established logarithms, data sets, and mathematical formulae is a time-honored practice of engineering seismologists.

Sensors. Not all earthquakes cause immediate and significant damage to affected structures. According to the USGA, the Greater San Francisco area experienced more than eighty earthquakes in 2011 alone, with none of those quakes registering higher than a 2.3 on the Richter scale. However, seismic activity on a small but frequent basis can cause long-term damage to structures. For example, seismic events can shift soil pressure on underground structures (such as pipes and foundations). Earthquake engineers are therefore highly reliant on sensor equipment, which enables them to gauge the effects of frequent seismic activity not only on above-ground structures but also on the ground itself.

To examine shifts in soil pressure caused by seismic activity, seismology engineers used an array of tactile

Fascinating Facts About Engineering Seismology

- There are approximately 500,000 detectable earthquakes in the world annually, 100,000 of which may be felt and only 100 that cause damage.

- Between 1975 and 1995, only the U.S. states of Florida, North Dakota, Iowa, and Wisconsin did not experience any earthquakes.

- The earliest known "seismograph" was introduced in 132 C.E. in China. It took the shape of a hollow urn (with a hidden pendulum inside) with dragon heads adorning the sides and frogs at the base directly under each dragon (which held a single ball). During an earthquake, a dragon would drop its ball into the mouth of the frog below, revealing the direction of the waves. The device once detected an earthquake four hundred miles away.

- Scientists cannot presently predict earthquakes; they can only calculate the potential for a quake to strike.

- The largest recorded earthquake in the world was a magnitude 9.5 quake in Chile in 1960.

- In 1931, there were 350 operating seismic stations in the world. In 2011, there existed more than 4,000.

- The largest recorded earthquake in the United States was a magnitude 9.2 quake in Alaska in 1964.

- A magnitude 8.8 earthquake in Chile in 2010 shifted that country's coastline approximately 1,640 feet toward the Pacific Ocean.

pressure sensors, which were originally designed for artificial intelligence systems but were later utilized for the purposes of designing car seats and brake pad systems.. The use of such equipment helps engineers study the long-term effects of seismic activity on water pipes, underground cables, and underground storage tanks.

Arguably one of the best-known types of seismic detector systems is the seismograph. The seismograph uses a pendulum-based system to detect ground motion from seismic activity. Originally, the modern seismograph was designed to detect only significant earthquakes and tremors, and it could be found only in stable environments (such as a laboratory). Today, however, there are many different types of seismographs; some may be placed underground, others can be used in the field, while others are so sensitive that they can detect distant explosions or minute tremors.

IMPACT ON INDUSTRY

Engineering seismology is a relatively new combination of civil engineering and seismology, along with other fields (such as geology, emergency management, and risk management). Its uses have proven invaluable, however, as many urban centers in earthquake-prone regions, such as Tokyo and San Francisco, have benefited from the careful application of safe building practices and disaster mitigation programs that are borne of engineering seismology. Japan, the United States, and Switzerland are among the leaders in earthquake engineering, along with Australia and New Zealand. Engineering seismology also involves a range of public and private organizations, such as the following:

Governments. National and regional governments play an important role in the application of the findings of engineering seismologists. The USGA, for example, offers a wide range of resources and services for studying seismic activity and the dangers it poses. Additionally, the Federal Emergency Management Agency (FEMA), the National Science Foundation, the National Institute of Standards, and the USGA combine their resources to operate the National Earthquake Hazards Reduction Program, which seeks to reduce property losses and human casualties caused by earthquakes through careful engineering seismology practices (including mapping seismically active areas and generating building codes).

Engineering Seismology Societies. Engineering seismologists share information and theories with their peers through professional associations and societies, many of which are global in nature. The International Association for Earthquake Engineering is one such network, holding worldwide conferences on engineering seismology every four years. This organization has branch societies in Japan, Europe, and the United States, each working locally but also contributing to the larger association.

Universities. Because of the contributions engineering seismology provides to the field of civil engineering, earthquake engineering continues to evolve within this educational discipline. Universities such as Stanford and the University of California, Berkeley, offer such programs. Many other universities feature coursework in environmental engineering, which includes earthquake engineering and seismology studies. Furthermore, a large number of universities house research laboratories, shake tables, and seismograph stations covering seismic activity throughout a given geographic area.

Consulting Firms. There is a considerable financial benefit to constructing a building, bridge, or other structure that will survive in a seismically active environment. Oil companies and other energy corporations, mining operations, and other businesses frequently seek the advice of private engineering consultants who offer seismic monitoring services. Construction companies also look to these consultants, seeking structural analyses and other services. The person who introduced the Richter scale, Charles Richter, founded a consulting firm upon his retirement from the California Institute of Technology in 1970.

CAREERS AND COURSEWORK

Students interested in engineering seismology should pursue and complete a bachelor's degree program in a related field, such as geology or civil engineering. They should also obtain a master's degree, preferably in a field of relevance to earthquake engineering, such as environmental engineering, structural engineering, geology, or seismology. Engineering seismologists' competitiveness as job candidates is improved greatly when they also earn a doctorate.

Engineering seismologists must receive training in the geosciences, including seismology, geology, and physics. These fields include courses in geodynamics, plate tectonics, statistical seismology, and mineral dynamics. They must also study civil engineering, structural design, and computer science (which must include training in computer modeling, digital mapping systems, and design software, which are essential in this arena). Furthermore, engineering seismologists must demonstrate excellent mathematical skills, particularly in geometry, algebra, and calculus. Finally, earthquake engineers must be trained in the use of many of the technical systems and devices that seismologists must use to monitor earthquakes.

SOCIAL CONTEXT AND FUTURE PROSPECTS

The study of earthquakes is a practice that dates back hundreds of years. Earthquake engineering specifically, however, represents an evolution toward a practical application of the study of seismic activity to the design and construction of large buildings, power plants, and other structures.

Engineering seismologists work closely with seismologists and civil engineers. On the former front, these engineers help design and operate detection equipment and systems to help explain seismic activity. This collaboration is critical, as improved knowledge of seismic activity can save lives and property.

For example, Japan has long utilized engineering seismology practices in its urban centers. The magnitude 8.9 earthquake in that country in March, 2011, did not devastate Tokyo because of strong building codes that, among other things, cause skyscrapers to sway with the region's seismic waves rather than stand in rigid fashion. Comparatively, the magnitude 7.0 Haiti earthquake in 2010 virtually flattened the country's capital, Port-au-Prince, and outlying areas, largely because Haiti did not have earthquake-safe building codes. Its buildings were built using insufficient steel and on slopes with no reinforcing foundation or support systems. One observer in Haiti reported that Port-au-Prince would likely not have survived even a magnitude 2.0, much less the 7.0 quake it did have.

The significance of the 2011 Japan disaster and of the rare 5.8 Virginia earthquake that struck the East Coast of the United States in August, 2011, continues to cast light on the need for earthquake engineering in structural design and construction. As regions with a history of major seismic activity (and those regions with the potential for such activity) continue to grow in size and population, engineering seismologists are likely to remain in high demand.

Michael P. Auerbach, B.A., M.A.

FURTHER READING

Chopra, Anil K. *Dynamics of Structures: Theory and Applications to Earthquake Engineering.* 3d ed. Upper Saddle River, N.J.: Prentice Hall, 2007. This book analyzes the theories of structural dynamics and applies the effects of earthquakes and seismology to them, including structural design and energy dissipation models.

Griffith, M. C., et al. "Earthquake Reconnaissance: Forensic Engineering on an Urban Scale." *Australian Journal of Structural Engineering* 11, no. 1 (2010): 63-74. This article describes the observations of a team of Australian aid workers who were dispatched to Indonesia to assess structural damages caused by a magnitude 7.6 earthquake there in 2009.

Saragoni, G. Rudolfo. "The Challenge of Centennial Earthquakes to Improve Modern Earthquake Engineering." *AIP Conference Proceedings*, 1020, no. 1 (July 8, 2008): 1113-1120. This article uses the 1906 San Francisco earthquake as a point of reference for reviewing the evolution of modern engineering seismology.

Slak, Tomaz, and Vojko Kilar. "Development of Earthquake Resistance in Architecture from an Intuitive to an Engineering Approach." *Prostor* 19, no. 1 (2011): 252-263. This article reviews the application of earthquake engineering concepts to the design of modern structures.

Stark, Andreas. *Seismic Methods and Applications.* Boca Raton, Fla.: BrownWalker Press, 2008. This book first reviews the principles of seismology and relevant geosciences and then proceeds to the practical application of these principles to structural design and engineering.

Villaverde, Roberto. *Fundamental Concepts of Earthquake Engineering.* New York: CRC Press, 2009. This book features a review of the history of the field of engineering seismology and includes some examples of how certain seismic wave types and other conditions are taken into account in building design.

See also: Applied Physics; Architecture and Architectural Engineering; Civil Engineering; Earthquake Engineering; Earthquake Prediction; Economic Geology; Engineering; Environmental Engineering; Urban Planning and Engineering.

ENVIRONMENTAL BIOTECHNOLOGY

FIELDS OF STUDY

Biology; microbiology; biochemistry; genetics; virology; agronomy; chemistry; toxicology; environmental monitoring; microbiology, earth and planetary sciences; ecology and evolutionary biology; environmental engineering; chemical and biomolecular engineering; atmospheric science.

SUMMARY

Environmental biotechnology, also known as biotechnical pollution control, is a rapidly developing science that uses biological resources to protect and restore the environment. It has significant implications and applications in both the prevention of air, soil, and water pollution and the restoration of contaminated environments.

KEY TERMS AND CONCEPTS

- **Bioaugmentation:** Form of bioremediation that involves increasing the number or activity level of microorganisms that assist in pollutant reduction.
- **Bioenrichment:** Form of bioremediation that involves adding nutrients or oxygen to environments to increase the breakdown of contaminants.
- **Biofilms:** An aggregated layer of microbial populations formed on aqueous (water) environments; sometimes referred to as microbial mats and known colloquially as slime.
- **Bioremediation:** Use of microorganisms, such as fungi and bacteria, and their enzymes to return a contaminated environment to its original condition.
- **Biostimulation:** Form of bioremediation in which the natural processes of degradation are encouraged through the introduction of certain stimuli, such as nutrients and additional substrates.
- **Biotechnology:** Exploitation and manipulation of living organisms to produce valuable and useful products or results.
- **End-of-Pipe Technology:** Treatment of waste and pollution that has already contaminated the air, soil, or water.

DEFINITION AND BASIC PRINCIPLES

Environmental biotechnology is a multidisciplinary science, a synthesis of environmental engineering, biochemistry, and microbiology. Those involved in environmental biotechnology generally concentrate on the development, application, adaptation, and management of biological systems and organisms to repair and prevent environmental damage caused by pollution and on the advancement of green technologies and sustainable development. Fundamentally, environmental biotechnology is the study of how microorganisms, plants, and their enzymes can assist in the restoration, remediation, preservation, and sustainable use of the world's natural environment. The primary role of environmental biotechnologists is to create a better balance between human development and the natural environment.

All fields of biotechnology have seen rapid growth and progress in the 1990's and 2000's. The field of environmental biotechnology, in particular, has benefited from advancements in genetic engineering and modern microbiological concepts, which offer both traditional and innovative solutions to different forms of contamination occurring in different mediums (soil, water, and air).

BACKGROUND AND HISTORY

The term "biotechnology" was being used as early as 1919 but did not occur frequently in scientific literature until the 1960's and 1970's, with the publication of the *Journal of Biotechnology*. Although the use of biotechnology within the medical and agricultural industry can be traced back many years, the purposeful use of biotechnology to mitigate environmental issues has a much shorter history.

Although people were familiar with the concept of biotechnology, for the most part, efforts were focused on the medical and agricultural industries. The arrival of the Industrial Revolution, however, rapidly altered and influenced the environment through the release of toxic pollutants into the waterways and soil. As people became wealthier, the demand for goods grew, and with the rise in industrial and agricultural production came the increase in the impact on the environment. Although the concept of using natural degradation processes was not new—for many years, communities had composted and relied on natural processes and microbes in the breakdown and treatment of sewage—environmental biotechnology was not yet considered a science.

In the 1960's, however, chemical pollution and its adverse effects came under significant public scrutiny. One of the landmark cases involved the chemical dichloro-diphenyl-trichloroethane (DDT), which was widely used as a pesticide but later was found to seriously affect bird populations and to cause cancer in humans. Other cases of chemical pollution and its adverse effects on people and the environment were becoming known, including mercury poisoning in Japan, Agent Orange in Vietnam, industrial sludge in the United States, and the devastating effects of oil spills. By the 1970's, it had become clear that the environment was becoming sullied and that contaminants were adversely affecting people. Concern regarding the environment led to the development of many laws and regulations in both developed and developing countries regarding the proper management of waste and pollution control. It was in the light of this political and social awareness that the field of environmental biotechnology emerged. Although environmental biotechnology originally focused on the treatment of wastewater, the field expanded to include areas of study such as soil contamination, solid-waste treatment, and air purification methods.

In 1992, during the United Nations Conference on Environment and Development in Rio de Janeiro, environmental biotechnology was recognized and embraced as a crucial tool for both repairing and preventing environmental and health issues caused by humans. Since this conference, the field of environmental biotechnology has advanced at a rapid rate and has grown to provide innovative approaches to the sustainable development and protection of the world's ecosystems.

How It Works

Traditional methods of waste removal, such as landfill and incineration, cannot cope with the sheer volume of waste created by human populations. This situation has increased the need to develop alternative environmentally sound treatments and techniques. Environmental biotechnology seeks to positively affect pollution control and waste management.

A rise in consumption since the 1990's has been accompanied by a corresponding increase in the release of pollutants into the environment. Some of these contaminants, particularly those that also are naturally occurring, can be digested, degraded, or removed from the soil and water through the action of microorganisms. However, some of these human-created pollutants rarely occur naturally, and the accumulation of such substances can have a serious ecological impact.

The use of biotechnology in the treatment of waste and pollution is not a new idea. For more than a century, many communities have relied on natural processes and microbes to break down and treat sewage. The fundamental aim of environmental biotechnology is to use organisms to control contamination and treat waste. In the process called bioremediation, microorganisms, including fungi and bacteria and their enzymes, are used to return a contaminated environment to its original condition. Naturally occurring biological degradation processes are purposely employed to remove contaminants from areas where they have been released. The use of such processes requires a solid scientific understanding of the contaminant, its impact, and the affected ecosystem.

The concept of environmental biotechnology depends on the notion that all living organisms, such as flora, fauna, bacteria, and fungi, consume nutrients for their survival and, in doing so, produce waste by-products. Not all organisms require the same nutrients nor react in the same manner, however. Some organisms, such as certain bacteria and microorganisms, flourish on chemicals and toxins that are actually poisonous or harmful to other organisms or ecosystems. The fact that some microorganisms and various strains of microbial species react differently to chemical toxins and environmental pollutants has advanced the concept of using genetic manipulation techniques.

Environmental biotechnology aims to provide a natural approach to tackling environmental issues, from identification of biohazards to restoration of industrial, agricultural, and natural areas affected by contamination. Central to the concept of environmental biotechnology is the ability to determine which contaminants are present, for how long, and in what quantity, and what recovery method is applicable. There are four basic concepts and approaches in the field of environmental biotechnology: bioremediation, prevention, detection and monitoring, and genetic engineering.

APPLICATIONS AND PRODUCTS

Environmental pollution can be a legacy of former industrial practices or a product of unsustainable present-day practices. One of the most serious environmental issues facing the world in the twenty-first century is the production of very large quantities of waste, the majority of which becomes landfill. Industrialized nations also produce significant quantities of chemicals that often end up in the soil and water. Environmental biotechnology seeks ways to combat the escalating ecological problems associated with such pollution and waste. Technologies that have been developed and implemented include bioremediation of water and soil, biomonitoring using biosensors and bioassays, and bioprocessing.

Bioremediation. Bioremediation is usually classified as either in situ or ex situ. In situ bioremediation entails treating the contamination in place and relies on the ability of the microorganisms to metabolize or remove the contaminants inside the naturally occurring system; ex situ remediation entails the polluted material being removed from the contaminated site and treated elsewhere and relies on some form of artificial engineering and input. The process of bioremediation can occur naturally or be encouraged through artificial stimulus. The process of attenuation occurs under natural conditions and incorporates the normal chemical, biological, and physical processes, such as aerobic and anaerobic degradation, that eliminate or reduce soil and water contaminants.

Biostimulation and Bioaugmentation. Biostimulation is a form of bioremediation in which the natural processes of degradation are encouraged through the introduction of certain stimuli, such as nutrients and additional substrates. Bioaugmentation is a form of bioremediation that involves increasing the activity of the microorganisms that assist in pollutant reduction or augmenting existing yet insufficient populations of microorganisms. Bioenrichment is another form of bioremediation that involves adding nutrients or oxygen to environments to increase the breakdown of contaminants.

Biodetectors. Biodetectors such as bioassays and biosensors are used to monitor, assess, and analyze

---

Fascinating Facts About Environmental Biotechnology

- Environmental biotechnology ranges from basic natural organic processes, such as fermentation and biodegradation, to enhanced and human-manipulated procedures, such as genetic engineering.
- Oil spills can be cleaned up using a bioremediation accelerator, which when sprayed on an oil sheen, creates a hydrocarbon-eating bacterium that breaks down hydrocarbons into water and carbon dioxide.
- Bioremediation can occur within soils, on and below the surface. Water and nutrients are mixed into the soil to facilitate bacterial growth. The bacteria used in bioremediation break down the carbon chains of harmful organic contaminants, leaving water and carbon dioxide.
- Scientists at the University of California, San Diego, have discovered an enzyme that, when manipulated, causes plants to absorb more carbon dioxide while retaining more water. This enzyme may address the dual problem of increased carbon levels and decreased water resources.
- Biotechnology is a $30 billion-a-year industry. There are more than 1,450 biotechnology companies in the United States alone.
- In the gold-mining industry in South Africa, bacteria are used to isolate gold from gold ore, thus producing less waste and avoiding the expenditure of a large amount of smelting energy.

---

biological material, and provide important information and data about the effects and concentration of pollutants in the natural environment. Bioassays are procedures or experiments in which the quantity of a contaminant is estimated by measuring its effect on living organisms. Although they can be relatively slow and expensive, they are essential for the assessment and prediction of real and potential effects of pollution on the natural environment.

Biosensors are devices that detect and measure minute amounts of or changes in concentration of chemical substances within an environmental area and translate that information into data. Because of their ability to detect even tiny quantities of targeted chemicals with greater speed and at less cost than bioassays, biosensors have become important tools for the monitoring and control of pollution levels, both before and after bioremediation measures are implemented.

Bioprocessing. Bioprocessing is a process that uses living cells or organisms to produce specific outcomes. Communities of microbial organisms perform a comprehensive range of bioprocesses within the natural environment, which can be exploited to benefit both the environment and industry. Bioprocesses used in environmental biotechnology include microbial enhanced oil recovery (MEOR), biological treatments of polluted air, biodesulfurization, conversion of pollutants into useful products such as fertilizers and green energy, and microbial exploration technology.

Biofilm Control. Biofilms, often referred to as slime, occur on the surface of aqueous environments and are caused by a complex accumulation of microorganisms. Biofilms are usually considered undesirable, as they are frequently associated with odors, infections, fouling, and corrosion, but they can be beneficial under some circumstances such as the treatment of wastewater.

Genetic Engineering and Manipulation. The advancement of genetics and genetic manipulation has had significant impact on environmental biotechnology. Research in molecular genetics has provided novel techniques for the detection and degradation of contaminants through the manipulation and enhancement of the microbes' ability to adapt themselves genetically to different pollutants. The ecologically useful and improved organisms are classified as genetically engineered microbes (GEM).

IMPACT ON INDUSTRY

Major Organizations. Many organizations worldwide are focusing on environmental biotechnology research. The International Society of Environmental Biotechnology is an umbrella organization that aims to enhance the progress and promotion of environmental biotechnology within industry and governments on an international scale. In Australia, the Australian Environmental Biotechnology Cooperative Research Centre focuses its research on biofilm control, microbial detection and control, and bioprocesses, and the Commonwealth Scientific and Industrial Research Organisation (CSIRO) focuses its research on the use of microorganisms

and biotechnological processes in the treatment of mining contamination and wastewater. In the United States, significant research on environmental biotechnology is undertaken at the Center for Environmental Biotechnology at the Lawrence Berkeley National Laboratory. Additionally, as world leaders in biotechnology and research, both the United States and the European Union (EU) have been instrumental in the evolution of environmental biotechnology, culminating in the establishment of the EU-U.S. Task Force on Biotechnology Research.

Government and University Research. Environmental biotechnology is an essential tool for the sustainability of the Earth's natural ecosystems and the balance between people and nature. Environmental pollution is a global issue, particularly the contamination of the world's oceans and the atmosphere. As such, many governments have initiated policies and regulations for waste removal and pollution control.

Although pollution is a global problem and requires international cooperation and control, the United States, China, India, and the European Union are home to a large proportion of the world's population and are the greatest contributors (by country) of pollution. Therefore, it is crucial that the governments of these countries research and implement environmental biotechnology, pollution control, and ecological restoration. The Chinese government has introduced programs to develop environmental technology, including the National Medium- and Long-Term Science and Technology Development Plan in 2006. The Indian government's Environmental Biotechnology Division was introduced to provide solutions for the abatement of contaminants, with the additional objectives of conserving endangered species and developing useful products from waste.

The U.S. government has also introduced initiatives such as the Environmental Remediation Sciences Program, formed in 2005 through the consolidation of the Natural and Accelerated Bioremediation Research (NABIR) program and the Environmental Management Science Program (EMSP), and administered by the U.S. Department of Energy's Office of Science. This program aims to address some of the United States' most pressing environmental recovery and contamination issues.

A growing number of international universities and institutes contribute significantly to environmental biotechnology. These include the Center for

Environmental Biotechnology at Arizona State University, the School of Biotechnology and Biomolecular Sciences and the Centre for Marine Biofouling and Bio-Innovation at the University of New South Wales, the Center for Environmental Biotechnology at the University of Tennessee, the Centre on Environmental Biotechnology at the University of Kalyani (India), and the University of Idaho's Environmental Biotechnology Institute.

CAREERS AND COURSE WORK

Students who wish to pursue a career in environmental biotechnology can have a degree in a number of fields, including environmental science, chemistry, microbiology, and biomolecular engineering. Many universities provide undergraduate and postgraduate degrees in environmental biotechnology. Upon course completion, students should have a solid understanding of environmental and contamination processes, and the theories and technologies used in environmental biotechnology to mitigate environmental damage.

Environmental biotechnology graduates can pursue such careers as environmental and bioengineering consultants and water-recycling and water resource managers in the private sector, nongovernmental organizations (NGOs), and specialized government organizations and agencies, and as researchers and professors at universities.

SOCIAL CONTEXT AND FUTURE PROSPECTS

In the early twenty-first century, the population of the world edged toward 7 billion, increasing the amount of pollution that reaches the land, water, and atmosphere. Many natural ecosystems are struggling to cope with the remnants of old toxic contamination and the influx of new contamination. The ecological, social, and economic costs of pollution are immeasurable, and environmental recovery is one of the most important problems facing the global community.

Conventional biotechnology processes and techniques have relied on end-of-pipe technologies, that is treatment of waste and pollution that has already contaminated the air, soil, or water. Although such methods are necessary, particularly in remediation of existing pollution, many environmental biotechnologists think that end-of-pipe methods should be regarded as last-resort efforts, rather than preferred

methods. As such, environmental biotechnology is moving from first-generation technology based on naturally occurring processes and microorganisms to second-generation technology based on high-tech anthropomorphic enhancement and manipulation of natural processes and microorganisms.

The future of environmental biotechnology lies in following an integrated environmental protection approach, with the fundamental goal being to control pollution before it enters the natural ecosystem and to recover already polluted areas. An essential step in this goal is to create controls and pass legislation to reduce the incidence of contamination. However, many researchers believe that the future of environmental engineering also will be closely aligned with the advancement and application of molecular and genetic methods. Decreasing or mitigating greenhouse gas pollution in the Earth's atmosphere is of vital importance to global health, so research in environmental biotechnology is focusing on biological organisms and processes that may help.

Christine Watts, Ph.D., B.App.Sc., B.Sc.

FURTHER READING

Cummings, Stephen, ed. *Bioremediation: Methods and Protocols.* New York: Humana Press, 2010. Experts in the field of environmental biotechnology present innovative and imaginative bioremediation techniques in pollution removal.

Evans, Gareth, and Judith Furlong. *Environmental Biotechnology: Theory and Application.* New York: John Wiley & Sons, 2003. A detailed examination of environmental biotechnology, focusing on present-day practices, the potential for biotechnological interventions, and microbial techniques and methods.

Illman, Walter, and Pedro Alvarez. "Performance Assessment of Bioremediation and Natural Attenuation." *Critical Reviews in Environmental Science and Technology,* 39, no. 4 (April, 2009): 209-270. A critical review of the state-of-the-art in performance assessment methods. Discusses future research directions in bioremediation, natural attenuation, chemical fingerprinting, and molecular biological tools.

Jördening, Hans-Joachim, and Josef Winter. *Environmental Biotechnology: Concepts and Applications.* Weinheim, Germany: Wiley-VCH, 2005. A solid foundation for students wishing to study environmental biotechnology. Examines in detail the microbiological treatment of waste and pollution in water, soil, and air.

Scragg, Alan. *Environmental Biotechnology.* 2d ed. New York: Oxford University Press, 2005. Examines the multitude of ways in which environmental biotechnology is applied in pollution control, environmental management, and removal of oil and minerals.

Thakur, Indu Shekhar. *Environmental Biotechnology: Basic Concepts and Applications.* New Delhi: I. K. International, 2006. A comprehensive examination of environmental processes and the many possible applications of environmental biotechnology, such as bioremediation, bioprocessing, and bioleaching.

WEB SITES

International Society for Biotechnology
http://www3.inecol.edu.mx/iseb

Lawrence Berkeley National Laboratory, Earth Sciences Division
Center for Environmental Biotechnology
http://esd.lbl.gov/CEB

U.S. Department of Energy, Office of Biological and Environmental Research
Environmental Remediation Sciences Program
http://www.er.doe.gov/ober/ERSD/ersphome.html

See also: Air-Quality Monitoring; Bioengineering; Ecological Engineering; Environmental Chemistry; Environmental Engineering; Genetic Engineering; Hazardous-Waste Disposal; Industrial Pollution Control; Landscape Ecology; Land-Use Management; Sanitary Engineering; Sewage Engineering; Water-Pollution Control.

ENVIRONMENTAL CHEMISTRY

FIELDS OF STUDY

Chemistry; chemical engineering; bioengineering; physics; physical chemistry; organic chemistry; biochemistry; molecular biology; electrochemistry; analytical chemistry; photochemistry; atmospheric chemistry; agricultural chemistry; industrial ecology; toxicology.

SUMMARY

Environmental chemistry is an interdisciplinary subject dealing with chemical phenomena in nature. Environmental chemists are concerned with the consequences of anthropogenic chemicals in the air people breathe and the water they drink. They have become increasingly involved in managing the effects of these chemicals through both the creation of ecologically friendly products and efforts to minimize the pollution of the land, water, and air.

KEY TERMS AND CONCEPTS

- **Anthrosphere:** Artificial environment created, modified, and used by humans for their purposes and activities, from houses and factories to chemicals and communications systems.
- **Biomagnification:** Increase in the concentration of such chemicals as dichloro-diphenyl-trichloroethane (DDT) in different life-forms at successively higher trophic levels of a food chain or web.
- **Carcinogen:** Chemical that causes cancer in organisms exposed to it.
- **Chlorofluorocarbon (CFC):** Organic chemical compound containing carbon, hydrogen, chlorine, and florine, such as Freon (a trademark for several CFCs), once extensively used as a refrigerant.
- **Dichloro-Diphenyl-Trichloroethane (DDT):** Organic chemical that was widely used for insect control during and after World War II but has been banned in many countries.
- **Ecological Footprint:** Measure of how much land and water a human population needs to regenerate consumed resources and to absorb wastes.
- **Greenhouse Effect:** Phenomenon in which the atmosphere, like a greenhouse, traps solar heat, with the trapping agents being such gases as carbon dioxide, methane, and water vapor.
- **Hazardous Waste:** Waste, or discarded material, that contains chemicals that are flammable, corrosive, toxic, or otherwise pose a threat to the health of humans or other organisms or a hazard to the environment.
- **Heavy Metal:** Any metal with a specific gravity greater than 5; usually used to refer to any metal that is poisonous to humans and other organisms, such as cadmium, lead, and mercury.
- **Mutagen:** Chemical that causes or increases the frequency of changes (mutations) in the genetic material of an organism.
- **Ozone:** Triatomic form of oxygen produced when diatomic oxygen is exposed to ultraviolent radiation (its presence in a stratospheric layer protects life-forms from harmful solar radiation).
- **Pesticide:** Chemical used to kill or inhibit the multiplication of organisms that humans consider undesirable, such as certain insects.

DEFINITION AND BASIC PRINCIPLES

Environmental chemistry is the science of chemical processes in the environment. It is a profoundly interdisciplinary and socially relevant field. What environmental chemists do has important consequences for society because they are concerned with the effect of pollutants on the land, water, and air that humans depend on for their life, health, and proper functioning. Environmental chemists are forced to break down barriers that have traditionally kept chemists isolated from other fields.

Environmental chemistry needs to be distinguished from its later offshoot, green chemistry. As environmental chemistry developed, it tended to emphasize the detection and mitigation of pollutants, the study of the beneficial and adverse effects of various chemicals on the environment, and how the beneficial effects could be enhanced and the adverse eliminated or attenuated. Green chemistry, however, focuses on how to create sustainable, safe, and nonpolluting chemicals in ways that minimize the ecological footprint of the processes. Some scholars define this field simply as sustainable chemistry.

The work of environmental chemists is governed by several basic principles. For example, the prevention principle states that when creating chemical products, it is better to minimize waste from the start than to later clean up wastes that could have been eliminated. Another principle declares that in making products, chemists should avoid using substances that could harm humans or the environment. Furthermore, chemists should design safe chemicals with the lowest practicable toxicity. In manufacturing products, chemists must minimize energy use and maximize energy efficiency; they should also use, as much as possible, renewable materials and energy resources. The products chemists make should be, if possible, biodegradable. Environmental chemists should use advanced technologies, such as computers, to monitor and control hazardous wastes. Finally, they must employ procedures that minimize accidents.

BACKGROUND AND HISTORY

Some scholars trace environmental chemistry's roots to the industrial revolutions in Europe and the United States in the eighteenth and nineteenth centuries. The newly created industries accelerated the rate at which chemicals were produced. By the late nineteenth century, some scientists and members of the general public were becoming aware of and concerned about certain negative consequences of modern chemical technologies. For example, the Swedish chemist Svante August Arrhenius recognized what later came to be known as the greenhouse effect, and a Viennese physician documented the health dangers of asbestos, which had become an important component in more than 3,000 products.

Modern industrialized societies were also generating increasing amounts of wastes, and cities and states were experiencing difficulties in discovering how to manage them without harm to the environment. World War I, often called the chemists' war, revealed the power of scientists to produce poisonous and explosive materials. During World War II, chemists were involved in the mass production of penicillin and DDT, substances that saved thousands of lives. However, the widespread and unwise use of DDT and other pesticides after the war prompted Rachel Carson to write *Silent Spring* (1962), which detailed the negative effects that pesticides were having on birds and other organisms. Many associate the

start of the modern environmental movement with the publication of Carson's book.

The warnings Carson issued played a role in the establishment of the Environmental Protection Agency (EPA) in 1971 and the EPA's ban of DDT in 1972. Reports in 1974 that CFCs were destroying the Earth's ozone layer eventually led many countries to halt, then ban their production. From the 1970's on, environmental chemists devoted themselves to the management of pollutants and participation in government policies and regulations that attempted to prevent or mitigate chemical pollution. In the 1990's, criticism of this command-and-control approach led to the formation of the green chemistry movement by Paul Anastas and others. These chemists fostered a comprehensive approach to the production, utilization, and termination of chemical materials that saved energy and minimized wastes. By the first

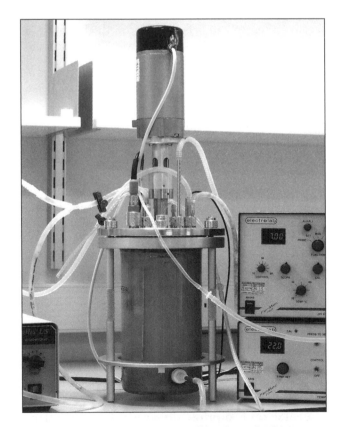

Small-scale bioreactor used for studying bacteria. Bioreactors are devices that maintain a specific environment to allow organisms or their derivative substances to undergo chemical processes. (Dr. Jeremy Burgess/Photo Researchers, Inc.)

decade of the twenty-first century, environmental chemistry had become a thriving profession with a wide spectrum of approaches and views.

HOW IT WORKS

Chemical Analysis and the Atmosphere. Indispensable to the progress of environmental chemistry is the ability to measure, quantitatively and qualitatively, certain substances that even at very low concentrations, pose harm to humans and the environment. By using such techniques as gravimetric and volumetric analysis, various types of spectroscopy, electroanalysis, and chromatography, environmental chemists have been able to accurately measure such atmospheric pollutants as sulfur dioxide, carbon monoxide, hydrogen sulfide, nitrogen oxides, and several hydrocarbons. Many governmental and nongovernmental organizations require that specific air contaminants be routinely monitored. Because of heavy demands on analytic chemists, much monitoring has become computerized and automatic.

Atmospheric particles range in size from a grain of sand to a molecule. Nanoparticles, for example, are about one-thousandth the size of a bacterial cell, but environmental chemists have discovered that they can have a deleterious effect on human health. Epidemiologists have found that these nanoparticles adversely affect respiratory and cardiovascular functioning.

Atmospheric aerosols are solid or liquid particles smaller than a hundred millimicrons in diameter. These particles undergo several possible transformations, from coagulation to phase transitions. For example, particles can serve as nuclei for the formation of water droplets, and some chemists have experimented with particulates in forming rain clouds. Human and natural biological sources contribute to atmospheric aerosols. The CFCs in aerosol cans have been factors in the depletion of the Earth's ozone layer. Marine organisms produce such chemicals as halogen radicals, which in turn influence reactions of atmospheric sulfur, nitrogen, and oxidants. As marine aerosol particles rise from the ocean and are oxidized in the atmosphere, they may react with its components, creating a substance that may harm human health. Marine aerosols contain carbonaceous as well as inorganic materials, and when an organic aerosol interacts with atmospheric oxygen, its inert hydrophobic (water-repelling) film is transformed into a reactive hydrophilic (water-absorbing) layer. A consequence of this process is that organic aerosols serve as a conduit for organic compounds to enter the atmosphere.

Carbon dioxide is an example of a molecular atmospheric component, and some environmental chemists have been devoting their efforts to determining its role in global warming, but others have studied the Earth's prebiotic environment to understand how inorganic carbon initially formed the organic molecules essential to life. Using photoelectrochemical techniques (how light affects electron transfers in chemical reactions), researchers discovered a possible metabolic pathway involving carbon dioxide fixation on mineral surfaces.

Water Pollution. Because of water's vital importance and its uneven distribution on the Earth's surface, environmental chemists have had to spend a great deal of time and energy studying this precious resource. Even before the development of environmental chemistry as a profession, many scientists, politicians, and citizens were concerned with water management. Various governmental and nongovernmental organizations were formed to monitor and manage the quality of water. Environmental chemists have been able to use their expertise to trace the origin and spread of water pollutants throughout the environment, paying special attention to the effects of water pollutants on plant, animal, and human life.

A particular interest of environmental chemists has been the interaction of inorganic and organic matter with bottom sediments in lakes, rivers, and oceans. These surface sediments are not simply unreactive sinks for pollutants but can be studied quantitatively in terms of how many specific chemicals are bonded to a certain amount of sediment, which in turn is influenced by whether the conditions are oxidizing or reducing. Chemists can then study the bioavailability of contaminants in sediments. Furthermore, environmental chemists have studied dissolution and precipitation, discovering that the rates of these processes depend on what happens in surface sediments. Using such techniques as scanning polarization force microscopy, they have been able to quantify pollutant immobilization and bacterial attachment on surface sediments. Specifically, they have used these methods to understand the concentrations and activities of heavy metals in aquatic sediments.

Hazardous Waste. One of the characteristics of advanced industrialized societies has been the creation of growing amounts of solid and liquid wastes, an important proportion of which pose severe dangers to the environment and human health. These chemicals can be toxic, corrosive, or flammable. The two largest categories of hazardous wastes are organic compounds such as polychlorinated biphenyls (PCBs) and dioxin, and heavy metals, such as lead and mercury. Environmental chemists have become involved in research on the health and environmental effects of these hazardous substances and the development of techniques to detect, monitor, and control them. For example, they have studied the rapidly growing technology of incineration as a means of reducing and disposing of wastes. They have also studied the chemical emissions from incinerators and researched methods for the safe disposal of the ash and slag produced. Because of the passage of the Resource Recovery Act of 1970, environmental chemists have devoted much attention to finding ways of reclaiming and recycling materials from solid wastes.

Pesticides. The mismanagement of pesticides inspired Carson to write *Silent Spring*, and pesticides continue to be a major concern of industrial and environmental chemists. One reason for the development of pesticides is the great success farmers had in using them to control insects, thereby dramatically increasing the quantity and quality of various agricultural products. By 1970, more than 30,000 pesticide products were being regularly used. This expansion in pesticides is what alarmed Carson, who was not in favor of a total ban of pesticides but rather their reduction and integration with biological and cultural controls.

Because pesticides are toxic to targeted species, they often cause harm to beneficial insects and, through biomagnification, to birds and other animals. Pesticide residues on agricultural products have also been shown to harm humans. Therefore, environmental chemists have become involved in monitoring pesticides from their development and use to their effects on the environment. They have also helped create pesticide regulations and laws. This regulatory system has become increasingly complex, costing companies, the government, and customers large amounts of money. The hope is that integrated control methods will prove safer and cheaper than traditional pesticides.

APPLICATIONS AND PRODUCTS

Anthrosphere. Environmental chemists are concerned with how humans and their activities, especially making and using chemicals, affect the environment. Building homes and factories, producing food and energy, and disposing of waste all have environmental consequences. Whereas some environmentalists study how to create ecologically friendly dwellings, environmental chemists study how chemical engineers should design factories that cause minimum harm to the environment. Specific examples of applications of environmental chemistry to industry include the creation of efficient catalysts that speed up reactions without themselves posing health or environmental hazards.

Because many problems arise from the use of hazardous solvents in chemical processes, environmental chemists try to develop processes that use only safe solvents or avoid their use altogether. Because the chemical industry depends heavily on petroleum resources, which are nonrenewable and becoming drastically diminished, environmental chemists study how renewable resources such as biomass may serve as substitutes for fossil fuels. They are also creating products that degrade rapidly after being discarded, so that their environmental impact is transient. Following the suggestions of environmental chemists, some companies are developing long-lasting, energy-saving batteries, and others are selling their products with less packaging than previously.

Hydrosphere. Throughout history, water has been essential in the development of human civilizations, some of which have declined and disappeared because of deforestation, desertification, and drought. Water has also been a vehicle for the spread of diseases and pollutants, both of which have caused serious harm to humans and their environment. Environmental chemists have consequently been involved in such applications as the purification of water for domestic use, the monitoring of water used in the making of chemicals, and the treatment of wastewater so that its release and reuse will not harm humans or the environment. For example, such heavy metals as cadmium, mercury, and lead are often found in wastewater from various industries, and environmental chemists have developed such techniques as electrodeposition, reverses osmosis, and ion exchange to remove them. Many organic compounds are carcinogens and mutagens, so chemists want to remove them from water. Besides such traditional methods as powdered activated carbon, chemists have used adsorbent synthetic polymers to attract insoluble organic compounds. Detergents in wastewater can contribute to lake eutrophication and cause harm to wildlife, and some companies have created detergents specially formulated to cause less environmental damage.

Atmosphere. In the twentieth century, scientists discovered that anthropogenic greenhouse gases have been contributing to global warming that could have catastrophic consequences for island nations and coastal cities. Industries, coal-burning power plants, and automobiles are major air polluters, and environmental chemical research has centered on finding ways to reduce or eliminate these pollutants. The Clean Air Act of 1970 and subsequent amendments set standards for air quality and put pressure on air polluters to reduce harmful emissions. For example, chemists have helped power plants develop desulfurization processes, and other scientists developed emission controls for automobiles.

The problem of global warming has proved difficult to solve. Some environmental chemists believe that capturing and storing carbon dioxide is the answer, whereas others believe that government regulation of carbon dioxide and methane by means of energy taxes will lead to a lessening of global warming. Some think that the Kyoto Protocol, an international agreement that went into effect in 2005, is a small but important first step, whereas others note that the lack of participation by the United States and the omission of a requirement that such countries as China and India reduce greenhouse gas emissions seriously weakened the agreement. On the other hand, the Montreal and Copenhagen Protocols did foster global cooperation in the reduction of and phasing out of CFCs, which should lead to a reversal in ozone layer depletion.

Agricultural and Industrial Ecology. Agriculture, which involves the production of plants and animals as food, is essential in ministering to basic human needs. Fertilizers and pesticides developed by chemists brought forth the green revolution, which increased crop yields in developed and developing countries. Some believe that genetic engineering techniques will further revolutionize agriculture. Initially, chemists created such highly effective insecticides as DDT, but DDT proved damaging to the environment. Activists

---

Fascinating Facts About Environmental Chemistry

- At life's beginning, single-celled cyanobacteria made possible the evolution of millions of new species; however, what modern humans are doing to the atmosphere will result in the extinction of hundreds of thousands of species.
- More than 99 percent of the total mass of the Earth's atmosphere is found within about 30 kilometers (about 20 miles) of its surface.
- Although 71 percent of the Earth's surface is covered by water, only 0.024 percent of this water is available as freshwater.
- According to a National Academy of Sciences study, legally permitted pesticide residues in food cause 4,000 to 20,000 cases of cancer per year in the United States.
- From 1980 to 2010, the quality of outdoor air in most developed countries greatly improved.
- From 1980 to 2008, the EPA placed 1,569 hazardous-waste sites on its priority list for cleanup.
- According to the U.S. Geological Survey, even though the population of the United States grew by 16 percent from 1980 to 2004, total water consumption decreased by about 9 percent.
- Because of global warming, in February and March, 2002, a mass of ice larger than the state of Rhode Island separated from the Antarctic Peninsula.
- The United States leads the world in producing solid waste. With only 4.6 percent of the world's population, it produced about one-third of the world's solid waste.
- Each year, 12,000 to 16,000 American children under nine years of age are treated for acute lead poisoning, and about 200 die.

---

then encouraged chemists to develop biopesticides from natural sources because they are generally more ecologically friendly than synthetics.

Industrial ecology is a new field based on chemical engineering and ecology; its goal is to create products in a way that minimizes environmental harm. Therefore, environmental chemical engineers strive to build factories that use renewable energy as much as possible, recycle most materials, minimize wastes, and extract useful materials from wastes. In general, these environmental chemists act as wise stewards of their facilities and the environment. A successful example of ecological engineering is phytoremediation, or the use of plants to remove pollutants from contaminated lands. Artificially constructed wetlands have also been used to purify wastewater.

IMPACT ON INDUSTRY

Industries are involved in a wide variety of processes that have environmental implications, from food production and mineral extraction to manufacturing and construction. In some cases, such as the renewable energy industries, the environmental influence is strong, but even in traditional industries such as utilities and transportation companies, problems such as air and water pollution have become corporate concerns.

Government and University Research. Since the beginning of the modern environmental movement, state and federal governments as well as universities have increased grants and fellowships for projects related to environmental chemistry. For example, the EPA's Green Chemistry Program has supported basic research to develop chemical products and manufacturing techniques that are ecologically benign. Sometimes government agencies cooperate with each other in funding environmental chemical projects; for instance, in 1992, the EPA's Office of Pollution Prevention and Toxics collaborated with the National Science Foundation to fund several green chemical proposals. These grants were significant, totaling tens of millions of dollars. Besides government and academia, professional organizations have also sponsored green research. For example, the American Chemical Society has established the Green Chemistry Institute, whose purpose is to encourage collaboration with scientists in other disciplines to discover chemical products and processes that reduce or eliminate hazardous wastes.

Industry and Business. Environmentalists and government regulations have forced leaders in business and industry to make sustainability a theme in their plans for future development. In particular, the U.S. chemical industry, the world's largest, has directly linked its growth and competitiveness to a concern for the environment. Industrial leaders realize that they will have to cooperate with officials in government and academia to realize this vision. They also understand that they will need to join with such

organizations as the Environmental Management Institute and the Society of Environmental Toxicology and Chemistry to minimize the environmental contamination that has at times characterized the chemical industry of the past.

Major Corporations. Top American chemical companies, such as Dow, DuPont, Eastman Chemical, and Union Carbide, have gone on record as vowing to use resources more efficiently, deliver products to consumers that meet their needs and enhance their quality of life, and preserve the environment for future generations. Nevertheless, these promised changes must be seen against the background of past environmental depredations and disasters. For example, Dow is responsible for ninety-six of the worst Superfund toxic-waste dumps, and Union Carbide shared responsibility for the deaths of more than 2,000 people in a release of toxic chemicals in Bhopal, India. Eastman Chemical, along with other industries, is a member of Responsible Care, an organization devoted to the principles of green chemistry, and the hope is that the member industries will encourage the production of ecologically friendly chemicals without the concomitant of dangerous wastes.

CAREERS AND COURSE WORK

Students in environmental chemistry need to take many chemistry courses, including general chemistry, organic chemistry, quantitative and qualitative analysis, instrumental analysis, inorganic chemistry, physical chemistry, and biochemistry. They also should study advanced mathematics, physics, and computer science. Although job opportunities exist for students with a bachelor's degree, there are greater opportunities for those who obtain a master's degree or a doctorate. Graduate training allows students to specialize in such areas as soil science, hazardous-waste management, air-quality management, water-quality management, environmental education, and environmental law. Because of increasing environmental concerns in industries, governments, and academia, numerous careers are possible for environmental chemistry graduates. They have found positions in business, law, marketing, public policy, government agencies, laboratories, and chemical industries. Some graduates pursue careers in such government agencies as the EPA, the Food and Drug Administration, the Natural Resource Conservation Services, the Forest Service, and the Department of Health and Human Services. After obtaining a doctorate, some environmental chemists become teachers and researchers in one of the many academic programs devoted to their field.

SOCIAL CONTEXT AND FUTURE PROSPECTS

In a world increasingly concerned with environmental quality, the sustainability of lifestyles, and environmental justice, the future for environmental chemistry appears bright. For example, analysts have predicted that environmental chemical engineers will have a much faster employment growth than the average for all other occupations. Environmental chemists will be needed to help industries comply with regulations and to develop ways of cleaning up hazardous wastes. However, other analysts warn that, in periods of economic recession, environmental concerns tend to be set aside, and this could complicate the employment forecast for environmental chemists.

Some organizations, such as the Environmental Chemistry Group in England, have as a principal goal the promotion of the expertise and interests of their members, and the American Chemical Society's Division of Environmental Chemistry similarly serves its members with information on educational programs, job opportunities, and awards for significant achievement, such as the Award for Creative Advances in Environmental Chemistry. These organizations also issue reports on their social goals, and documents detailing their social philosophy emphasize that environmental chemists should be devoted to the safe operation of their employers' facilities. Furthermore, they should strive to protect the environment and make sustainability an integral part of all business activities.

Robert J. Paradowski, M.S., Ph.D.

FURTHER READING

Baird, Colin, and Michael Cann. *Environmental Chemistry*. 4th ed. New York: W. H. Freeman, 2008. A clear and comprehensive survey of the field. Each chapter has further reading suggestions and Web sites of interest. Index.

Carson, Rachel. *Silent Spring*. 1962. Reprint. Boston: Houghton Mifflin, 2002. Originally serialized in *The New Yorker* magazine, this book has been honored as one of the best nonfiction works of the twentieth century. Its criticism of the chemical industry and the overuse of pesticides generated controversy, and most of its major points have stood the test of time.

Girard, James E. *Principles of Environmental Chemistry*. Sudbury, Mass.: Jones and Bartlett, 2010. Emphasizes the chemical principles undergirding environmental issues as well as the social and economic contexts in which they occur. Five appendixes and index.

Howard, Alan G. *Aquatic Environmental Chemistry*. 1998. Reprint. New York: Oxford University Press, 2004. Analyzes the chemistry behind freshwater and marine systems. Also includes useful secondary material that contains explanations of unusual terms and advanced chemical and mathematical concepts.

Manahan, Stanley E. *Environmental Chemistry*. 9th ed. Baca Raton, Fla.: CRC Press, 2010. Explores the anthrosphere, industrial ecosystems, geochemistry, and aquatic and atmospheric chemistry. Each chapter has a list of further references and cited literature. Index.

Schwedt, Georg. *The Essential Guide to Environmental Chemistry*. 2001. Reprint. New York: John Wiley & Sons, 2007. Provides a concise overview of the field. Contains many color illustrations and an index.

WEB SITES

American Chemical Society
Division of Environmental Chemistry
http://www.envirofacs.org

Environmental Protection Agency
National Exposure Research Library, Environmental Sciences Division
http://www.epa.gov/nerlesd1

Royal Society of Chemistry
Environmental Chemistry Group
http://www.rsc.org/membership/networking/interestgroups/environmental/index.asp

Society of Environmental Toxicology and Chemistry
http://www.setac.org

See also: Air-Quality Monitoring; Coastal Engineering; Ecological Engineering; Environmental Biotechnology; Environmental Engineering; Hazardous-Waste Disposal; Industrial Pollution Control; Recycling Technology; Sanitary Engineering; Sewage Engineering; Thermal Pollution Control; Water-Pollution Control.

ENVIRONMENTAL ENGINEERING

FIELDS OF STUDY

Mathematics; chemistry; physics; biology; geology; engineering mechanics; fluid mechanics; soil mechanics; hydrology.

SUMMARY

Environmental engineering is a field of engineering involving the planning, design, construction, and operation of equipment, systems, and structures to protect and enhance the environment. Major areas of application within the field of environmental engineering are wastewater treatment, water-pollution control, water treatment, air-pollution control, solid-waste management, and hazardous-waste management. Water-pollution control deals with physical, chemical, biological, radioactive, and thermal contaminants. Water treatment may be for the drinking water supply or for industrial water use. Air-pollution control is needed for stationary and moving sources. The management of solid and hazardous wastes includes landfill and incinerators for disposal of solid waste and identification and management of hazardous wastes.

KEY TERMS AND CONCEPTS

- **Activated Sludge Process:** Biological wastewater-treatment system for removing waste organic matter from wastewater.
- **Baghouse:** Air-pollution control device that filters particulates from an exhaust stream; also called a bag filter.
- **Biochemical Oxygen Demand (BOD):** Amount of oxygen needed to oxidize the organic matter in a water sample.
- **Catch Basin:** Chamber to retain matter flowing from a street gutter that might otherwise obstruct a sewer.
- **Digested Sludge:** Wastewater biosolids (sludge) that have been stabilized by an anaerobic or aerobic biological process.
- **Effluent:** Liquid flowing out from a wastewater-treatment process or treatment plant.
- **Electrostatic Precipitator:** Air-pollution control device to remove particulates from an exhaust stream by giving the particles a charge.
- **Nonpoint Source:** Pollution source that cannot be traced back to a single emission source, such as storm-water runoff.
- **Oxidation Pond:** Large shallow basin used to treat wastewater, using sunlight, bacteria, and algae.
- **Photochemical Smog:** Form of air pollution caused by nitrogen oxides and hydrocarbons in the air that react to form other pollutants because of catalysis by sunlight.
- **Pollution Prevention:** Use of conscious practices or processes to reduce or eliminate the creation of wastes at the source.
- **Primary Treatment:** Wastewater treatment to remove suspended and floating matter that will settle from incoming wastewater.
- **Sanitary Landfill:** Site used for disposal of solid waste that uses liners to prevent groundwater contamination and is covered daily with a layer of earth.
- **Secondary Treatment:** Wastewater treatment to remove dissolved and fine suspended organic matter that would exert an oxygen demand on a receiving stream.
- **Trickling Filter:** Biological wastewater-treatment process in which wastewater trickles through a bed of rocks with a coating containing microorganisms.

DEFINITION AND BASIC PRINCIPLES

Environmental engineering is a field of engineering that split off from civil engineering as the importance of the treatment of drinking water and wastewater was recognized. This field of engineering was first known as sanitary engineering and dealt almost exclusively with the treatment of water and wastewater. As awareness of other environmental concerns and the need to do something about them grew, this field of engineering became known as environmental engineering, with the expanded scope of dealing with air pollution, solid wastes, and hazardous wastes, in addition to water and wastewater treatment.

Environmental engineering is an interdisciplinary field that makes use of principles of chemistry, biology, mathematics, and physics, along with engineering sciences (such as soil mechanics, fluid mechanics, and hydrology) and empirical engineering correlations and knowledge to plan for, design, construct, maintain, and operate facilities for treatment of liquid and gaseous waste streams, for prevention of air pollution, and for management of solid and hazardous wastes.

The field also includes investigation of sites with contaminated soil and/or groundwater and the planning and design of remediation strategies. Environmental engineers also provide environmental impact analyses, in which they assess how a proposed project will affect the environment.

BACKGROUND AND HISTORY

When environmental engineering, once a branch of civil engineering, first became a separate field in the mid-1800's, it was known as sanitary engineering. Initially, the field involved the water supply, water treatment, and wastewater collection and treatment.

In the middle of the twentieth century, people began to become concerned about environmental quality issues such as water and air pollution. As a consequence, the field of sanitary engineering began to change to environmental engineering, expanding its scope to include air pollution, solid- and hazardous-waste management, and industrial hygiene.

Several pieces of legislation have affected and helped define the work of environmental engineers. Some of the major laws include the Clean Air Act of 1970, the Safe Drinking Water Act of 1974, the Toxic Substances Control Act of 1976, the Resource Recovery and Conservation Act (RCRA) of 1976, and the Clean Water Act of 1977.

HOW IT WORKS

Environmental engineering uses chemical, physical, and biological processes for the treatment of water, wastewater, and air, as well as in-site remediation processes. Therefore, knowledge of the basic sciences—chemistry, biology, and physics—is important along with knowledge of engineering sciences and applied engineering.

Chemistry. Chemical processes are used to treat water and wastewater, to control air pollution, and for site remediation. These chemical treatments include chlorination for disinfection of both water and wastewater, chemical oxidation for iron and manganese removal in water-treatment plants, chemical oxidation for odor control, chemical precipitation for removal of metals or phosphorus from wastewater, water softening by the lime-soda process, and chemical neutralization for pH (acidity) control and for scaling control.

The chemistry principles and knowledge that are needed for these treatment processes include the ability to understand and work with chemical equations, to make stoiciometric calculations for dosages, and to determine size and configuration requirements for chemical reactors to carry out the various processes.

Biology. The major biological treatment processes used in wastewater treatment are biological oxidation of dissolved and fine suspended organic matter in wastewater (secondary treatment) and stabilization

Air is injected and mixed into water in an aeration tank. This is the first step in the "activated sludge" water treatment process, in which both air and microorganisms are used to oxidize organic pollutants, producing a waste "sludge" of oxidized material. (Jonathan A. Meyers/Photo Researchers, Inc.)

of biological wastewater biosolids (sludge) by anaerobic digestion or aerobic digestion.

Biological principles and knowledge that are useful in designing and operating biological wastewater treatment and biosolids digestion processes include the kinetics of the biological reactions and knowledge of the environmental conditions required for the microorganisms. The required environmental conditions include the presence or absence of oxygen and the appropriate temperature and pH.

Physics. Physical treatment processes used in environmental engineering include screening, grinding, comminuting, mixing, flow equalization, flocculation, sedimentation, flotation, and granular filtration. These processes are used to remove materials that can be screened, settled, or filtered out of water or wastewater and to assist in managing some of the processes. Many of these physical treatment processes are designed on the basis of empirical loading factors, although some use theoretical relationships such as the use of estimated particle settling velocities for design of sedimentation equipment.

Soil Mechanics. Topics covered in soil mechanics include the physical properties of soil, the distribution of stress within the soil, soil compaction, and water flow through soil. Knowledge of soil mechanics is used by environmental engineers in connection with design and operation of sanitary landfills for solid waste, in storm water management, and in the investigation and remediation of contaminated soil and groundwater.

Fluid Mechanics. Principles of fluid mechanics are used by environmental engineers in connection with the transport of water and wastewater through pipes and open channels. Such transport takes place in water distribution systems, in sanitary sewer collection systems, in storm water sewers, and in wastewater-treatment and water-treatment plants. Design and sizing of the pipes and open channels make use of empirical relationships such as the Manning equation for open channel flow and the Darcy-Weisbach equation for frictional head loss in pipe flow. Environmental engineers also design and select pumps and flow measuring devices.

Hydrology. The principles of hydrology (the science of water) are used to determine flow rates for storm water management when designing storm sewers or storm water detention or retention facilities. Knowledge of hydrology is also helpful in planning and developing surface water or groundwater as sources of water.

Practical Knowledge. Environmental engineers make use of accumulated knowledge from their work in the field. Theoretical equations, empirical equations, graphs, nomographs, guidelines, and rules of thumb have been developed based on experience. Empirical loading factors are used to size and design many treatment processes for water and wastewater. For example, the design of rapid sand filters to treat drinking water was based on a specified loading rate in gallons per minute of water per square foot of sand filter. Also the size required for a rotating biological contactor to provide secondary treatment of wastewater was determined based on a loading rate in pounds of biochemical oxygen demand (BOD) per day per 1,000 square feet of contactor area.

Engineering Tools. Tools such as engineering graphics, computer-aided drafting (CAD), geographic information systems (GIS), and surveying are available for use by environmental engineers. These tools are used for working with plans and drawings and for laying out treatment facilities or landfills.

Codes and Design Criteria. Much environmental engineering work makes use of codes or design criteria specified by local, state, or federal government agencies. Examples of such design criteria are the storm return period to be used in designing storm sewers or storm water detention facilities and the loading factor for rapid sand filters. Design and operation of treatment facilities for water and wastewater are also based on mandated requirements for the finished water or the treated effluent.

APPLICATIONS AND PRODUCTS

Environmental engineers design, build, operate, and maintain treatment facilities and equipment for the treatment of drinking water and wastewater, air-pollution control, and the management of solid and hazardous wastes.

Air-Pollution Control. Increasing air pollution from industries and power plants as well as automobiles led to passage of the Clean Air Act of 1970. This law led to greater efforts to control air pollution. The two major ways to control air pollution are the treatment of emissions from fixed sources and from moving sources (primarily automobiles).

The fixed sources of air pollution are mainly the smokestacks of industrial facilities and power plants.

Devices used to reduce the number of particulates emitted include settling chambers, baghouses, cyclones, wet scrubbers, and electrostatic precipitators. Electrostatic precipitators impart the particles with an electric charge to aid in their removal. They are often used in power plants, at least in part because of the readily available electric power to run them. Water-soluble gaseous pollutants can be removed by wet scrubbers. Other options for gaseous pollutants are adsorption on activated carbon or incineration of combustible pollutants. Because sulfur is contained in the coal used as fuel, coal-fired power plants produce sulfur oxides, particularly troublesome pollutants. The main options for reducing these sulfur oxides are desulfurizing the coal or desulfurizing the flue gas, most typically with a wet scrubber using lime to preciptitate the sulfur oxides.

Legislation has greatly reduced the amount of automobile emissions, the main moving source of air pollution. The reduction in emissions has been accomplished through catalytic converters to treat exhaust gases and improvements in the efficiency of automobile engines.

Water Treatment. The two main sources for the water supply are surface water (river, lake, or reservoir) and groundwater. The treatment requirements for these two sources are somewhat different.

For surface water, treatment is aimed primarily at removal of turbidity (fine suspended matter) and perhaps softening the water. The typical treatment processes for removal of turbidity involve the addition of chemicals such as alum or ferric chloride. The chemicals are rapidly mixed into the water so that they react with alkalinity in the water, then slowly mixed (flocculation) to form a settleable precipitate. After sedimentation, the water passes through a sand filter and finally is disinfected with chlorine. If the water is to be softened as part of the treatment, lime, $Ca(OH)_2$, and soda ash, Na_2CO_3, are used in place of alum or ferric chloride, and the water hardness (calcium and magnesium ions) is removed along with its turbidity.

Groundwater is typically not turbid (cloudy), so it does not require the type of treatment used for surface water. At minimum, it requires disinfection. Removal of iron and manganese by aeration may be needed, and if the water is very hard, it may be softened by ion exchange.

Wastewater Treatment. The Clean Water Act of 1977 brought wastewater treatment to a new level by requiring that all wastewater discharged from municipal treatment plants must first undergo at least secondary treatment. Before the passage of the legislation, many large cities located on a river or along the ocean provided only primary treatment in their wastewater-treatment plants and discharged effluent with only settleable solids removed. All dissolved and fine suspended organic matter remained in the effluent. Upgrading treatment plants involved added a biological treatment to remove dissolved and fine suspended organic matter that would otherwise exert an oxygen demand on the receiving stream, perhaps depleting the oxygen enough to cause problems for fish and other aquatic life.

Fascinating Facts About Environmental Engineering

- In March, 1987, a barge loaded with municipal solid waste departed from Islip, New York, headed to a facility in North Carolina, where state officials turned it away. It traveled to six states and three countries over seven months trying to find a place to unload its cargo before it was allowed to return to New York.

- The activated sludge process for treating wastewater was invented in England in 1914. Interest in the activated sludge process spread rapidly, and it soon became the most widely used biological wastewater-treatment process in the world.

- The use of chlorine to disinfect drinking water supplies began in the late 1800's and early 1900's. It dramatically reduced the incidence of waterborne diseases such as cholera and typhoid fever.

- In the book *Silent Spring* (1962), Rachel Carson described the negative effect of the pesticide dichloro-diphenyl-trichloroethane (DDT) on birds. This book increased environmental awareness and is often cited as the beginning of the environmental movement.

- The first comprehensive sewer system in the United States was built in Chicago in 1850. The city level was raised 10 to 15 feet so that gravity would drain the sewers into the Chicago River, which emptied into Lake Michigan.

- A "solid waste" as defined by the U.S. Resource Conservation and Recovery Act may be solid, liquid, or semi-solid in form.

Solid-Waste Management. The main options for solid-waste management are incineration, which reduces the volume for disposal to that of the ash that is produced, and disposal in a sanitary landfill. Some efforts have been made to reuse and recycling materials to reduce the amount of waste sent to incinerators or landfills. A sanitary landfill is a big improvement over the traditional garbage dump, which was simply an open dumping ground. A sanitary landfill uses liners to prevent groundwater contamination, and each day, the solid waste is covered with soil.

Hazardous-Waste Management. The Resource Conservation and Recovery Act (RCRA) of 1976 provides the framework for regulating hazardous-waste handling and disposal in the United States. One very useful component of RCRA is that it specifies a very clear and organized procedure for determining if a particular material is a hazardous waste and therefore subject to RCRA regulations. If the material of interest is indeed a waste, then it is defined to be a hazardous waste if it appears on one of RCRA's lists of hazardous wastes, if it contains one or more hazardous chemicals that appear on an RCRA list, or if it has one or more of the four RCRA hazardous waste characteristics as defined by laboratory tests. The four RCRA hazardous waste characteristics are flammability, reactivity, corrosivity, and toxicity. The RCRA regulations set standards for secure landfills and treatment processes for disposal of hazardous waste.

Much work has been done in investigating and cleaning up sites that have been contaminated by hazardous wastes in the past. In some cases, funding is available for cleanup of such sites through the Comprehensive Environmental Response, Compensation, and Liability Act of 1980 (known as CERCLA or Superfund) or its amendment, the Superfund Amendments and Reauthorization Act (SARA) of 1986.

IMPACT ON INDUSTRY

Increased interest in environmental issues in the last quarter of the twentieth century has made environmental engineering more prominent. The U.S. Bureau of Labor Statistics shows environmental engineering as the eighth largest field of engineering, with an estimated 54,300 environmental engineers employed in the United States in 2008. The bureau projects a 31 percent rate of growth in environmental engineering employment through much of the 2010's, which is much higher than the average for all occupations. Environmental engineers are employed by consulting engineering firms, industry, universities, and federal, state, and local government agencies.

Consulting Engineering Firms. Slightly more than half of all environmental engineers in the United States are employed by firms that engage in consulting in architecture, engineering, management, and scientific and technical issues. Some engineering consulting companies specialize in environmental projects, while others have an environmental division or simply have some environmental engineers on staff. The U.S. environmental consulting industry is made up of about 8,000 companies, ranging in size from one-person shops to huge multinational corporations. Some of the largest environmental consulting firms are CH2M Hill, Veolia Environmental Services North America, and Tetra Tech. Two engineering and construction firms with large environmental divisions are Bechtel and URS.

Government Agencies. Environmental engineers are employed by government agencies at the local, state, and federal levels. The U.S. Environmental Protection Agency and state environmental agencies employ the most environmental engineers, but many other government agencies, such as the Army Corps of Engineers, the Bureau of Reclamation, the Department of Agriculture, the Department of Defense, the Federal Emergency Management Agency, and the Natural Resources Conservation Service also have environmental engineers on their staffs. At the local government level, environmental engineers are used by city and county governments, in city or county engineering offices, and in public works departments.

University Research and Teaching. Colleges and universities employ environmental engineers to teach environmental engineering and the environmental component of civil engineering programs. Civil engineering is one of the largest engineering specialties, and its former subspecialty, environmental engineering, is taught at numerous colleges and universities around the world.

Industry. A small percentage of environmental engineers work in industry, at companies such as 3M, Abbott Laboratories, BASF, Bristol-Myers Squibb, Chevron, the Dow Chemical Company, DuPont, and IBM.

CAREERS AND COURSE WORK

An entry-level environmental engineering position can be obtained with a bachelor's degree in environmental engineering or in civil or chemical engineering with an environmental specialization. However, because many positions require registration as an engineer in training or as a professional engineer, it is important that the bachelor's degree program is accredited by the Accreditation Board for Engineering and Technology (ABET). Students must first graduate from an accredited program before taking the exam to become a registered engineer in training. After four years of experience, the engineer in training can take another exam for registration as a professional engineer.

A typical program of study for an environmental engineering degree at the undergraduate level includes the chemistry, calculus-based physics, and mathematics that is typical of almost all engineering programs in the first two years of study. It also may include biology, additional chemistry, and engineering geology. The last two years of study will typically include hydrology, soil mechanics, an introductory course in environmental engineering, and courses in specialized areas of environmental engineering such as water treatment, wastewater treatment, air-pollution control, and solid- and hazardous-waste management.

Master's degree programs in environmental engineering fall into two categories: those designed primarily for people with an undergraduate degree in environmental engineering and those for people with an undergraduate degree in another type of engineering. Some environmental engineering positions require a master's degree. A doctoral degree in environmental engineering is necessary for a position in research or teaching at a college or university.

SOCIAL CONTEXT AND FUTURE PROSPECTS

Many major areas of concern in the United States and around the world are related to the environment. Issues such as water-pollution control, air-pollution control, global warming, and climate change all need the work of environmental engineers. These issues, as well as the need for environmental engineers, are likely to remain concerns for much of the twenty-first century.

Water supply, wastewater treatment, and solid-waste management all involve infrastructure, needing repair, maintenance, and upgrading, which are all likely to need the help of environmental engineers.

Harlan H. Bengtson, B.S., M.S., Ph.D.

FURTHER READING

Anderson, William C. "A History of Environmental Engineering in the United States." In *Environmental and Water Resources History*, edited by Jerry R. Rogers and Augustine J. Fredrich. Reston, Va.: American Society of Civil Engineers, 2003. Describes the development of environmental engineering in the United States, starting in the 1830's, through the growth of environmental awareness in the 1970's, and into the twenty-first century. Identifies and discusses significant pioneers in the field.

Davis, Mackenzie L., and Susan J. Masten. *Principles of Environmental Engineering and Science.* 2d ed. Boston: McGraw-Hill Higher Education, 2009. An introduction to the field of environmental engineering. Includes illustrations and maps.

Juuti, Petri S., Tapio S. Katko, and Heikki Vuorinen. *Environmental History of Water: Global Views on Community Water Supply and Sanitation.* London: IWA, 2007. Provides information on the history of the water supply and sanitation around the world.

Leonard, Kathleen M. "Brief History of Environmental Engineering: 'The World's Second Oldest Profession.'" *ASCE Conference Proceedings* 265, no. 47 (2001): 389-393. Describes the evolution of environmental engineering from its earliest beginnings.

Spellman, Frank R., and Nancy E. Whiting. *Environmental Science and Technology: Concepts and Applications.* 2d ed. Lanham, Md.: Government Institutes, 2006. Provides background basic science and engineering science information as well as an introduction to the different areas of environmental engineering.

Vesilind, P. Aarne, Susan M. Morgan, and Lauren G. Heine. *Introduction to Environmental Engineering.* 3d ed. Stamford, Conn.: Cengage Learning, 2010. A holistic approach to solving environmental problems with two unifying themes—material balances and environmental ethics.

WEB SITES

American Academy of Environmental Engineers
http://www.aaee.net

American Society of Civil Engineers
http://www.asce.org

U.S. Environmental Protection Agency
http://www.epa.gov

See also: Civil Engineering; Hazardous-Waste Disposal; Industrial Pollution Control; Landscape Architecture and Engineering; Landscape Ecology; Sanitary Engineering; Sewage Engineering; Water-Pollution Control.

ENVIRONMENTAL MICROBIOLOGY

FIELDS OF STUDY

Environmental microbiology; molecular biology; environmental science; biology; chemistry; biotechnology; botany; entomology; plant pathology; biochemistry; genetics; toxicology; environmental monitoring; chemical and biomolecular engineering; microbial ecology; soil science; biogeochemistry; modeling.

SUMMARY

The majority of the Earth's biomass consists of microorganisms, and everything consumed and created by microorganisms has a very significant impact on the surrounding environment and all other living organisms. Environmental microbiology focuses on the role of these microorganisms in both causing environmental deterioration and rectifying ecological degradation, while also considering microbial ecology in both natural and polluted environments. Environmental microbiology technology harnesses the natural ability of microorganisms to remove pollutants, such as crude oil and industrial waste, from the environment.

KEY TERMS AND CONCEPTS

- **Bioassessment:** Measurement of living organisms (their presence, condition, and number) as determination of water quality.
- **Biocriteria:** Narrative or numerical standards assessing the qualities required to support a desired condition, particularly in a body of water; also known as biological criteria.
- **Bioindicator:** Bacterium used to detect and respond to surrounding environmental conditions.
- **Bioremediation:** Use of microorganisms, such as fungi and bacteria, and their enzymes to return a contaminated environment to its original condition.
- **Biotechnology:** Exploitation and manipulation of living organisms to produce valuable and useful products or results.
- **Culture:** Artificial cultivation of microorganisms, usually under laboratory conditions.

- **Fossil Fuel:** Deposit within the Earth's crust of either solid (coal), liquid (oil), or gaseous (natural gas) hydrocarbons produced through the natural decomposition of organic material (plants and animals) over many millions of years.
- **Genomic Analysis:** Examination of an organism's complete DNA.
- **Metagenome Technology:** Field of microbiology that applies genomic analysis to entire microorganism communities, thus creating a culture-independent approach that avoids the need to isolate and culture individual microbes; also known as metagenomics.

DEFINITION AND BASIC PRINCIPLES

Environmental microbiology is the study of microbial community structure and physiology within the environment, whether soil, water, or air. At a fundamental level, environmental microbiology is the study of the role of microorganisms as both the cause and the remediation of environmental pollution and ecological degradation.

Environmental microbiology is a multidisciplinary science, blending environmental science and the biology of microscopic organisms. It involves the development, application, adaptation, and management of biological systems and microorganisms to repair and prevent environmental damage caused by pollution. This discipline investigates how microorganisms can be both detrimental and beneficial to the environment and human society, particularly in relation to the restoration and remediation of the world's natural environment (air, water, and soil). The field has benefited from advancements in genetic engineering and modern microbiological concepts, which have encouraged the development of traditional and innovative solutions to environmental contamination.

BACKGROUND AND HISTORY

Any discussion of environmental microbiology must touch on the history of microbial discovery. Although the technology involved in environmental microbiology is evolving and expanding, the first steps toward the establishment of this field began hundreds of years ago with the discovery of microorganisms. In the mid-seventeenth century, both Antoni van Leeuwenhoek and Robert Hooke observed nonliving and living microorganisms under self-made microscopes, thereby discovering the reason behind one of nature's conundrums—the decomposition of plants and animals. Although the identification of microbes (living things not visible to the naked eye) in 1675 was instrumental in advancing biological science, the true significance of the role of microbes in disease was not understood until some two hundred years later.

By the late nineteenth century, research conducted by Dutch microbiologist Martinus W. Beijerinck and Russian microbiologist Sergei Winogradsky had greatly advanced the understanding of microorganisms and their possible role in ecological processes. Beijerinck developed a method for isolating microorganisms called the enrichment culture technique, and Winogradsky enhanced scientific awareness of microbial diversity and discovered the autotrophic (self-feeding) ability of bacteria. He went on to develop the Winogradsky column culture technique, which showed that water in a lake contains a number of organisms performing various processes at different levels. The discipline of environmental microbiology was made possible by the work of these early scientists. Although traditional microbe research focused on the role of microbes in causing disease, Beijerinck and Winogradsky began to examine their role in disease prevention.

The roots of environmental microbiology also lie in urban waste management and treatment. The field originally focused on monitoring the movement of pathogens and treating them within natural and urban environments to protect municipal water quality and public health. As the world became more urbanized in the late nineteenth century, the incidence of communicable diseases such as typhoid fever and cholera increased. To combat the spread of diseases, cities and communities began to treat water with various filtration and disinfectant methods. For the most part, such approaches to water treatment were instrumental in the elimination of waterborne bacterial diseases in developed countries, and disinfection processes continue to be widely used.

As research continued into the 1960's, however, it became apparent that the viruses and protozoa parasites that caused waterborne diseases were much more resistant to the process of disinfection than were bacteria. Already treated urban water supplies were still plagued by uncontrolled outbreaks of giardia (which causes giardiasis), cryptosporidium (which causes cryptosporidiosis), and norovirus (which causes gastroenteritis). Although serious outbreaks within the developed world are rare, the continued occurrence of water pathogens has meant that the field of environmental microbiology still has a very strong focus on water quality and the treatment and control of water pathogens. Perhaps of most concern is the fact that some 10 to 50 percent of diarrheal illness is caused by waterborne microbial organisms not yet identified by science.

Before long, the effects of water quality on human health were not the only area of concern for environmental microbiology. By the 1960's, concern over the effects of chemicals in the natural environment had increased, and it became obvious that poor human health was not the only issue. Chemicals in the soil and water found their way into the human food chain through groundwater use and consumption, affecting not only human health but also animal and plant species using those same soil and water supplies. Significant chemical contamination, such as the massive *Exxon Valdez* oil spill in 1989, highlighted the need to investigate the potential for microorganisms in bioremediation. In the twenty-first century, the use of microorganisms in the treatment and control of pathogens and in bioremediation processes is a key feature of environmental microbiology.

HOW IT WORKS

Microorganisms are found in all areas of the biosphere, even environments of extreme temperatures, acidity, salinity, and darkness that are inhospitable to most other organisms. They account for the vast majority of the Earth's biomass. The entire number of microorganisms living on Earth is almost immeasurable, but it is estimated that more than 1 billion microorganisms live in just 1 gram of soil. Microorganisms are vital to the health and function of the planet and are responsible for a vast number of natural life processes. They are the most significant players involved in the synthesis and degradation of important molecules. They provide energy to other organisms, are responsible for most of the planet's photosynthesis, are able to fix nitrogen and recycle nutrients in ecosystems, and are also capable of causing lethal diseases for humans, flora, and fauna.

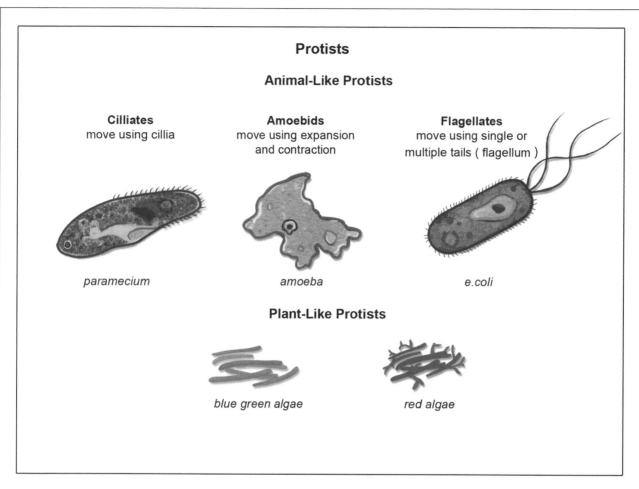

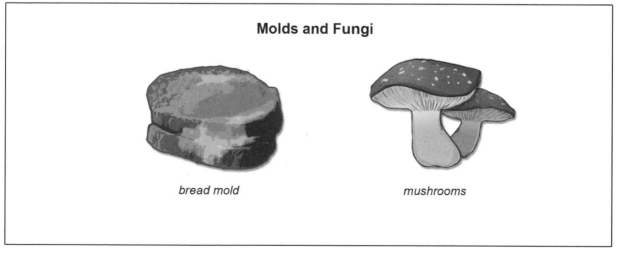

The microbial interactions of microorganisms such as protists and fungi are a primary focus of microbiologists.

Fundamentally, environmental microbiology is about harnessing the natural ability of microbes and their relationships with other microbes to address environmental issues and solve environmental problems. The main focus of environmental microbiology remains the use of microorganisms in the treatment and control of pathogens and in bioremediation processes. Modern-day technological advancements in genetic engineering and biotechnology have, however, greatly increased the applications of this scientific field. The increased applications for environmental microbiology are important in view of the ability of pathogens such as bacteria and viruses to evolve and emerge rapidly in an environment under increasing stress from the human population.

APPLICATIONS AND PRODUCTS

The application of genomics and molecular biology to environmental microbiology has allowed scientists to discover the vast diversity and complexity that exists in natural microorganism communities. Despite advanced human-developed technology, however, it is estimated that less than 2 percent of microorganisms have actually been described by science or have been grown in laboratory conditions. This means that scientists not only are unsure of all the possible effects of such organisms on human health and the environment but also have not even begun to understand or determine the many potential applications and products of such organisms.

New discoveries and principles of environmental microbiology can have vast potential in other areas of science such as biotechnology, pharmaceutics, biomedical research and engineering, the chemical and textile industries, bioremediation and wastewater treatment, pathogen control, mineral recovery, sustainable agriculture, and resource conservation. However, the majority of research in the applications for bioremediation, water quality, and the biotreatment of waste material and wastewater.

Bioremediation. Human society is increasingly exploiting the flexible appetite of microbes (particularly bacteria) to remediate environments containing contaminants such as industrial waste, crude oil, and polychlorinated biphenyls (PCBs). Bioremediation is usually classified as either in situ or ex situ and is defined as the use of microorganisms, such as fungi and bacteria, and their enzymes to return a contaminated environment to its original condition. In situ bioremediation entails treating the contamination in place and relies on the ability of the microorganisms to metabolize or remove the contaminants inside a naturally occurring system. Ex situ remediation involves removing the polluted material from the site and treating it elsewhere, and it relies on some form of artificial engineering and input. Although the process of bioremediation can occur naturally, it can be promoted through artificial human stimulus through the use of environmental microbiology technology.

For example, environmental microbiology technology is used in the remediation of crude oil spills in coastal areas. Although oil is a naturally occurring fossil fuel, it can cause major ecological damage when accidentally introduced into marine (and terrestrial) environments during oil spills. Large spills can be difficult to contain and collect, and the longer a spill remains in the environment, the greater the potential damage. Environmental microbiology techniques, specifically hydrocarbonoclastic bacteria (HCB), can be used to clean up oil spills. These bacteria are able to degrade hydrocarbons (in this case oil) and are therefore beneficial as bioremediators.

Waste Biotreatment and Water Quality Treatment. Biotreatment processes involve the environmentally friendly treatment of waste material, including wastewater, using living microorganisms. Scientific research has demonstrated that while wastewater processes are one of the most important functions of environmental microbiology and use a significant diversity of microorganisms, many of these microorganisms are yet to be classified or cultured. This indicates that the potential of microbes in biotreatment is not yet fully understood. Research on metagenome technology (metagenomics), however, is highlighting the diversity, structure, and functions of microorganisms. This research has looked at the role of microbes in processes such as nitrogen cycling, anaerobic ammonium oxidation, and methane fermentation in the treatment of wastewater, with a particular focus on the use of bacterial biofilms and bioreactors.

The World Health Organization (WHO) reported that 10 percent of the global health burden is related to poor-quality water and could be avoided. One of the most important applications of environmental microbiology, therefore, is in controlling waterborne pathogens and diseases. The treatment of drinking water is, of course, not new. Water disinfection processes were key in the fight against bacterial waterborne pathogens. The introduction of environmental microbiology techniques, however, has assisted in controlling pathogens and contaminants other than bacteria. In particular, microorganisms are used to decompose contaminants such as organic matter, nitrates, and phosphate in wastewater and to improve water quality. The removal of organic material is done through both bacterial aerobic and anaerobic decomposition, but the removal of ammonium and nitrates is much more complex. It requires both aerobic and anaerobic conversion to remove the pollutants and involves the bacterial conversion of ammonia to nitrite and nitrate to nitrogen gas, which can be safely released into the atmosphere.

Bioassessment and Bioindicators. In environmental microbiology, bioindicators are bacteria that are used to detect and respond to surrounding environmental conditions. Bioindicators can be any organism and are used in numerous industries and fields of science, including environmental microbiology. Bacteria, however, are considered to be superior precursors of human-caused environmental degradation, as they possess the highest surface area to volume ratio of all organisms.

IMPACT ON INDUSTRY

Major Organizations. The American Society for Microbiology is the world's largest professional life science organization. Although the society covers all areas of microbiology and releases numerous publications, including *Applied and Environmental Microbiology*, it is focused on advancing "microbiological sciences as a vehicle for understanding life processes and to apply and communicate this knowledge for the improvement of health and environmental and economic well-being worldwide." The International Society for Microbial Ecology is also one of the principal scientific societies involved in the emerging field of microbial ecology and its related disciplines. Organizations such as the World Health Organization are also concerned with environmental microbiology, particularly as it relates to environmental water quality and human health.

Government and University Research. A great many universities around the world are involved in environmental microbiology either directly or indirectly through biotechnology and microbial ecology. Significant research into environmental microbiology has been undertaken by universities such as Cornell and Stanford in the United States, the University of Ottawa in Canada, the University of Oxford and University of Birmingham in the United Kingdom, the University of New South Wales in Australia, and Peking University in China.

Many of the world's governments have been pursuing and implementing environmental microbiology research in an attempt to rectify and control environmental pollution and wastewater management. For example, the U.S. Environmental Protection Agency, particularly through the Microbiological and Chemical Exposure Assessment Research Division, has promoted significant research and remediation of environmental contaminants. Canada's National Research Council (NRC) has researched and developed environmental genomics, bioremediation of contaminated soil, and molecular techniques

---

Fascinating Facts About Environmental Microbiology

- Scientists with the U.S. Geological Survey added nutrients to contaminated soil to stimulate natural microbes that were actively consuming fuel-derived toxins in Hanahan, South Carolina, in 1992. After a year, contamination in residential areas was reduced by 75 percent.

- In 2010, the United Nations reported that worldwide, annually, 2 million tons of industrial and agricultural waste and sewage are discharged into waterways and more than 1.8 million children under the age of five die from water-related diseases.

- According to estimates, nearly 99.8 percent—or less than 1 percent—of microorganisms in the environment are nonculturable or not readily culturable.

- Deep in the Black Sea, scientists from the Max Planck Institute for Marine Microbiology have found bacteria that thrive on methane.

- Microorganisms exist in the human mouth. For instance, *Actinomyces naeslundii* is a bacterium that is associated with good oral hygiene.

- In the 1960's, Thomas Brock found *Thermus acquaticus*, an archaea organism, in the hot springs at Yellowstone National Park. The microbe's polymerase enzyme can duplicate DNA at high temperatures and has been used in genetic studies.

to assess environmental impacts and determine the diversity of microorganism species. Australia's leading scientific research center, the Commonwealth Scientific and Industrial Research Organization, has investigated bioremediation of contaminated sites, molecular detection and monitoring of microbial communities in natural and engineered environments, and the isolation and description of novel microorganisms. Additionally, the Chinese Academy of Sciences, China's prominent academic institution and broad research and development center in natural and technological science innovation, has undertaken significant research in the field of environmental microbiology, particularly in relation to water quality and bioremediation.

CAREERS AND COURSE WORK

Students who wish to pursue a career in environmental microbiology can come from a diverse number of fields but must have a strong grounding in basic microbiology, molecular biology, and environmental science. A solid grasp of genetic engineering technology and genetically modified organisms is also helpful.

Many universities provide undergraduate and postgraduate degrees in environmental microbiology. Upon course completion, students should have a solid understanding of the interactions between microorganisms in both natural and artificial environments, particularly contaminated aquatic and terrestrial ecosystems. Most courses focus teaching and research on both fundamental and applied features of biodegradation, bioremediation strategies, and the use of biosensors for toxicity assessment.

Graduates with environmental microbiology degrees who have done postgraduate research can enter careers in environmental and microbiology consulting, in water treatment and resource management, and bioremediation and bioassessment in the private sector, specialized government organizations and agencies, and universities and institutions undertaking teaching and research.

SOCIAL CONTEXT AND FUTURE PROSPECTS

The growing global population and its impact on the environment, particularly the supply of fresh water, make water treatment and bioremediation increasingly important. Therefore, future research will continue to focus on these aspects and applications of environmental microbiology. Less than 2 percent of the world's microbial species have been classified and cultured, which means there is significant unrealized potential for human uses of microorganisms. Many believe, however, that the emergence of metagenome technology holds the key to rapid advancement in the field of microbiology and in human understanding of life. Heralded as the most important advancement since the invention of the microscope, metagenomics applies genomic analysis (the examination of an organism's complete DNA) to entire microorganism communities, thereby avoiding the need to isolate and culture individual microbes.

Research has begun on applications of environmental microbiology to remove heavy metal pollution, destroy specific xenobiotics (foreign chemicals found in a living organism), treat water and air pollution caused by carbon dioxide and sulfur dioxide, and develop biodegradable plastics and other useful material.

Christine Watts, Ph.D., B.App.Sc., B.Sc.

FURTHER READING

Cummings, Stephen, ed. *Bioremediation: Methods and Protocols.* New York: Humana Press, 2010. Experts in the fields of environmental biotechnology and microbiology examine innovative and imaginative bioremediation techniques in pollution removal.

Hurst, Christon J., et al., eds. *Manual of Environmental Microbiology.* 3d ed. Washington, D.C.: ASM Press, 2007. A detailed examination of the role microbes play in planetary environments, with a focus on the basic principles of environmental microbiology, general methodologies, detection and impact of microbial activity within the environment, and detection and control of pathogens.

Illman, Walter, and Pedro Alvarez. "Performance Assessment of Bioremediation and Natural Attenuation." *Critical Reviews in Environmental Science and Technology* 39, no. 4 (April, 2009): 209-270. Reviews state-of-art performance assessment methods and discusses future research directions in bioremediation, natural attenuation, chemical fingerprinting, and molecular biological tools.

Liu, Wen-Tso, and Janet K. Jansson, eds. *Environmental Molecular Microbiology.* Norfolk, England: Caister Academic Press, 2010. Highlights the concepts and technology of environmental molecular microbiology, with contributions from international experts describing various technologies and their applications in environmental microbiology.

Maier, Raina M., Ian L. Pepper, and Charles P. Gerba, eds. *Environmental Microbiology*. 2d ed. Boston: Elsevier/Academic Press, 2009. Examines the fundamental concepts and principles related to environmental microbiology, with the addition of case studies to highlight relevant issues and solutions.

Mitchell, Ralph, and Ji-Dong Gu, eds. *Environmental Microbiology*. 2d ed. Hoboken, N.J.: Wiley-Blackwell, 2010. This revision of one of the most successful books on environmental microbiology takes a comprehensive look at the role of microbiological processes related to environmental deterioration, with a focus on the detection and control of environmental contaminants.

Rochelle, Paul A., ed. *Environmental Molecular Microbiology: Protocols and Applications*. New York: Springer-Verlag, 2001. This book provides a comprehensive collection of laboratory protocols, techniques, and applications in the field of environmental microbiology.

WEB SITES

American Society of Microbiology
http://www.asm.org

International Society for Microbial Ecology
http://www.isme-microbes.org

U.S. Environmental Protection Agency
EPA Microbiology
http://www.epa.gov/microbes

See also: Air-Quality Monitoring; Ecological Engineering; Environmental Chemistry; Environmental Engineering; Hazardous-Waste Disposal; Sewage Engineering; Soil Science; Thermal Pollution Control; Water-Pollution Control.

ENZYME ENGINEERING

FIELDS OF STUDY

Biology; biochemistry; biotechnology; chemistry; food science; genetics; medicine; microbiology; pharmacology.

SUMMARY

Catalysts accelerate the rate of chemical reactions without being essentially changed, and enzymes are biological catalysts that accelerate the rate of reactions that occur in living systems. Enzyme engineering identifies enzymes that have potentially useful catalytic activities and chemically or structurally modifies them to increase their activity, change their substrate specificity, change the types of reactions they catalyze, or change the properties of enzymes and the manner in which they are regulated. Engineered enzymes can generate completely novel molecules or new, improved ways to synthesize useful molecules.

KEY TERMS AND CONCEPTS

- **Active Site:** Pocket or cleft in an enzyme surrounded by amino acids that specifically bind the substrate and catalyze its conversion to the product.
- **Amino Acids:** Chemical building blocks of proteins that contain a nitrogen-containing amino group ($-NH_2$), a carboxylic acid (-COOH), and a functional group that varies between amino acids and determines their chemical properties.
- **Catalyst:** Substance that increases the rate of a reaction without being consumed or permanently changed in the process.
- **Cofactors:** Non-amino acid molecules that are associated with enzymes and necessary for enzymatic function.
- **Enzyme:** Proteins, groups of proteins, or ribonucleic acids (RNAs) that are produced by living organisms and catalyze biochemical reactions.
- **Protease:** Enzyme that degrades proteins.
- **Substrate:** Molecule or molecules physically engaged by an enzyme to accelerate the chemical reaction that consists of the conversion of the substrate or substrates to the product or products.

- **Transition State:** Intermediate chemical structure formed during the process of a chemical reaction that represents the highest energy state of the reaction and usually degenerates to form the product.

DEFINITION AND BASIC PRINCIPLES

Enzymes are widely used as catalysts in several industrial ventures, ranging from food to synthetic chemistry to many other industrial processes. However, enzymes often show insufficient substrate selectivity, poor stability, and catalytic activities that are not robust enough for industrial use. To remedy this shortcoming, enzyme engineering builds new enzymes or modifies existing enzymes to give them novel, useful properties, or the ability to catalyze valuable chemical reactions. Engineered enzymes come in several forms. Semisynthetic enzymes (synzymes) have specific amino acids that have been chemically modified. These modifications can significantly alter the activity, specificity, or properties of the enzyme. Directed evolution subjects the gene that encodes the enzyme to multiple rounds of mutation. The variant enzymes generated by these mutagenic genes are then sieved by some kind of selection scheme that identifies those mutant forms that display the desired characteristics or activities.

A cheaper strategy is rational design. Rational design uses detailed knowledge of the structure of proteins to identify those regions that are essential for its function and properties. By changing only those amino acids thought to be necessary for the modification of that function, the enzyme is potentially tailored for a new function with little investment.

Enzyme engineers use catalytic antibodies or abzymes. Antibodies are Y-shaped proteins made by vertebrate immune systems that bind to specific chemicals. Abzymes bind to chemicals and force them into the transition state of a chemical reaction, which accelerates the formation of the product from the reactants.

BACKGROUND AND HISTORY

Enzyme engineering arose only after advances in several other fields made it possible to determine the primary amino acid sequences and three-dimensional structure of enzymes and directly manipulate them at the molecular level. Swedish biochemist Pehr Victor Edman gave birth to protein sequencing in 1950, when he designed the Edman degradation reactions that can determine the primary amino acid sequence of proteins. In 1958, English biochemist John Cowdery Kendrew used X-ray crystallography to solve the three-dimensional structure of the muscle oxygen-storing protein myoglobin. In the 1970's, American biochemist Herbert Wayne Boyer and American geneticist Stanley Norman Cohen pioneered molecular cloning techniques that gave scientists the means to clone genes and insert them into bacteria for propagation.

The first studies in enzyme engineering examined the effects of mutations on enzyme active sites. Beta-lactamase, the enzyme used by bacteria to degrade beta-lactam antibiotic (penicillin, ampicillin, and amoxicillin) was one of the first enzymes examined

Chemical engineer adding enzymes into a bioreactor. (Science Source)

by enzyme engineering. In 1978, Canadian chemist Michael Smith and his colleagues invented site-directed mutagenesis, which gave biochemists a much better way to place targeted mutations into the genes that encode enzymes and thereby change their primary amino acid sequence. In 1986, the laboratories of Peter Schultz (University of California, Berkeley) and Richard Lerner (Research Institute of Scripps Clinic) made the first catalytic antibodies that could split ester bonds.

How It Works

Semisynthetic Enzymes. Enzymes that are modified by chemical means are known as semisynthetic enzymes. There are two main ways to produce semisynthetic enzymes: atom replacement or group attachment.

Atom replacement exchanges one atom within an enzyme for a different atom. Such replacements can modify enzyme activity or change the substrate specificity of the enzyme. Group attachment involves the use of particular chemical reagents to attach particular molecules to enzymes. Attaching additional molecules to enzymes can also markedly change enzyme activity and substrate specificity.

Directed Evolution. Directed evolution randomly changes amino acids in a protein without prior knowledge of the exact function of each amino acid. The first step, diversification, takes the gene that encodes the enzyme of interest and replicates it many times while using a copying machinery that is inherently error-prone. This introduces random mutations into the gene and creates a large collect of gene variants that are usually grown in bacteria. The second step, selection, tests or screens these enzyme variants for a desired property. Once the desired variants are identified, they undergo the third step, amplification, which replicates the identified variants and sequences them in order to determine which mutations produced the desired properties. Collectively, these three steps constitute one round of directed evolution, and the vast majority of such experiments require multiple rounds. The goal is to find those variant enzymes that show the most desired characteristics to the greatest extent. Directed-evolution studies suffer from the need to make huge numbers of mutants that produce no discernable effect, since up to 90 percent of all mutants made are uninformative.

Semirational Design. This enzyme engineering strategy employs sophisticated computer programs that assemble all the available structural information of the enzyme under study and predict how the mutations introduced into different locations within the enzyme might affect its activity. The enzyme engineer then notes the predicted changes that will potentially generate the desired property changes and uses this information to conduct targeted mutagenesis experiments. Targeted mutagenesis experiments introduce mutations into specific locations of a protein. Once these mutations are made, the variant enzyme with the engineered changes is tested to determine if it has the specific properties the enzyme engineer was hoping to produce in the enzyme. These approaches combine structural information with rational design. Two computer programs that make such predictions include Protein Sequence-Activity Relationship or ProSAR and Combinatorial Active-Site Saturation Test, otherwise known as CASTing.

Rational Design. If a great deal of structural information about the enzyme in question is available, then that structural information informs which amino acids should be changed. Many rational design attempts have not succeeded because of uncertainties regarding protein structure.

De Novo Design. A computer builds an enzyme around the transition state of a reaction from scratch. The computer begins by designing the active site by placing specific amino acids in strategic positions so that they efficiently bind the transition state of the chemical reaction and stabilize it. The program then constructs a protein backbone that supports and properly positions the active-site amino acids but still provides a coherent protein structure that is predictably stable under the desired conditions.

This particular strategy suffers from gaps in the ability to predict protein structure accurately and correlate this ideal structure with enzymatic activity. For example, two enzymes (retro-aldol enzyme and a Kemp elimination catalyst) were built completely from scratch by using computer programs. However, both enzymes required further optimization by directed evolution to achieve maximum activity.

Catalytic Antibodies. The immune system of some vertebrates makes Y-shaped proteins that specifically bind to and neutralize foreign substances that invade the body. Immunizing laboratory animals with stable analogues of the transition states

of various reactions directs the immune systems of those animals to synthesize antibodies that cannot only bind particular chemical reactants but force them into the transition state of the reaction, which subsequently forms the product.

APPLICATIONS AND PRODUCTS

Pharmaceutical Production. Beta-lactam and cephalosporin antibiotics are commonly prescribed to combat various illnesses. Both of these drugs kill bacteria by inhibiting the synthesis of the bacterial cell wall. Beta-lactam antibiotics include such widely recognized drugs as penicillin, ampicillin, and amoxicillin, whereas cephalosporin antibiotics include such popularly used antibiotics as Ceftin (cefuroxime), Kephlex (cephalexin), and Ceclor (cefaclor). Unfortunately, with repeated use, bacteria can become resistant to commonly used antibiotics, and making new, improved antibiotics is essential to treat some of the more recent and aggressive infectious diseases. To make new cephalosporin antibiotics, enzyme engineers have used enzymes called acylases to convert simple starting chemicals into various versions of these drugs. By engineering these acylase enzymes, pharmaceutical companies have been able to make new cephalosporin and beta-lactam antibiotics that have novel properties and can kill bacteria that are resistant to older drugs.

Enzymes as Medicines. When a person is cut, blood oozes from the damaged tissue. Fortunately, blood clotting (also known as coagulation) eventually stanches this blood flow. Blood clotting is an essential part of wound healing, but it is also a very highly regulated event. The formation of blood clots inside undamaged blood vessels clogs those vessels and leads to heart attacks if clots form inside the vessels that surround the heart, or a stroke, if they occur within vessels that surround the brain. The human body has ways to destroy unnecessary clots. An enzyme called tissue plasminogen activating factor (TPA) activates other enzymes in the body that degrade harmful clots. Commercially available, native TPA is called Alteplase, which has a half-life in the bloodstream of four to six minutes. Engineered forms of TPA are also clinically available. Reteplase, a shortened version of TPA (consists of 357 of the 527 amino acids of Alteplase), has a longer half-life (thirteen to sixteen minutes). Tenecteplase, which has two amino acid changes (substitutes asparagine[103] with a threonine and asparagine[114] with glutamine), has an even longer half-life of twenty to twenty-four minutes.

Engineered enzymes are also used in enzyme-replacement therapies. Several genetic diseases, known as lysosomal storage diseases, result from the inability to make functional versions of enzymes that degrade various biological molecules. The accumulation of these molecules kills brain cells and causes the death of the patient. Engineered enzymes used in enzyme-replacement therapies include Cerezyme (imiglucerase, used to treat Gaucher's disease), Naglazyme (galsulfase, used to treat mucopolysaccharidosis VI), Myozyme (alglucosidase alfa, used to treat Pompe disease), and Aldurazyme (laronidase, used to treat mucopolysaccharidosis I).

Enzyme Immobilization. By attaching enzymes to surfaces, embedding them in gel matrices, hollow fibers, or cross-linking them to each other, enzymes are immobilized on insoluble surfaces. This increases their stability, simplifies their recycling, and increases the tolerance of enzymes to high levels of substrate and products. Detergent enzyme preparations, such as Alcalase, immobilize the protease subtilisin by attaching it to insoluble particles. Attaching the enzyme to inert material increases its reuse as it degrades proteinaceous matter.

Making Enzymes Soluble in Organic Solvents. Enzymes usually work in water, but many reactions between organic chemicals occur in organic solvents. Although Russian chemist Alexander Klibanov showed that several enzymes are active in organic solvents, many enzymes are neither soluble in organic solvents nor work properly in such environments. Attaching a molecule called polyethylene glycol (PEG) to some enzymes makes them soluble and active in organic solvents and allows them to make things such as polyester, peptides (small proteins), esters (sweet-smelling things found in foods), and amides (nitrogen-containing compounds). Such modified enzymes also have clinical uses. For example, the enzyme asparaginase can kill cancer cells but is toxic, unstable, and some patients have severe allergies to it. PEG-treated asparaginase is not as toxic as the native enzyme, is much more stable, and does not cause allergy. PEG-asparaginase is used to treat tumors in humans.

Abzymes. A notable variety of reactions are catalyzed by catalytic antibodies that range from forming or breaking carbon-carbon bonds, rearrangements, hydrolysis of various bonds, transfer of chemical groups, and even an industrial reaction called the Diels-Alder reaction. However, abzymes are very expensive and tedious to make, and their catalytic activity is well below that of enzymes. Yet they do provide tailor-made catalysts when no other such reagent exists.

IMPACT ON INDUSTRY

Government and University Research. Regulation of research in academic venues that use genetically engineered organisms is specified by the National Institutes of Health (NIH) Guidelines for Research Involving Recombinant DNA Molecule. The Recombinant DNA Advisory Committee (RAC) is responsible for executing the regulations in the NIH guidelines. These guidelines typically regulate all work on pathogenic (disease-causing) organisms, prohibit genetic engineering of human beings without specific approvals for gene-therapy trials, and also prohibit the release of genetically engineered organisms into the environment without the express approval of government agencies such as the Environmental Protection Agency (EPA). However, research on specific enzymes receives little mention in these guidelines. Only the organisms that house the recombinant enzymes are regulated. Because the NIH is the largest funding agency in the United States, it sets the rules that are followed by other research-funding organizations.

Industry and Business. Any enzyme used to produce products that are sold as food or drugs comes under the regulatory auspices of the Food and Drug Administration (FDA). However, since enzymes do not usually end up in the finished product, manufacturers are not required to list them on product labels. Furthermore, if the enzyme is made by a genetically engineered organism, or if the enzyme itself is the result of enzyme engineering, the company is not required by law to notify the FDA. Engineered enzymes are subject to little regulatory oversight, and companies are free to experiment at will, provided that such enzymes do not become a part of the finished product.

Major Companies. Enzyme engineering has yet to make a significant mark on industry. Nevertheless, there are some companies that have made and marketed engineered enzymes.

Fascinating Facts About Enzyme Engineering

- Hemophilia is a genetic disease characterized by an inability to form blood clots. Because blood clotting requires the coordinated and sequential activity of a host of blood-based enzymes called clotting factors, hemophilia patients are treated with infusions of purified clotting enzymes. Unfortunately, some patients form antibodies against clotting factors, which abrogates the efficacy of such treatments. However, enzyme engineers have made a truncated version of clotting factor VIII (B-domain deleted recombinant factor VIII or BDDrFVIII) that is smaller than the native enzyme but just as active. This protein is not recognized by the immune system nearly as often as the native clotting factor VIII.

- Another clotting factor called factor VII (FVII) can bypass the need for other clotting factors if it is present in high enough concentrations. FVII is made in an inactive form that is only activated when the front tip of the protein is removed. Enzyme engineers have made a form of FVII that does not require activation called recombinant FVIIa (rFVIIa; the "a" stands for active). Infusions of rFVIIa can help clot the blood of hemophilia patients who have antibodies against other clotting factors and can also successfully treat patients who suffer from uncontrolled

bleeding from trauma, surgery, anticoagulant drugs, or pregnancy.

- Metabolic engineering manipulates enzymes that direct metabolic pathways. Such manipulation can generate organisms that synthesize industrially useful compounds or degrade environmental pollutants. Animal feed production uses a microorganism called *Methylophilus methylotrophus* to convert methanol to animal protein, but by inserting genes from the common intestinal bacterial *Escherichia coli* into *M. methylotrophus*, this metabolically engineered organism can make much more protein from the same initial mass of methanol, thus increasing the overall efficiency of animal protein production and decreasing production costs.

- Antibody-directed enzyme prodrug therapy (ADEPT) uses an enzyme to a human antibody that tightly binds to a tumor-specific surface protein. When these antibodies bind to cancer cells, the enzyme converts an anticancer drug that is present in an inactive form (a prodrug) into a cancer-killing chemical. Because the prodrugs are activated only at the surface of the tumors, they cause few side effects and effectively kill tumors. As of 2011, ADEPT is in clinical trials.

The largest maker of enzymes for the food industry is Novo Nordisk, which makes its enzymes from genetically engineered microorganisms. Few of the enzymes made by this company are engineered, although Novo Nordisk also produces several different types of insulin, some of which are engineered. Enzyme engineering research is a part of the research and development at Novo Nordisk.

Pharmaceutical companies that make engineered enzymes usually have individual enzymes that they produce and market. As of 2011, there is no clear industrial leader in the production of engineered enzymes for clinical use. Enzon Pharmaceuticals makes PEG-conjugated asparaginase (Oncaspar), which was approved by the FDA in 2006 as a treatment for tumors. This is a prime example of an engineered enzyme that is being used in a clinical setting. Reteplase (retavase), an engineered version of TPA, is made by Hospira Incorporated but is marketed by EKR Therapeutics.

Other companies, such as Genentech and Pfizer, have conducted extensive research on enzymes in order to make drugs that are inhibitors of important enzymes for clinical use.

Because enzyme engineering is rather labor-intensive and requires a great deal of time to discover something commercially useful, there are few engineered enzymes on the market. However, given the interest in this field it is only a matter of time before engineered enzymes begin to appear in larger numbers and companies begin to appear that specialize in enzyme engineering.

CAREERS AND COURSE WORK

Anyone who wishes to study enzyme engineering must have a good understanding of general, organic, and physical chemistry and biochemistry. Knowledge of calculus is also essential as is mastery of computers, since many structural studies of enzymes use somewhat sophisticated computer programs. Enzyme engineers also must have some mastery of the tools of molecular biology, gene cloning, and microbiology. Enzyme engineers will also need graduate training, since many of the tools used in enzyme engineering are simply beyond the typical undergraduate laboratory curriculum. A bachelor's degree in chemistry or biochemistry is necessary to work as a technician in the enzyme engineering field. A master's degree will also be sufficient for a technician, but to run an enzyme engineering lab, a Ph.D. in chemistry or biochemistry is required.

Enzyme engineers work either in academia or industry. In academia, they will run their own lab and train graduate students. Most of the research in academic settings is not applied, but theoretical. Academic enzyme engineers usually try to develop new technologies for enzyme engineering, or use enzyme engineering techniques to study various enzymes. In industry, enzyme engineering research is much more applied, since the goal is to optimize an enzyme for a specific synthetic process that saves time, money, and resources.

Enzyme engineering is almost certainly the next frontier in biochemistry. Many industries make copious use of enzymes already, and the need to tailor these enzymes to fit the needs of industrial uses is pressing. Enzyme engineers need to be good collaborators, visionary, and very patient, since most experiments require extensive trials before something interesting is discovered.

SOCIAL CONTEXT AND FUTURE PROSPECTS

Because modified enzymes can make certain products more cheaply, the public response to modified enzymes is generally positive. However, the genetically modified organisms (GMOs) that are used to produce these enzymes give many people pause, since the introduction of GMOs into the environment may have long-term consequences that are presently unrecognized. Strict government regulation that forbids the release of GMOs into the environment without approval allays most of these concerns, but some people are still troubled by the use of GMOs to make products that they eventually end up eating or using in some other manner.

Enzyme engineering is one of the up-and-coming fields in chemistry and biochemistry. Since the 1990's, the use of enzymes in industrial and academic chemistry has greatly increased. There are many advantages to using enzymes in that they can act outside cells and under mild conditions that minimize troublesome side effects, are environmentally innocuous, compatible with other enzymes, and are very efficient, though highly selective catalysts. The largest drawback of using enzymes is that the right enzyme is sometimes not available to catalyze the desired reaction. Enzyme engineering can eliminate this significant drawback.

Furthermore, as biochemists achieve a more profound understanding of protein structure, cheaper and faster ways of doing enzyme engineering, such as rational design, become more successful and practical. This will shorten the time required for enzyme engineering experiments and reduce its cost. Companies are already looking intently at enzyme engineering as a significant investment for their research and development departments.

Michael A. Buratovich, B.S., M.A., Ph.D.

FURTHER READING

Arnold, Frances H., and George Georgiou, eds. *Directed Enzyme Evolution: Screening and Selection Methods.* Totowa, N.J.: Humana Press, 2010. Laboratory protocol book that describes, in great detail with figures and graphs, some rather ingenious techniques for screening mutant clones of enzyme genes.

_____. *Directed Evolution Library Creation: Methods and Protocols.* Totowa, N.J.: Humana Press, 2010. Encyclopedic collection of protocols for generating libraries of randomly mutagenic enzyme genes in bacteria, with tables, graphs, and some figures.

Faber, Kurt. *Biotransformations in Organic Chemistry: A Textbook.* 5th ed. New York: Springer-Verlag, 2004. A very clear, useful textbook on the uses of enzymes in chemistry that includes a chapter on engineered enzymes.

Park, Sheldon J., and Jennifer R. Cochran, eds. *Protein Engineering and Design.* Boca Raton, Fla.: CRC Press, 2010. Covers the broader field of protein engineering—methods of developing altered proteins for novel applications—in two sections: one on experimental protein engineering and the other on computational design. Includes discussion of enzyme engineering using both rational and combinatorial approaches.

Scheindlin, Stanley. "Clinical Enzymology: Enzymes As Medicine." *Molecular Interventions* 7, no. 1

(February, 2007): 4-8. An absorbing and readable summary of the use of engineered enzymes in clinical diagnoses and treatments.

WEB SITES
International Enzyme Engineering Symposium
http://www.enzymeengineering.ege.edu.tr

National Institutes of Health
Recombinant DNA Advisory Committee
http://oba.od.nih.gov/rdna_rac/rac_about.html

See also: Chemical Engineering; Food Science; Metabolic Engineering; Pharmacology.

EPIDEMIOLOGY

FIELDS OF STUDY

Immunology; internal medicine and other medical specialties; demography; public health; sociology; biochemistry; bioethics; environmental science; genetics; microbiology; molecular biology; toxicology.

SUMMARY

Epidemiology is the branch of medical science that studies the occurrence of disease in human populations: How many people in the population have the disease? Are there common factors that distinguish people with the disease from those who are free of it? How is it spread? Why does it occur in a particular place? Are there associated risk factors? What causes the disease? What can be done to prevent or control it? Epidemiology is tasked with answering these questions, which will always have important public-health implications.

KEY TERMS AND CONCEPTS

- **Bias:** Flaw in a study's design or selection of subjects that distorts the results.
- **Confounding Variable/Factor:** Variable or factor that is related to both the risk factor and the disease or condition being studied and therefore distorts the relation between them.
- **Control Group:** Group of people who are similar to those in the study group but who do not have the disease or condition being studied or who did not receive the intervention being studied.
- **Dose-Response Curve:** Graph that plots changes in the dose of an agent against its effect.
- **Epidemic:** Occurrence of a disease that is clearly in excess of normal numbers.
- **Exposure:** Condition of being subjected to something that can be considered a determinant of disease.
- **Incidence:** Rate of occurrence of new cases of a disease during a specified time.
- **Morbidity:** Diseased state or condition.
- **Mortality Rate:** Death rate, usually discussed in relation to a particular morbidity.

- **Pandemic:** Epidemic that affects several countries or continents.
- **Pathogenic:** Having the ability to cause disease.
- **Prevalence:** Number of cases of a disease at a specific time.
- **Reservoir:** Carrier of a disease-causing organism that is not itself harmed by it.
- **Risk Factor:** Variable associated with a disease that increases the likelihood of developing that disease.

DEFINITION AND BASIC PRINCIPLES

Epidemiology is the study of variations in disease within population groups. The variations are rarely of only theoretical interest; rather they serve as the data source for devising measures to control or prevent health-related problems. Ideally, once a disease or health anomaly is identified, an investigation links it to risk factors and a cause is inferred. A range of activities can follow: design of biomedical research, communication of findings, and collaboration with other disciplines and public-health agencies. Epidemiologic studies can be expected to influence the policy and practice of public health.

Epidemiology derives from the term "epidemic," which traditionally described a rapidly spreading infectious disease. Epidemics have come instead to mean health-related conditions that occur in excess of expected numbers and may be related to environmental, socioeconomic, and other factors. The science of epidemiology considers a host of modern health issues within its purview.

BACKGROUND AND HISTORY

The history of epidemiology has been both consistent with and limited by the biological knowledge of the time, but its roots lie in the nineteenth century. In Victorian England and elsewhere, the explanation of choice for the cause of disease was miasma, a collective term, traceable to the ancient Greek physician Hippocrates, for foul atmospheric poisons. In the nineteenth century, a few pioneers attempted to make inroads into scientific methods in the face of widespread ignorance of disease causation. For example, the London Epidemiological

Society, chartered in 1850, consisted of a group of physicians whose mission was to investigate the nature of disease and prevent epidemics.

Several historical events—cholera in London, clinical study in Paris, and measles in the Faeroe Islands—are considered milestones in the development of epidemiologic studies.

Cholera in London. Miasma was believed responsible for London's periodic cholera epidemics, which occurred between 1831 and 1866. One expert of the time actually advocated removing noxious smells from the home by directing household waste to the water supply. John Snow, a physician widely known for administering chloroform to Queen Victoria during childbirth, reasoned that if symptoms of cholera were largely gastrointestinal, then ingesting something must cause it. He ultimately related variations in household and neighborhood mortality rates to the sources of their water supply. His resulting 1854 treatise indicting the infamous Broad Street pump is an epidemiologic classic.

Clinical Study in Paris. The early nineteenth century also saw the beginnings of clinical epidemiology studies. Pierre Louis recorded and published meticulous observations of the contemporary practice of bloodletting. His methods were mathematical, based on careful studies of patients with inflammatory disease. Among his conclusions: Bloodletting did not benefit patients with pneumonia, although it shortened the duration of the disease if done early and the patient survived.

Measles in the Faeroe Islands. Peter Panum was a young Danish doctor dispatched by his government in 1846 to study an outbreak of measles in the isolated Faeroe Islands, in the North Atlantic. Panum's deductions about the natural course of measles were derived largely from studying the islanders' way of life. In writing about his observations in a journal of the time, Panum emphasized that the physician who works in an unaccustomed environment must first study hygienic conditions that can affect the health of its population.

How It Works

Hill's Postulates. The tools of epidemiology are methodology, investigation, and inference. Sir Austin Hill was a professor of medical statistics at the University of London. His 1965 address to London's Royal Society of Medicine enumerated six criteria—strength,

consistency, specificity, temporality, biologic gradient, and plausibility—that should be used to infer cause from an association. None is absolutely required, but together they build a substantial case for establishing causation. First, Hill examined the strength of an association. His example was the high prevalence of scrotal cancer among chimney sweeps. Second, he looked at whether the association was consistently observed by separate investigators in different places, times, and circumstances. Third, he examined whether the association was limited to specific groups, sites, and disease, and whether there were any associations between those variables and another disease. Fourth, he looked at temporality,

Mike Fink, an epidemiology specialist for the Arizona Department of Health Services, separates the Culex genus mosquito, the carrier of the West Nile virus, from other mosquitos caught in a traps placed in suspected areas throughout the state at the department's lab August 5, 2004 in Phoenix, Arizona. (Getty Images)

noting that the cause of an association had to precede the effect. The fifth element was the biological gradient, or the dose-response curve, formed by the data. For example, the greater the number of hours of unprotected sun exposure, the higher the risk of cancer, if there is a causal relation. The sixth criterium was plausibility, whether the theorized cause made biological sense. This determination, of course, depends on the biological knowledge at the time it is made.

The tools of epidemiology are various types of studies and investigations. All studies depend heavily on biostatistics in their design and in testing the strength of the associations that they find. Most epidemiologic studies are observational; they look at diseases in their settings, estimate their prevalence, and point to risk factors. Data are collected and analyzed, outcomes are evaluated, and causal inferences are drawn when the evidence warrants it. The classic study types are case-control studies, cohort studies, cross-sectional studies, meta-analyses, and random controlled trials.

Case-Control Studies. Case-control studies are retrospective in that they look back in time, often involving the review of hospital charts of cases. The strategy compares people with a disease—cases—with a control group free of the disease but otherwise as similar as possible to the people in the cases. The goal is to identify disease risk factors. These studies are inexpensive and useful for studying rare diseases or those that take years to develop. They often rely on participants' memory, however, and bias in selecting cases or controls can mislead.

The first notable case-control study was reported in 1926 by Janet Lane-Claypon, who compared a large group of women with breast cancer with a group of healthy women. She found important differences, such as estrogen exposure, between the two groups.

Cohort Studies. Cohort studies are observational studies in which subjects, with and without an exposure of interest, are followed forward in time to determine the outcome. Investigators do not control the exposure or the treatment. Participants are chosen, sampled, or classified according to whether they were exposed to a treatment or risk factor.

The strengths of cohort studies lie in their ability to compute incidence rates and relative risks. They are the best way to examine the incidence and natural history of a disease. Confounding factors are

a particular problem, however. For example, one study showed an association between ice cream consumption and an increased risk of drowning. In this instance, the confounding factor was proximity to a beach community.

Cross-Sectional Studies. Often conducted through questionnaires or interviews, cross-sectional studies measure the presence or absence of a condition in a representative population at one particular time. The population consists of people of different ages and ethnic and socioeconomic groups, and questions include a number of possible exposures.

The studies cannot determine the timing of an exposure because the disease outcome and potential causes are ascertained at the same time; whether risk factor or outcome came first cannot be confirmed. Market research and political organizations frequently use this research method.

Meta-Analysis. Meta-analysis is a widely used method that employs statistical analysis to synthesize the findings of a number of independent studies. Meta-analyses assume that random variation accounts for differences in results from one study to another. These analyses require both statistical expertise and extensive clinical knowledge of the topic. Weaknesses of meta-analyses are implicit in the process. The included studies may have measured slightly different outcomes, they may have used different research designs, and there may have been selection bias in the criteria used to include a study.

Randomized Controlled Trials. Randomized controlled trials are large clinical trials, considered the gold standard for evaluating the effects of a medical intervention—often a single drug for a specific condition. A group of patients, the experimental group, is randomly assigned to receive the intervention; the control group receives a placebo or alternative treatment. The statistical assumption is that the groups differ only in whether they have received the intervention. The trial follows patients over time and compares the occurrence of prespecified outcomes in each group.

APPLICATIONS AND PRODUCTS

The application of epidemiology is public health—the control or prevention of disease in populations through some organized community-based or governmental effort. Products are also produced through privately run organizations, albeit under some governmental control. Most of the thousands of vaccines and prescription drugs that are listed on hospital formularies and stocked in American pharmacies owe their existence to epidemiologic research.

Vaccination. Vaccines are the quintessential product of applied epidemiology. Vaccination against smallpox has been one of the great public-health successes of the modern era. The last reported case of smallpox transmitted through human contact occurred in a Somali village in 1977.

Children in the United States are vaccinated against a growing list of diseases: pertussis (whooping cough), diphtheria, tetanus, measles, mumps, rubella, varicella (chicken pox), polio, hepatitis B, and pneumococcal disease. Annual influenza vaccination, depending on the prevalent viral strain, is recommended for most age groups. The H1N1 influenza pandemic of 2009 saw the rapid development of an effective vaccine; however, a vaccine for HIV remains ellusive.

Research is being directed to vaccines in other disease areas. A vaccine has been developed that prevents breast cancer in genetically predisposed mice, but much research must be done before a vaccine is ready for women. Because some strains of *Staphylococcus aureus* are resistant to methicillin, research may begin on a vaccine against this staph-causing bacterium. A study of the economic benefit of an *S. aureus* vaccine for newborns found it to be cost-effective.

The Framingham Heart Study. No better illustration exists of an epidemiologic study that has profoundly affected public health and generated a plethora of products in the process than the Framingham Heart Study, which is still operating.

The *Physician's Desk Reference* (PDR), which lies within easy reach of any physician who prescribes drugs, lists eighteen different classes of cardiovascular medication, each with a different mechanism of action or combination of mechanisms. An additional twelve classes of antihypertensive medication have a separate heading. Although no well-marked route exists between these drugs and the Framingham study, much of what is known about cardiovascular disease can be traced to it, and few epidemiology textbooks fail to mention the study.

Initiated in 1949 in Framingham, Massachusetts, a small town 20 miles from Boston, the study served as a model for all the cohort studies that came after it. The study enlisted and followed more than 5,000

Fascinating Facts About Epidemiology

- In the 1300's, the plague, or the Black Death, spread from the Middle East to Europe, where it killed nearly 30 million people, one-third of the population. Caused by the bacterium *Yersinia pestis*, the plague typically spreads from rodents or fleas to humans.

- The smallpox virus brought to the Americas by European settlers had a devastating effect on the Native American population, who had not developed any immunity to the disease. The mortality rate among Europeans was around 30 percent compared with about 50 percent for the Cherokee and the Iroquois, 66 percent for the Omaha and the Blackfeet, 90 percent for the Mandan, and 100 percent for the Taino Indians.

- In the 1918-1919 influenza pandemic, about 500 million people worldwide (one-third of the world's population) were infected, and more than 50 million people died.

- A large sample of California men and women whose scores on a standard test indicated major depression ate more than twice as much chocolate as adults whose scores were in the normal range. The higher the score, the deeper the depression and the greater the amount of chocolate consumed. Researchers did not speculate whether self-medication or chocolate-induced depression were factors.

- A unique form of social withdrawal termed *hikikomori*, known in Japan since the 1970's, has been described in English medical literature. Affected adolescent and young adult males, numbering in the thousands, withdraw into their parental homes for months or years, have no social contact, and neither attend school nor hold a job. No basis exists for any specific treatment.

- A cross-sectional study of more than seven hundred men and women, with the average age of sixty, found that visceral (abdominal) fat was associated with lower brain volume (as measured by brain scans). Generalized body fat had a weaker association with brain volume, and brain size was unrelated to any cardiovascular risk factors.

men and women between the ages of thirty and sixty-two who were examined and found free of coronary heart disease. The study served as a source of insight into the development of cardiovascular disease when infectious disease was the major epidemiologic concern.

Cardiovascular disease was a puzzle. The causes were unclear, the onset was not clinically observable, and the outcome could be suddenly lethal. Hypertension was considered a benign condition that was safely ignored. Gradually, serum cholesterol, high blood pressure, and cigarette smoking began to emerge as important associations with the development of heart disease. The very concept of "risk factor" comes from the Framingham study.

From the earliest reports, an elevated serum-cholesterol level differentiated those who developed heart disease from those who remained free of it. As a result, millions of people take drugs that lower cholesterol; at least one drug has been approved by the U.S. Food and Drug Administration to be administered to people with normal cholesterol levels. Research on new lipid-lowering agents is also under way.

The Diabetes Epidemic. Whether the incidence of type 2 diabetes is referred to as an epidemic or pandemic, it is increasing at an alarming rate. Among people who are at least twenty years old, the prevalence of diagnosed and undiagnosed cases has been estimated at 12.9 percent and rising.

This epidemic has spurred drug development. Some thirty brands of antidiabetic agents and insulin formulations are listed in the *Physician's Desk Reference*. Pharmaceutical companies are working to develop drugs that can compensate for loss of beta-cell function (insulin-secreting cells). Diabetes is expected to remain a major public-health concern for the foreseeable future.

IMPACT ON INDUSTRY

Wedded as it is to public health, epidemiology is largely a government enterprise, and consequently, its effects on industry are largely government mediated. For example, in the twentieth century, the research that conclusively linked cigarette smoking and lung cancer prompted increased government regulation of cigarette advertising.

The Food Industry. As the epidemiology of food-borne illness evolves, regulatory concern is increasingly directed at industries involved in food production. The traditional villains, undercooked meat and unpasteurized milk, have given way to contaminated eggs, fruits, and vegetables. New food-borne pathogens have emerged with new transmission routes, some international. From the farm to the table, keeping food safe requires industry to collaborate with regulatory agencies.

Better control of disease-causing *Escherichia coli* reservoirs in cattle, for example, might include improving the hygienic conditions in slaughterhouses, immunizing cattle against *E. coli* infection, and altering feed to make cattle more resistant to *E. coli* residence. Extensive use of antibiotics to promote animal growth is implicated in the prevalence of antibiotic-resistant pathogenic bacteria. Changing this practice would doubtless require government intervention.

Another pressure on the food industry is the epidemic of obesity. Epidemiologic studies have quantified a relationship between consumption of sugar-sweetened beverages (carbonated soft drinks, sports drinks, sweetened tea, and the like), long-term weight gain, and diabetes. Calls for changes in processed foods are increasing. The American Heart Association, the American Dietetic Association, and other associations have recommended limiting consumption of trans fats.

U.S. Governmental Agencies. The U.S. government contains many bureaus, institutes, and agencies that deal with epidemiology. A number of individual states also have departments or programs in this area. All are concerned with aspects of public health, and their activities and research findings filter down to affect a variety of industries.

The Centers for Disease Control and Prevention (CDC), founded in 1946, and part of the U.S. Department of Health and Human Services, is concerned with infectious and chronic diseases, injuries, and disabilities in and out of the workplace, as well as environmental health hazards. The CDC's Foodborne Diseases Active Surveillance Network (FoodNet) collaborates with industry, state health departments, and federal food regulatory agencies to monitor outbreaks of food-borne illness.

The National Institute of Environmental Health Sciences is one of twenty-seven institutes within the National Institutes of Health. The Epidemiology Branch of this institute applies epidemiologic methods to the study of environmental effects on human health. Its research into the long-term effects of lead exposure resulted in the removal of lead from paint and gasoline.

The Division of Cancer Epidemiology and Genetics, a branch of the National Cancer Institute, conducts research into the genetic and environmental determinants of cancer. Its projects include investigating municipal and privately supplied drinking water. Exposure to nitrates, arsenic, and chlorination by-products have been linked to several cancers, and studies are under way in a number of regions within the United States.

Worldwide Agencies. The primary international agency involved in epidemiology is the World Health Organization (WHO), which benefits more than 190 United Nations member states. The agency tracks global health trends, organizes and coordinates research efforts, provides technical support, and disseminates information on disease outbreaks. The WHO is geared to respond to pandemics and was actively involved in the 2009 H1N1 influenza pandemic. Under the guidance of the WHO, member states devised plans and stockpiled antiviral drugs.

Severe acute respiratory syndrome (SARS) holds the distinction of being the twenty-first century's first global epidemic of a new, life-threatening, and easily transmissible disease. The first appearance of the disease was in Guangdong Province, in southern China, in November, 2002. Because of the disease's rapid spread and severity, the WHO convened a global epidemiology conference just six months after its outbreak to gather together what was then known about SARS and to identify what still needed to be determined. Throughout the epidemic, the WHO coordinated a network of laboratories that shared information on developing a rapid diagnostic test and determining likely transmission routes. They determined that the virus has an animal origin, which leaves open the potential for another outbreak. Animals within the Guangdong Province still harbor the precursor virus, and animal markets in southern China still provide a venue for the initial animal-human contact, followed by human-human transmission. The WHO continues to monitor and guide efforts to develop a SARS vaccine.

CAREERS AND COURSE WORK

Epidemiology accommodates many interests. A college degree is a minimal requirement to work in the field, and some background in biostatistics is helpful. More than seventy colleges and universities across the United States have departments or entire schools devoted to epidemiology or public health. The Bloomberg School of Public Health at The Johns Hopkins University in Baltimore, Maryland, consists of ten departments, one of which is Epidemiology. Others are Biostatistics, Health Policy and Management, and Molecular Microbiology and Immunology.

Virtually all the government agencies concerned with epidemiology and public health (including the WHO) list job opportunities and training programs on their Web sites. For example, in addition to epidemiologists, the CDC employs a variety of specialists, including microbiologists, behavioral scientists, and public health advisors. Other positions, described as "mission support," resemble those found in any large corporation—accountants, public relations personnel, budget administrators, and Web developers.

Positions at the CDC and at other government agencies structure their requirements according to civil-service grades. Starting at the lowest grade, the minimum requirement for a position (for example, as a behavioral scientist) would be a bachelor's or higher degree in a related discipline from an accredited college or university, an equivalent combination of education and experience, or four years of appropriate experience. As the civil-service grade goes up, so do the requirements: education at the graduate level and beyond, specialized experience, and specific skills.

SOCIAL CONTEXT AND FUTURE PROSPECTS

The future focus of epidemiology is a frequent topic in epidemiology journals. Critics of purely academic epidemiology claim that it ignores the influence of society on population health and undervalues the part that community action can play in promoting health. This view stresses science's public-health mission, and it considers health and disease in the context of the entire human environment: society, economics, and ecology. Social epidemiology covers a range of topics that blend epidemiology and sociology.

Molecular/Genetic Epidemiology. The term "molecular epidemiology" was first used in the early 1980's. Rapidly emerging technology enabled molecular biomarkers to be added to traditionally collected data, first in the context of cancer research. The strategy was to identify molecular events and modify risk in vulnerable populations before cancer became a clinical entity, thus potentially preventing cancer.

The Human Genome Project extended the reach of research still further, with genome-wide association (GWA) studies. These studies examine common genetic variants that occur across the human genome to identify associations with observable disease traits (phenotypes).

Genome-wide association studies often take the form of molecular case-control studies, which compare differences in alleles between cases with the disease and controls. This genome-wide strategy has identified four genetic loci with variants that increase susceptibility to type 2 diabetes.

Epigenetics. The next frontier in epidemiology is likely to be epigenetics; the term "epigenetic epidemiology" has already appeared in the scientific literature. Epigenetics refers to changes in gene function or expression that do not change DNA sequences and that are heritable because the changes remain intact through cell division. For epidemiologists and other scientists, epigenetics provides another research pathway to environmental mechanisms that affect gene expression and cause disease.

Epigenetics has particularly excited the interest of cancer researchers, who see epigenetic mechanisms as contributing to cancer development. In contrast to genetic changes, epigenetic changes can potentially be reversed. Anticancer agents that counter epigenetic effects are already being tested in clinical trials.

Judith Weinblatt, M.A., M.S.

FURTHER READING

Byrne, Joseph P., ed. *Encyclopedia of Pestilence, Pandemics, and Plagues.* 2 vols. Westport, Conn.: Greenwood Press, 2008. An alphabetized treasure trove of information about anything and anyone connected with epidemiology, public health, and epidemic disease. Each entry has references.

Cockerham, William C. *Social Causes of Health and Disease.* Malden, Mass.: Polity Press, 2008. Highlights the confluence of epidemiology, public health, and medical sociology. Lifestyle, class, and ethnicity are considered in the context of health.

Desowitz, Robert S. *Who Gave Pinta to the Santa Maria? Torrid Diseases in a Temperate World.* San Diego, Calif.: Harcourt Brace, 1998. A renowned epidemiologist and lively writer tracks the spread of lethal tropical diseases into the Americas.

Dickerson, James L. *Yellow Fever: A Deadly Disease Poised to Kill Again.* Amherst, N.Y.: Prometheus Books, 2006. Focuses on yellow fever, which had a surprisingly virulent history in the early United States.

Gerstman, Burt B. *Epidemiology Kept Simple: An Introduction to Traditional and Modern Epidemiology.* 2d ed. Hoboken, N.J.: Wiley-Liss, 2003. Takes a comprehensive approach toward epidemiology and makes good use of concrete examples.

Gordis, Leon. *Epidemiology.* 4th ed. Philadelphia: Saunders Elsevier, 2009. Presents basic principles and important concepts with generous use of illustrations and examples.

Lock, Margaret, and Vinh-Kim Nguyen. *An Anthropology of Biomedicine.* Oxford, England: Wiley-Blackwell, 2010. Examines the assumption that human bodies are the same everywhere and presents the ways in which culture, history, politics, and environmental biology interact to change human biology.

WEB SITES

American Epidemiology Society
http://www.acepidemiology.org/societies/AES.shtml

Centers for Disease Control and Prevention
Epidemiology Program Office
http://www.cdc.gov/epo

National Cancer Institute
Division of Cancer Epidemiology and Genetics
http://dceg.cancer.gov

National Institute of Environmental Health Sciences
Epidemiology Branch
http://www.niehs.nih.gov/research/atniehs/labs/epi/index.cfm

Society for Epidemiologic Research
http://www.epiresearch.org

World Health Organization
Epidemiology
http://www.who.int/topics/epidemiology/en

See also: Food Preservation; Hazardous-Waste Disposal; Immunology and Vaccination; Parasitology; Pasteurization and Irradiation; Pathology; Sanitary Engineering.

EPOXIES AND RESIN TECHNOLOGIES

FIELDS OF STUDY

Organic chemistry; polymer chemistry; industrial chemistry; chemical engineering; civil engineering; automotive engineering; watercraft design; adhesives; advanced composite technology; molding; mold making; extrusion manufacturing; prosthetics; packaging; environmental chemistry; materials science; materials recycling.

SUMMARY

Epoxies and resins are chemical systems, as opposed to single compounds, that are used in a variety of applications. Their value derives from their polymerization into three-dimensional or cross-linked polymeric materials when the components are combined and allowed to react. Epoxies are so-named because the principal component is a reactive epoxide compound. The combination of the epoxy compound with a second material that promotes the polymerization reaction is called a resin. The term also applies generally to any polymerizing combination of materials that is not epoxy based. Epoxies and resins are used primarily in structural composite applications, in which the combination of a reinforcement material (usually a specialized fiber) bound within a solid matrix of polymerized resin provides the advantages of high strength, low weight, and unique design capabilities. Resins are also used in injection molding and other molding operations, extrusion and pultrusion, prototype modeling, and as high-strength adhesives.

KEY TERMS AND CONCEPTS

- **Amine:** Organic compound of nitrogen derived from ammonia (NH_3) by replacement of one or more hydrogen molecules by a hydrocarbon radical. Amines can be primary (RNH_2), secondary (R_2NH), or tertiary (R_3N).
- **Cross-Linking:** Bonding of two molecules to each other from separate polymerization chain reactions; also the extent to which intermolecular bonds have formed between the otherwise separate molecular chains of a polymer.
- **Epoxide:** Compound whose molecules contain a three-membered ring structure of two carbon atoms and one oxygen atom; normally formed through the incomplete oxidation of a carbon-carbon double bond.
- **Epoxy:** Resinous material made up, in part, of epoxide compounds.
- **Fiber-Reinforced Plastic (FRP):** Structural composite consisting of a fibrous support material encased in a solid matrix of polymerized resin.
- **Functional Group:** Group of atoms that exhibits specific chemical behavior in a molecule.
- **Glass Transition Temperature (Tg):** Temperature range at which a thermoplastic material begins to lose the characteristic behavior of fracturing conchoidally (like glass) and becomes plastic; also the temperature range at which a thermosetting material begins to decompose.
- **Polymer:** Compound formed by sequential repeating reactions between simpler molecules; from "poly," meaning many, and "mer," meaning form.
- **Polymerization:** Process of forming a polymer; a type of reaction that produces very large molecules by the intermolecular bonding of smaller molecules to each other in a repeating manner.
- **Thermoplastic:** Describing a material, usually a polymer, that softens or becomes plastic with heating.
- **Thermosetting:** Describing a material, usually a resin, that polymerizes and hardens with heating.

DEFINITION AND BASIC PRINCIPLES

In the field of polymers, resin refers to the material or blend of materials that is specifically prepared to undergo a polymerization reaction. In this type of reaction, molecules add together sequentially to form much longer and larger molecules. A polymerization reaction can proceed in a linear manner to form long-chain single molecules whose bulk strength derives from physical entanglement of the molecules. It also can proceed with branching to form large, multiple-branch molecules that derive their bulk strength from their sheer size and complex three-dimensional interlinking bonds between the molecules.

The particular combination of materials used to prepare the resin for polymerization is chosen according to the extent and type of polymerization desired. Monomers containing only one reactive site or two functional groups can form only linear polymers. Three-dimensional polymers require the presence of three or more functional groups or reactive sites in at least one of the resin components. The polymerization reaction can proceed as a simple addition reaction, in which the single monomer molecules simply add together by forming chemical bonds between the reactive sites or functional groups on different molecules.

Polymerization reactions are generally driven to completion with heating, although the heat produced by exothermic reactions must be controlled to prevent overheating, decomposition, and dangerous runaway reactions from occurring. The polymeric product of the reaction may be thermoplastic, becoming soft, or plastic, with heating. Thermoplastics are characterized by this change of behavior at the glass transition temperature, Tg. Below this temperature, the material is solid and fractures in the characteristic conchoidal manner of glass rather than along any regular planes that would denote a regular crystal structure. At the Tg, the material begins to deform rather than to fracture. The Tg is always stated as a fairly broad temperature range, and at its higher value, the material has no resistance to deformation that would result in fracture, although it may not yet be entirely liquefied.

A thermosetting resin produces a polymer that does not soften with further heating and exhibits conchoidal fracture behavior at all temperatures at which it is stable. Such polymers will undergo thermal decomposition (also called thermolysis) when heated, as their Tg is at a higher temperature than the temperature at which they break down.

Epoxies are a specific type of resin in which one of the components is an epoxide compound. A second component, typically an amine, reacts irreversibly with the epoxide functional group, causing its three-membered ring structure to open up. The intermediate form produced reacts in a chainlike manner with other epoxide molecules to form complex, three-dimensional polymeric molecules.

Various technologies, methods, and applications are encompassed by the field of epoxies and other resins. These range from molecular design

and testing in the chemical and material sciences laboratory, to injection molding and hand layup of fiber-reinforced plastics, and the repair of structures made from resin-based materials. The production of specific resin formulations on an industrial scale is particularly exacting because of the regulations governing certification of the materials for specific critical uses and concomitant purity requirements throughout the handling of the product. Specialized training and equipment is required for the safe production and transportation of the materials.

BACKGROUND AND HISTORY

Resins and their property of solidifying have been known and used since ancient times. During explorations of the New World and Asia, European explorers such as Christopher Columbus and Hernán Cortés found the indigenous peoples playing sports with balls that bounced and wearing clothing and footwear that had been waterproofed. The indigenous peoples used natural latex materials derived from plants in the making of these and other objects.

In the mid-nineteenth century, as the industrial sciences, especially chemistry, blossomed in Europe, these marvelous natural latex materials, known as caoutchouc and gutta percha, were imported and put to a variety of uses that capitalized on their unique properties. Gutta percha, for example, was used to make the corrosion-resistant coating and insulation for the first underseas telegraph cables laid across the English Channel between England and France. Other resins produced at this time were semisynthetic, chemical modifications of vegetable oils and latexes. The development of synthetic polymers such as Bakelite, especially after World War II, opened the way for untold applications. The unique and customizable properties of plastics and polymer resins served as the foundation of a very large and growing industry that has constantly sought new materials, new innovations, and new applications.

HOW IT WORKS

Resins and Chain Reactions. The term "resin" was originally used to refer to secretions of natural origin that could be used in waterproofing. It has since come to mean any organic polymer that does not have a distinct molecular weight. Typically, organic polymers form through sequential addition reactions between small molecules that then form much larger molecules through a chain reaction mechanism. Once initiated, the progress of such a reaction chain becomes entirely random. Any particular reaction chain will proceed and continue to add monomer molecules to the growing polymer molecule as long as it encounters them in an orientation that permits the additional step to occur. Typically, this happens several thousand times before a condition, such as an errant impurity, is encountered that terminates the series of reactions. The exact molecular and chemical identity of any individual polymer molecule is determined precisely by the number of monomer molecules that have been combined to produce that particular polymer molecule. However, within a bulk polymerization process, billions of individual reaction chains progress at the same time, in competition for the available monomers, and there is no way to directly control any of the individual reactions. As a result, any polymerized resin contains a variety of homologous molecules whose molecular weights follow a standard distribution pattern. In thermoplastic resins, this composition, consisting of what is technically a large number of different chemical compounds, is the main reason that the Tg is characterized by softening and gradual melting behavior over a range of temperatures rather than as the distinct melting point typical of a pure compound.

Polymerization Reaction Processes. Polymerization reactions occur in one of two modes. In one, monomer molecules add together in a linear head-to-tail manner in each single chain reaction. This occurs when only two atoms in the molecular structure function as reactive sites. In the other mode, there is more than one reactive site or functional group in each molecule. Polymerization reactions occur between reactive sites rather than between molecules. The presence of more than one reactive site in a molecule means that the molecule can take part in as many chain reaction sequences, with the resulting polymer molecules being cross-linked perhaps thousands of times and to as many different polymer chains. The result can, in theory, be a massive block of solid polymeric material composed of a single, large molecule.

Epoxy Resins and Cross-Linking. Polymerization and cross-linking bonds arise as the reactive site or sites of the molecules become connected by the formation of chemical bonds between them. As a bond

forms from the atom at one end of a reactive site, the atom at the other end becomes able to form a bond to the reactive site of another molecule. When the reactive site is an epoxide ring structure, the resulting resin is called an epoxy. Epoxy resins are two-part reaction systems, requiring the mixture and thorough blending of the epoxide compound and the catalyst, a second compound that initiates the ring opening of the epoxide. This is typically an amine, and the relative amount of amine to epoxide controls the rate at which the polymerization occurs. This represents essentially all the control that can be exercised over the progress of a polymerization reaction. It is therefore critical to control the relative amounts of epoxide and catalyst in an epoxy resin blend.

APPLICATIONS AND PRODUCTS

The value of epoxy and other resins is in their versatility; they are used in a wide variety of products and applications that have become central to modern society. Without epoxy resins and technology, many products would not exist, and modern society would be very different. Epoxy resins cure to a tough, resilient, and very durable solid that has high resistance to impact breakage, fracturing, erosion, and oxidation. They are also reasonably good thermal conductors that tolerate rapid temperature changes very well.

Aircraft. An excellent example of resin application is in aircraft technology, particularly in modern fighter jets. The fuselage and wing structures of many aircraft are constructed of fiber-reinforced plastics. The materials used in aircraft production must be able to tolerate drastic changes in temperature and pressure. For example, an aircraft may be stationed on the ground in a desert with surface temperatures in excess of 60 degrees Celsius, and less than one minute later, the aircraft may be in the air at altitudes where the air temperature is –35 degrees C or colder. That the structural materials of such an aircraft can repeatedly withstand abrupt changes of temperature and physical stresses says a great deal about the strength, toughness, and thermal properties of the epoxy resins used in its construction.

Electronic Devices. The thermal stability of epoxy resins is also evident in their use in the packaging material of modern integrated circuits, transistors, computer chips, and other electronic devices. The operation of these devices produces a great deal of heat because of the friction of electrons moving in the semiconductor material of the actual chip. Pushed to extremes, the devices can fail and burn out, but it is far more usual for the packaging material to adequately conduct and safely dissipate heat, allowing whatever process is running to continue uninterrupted. That may be something as trivial as some spare-time gaming, or as crucial as an emergency response call, the flight control program of an aircraft in the air, or an advanced medical procedure.

Structural Composite Applications. The applications of and the products produced from resins are numerous. The combinations of materials for the production of resins are essentially limitless, and each combination has specific qualities that make it suitable to particular applications. Thus, the varieties and possibilities in the field of epoxies and resin technologies are virtually limitless. A very significant area of application for epoxy resins and other types of resins is in the field of structural composites, particularly in fiber-reinforced plastics and as insulating or barrier foams. The particular application of a resin is determined as much by the desired properties of the product as by the properties of the polymerized resin. Resins that produce a hard, durable polymer such as those produced by epoxy resins are used in products of a corresponding nature. Resins that have good shape-retaining properties coupled with high compressibility, such as those used to produce urethane foams, are used in products such as furniture cushions, pillows, mattresses, floor mats, shoe insoles, and other applications in which the material provides protection from impact forces. Resins that exhibit high levels of expansion while forming a fairly rigid polymer with good thermal resistance are used in sealing and insulating applications, such as those for which urea-formaldehyde resin combinations are so useful.

Resin Production and Supply. A completely different set of technologies and applications is related to the supply and material processing of the resins themselves. Chemists and chemical engineers expend a great deal of effort and time in the development and testing of resins in order to identify new commercially valuable materials or to customize the properties of existing materials. When the new product leaves the laboratory for commercial applications, the system must then be established for the production and safe transport of the material from

Fascinating Facts About Epoxies and Resin Technologies

- The quest for a moldable material of chemical origin began with the need to find a replacement for ivory in the production of billiard balls. Ivory was considered to be the best material for billiard balls but was becoming scarce and expensive. In 1865, the company Phelan and Collender offered a reward of $10,000 in gold for a suitable substitute material.

- Nitrocellulose, or nitrated cellulose, was discovered accidentally by Christian Friedrich Schönbein in 1845 while he was carrying out chemical experiments with nitric acid and sulfuric acid in his kitchen. When he spilled a mixture of the two acids onto the linoleum floor, he used a clean cotton apron to mop up the spill. As it hung drying next to the hot stove, the apron burst into flame and vanished in a light puff of smoke. This was the origin of smokeless gun powder, called gun cotton.

- The first true synthetic polymer was a thermosetting resin compounded of phenol and formaldehyde called Bakelite. It was invented by Leo Hendrik Baekeland in 1907 in an effort to find a substitute for imported natural shellac.

- In theory, a single polymerization chain reaction initiated in a railway tank car full of an epoxy resin could turn the entire mass into a single, very large molecule.

- Variations in the composition of nitrocellulose led to its successful use as the basis of celluloid, the basic material of photographic film developed by George Eastman. The Kodak Brownie, a compact, portable camera (the body of which was made from Bakelite) was also introduced by Eastman, along with custom film-developing service.

- Epoxy resins represent some of the most advanced materials known, especially when used as the matrix material of a structural composite. In some modern aircraft, epoxy-based fiber-reinforced plastics have almost completely supplanted metals in structural components.

the supplier to the user. Systems and methods must also be established for the end user to prepare the intended products from the material. Resins for low-volume use can be packaged in cans and other small containers, while those for high-volume use may be transported by rail or in other types of large containers. Production methods must produce the resin material in a sufficiently pure state so that it will not polymerize en route. Methods of transport must also be such that the resin is protected against any contamination that could result in initiation of polymerization. This requires specialized applications in transportation technology. The end user of the resins will require the means to manipulate the resin, typically in spray-on applications or molding operations. The equipment used in the various molding operations also requires the creation of molds and forms appropriate to the product design. There is accordingly a very large sector of skilled support workers in industries and applications for resin usage.

IMPACT ON INDUSTRY

The twentieth century has been referred to as the "synthetic century," an indication of the impact that plastics have had on both industry and society. Up until the mid-nineteenth century, goods produced by industry were restricted to traditional materials of natural origin, such as woods, metals, minerals, and animal by-products, and to rubber compounds from plant sources. Gasoline and diesel internal combustion engines had not yet been developed, and the "science" of chemistry lacked any theory of atomic structure that would permit its practitioners to predict atomic behaviors or to understand molecular structures and their interactions. As growing populations demanded more and more commercial goods and advances were made in the modes of transportation (such as railroads), industrial manufacturing capability increased, creating a need for cheap, accessible materials to replace those from natural sources that were diminishing in supply, such as ivory and exotic hardwoods. New materials were also wanted for use in rapid production methods, such as hot pressing. These production methods would cut costs by eliminating the large numbers of skilled artisans and craftspeople performing piecework in production settings. The first truly successful artificial material that fit these needs was the nitrocellulose known as collodion, and the first truly successful commercial application of that material was in the production of celluloid. As the base material of convenient, flexible photographic film for small personal cameras, celluloid revolutionized the practice of photography and ushered in the age of industrial plastics in the last years of the nineteenth century.

Atomic Theory and Explosive Industrial Growth. Around the turn of the twentieth century, modern atomic theory was developed, and chemistry became a mainstream science through which new materials could be produced. Each new material engendered new applications, and each new application played to a demand for still newer materials, mostly derived from coal tar, of which a ready supply existed. The final key requirement was the discovery and development of polymerization. The first completely synthetic polymer, compounded from phenol and formaldehyde, was developed in 1907 by Belgian chemist Leo Hendrik Baekeland. It proved to be the elusive material needed to expedite the mass production of consumer goods. Soon, many other new materials were created from polymerization, which led to the development of the modern plastics industry. These versatile resin materials were used in a variety of applications, from the synthetic fibers used to make cloth to essential structural components of modern space and aircraft.

CAREERS AND COURSE WORK

The demand for products of resin technologies is likely to increase in accord with the expanding human population. The facility with which large quantities of objects can be produced by epoxy and resin technologies ensures the continued growth of the field as it keeps pace with the needs of the population. The need for new or improved qualities in the materials being used in resin products means a need for materials scientists with advanced training and knowledge in organic and physical chemistry.

An individual who chooses to make a career of resin technologies must learn the chemical principles of polymerization by taking courses in advanced organic chemistry, physical chemistry, analytical chemistry, reaction kinetics, and specialty polymer chemistry. He or she will also need courses in mathematics, statistics, and physics. Specialized fields of engineering related to epoxy and resin technologies include chemical engineering, mechanical engineering, and civil engineering (in regard to certain special infrastructure applications). Aircraft maintenance engineers and technicians must go through specific training programs in the use of resins and epoxies as they apply to aircraft structural maintenance standards. These are hands-on training programs focused on the physical use and applications of the materials rather than courses of instruction in chemical theory. No special training is required for the layperson to make use of the materials, which are sold in the automotive and marine supply sections of many retail outlets and by certain hobby and craft suppliers.

SOCIAL CONTEXT AND FUTURE PROSPECTS

A vast quantity of plastics is produced from resins. The strongest of these are the epoxy resins. In concert with the commercial and social benefits of epoxy and resin technologies are the logistical problems inherent in the materials themselves. The use of resin-based technologies carries with it the responsibility for the proper disposal of the used products. Thermoplastics are relatively easily managed because of their built-in ability to be reused. Because they can be rendered into a mobile fluid form simply by heating, used objects made from thermoplastic resins can be melted down and formed into new products. Thermosetting resins, however, cannot be reformed and must be processed for disposal in other ways. Thermoset plastics, such as the epoxy resins, are resistant to facile reprocessing as they are generally also impervious to solvents and all but the strongest oxidizing agents. Historically, and unfortunately, this has meant that the vast majority of goods made from thermosetting resins have been relegated to landfills, off-shore dumps, or just left as litter and refuse. Beginning in the late twentieth century, efforts began to be made to put such materials to other uses, the most common being simply to grind them up for use as bulk filling materials.

Epoxy and resin technologies and the plastics industry in general have had a huge impact on modern life since their inception, becoming essential to the infrastructure of modern society. Essentially every government and university research program, every industry and business sector, and every corporation that deals in material goods of any kind deals with resins and plastics in some way, and new ventures are established almost daily for the production of material goods designed specifically to be produced by epoxy or resin technologies.

Richard M. J. Renneboog, M.Sc.

FURTHER READING

Elias, Hans-Georg. *An Introduction to Plastics.* 2d ed. Weinheim, Germany: Wiley-VCH, 2003. Presents clear and usable descriptions of many aspects of the chemistry behind and the applications of resins and plastics, particularly in their manner of polymerization.
Fenichell, Stephen. *Plastic: The Making of a Synthetic Century.* New York: HarperBusiness, 1997. An extremely readable and entertaining account of the evolution of the plastics industry. Pays considerable attention to the social context and relevance of various materials.
Goodship, Vanessa. *Practical Guide to Injection Moulding.* Shrewsbury, England: Rapra Technology, 2004. Presents comprehensive information on the techniques and principles of injection molding, including troubleshooting of production machinery.
Green, Mark M., and Harold A. Wittcoff. *Organic Chemistry Principles and Industrial Practice.* 2003. Reprint. Weinheim, Germany: Wiley-VCH, 2006. Explains the principles of organic chemistry as related to epoxy resins and polymerization. Although the book is nicely presented in an understandable format, the level of chemistry is for the advanced student.
Lewis, Richard J., Sr., and Gessner G. Hawley. *Hawley's Condensed Chemical Dictionary.* 15th ed. Chichester, England: John Wiley & Sons, 2007. Entries include subjects such as trade names, chemical identities, and applications of diverse materials and methods.

WEB SITES

European Resin Manufacturers Associaton
http://www.erma.org.uk

Plastics Institute of America
http://www.plasticsinstitute.org/index.php

See also: Chemical Engineering; Plastics Engineering; Polymer Science.

ERGONOMICS

FIELDS OF STUDY

Biomechanics; rehabilitation engineering; sports engineering; prosthetics; mechanical engineering; industrial engineering; psychology; physiology; anthropometry; industrial design; kinesiology; engineering psychology; industrial hygiene; human factors engineering; physiology; psychology; environmental science; computer science.

SUMMARY

According to the International Ergonomics Association, "Ergonomics draws on the physical and life sciences, applying data on human physical and psychological characteristics to the design of machines, devices, systems, and environments as a means of improving the practicality, efficiency, and safety of human-machine relationships."

KEY TERMS AND CONCEPTS

- **Anthropometrics:** Statistical information about human body dimensions and product design applied to ergonomics.
- **Biomechanics:** Study of human body structure and function and the forces that affect the human body.
- **Cognitive Ergonomics:** Mental processes of humans such as memory, motor response, and reasoning as related to interactions with elements of the system such as mental workload, performance, and reliability.
- **Engineering Psychology:** Application of psychological factors related to design and use of equipment.
- **Human Factors Engineering:** Application of human characteristics, such as arm length, to design equipment or products to fit the human while enhancing work efficiency and limiting stress.
- **Human Performance Engineering:** Approach to solving problems by individuals or teams that analyzes performance gaps, defines interventions, and evaluates outcomes.
- **Job Risk Factors:** Conditions and job demands in the workplace that pose a risk to the worker.

- **Macroergonomics:** Broad system review of human worker and technology interface with a goal of optimal worker productivity, job satisfaction and commitment, and health and safety.
- **Occupational Health and Safety:** Programs, including ergonomics, that support and foster health and safety in the workplace.
- **Organizational Ergonomics:** Optimization of human-technology systems and organizational policies and procedures to include teamwork, cooperative participation, and community ergonomics.
- **Participatory Ergonomics:** Active participation of workers in the development and evaluation of a work-related ergonomics programs.
- **Physical Ergonomics:** Physiological, anatomical, and biomechanical characteristics of the human body related to physical activity.
- **Work-Related Musculoskeletal Disorders:** Physical conditions that result from work-related stress or injury.

DEFINITION AND BASIC PRINCIPLES

Ergonomics, a holistic multidisciplinary science, draws from the fields of engineering, psychology, physiology, and computer and environmental sciences. These sciences define ergonomics as optimizing effectiveness of human activities while improving the quality of life with safety, comfort, and reduced fatigue.

The term ergonomics dates to the mid-1800's, but credit for applying the term generally goes to Hywel Murrell, a British chemist. Ergonomics derives from the Greek word *ergon*, meaning "work," and *nomos*, meaning "law." Ergonomics studies work within the natural laws of the human body.

The International Ergonomics Association promotes a systematic approach to the ergonomic process, to incorporate human factors and human performance engineering and address problems in design of machines, environments, or systems. This can improve efficiency and safety of the human-machine relationship. The basic steps in the ergonomic process include organization of the process, identification and analyzation the problem, development of a solution, implementation of the solution, and evaluation of the result.

BACKGROUND AND HISTORY

The types of work and settings have changed over the centuries. Humans have consistently been aware of the need for a good fit between work tools and the human body. While he was a medical student at Parma University in Italy, Bernardino Ramazzini recognized that workers suffered certain diseases. In 1682, he focused on worker health concerns. His scholarly collection of observations, *De Mortis Artificum Diatriba (Diseases of Workers)*, published in 1700, detailed conditions associated with specific work environments and factors such as prolonged body postures and repetitive motion. His work earned him the title "Father of Occupational Medicine."

The term "ergonomics" is attributed to Hywel Murrell, a chemist who worked with the British Army Operational Research Group during World War II. In 1949, he served as leader of the Naval Motion Study Unit. He invited people with like interests in human factors research to meet with him, forming the Ergonomics Research Society. He remained active in academia until his death in 1989 at age seventy-six. Murrell's specialties included skill development and use and fatigue and aging. He was interested in applications of psychology and ergonomics in day-to-day situations. Murrell authored the first textbook on ergonomics, *Ergonomics: Man in His Working Environment*.

In the industrial era, tools and machines were developed to increase productivity. These put a new strain on the relationship between work and the human body. Between World War I and II, classic work was accomplished by the British Industrial Fatigue Research Board on the impact of environmental factors and human work performance. By the time World War II had begun, worker safety became a primary concern, leading the way for the science of ergonomics.

HOW IT WORKS

Ergonomics, the science of adapting the workplace environment to the work and workers, seeks to maintain worker safety. The goal of industry employers is to keep workers well and comfortable while functioning efficiently on the job. This can be best accomplished by providing safe working conditions to prevent work-related injuries.

National Institute for Occupational Safety and Health's Seven-Step Approach. Businesses can assume a reactive or proactive approach. The National Institute for Occupational Safety and Health (NIOSH) has defined a seven-step program for evaluating and addressing potential musculoskeletal problems in the workplace. First is finding worker complaints of pains or aches and defining jobs that require repetitive movement or forceful exertion. Management must then commit to addressing the problem with input from the worker. Participatory ergonomics is important in encouraging workers to help define problems and solutions to work-related stress. Key is education and training about the potential work-related risks and musculoskeletal problems from defined jobs. Using attendance, illness, and medical records, management should investigate high-risk jobs, where injury is most common. Leadership must analyze job descriptions and functions to see if risky work-related tasks can be eliminated. Management should support health care intervention that emphasizes early detection and treatment to avoid work-related impairment and disability. Finally, management should use this information to minimize work-related musculoskeletal risks when creating new jobs, policies, and procedures.

Physical Ergonomics. Ergonomics can be subdivided into several disciplines: physical, cognitive, and organizational. Physical ergonomics is the body's response to physical workloads. It addresses physiological and anatomical characteristics of humans as related to physical activity. Biomechanics and anthropometrics fall into this category. This discipline

The Contoured Ergonomic Keyboard from Kinesis lines the keys up in a grid, and splits the keys into two widely separated concave areas, one for each hand. (AP Photo)

is concerned with safety and health and encompasses work postures, repetitive movements, vibration, materials handling, posture, workspace layout, and work-related musculoskeletal disorders. Common injuries in an office setting result from computer use (keyboard, mouse, and viewing the monitor).

Cognitive Ergonomics. Cognitive ergonomics deals with human mental processes and capabilities at work, such as reasoning, perception, and memory, as well as motor response. Topics related to cognitive ergonomics include work stress and mental workload, decision making, performance, and reliability. Computer-human interaction and human training are sometimes listed here.

Organizational Ergonomics. Organizational ergonomics addresses sociotechnical systems of the organization and its policies, procedures, processes, and structures. Concepts in this subdiscipline could include work design and hours, job satisfaction, time management, telecommuting, ethics, and motivation, as well as teamwork, cooperation, participation, and communication.

APPLICATIONS AND PRODUCTS

Ergonomics can be applied to work in any setting with the goal of achieving efficiency and effectiveness while maintaining worker comfort and safety. The principles of ergonomics have been applied to many industries, including aerospace, health care, communications, geriatrics, transportation, product design, and information technology.

Office Workers. Global industry requires office workers to use computers every day. Product orders are taken by workers via phone and the Internet. Office workers spend time tied to phones and computers, while sitting in one place. These workers are subject to work-related injury and stress created by continuous computer and phone use.

Ergonomic experts have taken the principles of human factors engineering to improve the work environment for computer users. The placement and maintenance of the computer monitor will affect the user's eyes and musculoskeletal system. The monitor should be clean with brightness and contrast set for the comfort of the user. Placing the monitor directly in front of the user will minimize neck strain. The monitor should be set one arm's length away, tilted back by 10 to 20 degrees, and positioned away from windows or direct lighting to reduce glare.

Office workers often sit for extended periods while working, which is stressful on legs, feet, and the intervertebral discs of the spine. Pooling of the blood in the feet and ankles can cause swelling and place stress on the heart. Employers should encourage workers to alternate between standing and sitting. Ergonomic chairs are designed to relieve the pressure placed on the back while sitting for extended periods. Arm rests should be adjusted so arms rest at the side of the body, allowing the shoulder to drop to a natural, relaxed position.

Many ergonomic ailments occur in the soft tissues of the wrist and forearm, as continuous computer use subjects workers to repetitive motions and sometimes awkward positioning. Computer mouses are ergonomically designed to minimize worker injury, and the no-hands mouse uses foot pedals to navigate. Ergonomically friendly computer keyboards are also available.

Health Care. In the health care field, ergonomics is useful in designing products for conditions such as

---

Fascinating Facts About Ergonomics

- October is National Ergonomics Month.
- Ergonomics is sometimes called human engineering.
- In 2007, the National Institute for Occupational Safety and Health (NIOSH) stated that 32 percent of the workers' compensation claims in the construction industry could have been prevented by proper ergonomic procedures.
- Repetitive strain injuries cost U.S. businesses more than $1 billion annually.
- Five common aspects of ergonomics include safety, comfort, productivity or performance, ease of use, and aesthetics.
- Some ergonomically designed keyboards are available with a rest-time indicator, which encourages the user to take a break from the computer.
- Although laptop computers are portable and convenient, work-related musculoskeletal injuries can still occur if one does not observe proper ergonomic principles while using a laptop.
- Cumulative trauma disorders, caused by repetitive strain or motion injuries and work-related musculoskeletal disorders, are the largest cause of occupational disease in the United States.

arthritis, carpal tunnel syndrome, and chronic pain. Ergonomic applications for persons with arthritis, some 46 million adults in the United States, include appliances with larger dials that can be grasped more easily, levers rather than door knobs, and cars with keyless entry and ignition. Larger controls on the dashboard and thicker steering wheels can be more easily grasped.

Health care workers are at risk for work-related musculoskeletal injuries, such as back or muscle strain, without adequate ergonomics. This is true in nursing homes where nursing assistants must lift patients with impaired mobility. These workers can benefit from ergonomically designed patient-handling equipment and devices such as belts and portable hoists to lift patients.

Dentists are at risk for work-related musculoskeletal disorders. They experience repetitive hand movements, vibrating tools, and fixed and awkward posturing. Neck, back, hand, and wrist injuries are common. Ergonomic equipment is available for dentists, including specially designed hand instruments, syringes and dispensers, lighting, magnification tools, and patient chairs.

Transportation. Ergonomics has applications useful to anthropometry. A 2001 study in the United Kingdom found the airline industry did not provide adequate space for passengers in the economy-class sections. The study focused on seating standards and the passengers' ability to make a safe emergency exit. They found that the economy-class seats did not have enough space to brace for an emergency landing and that even the seats themselves could delay a safe exit. The study also stated that the existing seating would accommodate only up to the 77 percentile of European travelers based on the buttock-knee length dimension.

Other applications include ergonomic food carts and passenger delivery, crew rest seats, and ergonomic design for first-class, business, and economy passenger seats. Cockpit design is important for pilot comfort and safety and to minimize fatigue.

Many competitive manufacturers in the automotive industry have employed ergonomics in designing cars for comfort, safety, and efficiency. Examples include options for driver seat position to accommodate variation in body size and allow the steering wheel and backrest to be ergonomically positioned. Also noted are passenger-seat comfort and safety,

placement of controls, and an option for cell phone placement. These considerations can lessen work stress considerably when someone drives as part of his or her job.

Communication Technology. Cell phones have been a plus for industry and individuals but can come at a price. Shoulder and neck pain may be related to cell phone use. Many people will cradle the phone between their head and shoulder when talking. Ergonomic solutions exist to decrease user strain and pain. Headsets keep hands and head free of awkward posturing. Frequently changing sides can help if a headset is not available. Keying into the phone's address book one's most-dialed numbers can decrease repetitive movement of the fingers. Using cell phone technology with ergonomic design can reduce the daily and cumulative stress of cell phone use.

Aging. With an increasing aging population, industries are applying ergonomic solutions to meet senior needs. Gerontechnology addresses the need for work and leisure, comfort and safety in older adults. Some automobiles have larger and simple dashboard controls. Many tools for use in kitchen or garden have special adaptive handles for less strain on the hands and muscles. Phones are equipped with different levels of tone for varying hearing issues; digits are larger and easier to push. Bathrooms are equipped with safety handles and equipment that allow independence.

Many seniors still hold jobs and make good employees. Employers need ergonomically designed workplaces to accommodate the physical and cognitive changes of normal aging. Seats may need to be firmer and higher to allow for decreased joint flexibility. Good lighting is important for safe work. Restrooms with modifications may be necessary. Flexible-schedule availability will assist with worker fatigue. By redesigning the work environment for aging workers, the risk of illness and injury can be diminished and performance improved.

IMPACT ON INDUSTRY

Worker safety and health is critical for employers in all industries. The science of ergonomics impacts the quality of life and job satisfaction for workers while affecting the financial success of the employer. Many organizations have dedicated programs and research to support this growing field.

Government and University Research. National Institute for Occupational Safety and Health (NIOSH) offers copies of ergonomic-related studies in downloadable files. This government-agency Web site contains links to many published research articles on various aspects of ergonomics. The landmark 1997 study about musculoskeletal disorders and workplace factors is one that is available to any business or individual. Others include the research of violence in the workplace and musculoskeletal pain with regard to nursing-home workers published in the journal *Occupational and Environmental Medicine.* This was the first study of the hazard of workplace violence as linked to work-related musculoskeletal disorders. The results showed that the incidence of musculoskeletal pain increased from 40 percent in those workers that were not assaulted to 70 percent in victims of workplace violence.

T. H. Tveito of the Harvard School of Public Health published findings on workers and low-back pain in the November 2010 issue of *Disability and Rehabilitation.* This study indicated that employers who wanted to retain workers with low-back pain needed to focus on worker communication, pacing work, and options for altered job routines.

Cornell University has a unique Web site called Cornell University Ergonomic Web, where visitors can read research findings from students and professors who participate in the Cornell Human Factors and Ergonomics Research Group. Topics range from computer workstation guidelines to hospital ergonomics.

Industry and Business. Most industries and businesses address worker safety as required by law. However, the Material Handling Industry of America (MHIA) went further by publishing an article entitled "Ergonomics = Good Business Practice." This work describes many reasons why industries should be proactive in addressing ergonomic concerns to improve worker safety and morale. The author provides startling statistics, such as the fact more than 5,000 American workers died on the job in 2008. The U.S. Bureau of Labor Statistics recorded about 3.7 million cases of worker illness and injury that same year, mostly strains and sprains. The article promotes ergonomics as the answer to many of these industry-specific worker events, including cranes, ergonomic shelving, lift tables, workbenches, and adjustable platforms to handle materials safely without unnecessary bending or twisting.

The MHIA has a subgroup of its industry dedicated to the study and implementation of ergonomics. The Ergonomic Assist Systems and Equipment (EASE) reviews standards and serves as a resource to other industry groups.

Major Corporations. In 1991, the Intel Corporation, known for computers and computer chips, recognized that a significant amount of its workers' injuries and illnesses was related to ergonomics. They hired ergonomic professionals to provide leadership for a interdisciplinary team of occupational health professionals, ergonomists, and management. They worked at all levels to provide a companywide training program and employed twenty-eight full-time ergonomists globally. Much of the training is Web based, and the company credits its success to proactive leadership and dedicated workers who practice ergonomic safety. The lost-day case rate has dropped 95 percent.

Intel has continued its dedication to worker safety through ergonomics. In 1996, the company won the Outstanding Achievement Award from the Institute of Industrial Engineers for their work in ergonomics and productivity. Three years later they were given the Outstanding Office Ergonomics Award from the Center for Office Technology, and in 2001 Intel received the National Safety Council's Green Cross for Safety.

CAREERS AND COURSE WORK

The health and safety of workers continues to be a primary concern for employers. The job opportunities for persons interested in ergonomics are varied and depend on the role desired. As with other professions, some jobs require formal education, while others provide on-the-job training. Other jobs require certification or special training in the area of interest.

Jobs related to ergonomics include the role of ergonomist, who has special knowledge and skills in the science of ergonomics, designing the workplace to fit the worker. Ergonomists typically have the minimum of a bachelor's degree in industrial or mechanical engineering, industrial design, psychology, or health care sciences and often a master's or doctoral degree in a related area such as human factors engineering. The International Ergonomics Association (IEA) encourages all ergonomists to become board certified.

Health care professionals such as occupational therapists have also become interested in the growing field of ergonomics.

SOCIAL CONTEXT AND FUTURE PROSPECTS

The IEA attests to the fact that ergonomics is an international concern that affects the global economy. IEA is composed of forty-two organizations worldwide run by a council with representatives from these groups. IEA supports ergonomic efforts in developing countries and keeps a directory of educational programs in some forty-five countries. IEA produced the standard guidelines for industry ergonomics and established a certification program called Ergonomics Quality in Design (EQUID).

The goals of IEA include advancing ergonomics to the international level and enhancing the contribution of the discipline of ergonomics in a global society.

Marylane Wade Koch, M.S.N., R.N.

FURTHER READING

Bhattacharya, Amit, and James D. McGlothin, *Occupational Ergonomics: Theory and Applications.* New York: Marcel Dekker, 1996. Provides a comprehensive look at basic ergonomic principles, including physiology of body movement; practical applications for the workplace; medical implications; and case studies in various industries.

Dul, Jan, and Bernard Weerdmeester. *Ergonomics for Beginners: A Quick Reference Guide.* 3d ed. Boca Raton, Fla.: CRC Press, 2008. A reference for basic ergonomic principles with updated applications for the growing communications-technology workplace, International Organization for Standardization (ISO) ergonomics standards, and human-centered workplace design.

Eastman Kodak Company. *Kodak's Ergonomic Design for People at Work.* 2d ed. Hoboken, N.J.: John Wiley & Sons, 2003. Written for people who may not be trained in ergonomics but want to reduce the incidence of workplace injury through understanding basic principles and reviewing ISO standards; illustrations.

Marras, William S., and Waldemar Karwowski, eds. *Interventions, Controls, and Applications in Occupational Ergonomics.* 2d ed. Boca Raton, Fla.: CRC Press, 2006. A complete resource book to help the reader understand every aspect of ergonomics from the basic ergonomic processes to the future of ergonomics and human work.

National Institute for Occupational Safety and Health. *Elements of Ergonomics Programs.* Cincinnati, Ohio: Alexander L. Cohen, et al. Available at http://www.cdc.gov/niosh/docs/97-117. Looks at what components are necessary for ergonomics programs.

Stanton, Nelville A., et al., eds. *Human Factor Methods: A Practical Guide for Engineering and Design.* Burlington, Vt.: Ashgate Publishing, 2005. Designed to serve as an ergonomics how-to manual for students and practitioners, with examples, flowcharts, and case studies.

WEB SITES

Cornell University Ergonomics Web
http://ergo.human.cornell.edu

Human Factors and Ergonomics Society
http://www.hfes.org/web/Default.aspx

International Ergonomics Association
http://www.iea.cc

National Institute for Occupational Safety and Health
Ergonomics and Musculoskeletal Disorders
http://www.cdc.gov/niosh/topics/ergonomics

Office of Research Services: Division of Occupational Health and Safety
Ergonomics for Computer Workstations
http://dohs.ors.od.nih.gov/ergo_computers.htm

See also: Biomechanical Engineering; Biomechanics; Computer Science; Kinesiology; Mechanical Engineering.

EROSION CONTROL

FIELDS OF STUDY

Soil science; soil conservation; hydrology; hydrogeology; geology; agronomy; watershed management; water quality; air quality control.

SUMMARY

Preventing or eliminating the erosion of soil and rock protects water and air quality and the integrity and usability of public and private lands. As a result, erosion control requires both engineering and land-use management techniques. Erosion control is often seen as primarily an agricultural problem because the land area affected by farming practices is large. However, erosion control is also necessary in coastal regions and areas used for forestry, transportation, development, and recreational purposes because natural or human actions can change landscapes or soil covers and result in increased erosion.

KEY TERMS AND CONCEPTS

- **Conservation Tillage:** Sequence of farming practices that reduces loss of soil or water.
- **Desertification:** Gradual conversion of productive land into desert through loss of topsoil from abusive land-management practices.
- **Nonpoint Source Pollution:** Pollution from dispersed natural and human sources accumulated and carried by rainfall or snowmelt moving over and through the ground.
- **Overgrazing:** Pressure from travel and vegetation removal by grazing animals that results in exposure of bare soil surfaces.
- **Slope Stabilization:** Techniques or structures used to prevent movement of areas adjacent to excavations or in natural slopes forming cliffs, valley sides, and reservoirs that can have significant impacts.

DEFINITION AND BASIC PRINCIPLES

Erosion control is the practice of preventing the movement of soil or rock by the action of wind or water. Uncontrolled erosion by either natural or human actions can cause water or air pollution or damage to property. Erosion is a natural result of the action of water or wind; however, human activity can accelerate this process by removing protective vegetation or creating instabilities in existing soil and rock structures.

Wind erosion acts by selectively transporting soil particles; in other words, the higher the wind velocity, the larger the soil particle that can be transported. Water erosion works both as precipitation hits the ground surface and as water flows over the land. Erosion-control techniques and structures work to reduce the potential for soil or rock transport either by reducing the exposure of land surfaces to the effects of wind and water or by modifying the landscape or installing structures that increase the stability of the landform. For example, mulching, revegetation, and the application of geotextiles are all approaches that work to reduce the vulnerability of soil surfaces to erosion. Gabions, retaining walls, and terracing are examples of structures or modifications to landforms.

Lack of erosion control can result in declines in soil productivity for agricultural purposes and loss of land stability for land development and uses. Examples of the large-scale effects of erosion include landslides and desertification.

BACKGROUND AND HISTORY

Before 1920, erosion was noted during farming practices but not necessarily considered reparable, and eroded lands were abandoned in favor of new fields. This mind-set had its roots in the colonial days of the United States, where potential new farmland stretched to the horizon and was freely available to those willing to clear the land. Decades of intensive farming practice led to the loss of soil productivity and then the soil itself to water and wind erosion, with dramatic examples of wind erosion occurring during the Dust Bowl era of the 1930's.

In 1935, the United States Congress established the Soil Conservation Service (now the Natural Resources Conservation Service) to provide for ongoing work to conserve the nation's soils. This act established soil conservation as a national priority independent of agricultural programs, which had far-reaching advantages borne out by more recent concerns about erosion related to storm-water runoff, coastal wave and storm action, and river movement.

Erosion-control practices are required in many different circumstances. Agricultural practices for animal husbandry or to produce food crops can have significant adverse impacts on soil health if soil conservation techniques are not applied. Land development for residential housing or commercial and industrial facilities can have short-term erosion impacts during construction or long-term effects from rainwater or snowmelt on the sites. Recreational uses near lakes, streams, or oceans can affect the stability of the shoreline. Transportation systems may create unstable slopes with increased potential for landslides from cuts through hills for roads and rail lines.

HOW IT WORKS

Agriculture. Erosion-control practices in agriculture focus primarily on nonstructural techniques to retain and improve soil productivity. The intent of erosion-control measures is not only to protect the soil from raindrop impacts or wind transport but also to increase the infiltration capacity of the soil to lower the amount of water that runs off over land. One way that infiltration capacity is increased is to reduce the length of the slope that is available for water to travel by using terraces or contour plowing. Another way to increase infiltration is to slow the movement of water and physically protect the soil surface with mulch or vegetation, practices that also increase soil fertility by increasing organic content and biological activity.

Coastal Zone Management. Wave action and storm surges in coastal environments cause beach and headland erosion that is typically controlled through revegetation of coastal dunes and barrier islands or hardened structures like riprap, gabions, and retaining walls. An understanding of beach dynamics is needed, particularly of the seasonal transport of sand on- and offshore or the transverse migration of sediments along shorelines. Hard structural erosion-control features must be evaluated and designed to ensure that erosion is not increased elsewhere as an unintended consequence.

Transportation. The development and maintenance of transportation systems can increase erosion potential in landscape by altering the stability of landforms through either physically changing the landscape or by increasing water runoff. For example,

road cuts can increase the potential for erosion and mass wasting by creating new steep slopes in previously stable hillsides. Unless control measures are used, for example, gabions, geotextiles, or terraces, erosional forces will work to reduce the steep slope to a more stable form. This erosive action can take the form of catastrophic landslides. Increased water runoff is also generated by road systems because of the impermeable nature of most road surfaces. The runoff, if not addressed, can increase erosion of road beds or nearby stream drainages, which can increase road-maintenance requirements or property damage.

Storm-Water Management. Residential, commercial, industrial, and public-facility uses generate erosion-control needs during both the development phase and long-term occupation of the site. Site preparation include activities such as tree and vegetation clearing and grading that expose soils and subsoils to erosion and alter the topography of the site. Erosion-control features during construction include phasing the work to minimize the amount of land exposed at any particular time, rapid revegetation of the site, and temporary approaches such as mulching.

Long-term land uses can also increase erosion potential if rainfall or snowmelt is not adequately accommodated on the site. In a similar manner to road systems, the impermeable surfaces created by roofs, driveways, parking areas, and walkways means that overland flow is increased and the infiltrative capacity of the site is reduced. Increased off-site erosion can result if excess overland flow is discharged to adjacent properties or the road system.

Recreation Management. The effects of recreational activities on landscapes may be dealt with through other disciplines and programs mentioned above. Effects of recreation often include soil compaction and vegetation removal through foot or vehicle travel. When the effects are limited in area, the overall effect on erosion potential can be slight; however, when the effects are widespread or occur on steep slopes or near bodies of water, the erosion potential can be high. Mitigating approaches may be to limit the area of travel, for example, designated trails or sites for activities, which will allow revegetation to occur naturally or through replanting. In some instances, the volume of traffic may be so great that structural approaches are needed. For example, a shoreline hardened with a retaining wall can allow

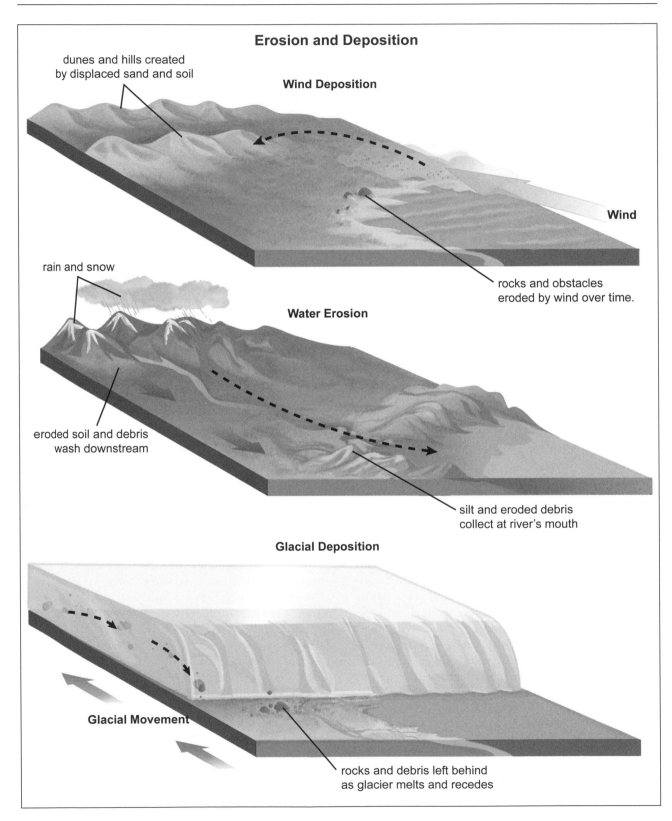

Erosion and Deposition

Wind Deposition

dunes and hills created
by displaced sand and soil

Wind

rocks and obstacles
eroded by wind over time.

rain and snow

Water Erosion

eroded soil and debris
wash downstream

silt and eroded debris
collect at river's mouth

Glacial Deposition

Glacial Movement

rocks and debris left behind
as glacier melts and recedes

increased access for fishing when foot traffic would heavily impact a naturally vegetated shoreline.

APPLICATIONS AND PRODUCTS

Agriculture. Many agricultural erosion-control measures emphasize changing farming practices rather than necessarily installing a structure or feature on the site. For example, conservation tillage works to minimize soil tilling to leave as much crop residue in place as possible on the surface of the soil. This approach requires that the crop residue is not plowed into the soil at the end of the growing season and the new crop is planted in rows plowed through the residue at the beginning of the next season. The crop residue acts as a mulch to help stabilize the soil, retain soil moisture, and increase the soil's organic content. In this approach, chemical weed control is often used to eliminate undesirable plants or the crop from the previous year. Alternative methods of controlling weeds are available if a reduction in use of herbicides is needed.

Other nonstructural approaches to erosion control on agricultural land include contour plowing and crop or pasture rotation. Crop rotation is a practice where a sequence of crops are cultivated on the land over a series of seasons or years. Rotating crops improves soil structure and makes it more resistant to erosion, particularly when paired with conservation-tillage techniques. In addition, varying the crops grown can reduce reliance on herbicides and pesticides by, for example, changing the crop to one that creates field conditions that are less favorable to the weeds or pests of concern.

Structural erosion-control measures for agriculture include terraces and large windbreaks. Although windbreaks are not constructed, they usually are composed of trees planted in rows perpendicular to the prevailing wind direction. They do require planning for location and long-term maintenance to ensure continued effectiveness. Terraces are constructed to created areas of flat land in an otherwise sloped terrain. Terraces retain water behind a dyke or berm at the edge of the terrace and can help manage water on a site after heavy rains. Terraces cost more than other erosion-control measures in terms of labor and equipment to construct but can be effectively farmed with large equipment when properly designed, located, and constructed.

Coastal Zone Management. Erosion control in coastal zones can be necessary because the action of wind and waves erodes coastal lands so that it threatens structures or features such as roads, buildings, or navigation channels. Erosion control can also be needed when structures encroach into areas that are vulnerable to erosion or are otherwise unstable landforms. Examples include the construction of buildings on barrier islands or on exposed headlands.

Nonstructural erosion-control methods in coastal environments can be modeled after natural features such as coastal-dune environments or saltwater marshes to create buffers that absorb the energy of waves or storm surges. These approaches include limiting access for foot and vehicle traffic in areas vulnerable to erosion and revegetating with native plants. The plants trap wind-borne sediment and sand and anchor coastal soils with their root systems. Revegetation measures may include the use of geotextiles or mulches to protect the soil surface while the vegetation becomes established.

Structural erosion-control measures in coastal environments include riprap, gabions, and retaining walls. Riprap is rock placed to protect the toe of a slope or the base of a vertical wall. The rock absorbs the energy of wave action and prevents the undermining of existing structures. Gabions are large cages filled with rock, often the same rock that would be used for riprap, that armor the shoreline. Gabions can be stacked to create a retaining wall and can be used in place of riprap where space is limited. Gabions have an advantage over other armoring approaches in that the rock-filled cages drain water freely and are flexible enough to accommodate ground movement. In general, retaining walls need to be properly designed and installed to ensure that the pressure of the ground and subsurface water behind the wall can be withstood for the long term by creating weep holes for moisture to escape and using devices to prevent the soil load from transferring to the wall. Retaining walls can be constructed using either non-proprietary methods or proprietary products such as interlocking concrete blocks and grid materials.

Transportation. Transportation facilities, in particular roads and railways, are often constructed through hilly terrain using cuts into steep slopes. The resulting wall often creates an instability in the landscape that needs to be addressed to avoid damage to facilities and delays in shipping and travel.

Slope-stabilization techniques are commonly used in road cuts to prevent rock fall and landslides. These techniques include structural solutions (such as gabions, retaining walls, and spray-on concrete) and less structural approaches incorporating geotextiles and revegetation.

In addition, transportation facilities need to be designed and constructed to ensure that unintended erosion impacts are not created by runoff of storm water from impermeable surfaces. In these circumstances, facility designs need to include consideration of how storm water can be infiltrated.

Storm-Water Management. Construction and other site-development activities can significantly increase erosion potential on sites by removing vegetation and changing the topography by grading. Erosion control on construction sites can be achieved either temporarily during construction or for the long-term operation of the site after construction is complete. Temporary measures include phasing construction activities to minimize how much of the soil on the site is exposed at any one time and use of stabilizing materials such as mulches or geotextiles. Products such as sedimentation fences or storm-drain filters are not erosion-control devices, instead they control the movement of sediments that have already eroded to ensure they do not create off-site impacts to roads or adjacent lands.

Long-term storm-water management on developed sites includes the use of structural, such as retaining walls, or nonstructural approaches, such as revegetation, to stabilize sites and prevent erosion as the site is used.

Recreation Management. Recreation activities can be either dispersed or focused in terms of impacts to a site's erosion potential. Activities such as hiking or canoeing can have small impacts on sites if traffic is light. However, heavily used trails or boat launches can become significantly eroded. Other activities, for example off-road recreational vehicles, can create significantly de-vegetated areas and bare compacted soils vulnerable to wind and water erosion even with light usage. Erosion-control measures related to recreation areas will vary by need and facility. For example, boat launches for light craft in a rural area may consist of a graveled surface along the bank, whereas heavily used launches for heavy craft may consist of a concrete ramp extending into the water.

Fascinating Facts About Erosion Control

- Dust storms occurring during the 1930's in the Dust Bowl blackened skies from the Great Plains to as far east as Washington, D.C., and left more than a half million people homeless.
- Hurricane Hugo hit the east coast of the United States north of Charleston, South Carolina, in September, 1989. The nearly 20-foot-high storm surge inundated much of the barrier island system, which was less than 10 feet above sea level, and caused almost $6 billion in damages.
- The Grand Canyon is a dramatic example of naturally occurring erosion. It is more than a mile deep and fifteen miles wide.
- Development in hilly terrain and on steep hillsides increases the potential for landslides through a combination of factors, including the increased load on the slope from the weight of structures while increasing erosion potential by directing flow from roofs and pavement to surface drainages, and the loss of vegetation, the root systems of which anchor soils.
- Areas with high annual precipitation rates can be more vulnerable to landslides. For example, the state of Oregon averages more than $10 million in damages from landslides each year.
- The term "gabion" is derived from the Italian word for big cage, *gabbione*. Gabions have historically been used to create protective barriers for troops and equipment attacking a fortress or walled city.

Erosion control along river channels historically emphasized engineering solutions such as riprap, retaining walls, and channelization where the stream channel was lined with concrete. Because of the unintended erosion and flooding effects experienced as a result of these approaches, more recent stream-erosion projects have mimicked natural systems by reestablishing riparian vegetation and features such as floodplains and meanders. These approaches include removing barriers to river movement and using products such as coconut fiber bales and geotextiles to help revegetate areas and improve infiltration.

IMPACT ON INDUSTRY

Erosion-control requirements affect a wide variety of public and private programs. Public agencies include environmental protection agencies concerned with protecting and enhancing water and air quality, public works or road departments responsible for constructing and maintaining roads and other transportation facilities, and agencies working on preparation for and the aftermath of natural hazards. Organizations in the private sector include technical consultants, land developers, construction contractors, and manufacturers of products used in erosion-control measures. The following provides examples of organizations working with erosion-control measures or regulations.

Government. The United States Environmental Protection Agency (EPA) administers the federal Clean Water Act, which requires protection of water quality. This act establishes water-quality standards, the foundation for storm-water management requirements and, through section 319, funds implementation projects to demonstrate the effectiveness of measures to control pollution from nonpoint sources.

The Natural Resources Conservation Service (NRCS), which is part of the United States Department of Agriculture, conducts assessments and research into the effectiveness of various soil-conservation approaches. The NRCS works with universities in each state to provide technical assistance on erosion control to land owners. The United States Geological Survey researches and compiles information on geologic hazards in each state, including landslide hazards. State agencies administer the Coastal Zone Management Program to respond to erosion-control needs in coastal environments. State and local agencies administer programs for public works and land-use management.

Industry and Business. Geotechnical and engineering firms provide professional services related to working in areas of geologic hazards such as areas prone to landslides or coastal environments or services related to design and construction of erosion-control structures or nonstructural installations. Land developers and construction contractors are responsible for ensuring that appropriate erosion-control measures are implemented during projects. Manufacturers and suppliers provide the products and materials required to construct or install erosion-control measures. Products include retaining-wall components, rock, geotextiles, or plants for revegetation.

CAREERS AND COURSE WORK

Course work that supports careers in erosion control or soil conservation are based in the physical and applied sciences. Degree programs include soil science, agronomy, hydrology, hydrogeology, and geology. Civil engineering is another alternative degree program, although course work from the other disciplines is required in order to create a focus on erosion control. A solid foundation of course work in soil science, geology, and hydrogeology is required to understand the basic physical processes involved in this subject in addition to course work in mathematics, physics, and engineering.

Students may focus on the engineering, land-use management, or regulatory approach to erosion control. A career oriented toward engineering or structural solutions to erosion-control requires an engineering degree, including course work in physics, mathematics, and engineering dynamics. This career may also require professional certifications, such as being certified as a Professional Engineer, which potentially requires additional study for the examination that must be passed to achieve the certification. A career focusing on land-use management or a regulatory approach to erosion control may be founded on any of the degree programs listed above and includes additional course work in land use and public-policy analysis.

A bachelor or master of science degree is typically required for career opportunities designing and implementing erosion-control practices. A degree is typically not required for a career as a construction contractor installing or building erosion-control structures or features, such as retaining walls or regrading and revegetation projects. However, certain professional certifications or licenses may be required for construction contractors depending on the jurisdiction.

SOCIAL CONTEXT AND FUTURE PROSPECTS

Erosion-control requirements affect many sectors of society and have become integrated into a wide variety of occupational disciplines. Given the potential for significant damage if erosion-control measures fail—for example, in areas with steep slopes or coastal zone environments—the emphasis on and expectations for high levels of success increase with increasing development pressures. Increasing development and market pressures can also affect food production in that farmers seek to maximize the productivity of their land while reducing costs. Soil-conservation measures can increase productivity while reducing the need for some herbicides and pesticides. Soil-conservation measures that increase the moisture-holding capacity of sites and make those sites less dependent on irrigation are also important for agriculture and urban or suburban areas where water supplies may be stressed by demands for domestic use.

Barbara J. Rich, B.S., M.A.

FURTHER READING

Baxter, Roberta. "Maintaining Vertical: Techniques for Slope Stabilization." *Erosion Control* 11, no. 2 (March/April, 2004). Provides an overview of slope stabilization.

Bell, F. G. *Basic Environmental and Engineering Geology.* Caithness, Scotland: Whittles Publishing, 2007. This book discusses geologic hazards, including soil erosion, in the context of the built environment and its effects on natural resources.

Coetzee, Ken. *Caring for Natural Rangelands.* Scottsville, South Africa: University of KwaZulu-Natal Press, 2005. Includes practical approaches for dealing with soil erosion on rangelands.

England, Gordon. "Implementing LID for New Development." *Stormwater* 11, no. 5 (July/August 2010). Covers the issues involved in low-impact development related to storm-water management.

Leposky, Rosalie E. "Retaining Walls: What You See and What You Don't." *Erosion Control* 11, no. 1 (January/February, 2004). Discusses manufactured materials and their use in stabilizing soil and managing groundwater to prevent mass wasting.

Lundgren, Lawrence W. *Environmental Geology.* 2d ed. Upper Saddle River, N.J.: Prentice Hall, 1999. Comprehensive overview of natural and anthropogenic causes of soil erosion, including discussion of land use, risk assessment, and surface water management.

Thompson, J. William, and Kim Sorvig. *Sustainable Landscape Construction: A Guide to Green Building Outdoors.* Washington, D.C.: Island Press, 2008. Provides information on land-development practices to preserve and restore vegetation and soil during and after construction.

WEB SITES

Erosion Control Network
http://www.erosioncontrolnetwork.com

International Erosion Control Society
http://www.ieca.org

Natural Resources Conservation Service
http://www.nrcs.usda.gov

See also: Agricultural Science; Civil Engineering; Hydrology and Hydrogeology; Land-Use Management; Soil Science.

EXPLOSIVES TECHNOLOGIES

FIELDS OF STUDY

Chemistry; structural engineering; metallurgy; hazardous materials handling and transportation.

SUMMARY

Explosives, or the more generalized term "high-energy materials," are products that convert chemical energy into physical force through heat, blast, and compression. Explosives are useful in both creation and destruction. They are a key tool used by modern military forces and provide both destructive and propellant forces. A number of industries use explosives for both the separation of materials and the compression and shaping of processed products. There are also commercial applications for explosives, most notably in pyrotechnics and fireworks for both entertainment and commercial displays.

KEY TERMS AND CONCEPTS

- **Brisance:** Measurement of the shock output of an explosive product based on its ability to shatter a standardized material; from the French word for "break."
- **Deflagration:** Generating an explosive reaction by use of heat or fire; characteristic of low explosives.
- **Detonation:** Generating an explosive reaction by shock; characteristic of high explosives.
- **High Explosive:** Explosive substance that has a high rate of combustion, creating a large amount of blast.
- **Low Explosive:** Explosive substance with a low rate of combustion, creating a relatively small blast.
- **Nitroglycerin:** First advanced chemical explosive and the basis for all modern explosives developed in the last 150 years.
- **Oxidizer:** Product that promotes the propagation of oxygen consumption in an explosion, allowing the explosive product to achieve its most efficient consumption.

DEFINITION AND BASIC PRINCIPLES

Explosives are chemical compounds that contain high levels of stored chemical energy that release when the stored form is transformed by a catalyst into heat (often accompanied by visible flame) and blast (the sudden overpressurization of the atmosphere around the explosion). While some early explosives were very unstable in their resting form and prone to conversion by the smallest catalyst, modern explosives are relatively safe to handle and will explode only when initiated by specific means of detonation. The relative safety of explosives makes them ideal for uses ranging from constructive to lethal.

With the exception of nuclear reactions, explosives are chemical compounds that are not individually prone to detonation. When combined with other elements, however, the reaction between the chemicals, initiated by an outside promotion, causes the materials to break down their molecular bonds and release their stored potential energy in the form of blast and heat. Nuclear explosions are again an exception in that, besides heat and blast, they also release a large amount of radioactive particles not present in conventional explosions.

Unlike many products that burn or explode when exposed to heat or pressure, explosives are examples of the static potential of chemical energy in that they release their energy only when determined by the user. Instead of uncontrollable reactions, explosives can be manufactured, stored, and used when needed, a safe and predictable application for a potentially dangerous product. Because of their potential to convert mass into heat and energy, explosives are by definition a very hazardous product. The usefulness of their properties, however, means that explosives are constantly needed—and improved. Consequently, more powerful explosives constantly appear, and the methods for using them safely also keep pace.

BACKGROUND AND HISTORY

The earliest explosive, gunpowder, first appeared in ninth-century China. It was a combination of sulfur, saltpeter, and charcoal. The Chinese used gunpowder for limited military purposes, but its primary purpose was to provide pyrotechnics and fireworks for ceremonial and religious rites. By the thirteenth century, gunpowder was being widely used in European firearms and cannons. Because it was relatively weak, gunpowder had few industrial or commercial uses. In the 1840's, more potent and complex explosives, such as guncotton (cotton fibers seeped with nitric acid) and nitroglycerin, appeared and were used industrially, such as for mining. Both guncotton and nitroglycerin proved susceptible to spontaneous detonation, however, until mixed with a stabilizing ingredient. Swedish scientist and inventor Alfred Nobel was the first to create a safe, stable, and storable explosive in 1867 when he introduced dynamite: nitroglycerin stabilized by absorption of sawdust and containment in a waterproof wrapper.

These early products soon expanded into a wide range of acid-based explosives in the late nineteenth century. Cordite, an explosive commonly used in naval warfare, was a prime example. Unlike the relative instability of guncotton and nitroglycerin, cordite was inflammable, vibration resistant, and waterproof. Trinitrotoluene, better known as TNT, was another benchmark explosive. Very stable and of a consistency that allows it to be formed into virtually any shape, TNT remains in common use more than a century after its invention.

HOW IT WORKS

Types of Explosives. Because of the myriad uses for explosive material, the term "explosives" is a general term and not always specific enough. The terminology for high-energy materials relates to the material's use and application. Explosive, as a general term, applies to any product that explodes, but it really refers to a substance intended to alter the form or physical properties of whatever is in its immediate vicinity when it explodes. A specifically placed explosive charge or an explosive used as a warhead in a shell or missile fits this definition. Explosive reactions can also be harnessed as a propellant to move projectiles toward a target. Simple explosives, like gunpowder used in a firearm, can be used as propellants, but more complex explosives are also used as fuel in missiles and rockets, where the propagation of energy must be sustained over an extended period. Propellants are also used to lift fireworks into the air and as pyrotechnics; this is the colorful burst of light that appears when the firework detonates. In addition to lifting the firework, the formulation of the bursting charge determines the size, pattern, and even the color of the resulting explosion.

Categories of Explosives. Low explosives have a relatively slow burning time and tend to generate more heat than blast. Low explosives, such as gunpowder, are commonly used in firearms because their chemical release is contained and will not burst the gun in which they are used. High explosives, which burn quickly and typically release more blast than heat, are used for open-air purposes because the blast needs to dissipate soon after explosion to avoid unnecessary or undesirable consequences. Instead of burning the application on which the high explosives are used, the blast is intended to shift, move, or remove something. The use of explosives to create mines and tunnels is a good example, as is the use of explosives to level buildings where conventional demolition is too dangerous or time consuming.

Characteristics of Explosives. How explosives transform into matter and energy is also an element of their type. Low explosives consume their mass through the process of deflagration. The initial ignition of the low explosive creates a concentrated charge of flame that breaks down the molecular bonds of the explosive matter, which in turn generates more flame and breaks down the next sequence of molecular bonds. Because the process must move from molecule to molecule, the resulting explosion takes more time and the low explosives do not dispel all of their stored energy at once, producing less blast. The generation of flame over mass is useful in applications that rely on heat, such as using explosives to burn through or sever metal parts. Because flame consumes materials in an irregular fashion, however, low explosives, especially gunpowder, often leave unburned residue as an unwanted byproduct. High explosives, on the other hand, consume their mass through the process of detonation. The molecular bonds of the explosive material are rapidly broken down by a shock wave passing through it. Because the detonation happens almost simultaneously, the resulting explosion largely occurs before flame can develop, and most of the explosive potential is released in the form of blast. This is particularly useful in applications where force transference is the goal, but damage by fire is not desirable. The variations in explosive characteristics permit a wider range of options in selecting the best explosive for a particular purpose. For many military applications, for instance, fire is a useful byproduct of an explosion, so low explosives are best. For purposes that prefer shock over flame, an explosive with a high brisance, or shattering, effect is the ideal option. Brisance is measured and standardized by the speed in which the explosion reaches its maximum force; the faster the propagation of force, the higher the brisance number. Maximum brisance is usually achieved with the aid of an oxidizer, an ingredient that burns at the start of the detonation to accelerate the charge to its maximum heat level at a rate faster than what the explosive would achieve on its own.

Initiation of Explosives. One of the early problems with explosives like nitroglycerin was their instability, which led to accidental explosions, property damage, and deaths. The development of later explosives removed this problem by creating stable explosives that could be safely stored, transported, and handled until detonated. Instead of casual handling, newer explosives required detonation by methods determined by the specialized purpose of the explosives themselves. Many military explosives are detonated by impact with a target that causes compression and initiation of the explosion. Friction is also a detonation method, whereby the heat generated by the movement of the explosion causes the explosion. For precisely timed explosions, electrical charges can also detonate high explosives. The electrical charge method allows the use of a timer mechanism or the initiation of a precisely timed sequence of explosions.

APPLICATIONS AND PRODUCTS

Military Use of Explosives. The military use of explosives has been the primary driving factor in explosives research, as rival militaries attempt to create bigger and better explosions to gain military advantage of a potential enemy. Starting with crude firearms and cannons in the thirteenth century, gunpowder transformed the battlefield by transferring the emphasis of military power from man and animal muscle power to chemical power. Firearms permitted soldiers to attack and defend an area much larger than a hand-operated weapon, such as a spear or bow. The military applications of gunpowder were limited, however, by the relative weakness of gunpowder itself. Only a small percentage of the black powder actually propelled the bullet, while the rest remained in the gun as residue or went out the barrel in a large cloud of dense smoke. To improve their firearms, scientists in the late nineteenth century devised improved propellants, termed "smokeless" powders, that generated virtually no smoke, no residue, and used much more of the powder to propel the bullet downrange.

The invention of more complex and potent explosives in the late nineteenth century transformed war forever. The more lethal explosives, coupled with the new internal combustion engine, led to the tools of all modern armies: machine guns, grenades and rockets, heavy artillery, and tanks. Individual soldiers carried more personal firepower with each succeeding generation, and the accompanying heavy weapons also possessed greater ability to distribute explosives around the battlefield. Explosives also caused a decline in defensive warfare. Before gunpowder, a defender inside a stout fortification or castle could withstand the assault of a force several times the size of the defenders. Gunpowder reversed the advantage, as explosives, in sufficient

---

Fascinating Facts About Explosives Technologies

- In Greek, dynamite means "connected with power."
- The Chinese, the first innovators of explosives, used fireworks during religious ceremonies, believing the noise and flame produced by gunpowder drove away evil spirits.
- In the 1950's, the nuclear scientists at the Los Alamos Laboratory devised Project Orion, a proposed interplanetary vehicle propelled through space by riding the shockwave of a nuclear device exploded a distance behind the vehicle. Project Orion never got off the drawing board.
- Alfred Nobel, the inventor of dynamite, donated his vast fortune to the creation of the Nobel Prize, five international prizes, one of which was the peace prize, which would go to someone who "shall have done the most or the best work for fraternity between nations, for the abolition or reduction of standing armies and for the holding of peace congresses."
- The 21,000-pound GBU-43/B Massive Ordnance Air Blast Bomb (MOAB, also known as the Mother of all Bombs) is the largest and most powerful conventional explosive bomb in the world.
- In December, 1917, an ammunition ship, the *Mont Blanc*, exploded in the harbor at Halifax, Nova Scotia, with a force of roughly 3 kilotons, the largest conventional explosion in history. The damage from the explosion devastated most of the city and caused more than 2,000 deaths.

amounts, could defeat any fixed defensive position, causing armies to pursue more mobile strategies. Explosives also went to sea, resulting in modern naval weapons such as the large-caliber guns on battleships, torpedoes, and mines. Aerial warfare also developed a wide range of explosive weapons, primarily warheads for bombs and missiles aimed at targets on the ground. Small hand grenades were hurled by pilots during World War I, and by World War II a single B-17 Flying Fortress bomber could carry up to four tons of bombs, while a Cold War-era B-52 Stratofortress could carry up to thirty-five tons. Modern cruise missiles carry a range of potential warheads, including low explosive fragmentation, high explosive blast, and nuclear warheads.

Civilian Use of Explosives. Explosives have found their way into a wide range of civilian applications. The earliest and perhaps best known is in the pyrotechnics industry. Fireworks displays are a common entertainment feature of many holidays and special events, and the evolution of explosives parallels the growing complexity and scale of modern fireworks companies. The railroad industry embraced the use of explosives: Confronted by mountains that defied construction, explosives were used to tunnel through them instead. The appearance of dynamite, the first safe and stable commercial explosive, appeared just as the expanding railroad system began moving into the mountainous western regions of North America, and its ability to create tunnels made explosives an important tool. The mining industry, also interested in boring holes through mountains, began to use explosives on a large scale. The blast component of explosives proved very valuable in creating tunnels, but also in crushing rock formations to ease the removal of the trace ores within.

The most visible industrial use of explosives is by the construction industry for the implosion of large structures. In circumstances where the traditional demolition of a building is too dangerous, time-consuming, or expensive, carefully placed explosives can topple a building so that it falls within a very precise footprint, preventing damage to nearby structures. The ability of explosives to topple buildings has, unfortunately, also attracted its use by criminal and terrorist organizations. Bombings, ranging from complex remote detonated bombs to crude "truck bombs," have become a common feature of life in unstable and contested regions of the world.

On a more benign level, explosives also play an unseen role in very common items. Small explosives charges, for instance, cause the rapid inflation of an automobile's air bag in case of a collision. On the scientific frontier, explosives in the form of propellants are the only means of lifting large cargoes into space, and heavy rockets consume tons of explosive material to maintain the International Space Station and the array of satellites in orbit.

IMPACT ON INDUSTRY

Corporate Control. Because of its importance to national security, explosives production usually resided within only a limited number of easily monitored producers during most of the industrial age. Governments ran explosive factories themselves or, more usually, entrusted the production of explosives to large corporations that enjoyed virtual monopolistic control of the market and government contracts. In the United States, the DuPont company was the leading producer of explosives for both the U.S. military and industrial purposes. Founded in Delaware in 1802 by Eleuthère Irénée du Pont, who fled France to avoid the French Revolution, the company was a major producer of explosives and munitions throughout the nineteenth and early twentieth centuries. Although still a major component of the chemical industry, DuPont lost its control of the explosives industry to government production during World War I and II—and to producers who introduced new and more potent explosives into the market.

Research. As control of the explosives industry passed into the hands of a wider variety of manufacturers, the impetus to capture a larger portion of the market led to increased research into improving the characteristics, safety, and usefulness of high-energy materials. Instead of a one-size-fits-all approach, research has led to explosives for very specific purposes in very specific places. A good example is the mining industry. In open-air pit mining, traditional explosives are suitable because the blast can vent into open air. In shaft mining, however, the heat, blast, and gas are contained within the mine. Consequently, specific explosives that produce shorter blast areas and less heat (which can ignite other naturally occurring gases in the mine) have been developed. Explosives that produce less toxic gases and byproducts have appeared, permitting their use in traditional roles while creating less negative impact on the environment.

Government Regulation. Because of the expansive nature of the explosives industry and the potential damage from the irresponsible manufacture and distribution of explosives, explosives are heavily regulated throughout the industrialized world. In the United States, explosives are regulated by the Department of Justice through its Bureau of Alcohol, Tobacco, Firearms, and Explosives, known as the ATF. Congress granted the ATF the power to regulate explosives under the terms of the Organized Crime Control Act of 1970. Through its Explosives Industries Programs Branch, the ATF coordinates and oversees the production of explosive materials at more than 11,000 licensed manufacturers in the United States, ranging from small pyrotechnics companies to large defense contractors. The production of explosives also requires close cooperation between the ATF and the Environmental Protection Agency (EPA) to ensure the safe disposal of hazardous by-products and degraded explosive material.

CAREERS AND COURSE WORK

Because of their complex chemical structures and potentially devastating potential, explosives technology requires very advanced and specific training and course work. Chemistry is the primary field of study leading to careers in the explosives industry, with jobs tending toward the development of new products that can be applied to new purposes. When using explosives in an occupational setting, knowing how explosives will react and change the environment in which they are used is essential. Structural engineering, for instance, provides knowledge on how explosives can affect or even destroy a structure or building. Conversely, structural engineering can also provide training on how to build structures resistant to explosive contact. If one is interested in propellants, a background in physics would prove useful, as would aerodynamics in the pyrotechnics industry. The potential hazards associated with explosives also require specialized training. The transport, storage, and disposal of explosives demand the adherence to strict safety protocols and the knowledge of how to use specialized equipment.

SOCIAL CONTEXT AND FUTURE PROSPECTS

Explosives have a long past and a broad future. As a key element of industrialization over the past few centuries, explosives will continue to play the same role as other world economies continue to develop. As people's understanding of the physical world continues to change, new developments in explosives technology will lead to more potent and potentially lethal products for both civilian and military applications. Just as explosives joined with the internal combustion engine to revolutionize warfare in the past, explosives have begun joining with electronics to become smaller, smarter, and more lethal. Coupled with electronic aiming mechanisms, common explosive bombs became "smart" bombs capable of steering themselves to a target. As targeting got even more precise, military explosives reversed the trend of ever-larger explosive charges to become smaller and more discriminate, destroying the targeted adversary but leaving nearby civilian life and property unharmed, avoiding unwanted collateral damage. Instead of dropping tons of bombs by a B-52, the U.S. Air Force, for instance, has developed the diminutive 250-pound GBU-39 Small Diameter Bomb big enough to destroy a target yet small enough to avoid excessive blast or fire damage around the target.

Steven J. Ramold, Ph.D.

FURTHER READING

Agrawal, Jai Prakash. *High Energy Materials: Propellants, Explosives and Pyrotechnics.* Weinheim, Germany: Wiley-VCH, 2010. A comprehensive overview of explosives, both past and present, with an emphasis on their chemical construction and use.

Akhavan, Jacqueline. *The Chemistry of Explosives.* 2d ed. London: Royal Society of Chemistry, 2004. Similar to the Agrawal book, but with an emphasis on explosives developed since World War II.

Fant, Kenne. *Alfred Nobel: A Biography.* New York: Arcade, 1993. The best biography of the controversial inventor who created a product that changed the world, but who was haunted by its consequences.

Griffin, Roger D. *Principles of Hazardous Materials Management.* 2d ed. Boca Raton, Fla.: CRC Press, 2009. The book contains a lengthy discussion of the handling and transport of explosives, detailing how to move and employ safely even the most dangerous materials.

Persson, Per-Anders, Roger Holmberg, and Jaimin Lee. *Rock Blasting and Explosives Engineering.* Boca Raton, Fla.: CRC Press, 1994. A good example of how the results of explosives are as important as the explosives themselves, this study explains the use of explosives in the mining industry.

WEB SITES

Bureau of Alcohol, Tobacco, Firearms, and Explosives
http://www.atf.gov

Institute of Makers of Explosives
http://www.ime.org

International Association of Bomb Technicians and Investigators
https://iabti.org

International Pyrotechnics Society
http://www.intpyro.org/Society.aspx

See also: Applied Physics; Hazardous-Waste Disposal; Metallurgy; Pyrotechnics.

FIBER-OPTIC COMMUNICATIONS

FIELDS OF STUDY

Physics; chemistry; solid-state physics; electrical engineering; electromechanical engineering; mechanical engineering; materials science and engineering; telecommunications; computer programming; broadcast technology; information technology; electronics; computer networking; mathematics; network security.

SUMMARY

The field of fiber optics focuses on the transmission of signals made of light through fibers made of glass, plastic, or other transparent materials. The field includes the technology used to create optic fibers as well as modern applications such as telephone networks, computer networks, and cable television. Fiber optics are used in almost every part of daily life in technologies such as fax machines, cell phones, television, computers, and the Internet.

KEY TERMS AND CONCEPTS

- **Attenuation:** Loss of light power as the signal travels through fiber-optic cable.
- **Bandwidth:** Range of frequencies within which a fiber-optic transmitting device can transmit data or information.
- **Broadband:** Telecommunications signal with a larger-than-usual bandwidth.
- **Dispersion:** Spreading of light-signal pulses as they travel through fiber-optic cable.
- **Endoscope:** Fiber optic medical device that is used to see inside the human body without surgery.
- **Fiber-Optic Cable:** Cable consisting of numerous fiber-optic fibers lined with a reflective core medium to direct light.
- **Fiberscope:** First device that was able to transmit images over a glass fiber.
- **Light-Emitting Diodes (LED):** Light source sometimes used in fiber-optic systems to transmit data.

- **Receiver:** In fiber-optics systems, a device that captures the signals transmitted through the fiber-optic cable and then translates the light back into electronic data.
- **Semiconductor Laser:** Laser sometimes used in fiber-optic systems to transmit data.
- **Transmitter:** In fiber optic systems, a device that codes electronic data into light signals.

DEFINITION AND BASIC PRINCIPLES

The field of fiber optics focuses on the transmission of signals made of light through fibers made of glass, plastic, or other transparent media. Using the principles of reflection, optical fibers transmit images, data, or voices and provide communications links for a variety of applications such as telephone networks, computer networks, and cable television.

BACKGROUND AND HISTORY

The modern field of fiber optics developed from a series of important scientific discoveries, principles, technologies, and applications. Early work in use of light as a signal by French engineer Claude Chappe, British physicist John Tyndall, Scottish physicist Alexander Graham Bell, and American engineer William Wheeler in the eighteenth and nineteenth centuries laid the foundation for harnessing light through conductible materials such as glass. These experiments also served as proof of the concept that sound could be transmitted as light. The failures in the inventions indicated further areas of work before use in practical applications: The main available light source was the Sun, and the light signal was reduced by travel through the conductible substance. For example, in 1880 Bell created a light-based system of sound transmission or photophone that was abandoned for being too affected by the interruption of the light transmission beam. In the 1920's, the transmission of facsimiles (faxes) or television images through light signals via glass or plastic rods or pipes was patented by Scottish inventor John Logie Baird and American engineer Clarence Hansell. The fiberscope, developed in the 1950's, was able to transmit low-resolution images of metal welds over a glass fiber. In the mid-1950's, Dutch scientist Abraham Van Heel reported a method of gathering fibers into bundles and coating them in a clear coating or cladding that decreased interference between the fibers and reduced distortion effects from the outside. In 1966, English engineer George Hockham and Chinese physicist Charles Kao published a theoretical method designed to dramatically decrease the amount of light lost as it traveled through glass fibers. By 1970, scientists at Corning Glass Works created fibers that actualized Hockham and Kao's theoretical method. In the mid-1970's, the first telephone systems using fiber optics were piloted in Atlanta and Chicago. By 1984, other major cities on the Eastern seaboard were connected by AT&T's fiber-optic systems. In 1988, the first transatlantic fiber-optic cable connected the United States to England and France.

By the late 1980's, fiber-optic technology was in use for such medical applications as the gastroscope, which allowed doctors to look inside a patient and see the image transmitted along the fibers. However, more work was still needed to allow effective and accurate transmission of electronic data for computer work.

In the mid-twentieth century, the use of fiber optics accelerated in number of applications and technological advances. Scientists found a way to create a glass fiber coated in such a way that the light transmitted moved forward at full strength and signal. Coupled with the development of the semiconductor laser, which could emit a high-powered, yet cool and energy-efficient, targeted stream of light, fiber optics quickly became integrated into existing and new technology associated with computer networking, cable television, telephone networks, and other industry applications that benefited from high-speed and long-distance data transfer.

HOW IT WORKS

The major elements required for fiber-optics transmission include: long flexible fibers made of transparent materials such as glass, plastic, or plastic-clad silica; a light-transmittal source such as a laser of light-emitting diode (LED); cables or rods lined with a reflective core medium to direct light; and a receiver to capture the signal. Many systems

Fiber optics. (Phillip Hayson/Photo Researchers, Inc.)

also include a signal amplifier or optoelectronic repeater to increase the transmission distance of a signal. Electronic data is coded into light signals using the transmitter. The light signals then move down the fibers bouncing off the reflective core of the fibers to the receiver. The receiver captures the signals and then translates the light back into electronic data. This process is used to transmit data in the form of images, sound, or other signals down the rods at the speed of light.

APPLICATIONS AND PRODUCTS

Information Transmittal. Fiber-optics technology revolutionized the ability to transfer data between computers. Networked computers share and distribute information via a main computer (a server) and its connected computers (nodes). The use of fiber optics exponentially increases the data-transmission speed and ability of computers to communicate. In addition, fiber-optics data transfer is more secure than lines affected by magnetic interference. Industries that use information and data transmission through networks include banking, communications, cable television, and telecommunications. Fiber-optic information transmission has advantages over copper-cable transmission in that it is relatively easy to install, is lighter weight, very durable, can transmit for long distances at a higher bandwidth, and is not influenced by electromagnetic disruptions such as lightning or fluorescent lighting fixture transformers.

Modern Communications. The use of fiber-optics technology in telephone communication has increased the capacity, ease, and speed of standard copper-wired phones. The quality of voices over the phone is improved, as the sound signal is no longer distorted by distance or is subject to time delay. Fiber-optic lines are not affected by electromagnetic interference and are less subject to security breaches related to unauthorized access to phone calls and data transfer over phone lines. Additionally, fiber-optic cables are less expensive and easier to install than copper wire or coaxial cables, and since the 1980's they have been installed in many areas. Fiber-optic cabling can be used to provide high-speed Internet access, cable television, and regular telephone service over one line. In addition to traditional phone lines and home-based services, fiber-optic links between mobile towers and networks also allow the use of smart phones, which can be used to send and receive e-mails, surf the Internet, and have device-specific applications such as Global Positioning Systems.

Manufacturing. The increased globalization of the manufacturing of goods requires information, images, and data to be transmitted quickly from one location to another (known as point-to-point connections). For example, a car may be assembled in Detroit, but one part may be made in Mexico, another in Taiwan, and a third in Alabama. The logistics to make sure all the parts are of appropriate quality and quantity to be shipped to Detroit for assembly are coordinated through networked computer systems and fiber-optic telephone lines. In addition, the ability to use fiber optics to capture and transmit images down a very small cable allows quality-control personnel to "see" inside areas that the human eye cannot. As an example, a fiberscope can be used to inspect a jet engine's welding work within combustion chambers and reactor vessels.

The Internet. According to the United Nations' International Telecommunication Union (ITU), the number of Internet users across the world met the two billion mark in January 2011. Much like standard telephone service, the capacity, ease, and speed to the Internet has been greatly increased by the replacement of phone-based modem systems, cable modems, and digital subscriber line (DSL) by fiber-optics wired systems. Although fiber-optic connections directly to homes in the United States are not available in all areas, some companies use fiber-optic systems down major networking lines and then split to traditional copper wiring for houses.

Medicine. Fiber optics have significantly altered medical practice by allowing physicians to see and work within the human body using natural or small surgical openings. The fiberscopes or endoscopes are fiber-optics-based instruments that can image and illuminate internal organs and tissues deep within the human body. A surgeon is able to visualize an area of concern without performing large-scale exploratory surgery. In addition to viewing internal body surfaces, laproscopic surgery using fiber-optic visualization allows the creation of very small cuts to target and perform surgery reducing overall surgical risks and recovery time in many cases. Beyond the use of endoscopes, fiber-optic technology has been used to update standard medical equipment so that it may be used in devices that emit electromagnetic fields. As an example, companies have developed a fiber-optic pulse oximeter to be used to measure heart rate and oxygen saturation during magnetic resonance imaging (MRI).

Broadcast Industry. The broadcast industry has moved much of its infrastructure to fiber-optics technology. This change has also allowed the creation and transmission of television signals with increased clarity and picture definition known as high-definition television (HDTV). The use of fiber optics and its increased data-transmission ability was key in 2009, when all television stations changed from analogue to digital signals for their content broadcasts.

Military. The military began using fiber optics as a reliable method of communications early in the development of the technology. This quick implementation was due to recognition that fiber optics cables were able to withstand demanding conditions and temperature extremes while still transferring information accurately and quickly. Programs such as the Air Force's Airborne Light Optical Fiber Technology (ALOFT) program helped move fiber-optic technology along even as it served as proof of concept: fiber-optic signal transmission could transmit data reliably even in outer space. Beyond communications, the military uses fiber-optic gyroscopes (FOGs) in navigation systems to direct guided missiles accurately. Additionally, fiber optics have been used to increase the accuracy of rifle-bullet targeting by using sensitive laser-based fiber-optic sensors that adjust crosshairs on the scope based on the precise measurement of the barrel's deflection.

Traffic Control. According to the United States Department of Transportation, traffic signals that are not synchronized result in nearly 10 percent of all traffic delays and waste nearly 300 million vehicle-hours nationwide each year. Fiber optics has been used as part of intelligent transport systems to help coordinate traffic signals and improve the flow of cars via real-time monitoring of congestion, accidents, and traffic flow. Beyond traffic congestion, some cities capture data on cars running red lights, paying tolls, and the license plates moving through toll roads, tunnels, and bridges.

IMPACT ON INDUSTRY

The total value of the fiber-optic communications industry is difficult to estimate as different aspects of the industry are divided and captured under different financial projections. For example, according to market-research firm BCC Research, the total estimated global market value for fiber-optic connectors in 2010 was an estimated $1.9 billion with an annual average growth rate of 9.6 percent. These projections are different from the total estimated global market value for fiber-optic circulators, which, as of 2011, is forecast to increase annually at 14.29 percent according to technology-forecasting firm ElectroniCast. Further projections are divided into categories such as medical fiber optics, glass and fiber manufacturing, and fiber optic components.

The fiber-optics industry has a global presence; however, the United States is a major player, dominating the optical cable and fiber markets internationally. According to market-research firm First Research, the U.S. glass and fiber-optic manufacturing industry includes about 2,000 companies with combined annual revenue of $20 billion. The majority of the market is concentrated on companies such as Corning and PPG Industries, and 80 percent of the market is captured by the fifty largest companies. Japan and the countries of the European Union are also instrumental in the fiber-optic communications industries as both manufacturers and consumers.

Government and University Research. The United States federal government funds fiber-optics research through branches such as the Department of Defense (DOD). One main category of basic research funded through the DOD includes lasers and fiber optics in communications and medicine. This research funding encourages joint ventures between universities and corporations such as the Lockheed Martin-Michigan Technological University project to develop a fiber-optic-based circuit board manufacturing process.

Industry and Business. Research and development programs in the industry and business sector continue to search for the next technological innovation in fiber-optics access and technology. Upcoming updates in speed and improved processing from industry will soon be seen in the next generation of fiber-optic cables. In addition, businesses are working to connect phone services, the Internet, and cable directly to an increasing number of private homes via fiber-optic cables.

Military. The military continues to fund research and implement new fiber-optics technologies related

---

Fascinating Facts About Fiber-Optic Communications

- As of 2011, all new undersea cables are made of optical fibers.
- Airplanes use fiber-optic cabling in order to keep overall weight down and increase available capacity.
- Fiber optics were an integral part of the television cameras sent to film the first Moon walk in 1969.
- Industry analysts predict that sometime in the early twenty-first century 98 percent of copper wire will have been replaced by fiber-optic cable.
- A fiber-optic fiber is thinner than a human hair.
- A fiber-optics system is capable of transmitting more than the equivalent of a twenty-four-volume encyclopedia worth of information in one second.

to several applications. Monitoring systems based on fiber optics are being developed to detect chemical weapons, explosives, or biohazardous substances based on a specific wavelength emitted by the substance or device. The military has also integrated use of smart phones into its operation activities with applications such as BulletFlight, which helps snipers determine the most effective angle from which to fire and input data to account for changes based on altitude and weather conditions.

CAREERS AND COURSE WORK

There are many careers in the fiber-optics industry and entry-level requirements vary significantly by position. Given the wide spectrum of difference between the careers, a sampling of careers and course work follows.

Professional, management, and sales occupations generally require a bachelor's degree. Technical occupations often require specific course work but not necessarily a bachelor's degree. However, it is easier to obtain employment and gain promotions with a degree, especially in larger, more competitive markets. Advanced schooling usually is required for supervisory positions—including technical occupations—which have greater responsibility and higher salaries. These positions comprise about 19 percent of the fiber-optics communications industry careers.

Engineering roles in the fiber-optics industry range from cable logistics and installation planning to research and development positions in fiber optics and lasers. Positions may be found in universities, corporations, and the military. Engineers may specialize in a particular area of fiber optics such as communication systems, telecommunications design, or computer network integration with fiber-optic technology. Education requirements for entry-level positions begin with a bachelor's degree in engineering, computer science, or a related field.

Telecommunications equipment installers and repairers usually acquire their skills through formal training at technical schools or college, where they major in electronics, communications technology, or computer science. Military experience in the field, on-the-job training with a software manufacturer, or prior work as a telecommunications line installer may also provide entry into more complicated or complex positions.

Optics physicists work in the fiber-optics industry in research and development. The role of the optics physicist is to develop solutions to fiber-optics communications quandaries using the laws of physics. Most optics physicists have a doctorate in physics, usually with a specialization in optics. They also tend to spend several years after obtaining their doctorate performing academic research before moving to industry positions.

Specialized roles in computer software engineering and networking in the fiber-optic telecommunications industry also exist. Much like the engineering roles, individuals may specialize in a particular area of fiber optics such as networking, communication systems, telecommunications design, data communications, or computer software. Education requirements for entry-level positions begin with a bachelor's degree with a major in engineering, computer science, or a related field.

SOCIAL CONTEXT AND FUTURE PROSPECTS

Fiber-optic communications technologies are constantly changing and integrating new innovations and applications. Some countries, such as Japan, have fully embraced use of fiber optics in the home as well as in business; however, the investment in infrastructure is not as fully actualized in other areas. The consumer demand for faster, better access to the Internet and related data-transmittal applications is driving the move from standard copper wiring to fiber optics. New types of fibers will increase fiber-optic application beyond telecommunications into more medical, military, and industrial uses. Though wireless technology use could negatively check industry growth, the strong consumer demand and increasing number of fiber-optics applications suggest that the fiber-optic industry will continue to grow. However, the industry may have more moderate growth as the telecommunication industry experiences decreased growth. This was seen during the economic recession of 2008 to 2010, as consumers held off upgrading from copper cabling to fiber optics. According to a report by Global Industry Analysts, the recession's impact on fiber-optic cabling ended in 2011. Overall, the report anticipates significant growth in the industry as more fiber-optic cable networks are installed and businesses, consumers, and telecom providers invest in advanced tools to facilitate the new networks. Employment in the wired telecommunications industry is expected to decline by 11 percent during the period from 2008 to 2018; however, telecommunications jobs focused on fiber optics are expected to rise.

Dawn A. Laney, B.A., M.S., C.G.C., C.C.R.C.

FURTHER READING

Allen, Thomas B. "The Future Is Calling," *National Geographic* 200, Issue 6 (December 2001): 76. An interesting and well-written description of the growth of the fiber-optics industry.

Belson, Ken. "Unlike U.S., Japanese Push Fiber Over Profit." *The New York Times.* October 3, 2007. Compares the United States' and Japan's approach to updating infrastructure with fiber-optic cabling.

Crisp, John, and Barry Elliott. *Introduction to Fiber Optics.* 3d ed. Burlington, Mass.: Elsevier, 2005. An excellent text for anyone, of any skill level, who wants to learn more about fiber optics from the ground up. Each chapter ends with review questions.

Goff, David R. *Fiber Optic Reference Guide: A Practical Guide to Communications Technology.* 3d ed. Burlington, Mass.: Focal Press, 2002. An excellent review of the history of fiber optics, the basic principles of the technology, and information on practical applications particularly in communications.

Hayes, Jim. *FOA Reference Guide to Fiber Optics: Study Guide to FOA Certification.* Fallbrook, Calif.: The Fiber Optic Association, 2009. A useful guide that details the design and installation of fiber optic cabling networks including expansive coverage of the components and processes of fiber optics.

Hecht, Jeff. *City of Light: The Story of Fiber Optics.* Rev. ed. New York: Oxford University Press, 1999. A readable history of the development of fiber optics.

WEB SITES

Fiber Optic Association
http://www.thefoa.org

International Telecommunication Union (ITU)
http://www.itu.int

U.S. Bureau of Labor Statistics
Career Guide to Industries: Telecommunications
http://www.bls.gov/oco/cg/cgs020.htm

See also: Computer Engineering; Computer Networks; Electrical Engineering; Mechanical Engineering; Telecommunications; Telephone Technology and Networks; Television Technologies.

FIBER TECHNOLOGIES

FIELDS OF STUDY

Chemistry; physics; mathematics; chemical engineering; polymer chemistry; agriculture; agronomy; mechanical engineering; industrial management; waste management; business management

SUMMARY

Fibers have been used for thousands of years, but not until the nineteenth and twentieth centuries did chemically modified natural fibers (cellulose) and synthetic plastic or polymer fibers become extremely important, opening new fields of application. Advanced composite materials rely exclusively on synthetic fibers. Research has also produced new applications of natural materials such as glass and basalt in the form of fibers. The current "king" among fibers is carbon, and new forms of carbon, such as carbon nanotubes, promise to advance fiber technology even further.

KEY TERMS AND CONCEPTS

- **Denier:** A unit indicating the fineness of a filament; a filament 9,000 meters in length weighing 1 gram has a fineness of 1 denier.
- **Roving:** Nonwoven fiber fabric whose strands all have the same absolute orientation.
- **Warp Clock:** A visual guide to the orientation of the warp of woven fiber fabrics, used in the laying-up of composite materials to provide a quasi-isotropic character to the final product.

DEFINITION AND BASIC PRINCIPLES

A fiber is a long, thin filament of a material. Fiber technologies are used to produce fibers from different materials that are either obtained from natural sources or produced synthetically. Natural fibers are either cellulose-based or protein-based, depending on their source. All cellulosic fibers come from plant sources, while protein-based fibers such as silk and wool are exclusively from animal sources; both fiber types are referred to as biopolymers. Synthetic fibers are manufactured from synthetic polymers, such as nylon, rayon, polyaramides, and polyesters. An infinite variety of synthetic materials can be used for the production of synthetic fibers.

Production typically consists of drawing a melted material through an orifice in such a way that it solidifies as it leaves the orifice, producing a single long strand or fiber. Any material that can be made to melt can be used in this way to produce fibers. There are also other ways in which specialty fibers also can be produced through chemical vapor deposition. Fibers are subsequently used in different ways, according to the characteristics of the material.

BACKGROUND AND HISTORY

Some of the earliest known applications of fibers date back to the ancient Egyptian and Babylonian civilizations. Papyrus was formed from the fibers of the papyrus reed. Linen fabrics were woven from flax fibers. Cotton fibers were used to make sail fabric. Ancient China produced the first paper from cellulose fiber and perfected the use of silk fiber.

Until the nineteenth century, all fibers came from natural sources. In the late nineteenth century, nitrocellulose was first used to develop smokeless gunpowder; it also became the first commercially successful plastic: celluloid.

As polymer science developed in the twentieth century, new and entirely synthetic materials were discovered that could be formed into fine fibers. Nylon-66 was invented in 1935 and Teflon in 1938. Following World War II, the plastics industry grew rapidly as new materials and uses were invented. The immense variety of polymer formulations provides an almost limitless array of materials, each with its own unique characteristics. The principal fibers used today are varieties of nylons, polyesters, polyamides, and epoxies that are capable of being produced in fiber form. In addition, large quantities of carbon and glass fibers are used in an ever-growing variety of functions.

HOW IT WORKS

The formation of fibers from natural or synthetic materials depends on some specific factors. A material must have the correct plastic characteristics that allow it to be formed into fibers. Without exception, all natural plant fibers are cellulose-based, and all fibers from animal sources are protein-based. In some cases, the fibers can be used just as they are taken from their source, but the vast majority of natural fibers must be subjected to chemical and physical treatment processes to improve their properties.

Cellulose Fibers. Cellulose fibers provide the greatest natural variety of fiber forms and types. Cellulose is a biopolymer; its individual molecules are constructed of thousands of molecules of glucose chemically bonded in a head-to-tail manner. Polymers in general are mixtures of many similar compounds that differ only in the number of monomer units from which they are constructed. The processes used to make natural and synthetic polymers produce similar molecules having a range of molecular weights. Physical and chemical manipulation of the bulk cellulose material, as in the production of rayon, is designed to provide a consistent form of the material that can then be formed into long filaments, or fibers.

Synthetic Polymers. Synthetic polymers have greatly expanded the range of fiber materials that are available, and the range of uses to which they can be applied. Synthetic polymers come in two varieties: thermoplastic and thermosetting. Thermoplastic polymers are those whose material becomes softer and eventually melts when heated. Thermosetting polymers are those whose the material sets and becomes hard or brittle through heating. It is possible to use both types of polymers to produce fibers, although thermoplastics are most commonly used for fiber production.

The process for both synthetic fibers is essentially the same, but with reversed logic. Fibers from thermoplastic polymers are produced by drawing the liquefied material through dies with orifices of the desired size. The material enters the die as a viscous liquid that is cooled and solidifies as it exits the die. The now-solid filament is then pulled from the die, drawing more molten material along as a continuous fiber. This is a simpler and more easily controlled method than forcing the liquid material through the die using pressure, and it produces highly consistent fibers with predictable properties.

Fibers from thermosetting polymers are formed in a similar manner, as the unpolymerized material is forced through the die. Rather than cooling, however, the material is heated as it exits the die to drive the polymerization to completion and to set the polymer.

Other materials are used to produce fibers in the manner used to produce fibers from thermoplastic polymers. Metal fibers were the first of these materials. The processes used for their production provided the basic technology for the production of fibers from polymers and other nonmetals. The best-known of these fibers is glass fiber, which is used with polymer resins to form composite materials. A somewhat more high-tech variety of glass fiber is used in fiber optics for high-speed communications networks. Basalt fiber has also been developed for use in composite materials. Both are available commercially in a variety of dimensions and forms.

Production of carbon fiber begins with fibers already formed from a carbon-based material, referred to as either pitch or PAN. Pitch is a blend of polymeric substances from tars, while PAN indicates that the carbon-based starting material is polyacrylonitrile. These starting fibers are then heat-treated in such a way that essentially all other atoms in the material are driven off, leaving the carbon skeletons of the original polymeric material as the end-product fiber.

Boron fiber is produced by passing a very thin filament of tungsten through a sealed chamber, during which the element boron is deposited onto the tungsten fiber by the process of chemical vapor deposition.

APPLICATIONS AND PRODUCTS

All fiber applications derive from the intrinsic nature of the material from which the fibers are formed. Each material, and each molecular variation of a material, produces fibers with unique characteristics and properties, even though the basic molecular formulas of different materials are very similar. As well, the physical structure of the fibers and the manner in which they were processed work to determine the properties of those fibers. The diameter of the fibers is a very important consideration. Other considerations are the temperature of the melt from which fibers of a material were drawn; whether the fibers were stretched or not, and the degree by which they were stretched; whether the fibers are hollow, filled, or solid; and the resistance of the fiber material to such environmental influences as exposure to light and other materials.

Structural Fibers. Loosely defined, all fibers are structural fibers in that they are used to form various structures, from plain weave cloth for clothing to advanced composite materials for high-tech applications. That they must resist physical loading is the common feature identifying them as structural fibers. In a stricter sense, structural fibers are fibers (materials such as glass, carbon, aramid, basalt, and boron) that are ordinarily used for construction purposes. They are used in normal and advanced composite materials to provide the fundamental load-bearing strength of the structure.

A typical application involves "laying-up" a structure of several layers of the fiber material, each with its own orientation, and encasing it within a rigid matrix of polymeric resin or other solidifying material. The solid matrix maintains the proper orientation

Kevin Ausman, executive director of the Rice University center for biological and environmental nanotechnology, holds a bottle of carbon nanotubes. Nanotechnology is supposed to make computers small enough to implant into a wrist and supply materials that strengthen and lighten bridges and airplanes. (AP Photo)

of the encased fibers to maintain the intrinsic strength of the structure.

Materials so formed have many structural applications. Glass fiber, for example, is commonly used to construct different fiberglass shapes, from flower pots to boat hulls, and is the most familiar of composite fiber materials. Glass fiber is also used in the construction of modern aircraft, such as the Airbus A-380, whose fuselage panels are composite structures of glass fibers embedded in a matrix of aluminum metal.

Carbon and aramid fibers such as Kevlar are used for high-strength structures. Their strength is such that the application of a layer of carbon fiber composite is frequently used to prolong the usable lifetime of weakened concrete structures, such as bridge pillars and structural joists, by several years. While very light, Kevlar is so strong that high-performance automotive drive trains can be constructed from it. It is the material of choice for the construction of modern high-performance military and civilian aircraft, and for the remote manipulators that were used aboard the space shuttles of the National Aeronautics and Space Administration. Kevlar is recognizable as the high stretch-resistance cord used to reinforce vehicle tires of all kinds and as the material that provides the impact-resistance of bulletproof vests.

In fiber structural applications, as with all material applications, it is important to understand the manner in which one material can interact with another. Allowing carbon fiber to form a galvanic connection to another structural component such as aluminum, for example, can result in damage to the overall structure caused by the electrical current that naturally results.

Fabrics and Textiles. The single most recognized application of fiber technologies is in the manufacture of textiles and fabrics. Textiles and fabrics are produced by interweaving strands of fibers consisting of single long fibers or of a number of fibers that have been spun together to form a single strand. There is no limit to the number of types of fibers that can be combined to form strands, or on the number of types of strands that can be combined in a weave.

The fiber manufacturing processes used with any individual material can be adjusted or altered to produce a range of fiber textures, including those that are soft and spongy or hard and resilient. The range of chemical compositions for any individual polymeric material, natural or synthetic, and the range of available processing options, provides a variety of properties that affect the application of fabrics and textiles produced.

Clothing and clothing design consume great quantities of fabrics and textiles. Also, clothing designers seek to find and utilize basic differences in fabric and textile properties that derive from variations in chemical composition and fiber processing methods.

Fibers for fabrics and textiles are quantified in units of deniers. Because the diameter of the fiber can be produced on a continuous diameter scale, it is therefore possible to have an essentially infinite range of denier weights. The effective weight of a fiber may also be adjusted by the use of sizing materials added to fibers during processing to augment or improve their stiffness, strength, smoothness, or weight. The gradual loss of sizing from the fibers

accounts for cotton denim jeans and other clothing items becoming suppler, less weighty, and more comfortable over time.

The high resistance of woven fabrics and textiles to physical loading makes them extremely valuable in many applications that do not relate to clothing. Sailcloth, whether from heavy cotton canvas or light nylon fabric, is more than sufficiently strong to move the entire mass of a large ship through water by resisting the force of wind pressing against the sails. Utility covers made from woven polypropylene strands are also a common consumer item, though used more for their water-repellent properties than for their strength. Sacks made from woven materials are used worldwide to carry quantities of goods ranging from coffee beans to gold coins and bullion. One reason for this latter use is that the fiber fabric can at some point be completely burned away to permit recovery of miniscule flakes of gold that chip off during handling.

Cordage. Ropes, cords, and strings in many weights and winds traditionally have been made from natural fibers such as cotton, hemp, sisal, and manila. These require little processing for rough cordage, but the suppleness of the cordage product increases with additional processing. Typically, many small fibers are combined to produce strands of the desired size, and these larger strands can then be entwined or plaited to produce cordage of larger sizes. The accumulated strength of the small fibers produces cordage that is stronger than cordage of the same size consisting of a single strand. The same concept is applied to cordage made from synthetic fibers.

Ropes and cords made from polypropylene can be produced as a single strand. However, the properties of such cordage would reflect the properties of the bulk material rather than the properties of combined small fibers. It would become brittle when cold, overly stretchy when warm, and subject to failure by impact shock. Combined fibers, although still subject to the effects of heat, cold, and impact shock, overcome many of these properties as the individual fibers act to support each other and provide superior resistance.

IMPACT ON INDUSTRY

Industries based on fiber technologies can be divided into two sectors: those that produce fibers and those that consume fibers; both are multibillion dollar sectors.

Fiber production industries are agricultural and technical. Forestry and other agricultural industries produce large amounts of cellulosic fiber each year. Production of cellulosic fiber peaked at some 3 million metric tons in 1982, representing 21 percent of the fiber market share, but this heavy production has steadily declined since 1982. By 2002, cellulosic fiber production had decreased to 6 percent of the world fiber-market share. Production in Eastern Europe dropped from 1.1 million metric tons to just 92,000 metric tons, and production in Asia increased by 660,000 metric tons over the same period, accounting for fully 69 percent of global production in 2002. Most of the cellulosic fiber that is produced by the forestry industry is used in the manufacture of paper and paper products, while the other cellulosic fiber types—cotton, hemp, sisal, and manila—are used primarily for fabrics, textiles, and cordage.

Synthetic fiber production is the principal driving force behind the decline in cellulosic fiber production. Industrial polymerization processes provide a much greater variety of fiber-forming materials with a lower requirement for process control. In the period from 1982 to 2002, manufactured fiber production increased by 155 percent, with the rise attributed to synthetic fiber as the production of cellulosic fiber decreased. Fibers from synthetic materials account for 94 percent of total global fiber production.

A great deal of research has been expended in the field of polymerization chemistry to identify effective catalysts and processes for the reactions involved, so it has become ever easier to control the nature of the materials being produced. Polymerization reactions have become controllable such that their products now span a much narrower range of monomer weights. This specificity of control has enabled the production of more specific fiber materials, which has in turn driven research and development of more specific uses of those fibers.

The greatest increase in synthetic fiber materials has been an increase in the polyesters, which now account for almost two-thirds of total synthetic fiber production. It must be remembered that terms such as "polyester" and "polyamide" refer to broad classes of compounds and not to specific materials. Each class contains untold thousands of possible variations in molecular structure, both from the chemical identity of the monomers used and from the order in which they react during polymerization.

Technical industries that produce fibers from both natural and synthetic materials make use of several classes of machinery. One class of machinery (for example, pelletizers and masticators) manipulates raw materials into a form that is readily transported and usable in the fiber-forming process. Another class of machinery (for example, injection molders and spinnerets) carries out the fiber-forming process, while still another class of machinery (for example, sizers and mercerizers) uses the raw fiber to provide a finished fiber product ready for consumers.

Industries that consume finished fibers include weaving mills that produce fabrics and textiles from all manner of fibers, including glass, basalt, and carbon. These fibers are then used accordingly, for clothing, composites, or other uses. Cordage manufacturers use finished fibers as an input feedstock to produce every type of cordage, from fine thread to coarse rope.

A more recent development in the fiber industry is the utilization of recycling wastes as feedstock for fiber-forming processes. Synthetic materials such as nylon, polyethylene, and polyethylene terephalate (PET), which have historically been condemned to landfills, are now accepted for reprocessing through recycling programs and formed into useful fibers for many different purposes. The versatility of the materials themselves lends to the development of new industries based on their use.

CAREERS AND COURSEWORK

Careers in textile production depend on a sound basic education in chemistry, physics, mathematics, and materials science. Students anticipating such a career should be prepared to take advanced courses in these areas at the college and university level, with specialization in organic chemistry, polymer chemistry, and industrial chemistry. Postsecondary programs specializing in textiles and the textile industry also are available, and provide the specialist training necessary for a career in the manufacture and use of fibers and textiles. The chemistry of color and dyes is another important aspect of the field, representing a distinct area of specialization.

Composite-materials training and advanced composite-materials specialist training can be obtained through only a limited number of public and private training facilities, although considerable research in this field is carried out in a number of universities and colleges. Private industries that require this kind of specialization, particularly aircraft manufacturers, often have their own patented fabrication processes and so prefer to train their personnel on-site rather than through outside agencies.

SOCIAL CONTEXT AND FUTURE PROSPECTS

One could argue that the fiber industry is the principal industry of modern society, solely on the basis that everyone wears clothes of some kind that have been made from natural or synthetic fibers. As this is unlikely ever to change, given the climatic conditions that prevail on this planet and given the need for protective outerwear in any environment, there is every likelihood that there will always be a need for specialists who are proficient in both fiber manufacturing and fiber utilization.

Richard M. Renneboog, M.Sc.

FURTHER READING

Fenichell, Stephen. *Plastic: The Making of a Synthetic Century.* New York: HarperCollins, 1996. A well-researched account of the plastics industry, focusing on the social and historical contexts of plastics, their technical development, and the many uses for synthetic fibers.

Morrison, Robert Thornton, and Robert Nielson Boyd. *Organic Chemistry..* 5th ed. Newton, Mass.: Allyn & Bacon, 1987. Provides one of the best and most readable introductions to organic chemistry and polymerization.

Selinger, Ben. *Chemistry in the Marketplace.* 5th ed. Sydney: Allen & Unwin, 2002. The seventh chapter of this book provides a concise overview of many fiber materials and their common uses and properties.

Weinberger, Charles B. "Instructional Module on Synthetic Fiber Manufacturing." Gateway Engineering Education Coalition: 30 Aug. 1996. This article presents an introduction to the chemical engineering of synthetic fiber production, giving an idea of the sort of training and specialization required for careers in this field.

WEB SITES

University of Tennessee Space Institute
http://www.utsi.edu/research/carbonfiber

U.S. Environmental Protection Agency
http://www.epa.gov

See also: Agricultural Science; Bioprocess Engineering; Fiber-Optic Communications; Forestry; Plastics Engineering; Polymer Science; Structural Composites.

FISHERIES SCIENCE

FIELDS OF STUDY

Ecology; biology; chemistry; marine biology; limnology; ichthyology; oceanography; ethology; hydrology; anthropology; law; economics; political science; genetic engineering; mechanical engineering; electrical engineering; systems engineering; veterinary medicine; ethology.

SUMMARY

Fisheries science is an interdisciplinary study concerned with the hunting and farming of aquatic organisms in oceans and bodies of fresh- or saltwater as part of the ongoing effort to feed the world's population. Until recently, most aquatic organisms from fisheries were wild creatures. Farming and the use of genetic-engineering techniques to produce desirable domesticated aquatic food organisms are widespread, but the majority of fisheries science still involves the management of fish stocks for sustainability of the resource.

KEY TERMS AND CONCEPTS

- **Allowable Biological Catch (ABC):** Maximum amount of fish stock that can be harvested without adversely affecting recruitment or other biological components of the stock.
- **Benthos:** All those animals and plants living on or in sediments at the bottom of the sea; benthic animals are usually described by their position in the sediment relative to the surface, and their size, either living within (burrowing in) the sediments, or living at, near, or on the sediment surface.
- **Biomass:** Amount or mass of some organism, such as fish.
- **Bycatch:** Fish, other than the primary target species, that are caught incidental to the harvest of the primary target species.
- **Derby:** Fishery in which the total allowable catch is fixed and participants do not have individual quotas; participants attempt to maximize their harvest as quickly as possible, before the fishery is closed for the season.
- **Epipelagic:** Pertaining to the community of suspended organisms inhabiting an aquatic environment between the surface and a depth of 200 meters.
- **Individual Fishing Quota (IFQ):** Fishery management tool that allocates a certain proportion of the total allowable catch to individual vessels, fishermen, or other eligible recipients based on initial qualifying criteria.
- **Input Controls:** Fishery management measures that seek to limit the amount or effectiveness of effort in a fishery, including licenses limiting numbers of fishermen, gear restrictions, and time limits on fishing activities.
- **Output Controls:** Fishery management measures (including total allowable catch and individual fishing quotas) intended to limit the amount of catch or harvest in a fishery.
- **Pelagic:** Of or pertaining to the open waters of the sea.
- **Recruitment:** Number or percentage of fish that survive from birth to a specific age or size.
- **Total Allowable Catch (TAC):** Total catch permitted to be caught from a stock in a given period, typically a year; usually this amount is less than the allowable biological catch.

DEFINITION AND BASIC PRINCIPLES

Fisheries science is concerned with the continued extraction of aquatic organisms from marine, brackish, and freshwater environments for subsistence, commercial, or recreational purposes. As such, fisheries science necessarily involves issues of yield (the numbers of fish harvested from a given stock) and sustainability (the numbers of fish that must not be harvested if the particular fish stock is to demonstrate continued productivity). This seemingly simple biological situation, however, is made much more complex by ecological, economic, and social considerations.

No aquatic species—not even farmed fish—exists in isolation, unconnected to the food web of its surrounding ecosystem. To the extent that fishing or aquaculture alters the population of one species, it also inevitably alters the populations of other species in the same environment. The health of a fishery is dependent as much on the health of the entire ecosystem as it is on any particular species. Sustainability, then, must ultimately take into account not just numbers of a particular fish stock harvested or left behind but also the health of that fish stock's entire ecosystem.

Fisheries science is further complicated by the fact that fish, particularly in the sea, have long been assumed to be a limitless resource, available to all for the taking. Depleted stocks and increasing competition for declining numbers of fish have led to an understanding that even fish in the sea are also a limited resource. This has led to the search for new ways to manage marine fisheries and to produce more fish aquaculturally through fish farming of various types. Basic, applied, and developmental research in fisheries science is deeply involved with aquaculture and fisheries management. These are fields in which biological sustainability and productivity must be balanced with the economic productivity of those employed in the fishing and fish-farming industries and with the social productivity of communities who have long been dependent on the harvest of aquatic organisms.

BACKGROUND AND HISTORY

Although evidence of aquaculture (particularly the raising of fish in ponds or estuaries) goes back at least 5,000 years, and evidence of the hunting and gathering of aquatic organisms goes back significantly more than 100,000 years. It was only in the twentieth century that fisheries science developed out of the biological study of aquatic organisms. The development of the discipline can be traced to the fact that, although there had been occasional spot depletions of fish stocks (particularly in freshwater environments or among large sea mammals such as whales), widespread depletions of fish stocks did not become a concern until the twentieth century.

These depletions and eventual collapses of fisheries resulted from technological advances that increased both the efficiency in how a given fish stock was harvested and the rate at which that stock was exploited. Through the first half of the twentieth century these technological advances were primarily mechanical: More powerful steam and then diesel engines allowed larger and sturdier fishing vessels, hauling larger and stronger fishing gear, to travel farther and, along with better refrigeration, stay at sea longer. During the second half of the twentieth century, the technological advances were primarily

electronic: sonar and the Global Positioning System (GPS), along with many other informational technologies, made it possible to find the fish more accurately and extract them.

Over the course of the twentieth century, each new technological advance allowed fishers to eat deeper into the natural capital of a given fish stock. Short-term-oriented market forces, driving an extractive technological arms race within the context of open-access fisheries, often reached the point that the fish stocks in question collapsed entirely. The need to manage fish stocks more rationally became increasingly obvious, and fisheries science has begun to serve this need.

HOW IT WORKS

The core of fisheries science is biological. It is concerned with understanding the life cycles of individual aquatic organisms, including growth rates, ages of sexual maturity, longevity, predation, and mortality. How these factors affect estimates of fish-stock population sizes and long-term population management of fish stocks are also all of key importance in fisheries science. Fisheries science also requires practical understanding of fish tagging and fish marking, the particular fishing gear used in specific fisheries, habitat improvement and bioremediation, fishways, screens, and guiding devices, the role of hatchery-raised fish and stocking, and small-pond, floating-enclosure, or net-pen management (these last being particularly important in aquacultural contexts).

Whether in aquacultural or traditional fishing approaches, however, fisheries scientists find they must go beyond a simply biological understanding of fish stocks. Fisheries science is profoundly influenced by the context of the cultural and legal framework within which fisheries management takes place. An important part of this framework is the common-law doctrine of the "public trust," particularly the idea that the resources of the rivers and seas within a nation's jurisdiction belong to the people, and the government holds them in trust for the public.

Fisheries as a Public Trust. The idea of the public trust has its roots in the principle of ancient Roman law that things such as the air, running water, the sea, and the shores of the sea are incapable of private ownership. In English law, and later in U.S. law, the fish and wild beasts were added to the list of "common

From left, Matt Blommel, a fisheries technician, Bob Greenlee, a district fisheries biologist, and Scott Herrmann, a fisheries biologist, measure and weigh blue catfish caught while electro-fishing on the James River on Thursday, April 24, 2008. (AP Photo)

property" that the government holds in trust for the public. Resources relating to the public good cannot be given away to private interests because public trust resources are inalienable—that is, they cannot be given over or transferred by the government to other entities or individuals. The government also must, by law, exercise continuing stewardship responsibility and authority over public trust resources, including fisheries.

In conjunction with improved extractive technologies and largely unregulated market forces, the open-access nature of fisheries in many cases resulted in the degradation of those fisheries. It thus became increasingly clear that government, as trustee of the fisheries, ought to exercise its stewardship authority over these public trust resources. In the United States, this stewardship responsibility is seen most clearly in the Magnuson-Stevens Fishery Conservation and Management Act of 1976 (which extended to 200 miles offshore the nation's jurisdiction over the sea as an "exclusive economic zone" for fishing) and the Sustainable Fisheries Act of 1996 (which called for further analysis of sustainability and closed to fishing many depleted fisheries, in hopes of their recovery).

The places where aquatic organisms (whether wild or farmed) live and from which they are to be

harvested or protected from harvest will likely remain waters held primarily in public trust. As a result, the fisheries scientists will continue to serve most often in the role of knowledge expert, assessing fish stocks and advising government, industry, and the public how best to use the fish stocks that are the common property of a nation's citizens.

Assessing Estimated Population Size and Ecological Risk: Traditional Fisheries. To estimate yield, to understand basic changes in population number and composition, and as a basis for sound management, the fisheries scientist must be able to estimate fish population reliably. Like much of the rest of fisheries science, population assessment is based in mathematical and systems-analysis approaches. Although opportunities do exist for direct counting of fish, most statistics on fish populations are estimates. Some examples of methods for estimating population include area density, mark-recapture or "single census," and catch-effort.

The area-density approach to estimating population involves counting the number of animals in a series of sample strips or plots distributed randomly or systematically throughout the total environmental area in which the fish stock population is to be determined. The sample count, once taken, is expanded to an estimate of the entire population by multiplying the aggregate sample count by a particular fraction (total area divided by the sum of sample areas). Because a subarea can also be sampled for time instead of space, the same equations applied to area-density can also be applied to partial-time coverage.

In the mark-recapture method of estimating fish population, a sample of fish is collected, marked, and released, and then at a later time a second or recapture sample is taken, which includes both marked and unmarked fish. This approach is based on the assumption that the proportion of marked fish recovered is to the total catch in the second sample as the total number of marked fish released is to the total fish population.

The catch-effort method of estimating fish population depends on the premise that all individual fish in a sample have the same chance of being caught and that the effort made by fishermen to catch the fish is constant. This approach works best when the population is closed, when chance of capture and constancy of effort remains unchanged from sample to sample, and when there are enough fish removed so that the population shows a decline.

Assessing Estimated Ecological Risk: Aquaculture. In aquaculture, population is much more controlled, so assessment emphasizes questions of ecological risk rather than determination of population. Because of the intensive nature of fish-farming practices, these questions of ecological risk include near-field and far-field effects of increased organic loading (mainly from uneaten fish feed, fish fecal material, and decomposing dead farm fish), increased inorganic loading (mainly nitrogen, phosphorus, trace elements, and vitamins in fish excretory products and uneaten feed), residual heavy metals (mainly zinc compounds in uneaten feed and fish feces), the transmission of disease organisms (often increased because of the high density of fish per volume of water in net-pen enclosures), and residual therapeutants (from biomedical treatments of the fish performed in response to the increased transmission of disease organisms resulting from crowded net-pen conditions).

Other ecological risks to be assessed for aquaculture include biological interaction of farmed fish that have escaped into wild populations and vice versa (with the potential not only for breeding but also for cross-infection with parasites and pathogens), the impact on marine habitat of fish-farm enclosures (including entanglements involving nets, anchors, and moorings), control of natural predators in the fish-farm environment, and increasing pressure on shoaling small pelagic fish populations to be fed to the farmed fish.

APPLICATIONS AND PRODUCTS

The primary role of most fisheries scientists continues to be consultative. That consultative role usually falls into three categories: assessor (the person responsible for making reliable estimates of fish populations, predicting what harvest levels those populations can support, and the environmental impacts of both fishing and fish-farming approaches), adviser (the person responsible for communicating to government, industry, and public all findings relevant to the health of a fishery and its environment), and educator (the person responsible for increasing governmental, industrial, and popular awareness of both the economic and ecological importance of aquatic organisms).

Some fisheries scientists, however, already find themselves involved in efforts with direct applications and products. These include new ways of growing or harvesting aquatic organisms, some as revolutionary as genetic modification of farmed fish, some as evolutionary as the development of more efficient fishing gear.

Aquatic and Marine Food. Fisheries scientists, as experts in the health of individual fish species and overall fishery populations, help industry provide high-quality seafood products for the consumer—an important activity, considering that seafoods are one of the world's primary sources of high-quality protein. Fisheries scientists are involved in product development, physicochemical principles, and process

---

Fascinating Facts About Fisheries Science

- Wild fisheries generated more than $63 billion in household incomes worldwide in 2009.
- Of the world's total fish stocks, 80 percent are either fully exploited or have collapsed.
- One billion residents of Asia depend on seafood for the majority of their animal protein.
- One in five Africans depends on seafood for the majority of his or her animal protein.
- The United States, Japan, and the countries of the European Union are net importers of seafood.
- Between 1986 and 2009, the world population of hammerhead sharks declined by 89 percent—largely from "finning" to make soup.
- Between 1965 and 2009, the Russian sturgeon population, exploited for caviar, declined by 90 percent.
- Overfishing by bottom trawlers along the New Zealand coast has resulted in an 80 percent decline in orange roughy numbers.
- Ghost fishing—nets, lines, and traps that continue to catch fish despite having been lost or abandoned—is a particular problem of derby fisheries.

---

technology for aquatic food and marine bioproduct utilization, as well as in examining and improving aquatic and marine products and manufacturing processes.

In the aquacultural context, fisheries science helps to increase the aquacultural contribution to the food supply while also developing new methods of production and improved cultural practices for selective species. These production and cultural concerns include environmental, ecological, and disease considerations, selective breeding, feeding, processing, and marketing.

Aquatic and Marine Nonfood Products. In addition to foodstuffs, however, many other products are derived from aquatic and marine organisms. Fish oil, which contains omega-3 fatty acids and anti-inflammatory eicosanoids, are important for a healthy diet. Fish meal, a high-protein food supplement used in aquaculture, is a by-product of rendering and processing fish for fish oil. Fish emulsion, a fertilizer, is produced from the fluid remainder of fish already processed for fish oil and fish meal. Fish skins, swim bladders, and bones are boiled to produce fish glue for specialized uses, while isinglass, a form of collagen obtained from dried swim bladders, is used for clarifying and refining wine and beer, for preserving eggs, and for conserving parchment. Kelp is steadily growing in popularity as a fertilizer and has long been a major source of iodine. Traditional royal or Tyrian purple is derived from sea snails of the *Murex* genus. Pearls and mother-of-pearl are key components of lustrous jewelry.

Fisheries scientists continue to explore aquatic and marine bioresources for pharmaceuticals, nutraceuticals, and novel biomaterials as well as investigate distribution and biodiversity of marine organisms important to industrial utilization. In finding innovative uses for aquatic and marine products and thereby increasing the value of specific fish stocks, fisheries scientists also make more likely the prospect that a given stock will be more sustainably harvested over the long term.

Aquaculture. In the aquaculture context, fisheries scientists not only examine the impacts of aquaculture practices on the environment (including habitat alteration, release of drugs and chemicals, interaction of cultured and wild organisms, and related environmental and regulatory issues) but also propose and evaluate methods for reducing or eliminating those impacts, including modeling, siting, and monitoring of aquaculture facilities and the use of polyculture and water-reuse systems. Fisheries scientists propose the design criteria, provide operational analysis, and develop management strategies for selected species in water-reuse systems.

Fisheries scientists provide in-depth understanding of the natural and social ecology of aquaculture ecosystems, applying principles of systems ecology to the management of the world's aquaculture ecosystems. They also evaluate the nature, causes, and spread of diseases limiting the success of freshwater and marine aquaculture projects. They provide diagnoses of diseases affecting hatchery management and other aquacultural contexts, as well as define appropriate prevention, control, and treatment strategies for those diseases.

Traditional Fisheries. For traditional or wild fisheries, much fisheries science involves advising and educating government, industry, and the public on the biology of aquatic-resource animals, as well as assessing fisheries' populations, stock abundance, and overall ecological health. Fisheries scientists also help frame the debate concerning aquatic resource management, conservation legislation, rehabilitation of depleted fisheries, and socioeconomic considerations involved in national and international fishery issues, practices, patterns, and public policy.

Fisheries scientists also are involved in the development of new fish-catching methods and technologies, including the development and assessment of electronic enhancements to fishing and fishing-vessel operation that have increased fishing power. Fisheries science contributes to the application of these new methods to scientific sampling, to commercial harvesting, to recreational and subsistence fishing. Fisheries scientists not only contribute to advancements and innovations in fishing-gear construction, maintenance, and operation, but also to the evaluation—through empirical, theoretical, model scaling, and statistical-analysis techniques—of the behavior and performance of fish-capture systems.

IMPACT ON INDUSTRY

The best current estimates conclude that global fisheries contribute between $225 billion and $240 billion per year to the worldwide economy. Marine fisheries alone provide fifteen percent of the animal protein consumed by humans annually. Despite the importance of fisheries, however, management of fisheries worldwide lags far behind international guidelines recommended to minimize the effects of exploitation. Many researchers agree that conversion of scientific advice into policy, through a participatory and transparent process, is at the core of achieving fisheries' sustainability built on science-based management.

Achieving such sustainability is no easy task, however. Conservation and economic growth often tend to work against each other, at least in the short term. According to a recent National Oceanic and Atmospheric Administration (NOAA) study, in 2009 heavy and effective controls on American fishermen's efforts (focused on limiting fishermen's days at sea and barring access to certain fishing grounds at certain times of the year) drove down landings of fish by 6 percent in the United States—but also moved the nation's fisheries closer to sustainability for the tenth straight year.

A particular challenge that faces fisheries scientists in their role as advisers to government, industry, and the public is that prediction of fish stocks in fisheries and of risk assessment in aquaculture tend to be complicated by the fact that most of the factors involved are interactive. It is therefore important that the fisheries scientist, in communicating findings and making recommendations, ought to qualify and quantify the uncertainty associated with both the findings and the recommendations.

Despite the inherent uncertainties, important actions do occur as a result of the advice of fisheries scientists. For example, more and more of the world's major fisheries are no longer open access. Derby or race-for-fish situations—in which there is a total allowable catch in a fishery but no limitation on fishing by any individual fisherman—are being phased out in most major fisheries. Input controls limiting fisher effort and output controls limiting catch or harvest have become increasingly important tools for managing fisheries as a public trust resource.

Also on the advice of fisheries scientists, the duty to prevent overfishing, attention to marine resources used for noncommercial purposes, and the broader ecological context of fisheries have all increasingly become legally enforceable obligations. This more responsible approach has in many cases resulted in the reduction of bycatch and fish discarding as well as reduced the adverse impacts of fishing activity on critical fish habitat.

Now widely implemented in U.S. fisheries, the catch-share system is a transferable version of individual fishing quota, which grants fishermen in sector cooperatives an assigned share of a total allowable catch for each species and encourages fishermen to buy, sell, or trade shares with colleagues or outside investors. The catch-share approach, in removing the race for fish, has reduced the incentive to fish during unsafe conditions and to buy ever-larger vessels and more equipment (overcapitalization). Under this approach, fresh fish are available to consumers over longer periods during the year, and fishermen have also been able to maintain higher-quality product while reducing bycatch and gear conflicts.

Economic and socioeconomic downsides to the approach do exist, however. Fish imported into the United States from unsustainable Asian and European fisheries are cheaper than sustainably caught U.S. fish. There are also important concerns about the equity of gifting a public trust resource, the elimination of vessels and reductions in crews, and the consolidation of shares (and therefore economic power) in fewer hands and into larger businesses. The effects of this last not only on jobs but also on the fishing culture the industry has long made possible in many locales.

Government and University Research. As part of the public trust nature of the fisheries involved, many U.S. governmental entities make use of the advice provided by fisheries scientists—departments and agencies as diverse as NOAA, the United States Geological Survey (USGS), the Food and Drug Administration (FDA), the Department of Commerce and the National Marine Fisheries Service (NMFS), the National Academy of Science and National Research Council, the eight regional fisheries management councils, and the scientific and statistical committees that provide scientific advice to them. Individual state and even county departments of fish and game make use of such expertise, as do supranational entities like the United Nations Food and Agriculture Organization (FAO), the International Council for the Exploration of the Sea (ICES), the North Atlantic Fisheries Organization (NAFO), and the International Pacific Halibut Commission (IPHC).

Universities and related research institutions employ nearly as many fisheries scientists as do governments. Massachusetts' Woods Hole Oceanographic Institution and its marine biological laboratory, several campuses in the University of California system, the University of Maryland system, the University of Florida system, and the University of Alaska system all have substantial research programs in fisheries science.

Industry, Business, and Nongovernmental (NGO) Sectors. Although most participants in the fishing industry continue to rely on statistics produced by governmental agencies and university researchers, fisheries scientists are also being employed in increasing numbers by industry groups ranging from aquaculture firms such as AquaBounty (creators of the AquAdvantage line of genetically modified salmon), to fishermen's associations, to large fish-processing concerns such as the At Sea Processor's Association.

In response to the changing landscape in the fishing and aquaculture industries, nongovernmental organizations such as the Center for Science in the Public Interest, the Alliance for Natural Health, the World Wildlife Fund, People for the Ethical Treatment of Animals (PETA), the Marine Stewardship Council, Earth Island Institute, and many others are also increasingly finding a need for fisheries scientists and their advice.

CAREERS AND COURSE WORK

Career titles in fisheries science involve both traditional wild fisheries and aquaculture. They include aquarium director, biologist, conservation ecologist, conservation officer, environmental consultant, fish culturist, fish-processing manager, fisheries manager, fisheries technician, fisheries biologist, fisherman, hatchery manager, lab technician, museum or aquarium curator, marine biologist, natural-resource specialist, nature interpreter, lake or pond manager, researcher, vessel captain, and wildlife manager.

Courses in ecology, biology, chemistry, and mathematics are foundational for students wishing to pursue careers in fisheries science. A fisheries technician, fish-hatchery manager, fish-processing manager, lab technician, or nature interpreter may need little more than this background, together with some basic skills in mechanical, electrical, or systems engineering.

The pursuit of the master's and doctoral degrees—which are often the necessary minimum qualification for more advanced academic, governmental, or industrial career in fisheries science—generally requires more specialized course work beginning at the upper division of undergraduate studies or the beginning of graduate studies. More specialized courses may include genetics, marine biology, limnology, ichthyology, oceanography, and veterinary medicine.

Although fisheries science is primarily biological at its root, it is also strongly interdisciplinary and advisory, so background in fields such as hydrology, anthropology, law, economics, and political science can also prove very helpful.

SOCIAL CONTEXT AND FUTURE PROSPECTS

As more and more of the world's fisheries become overfished or fished out, and wild fish stocks are threatened globally, there is increased pressure to make the changeover from the mechanized hunting-gathering approach of commercial fishing to aquaculture. This must be done in decades, rather than the millennia involved in the analogous changeover to agriculture. Concerns about potential ecological risks, ranging from aquaculture's inherently dense and intense practices to genetic modification of fish species, have also increased as the rapidity of this changeover has increased.

One of the most pivotal roles of fisheries science in the future will be the assessment of how much of the near-shore oceans should remain wild and how much should be intensively farmed. This will involve not only the assessment of ecological risk associated with the growing aquaculture industry but also the determination of how fully and how rapidly depleted or damaged fisheries can recover. Marine reserves, aquatic refuges, and bioremediation all have roles to play here, as do concerns associated with reef die-offs, pollution and climate change, and economic and political pressures on the remaining 20 percent of fisheries worldwide that are not yet fully exploited.

During the twentieth century, the survival or extinction of many populations of aquatic organisms will depend on how the growing population of a terrestrial organism with a fondness for seafood—the human species—interacts with its global environment. Fisheries scientists, in their roles as assessors and advisers, are key to helping shape that interaction.

Howard V. Hendrix, B.A., M.A., Ph.D.

FURTHER READING

Clover, Charles. *The End of the Line: How Overfishing Is Changing the World and What We Eat.* Berkeley: University of California Press, 2008. An investigative journalist passionately (and sometimes polemically) writes about his personal research into the crisis in global fisheries.

Committee to Review Individual Fishing Quotas, et al. *Sharing the Fish: Toward a National Policy on Individual Fishing Quotas.* Washington, D.C.: National Academy Press, 1999. Report of committee empaneled by Congress to report on individual fishing quotas.

Everhart, W. Harry, and William D. Youngs. *Principles of Fisheries Science.* 2d ed. Ithaca, N.Y.: Cornell University Press, 1981. Classic introductory textbook on the subject, providing a broad overview of the field.

Iudicello, Suzanne, Michael Weber, and Robert Wieland. *Fish, Markets, and Fishermen: The Economics of Overfishing.* Washington, D.C.: Island Press, 1999. Emphasizes the importance of human economic behavior in the exploitation of fisheries and their management.

McGoodwin, James R. *Crisis in the World's Fisheries: People, Problems, and Policies.* Stanford, Calif.: Stanford University Press, 1990. An anthropological perspective on fisheries, emphasizing indigenous and small-scale fishers.

Molyneaux, Paul. *Swimming in Circles: Aquaculture and the End of Wild Oceans.* New York: Thunder's Mouth Press, 2007. A thoroughly researched, critical journalistic account of salmon aquaculture in Maine and shrimp aquaculture in Mexico.

Walters, Carl J., and Steven J. D. Martell. *Fisheries Ecology and Management.* Princeton, N.J.: Princeton University Press, 2004. Upper division and graduate textbook for classes in fisheries assessment and management, with an emphasis on quantitative modeling methods.

WEB SITES

American Fisheries Society
http://www.fisheries.org

Food and Agriculture Organization of the United Nations
Fisheries and Aquaculture Department
http://www.fao.org/fishery/en

National Oceanic and Atmospheric Administration
National Marine Fisheries Service
http://www.nmfs.noaa.gov

National Oceanic and Atmospheric Administration
Office of Sustainable Fisheries
http://www.nmfs.noaa.gov/sfa/statusoffisheries/SOSmain.htm

See also: Anthropology; Electrical Engineering; Mechanical Engineering; Oceanography; Veterinary Science.

FLOOD-CONTROL TECHNOLOGY

FIELDS OF STUDY

Hydrology; civil engineering; fluvial geomorphology; watersheds; climatology; meteorology; coastal erosion; stream hydraulics; flood forecasting; dam building; physical geography.

SUMMARY

Flood-control technology deals with the myriad techniques that can be employed to deal with water that overflows stream banks, thereby leading to deaths and injuries, property and crop damage, and severe erosion. The magnitude of the flood damage varies considerably, as it depends on the duration of the storm and the amount of precipitation. Each watershed has different physical features, such as size, slope, basin relief, impoundments, flood history, soil types, and drainage characteristics.

KEY TERMS AND CONCEPTS

- **Flash Flood:** Occurs when storms cause a river to rise rapidly.
- **Flood Control:** Refers to various measures that could reduce flood damage.
- **Flood Frequency:** Average amount of time between floods that are the same or greater than a selected magnitude.
- **Floodplain:** Low-lying area of varying width that could be on one or both sides of a river.
- **Flood Probability:** Statistical determination that flood events of a particular size would be less or the same during a particular period of time.
- **Flood Stage:** Occurs when the river overflows its banks and moves onto the floodplain.
- **Flood Wall:** Structure designed to reduce flooding.
- **Floodway:** Large channel built to divert floods away from populated areas.
- **Hydrograph:** Graphic representation of the amount of water passing a given point.

DEFINITION AND BASIC PRINCIPLES

Flood-control technology includes various structural and nonstructural measures that play a substantial role in mitigating flood damages. The methods vary from place to place, given the enormous heterogeneity of stream discharge, precipitation frequency and amount, and watershed factors that include slope, soil permeability, infiltration characteristics, degree of urbanization, varying land-management practices, governmental interest, and land-ownership practices.

The structural or physical measures that are employed include levees or flood walls made of earth or concrete, channel modifications such as deepening or widening the stream itself, building dams or reservoirs to hold additional water, watershed-improvement techniques that include reforestation and planting of vegetative cover in bare areas, flood proofing in lower-risk floodplain areas that include dikes, elevating the buildings, and waterproofing.

The nonstructural or preventative measures that could be adopted include floodplain regulations such as zoning laws that stipulate where development may or may not occur, building codes that put restrictions on basements and the permissible amount of impervious cover allowed on the site, open-space requirements for large-scale developments, and designing flood warning systems to provide advance notice about impending problems.

BACKGROUND AND HISTORY

The beginnings of speculation about water movement (the hydrologic cycle) began in ancient times by such renowned philosophers as Homer, Thales, Plato, and Aristotle in Greece, and later on by Lucretius, Seneca, and Pliny in ancient Rome. The field of hydrometry that pertains to stream-flow measurement began in the seventeenth century and improved techniques increased with each century. For example, flow formulas, measuring instruments, and stream-gauging procedures became better established in the eighteenth and later centuries. Some of the many advances include civil engineer Theodore G. Ellis's and surveyor William Gunn Price's stream current meters in 1870 and 1885, respectively, Irish engineer Robert Manning's flow formula in 1891, civil engineer Allen Hazen's comments on increasing the use of statistics in flood studies in 1930, and German statistician E.J. Gumbel's suggestion for using frequency analysis for floods in 1941.

Major changes at the governmental level in the United States assumed increasing importance as several hydrologic agencies were created in the nineteenth century: the Army Corps of Engineers in 1802, the Weather Bureau in 1870 (now called the National Weather Service), the U.S. Geological Survey, and the Mississippi River Commission in 1879.

How It Works

Floods. In the hydrologic cycle, atmospheric precipitation falls to Earth and is either evapotranspired back to the atmosphere, absorbed by vegetation, or moved downslope as overland flow that eventually becomes large enough to form stream channels. Water generally stays within the channel for most of the year. However, if the precipitation is heavy enough, the channel cannot transport all of this water and the stream overflows its banks, thereby creating floods.

Floodplains are low areas that can be found on one or both sides of a stream and are common in humid regions. Small watersheds with steep slopes often have flash floods that move very quickly with enough turbulence to damage buildings and vehicles easily. Since the flash floods occur so rapidly, people cannot be readily warned and drownings can occur.

Stream Discharge. In the United States, stream flow is measured in cubic feet per second at most gauging stations operated by the U.S. Geological Survey (USGS). Most of the larger streams have higher discharges as they flow downstream. There are some exceptions, such as the Colorado River in the arid southwestern United States, where the demand for irrigation water and public supply literally dries up the stream at its mouth.

Flood Frequency. It would be wonderful if flood prediction were reasonably exact. However, the probability of knowing when floods would occur turns out to be a markedly difficult task as it involves frequency analysis, probability theory, data that need to be homogeneous and independent, and selected assumptions that existing stream-flow data would be similar to future flows. This fallacious assumption implies that there will be no changes in future land use and climate.

The longer the period of record, the better the estimation of future flood flows. Accordingly, a "100-year flood" does not mean that the next one of that magnitude will occur 100 years in the future. It does mean that there is a 1 percent chance that a 100-year

flood could occur in any year. Accordingly, a 100-year flood could occur this year and again the following year, although the odds are against it. In hindsight, it would have been better if the term "100-year flood" would have been expressed as a "1-in-100 chance flood" in order to reduce public confusion.

Floodplain Development. The stream channel itself occupies a relatively narrow area within a floodplain. The channel is bordered by low-lying land that is called a floodway. As the distance away from the river increases, the slope of the land gradually increases and a slightly higher landscape called a floodway fringe is created. A cross-sectional profile of this floodplain would then show a channel with a floodway on one or both sides of the channel and a floodway fringe farther away from the stream.

In a nonregulated development scenario that was common in earlier years and still exists in some areas, residential and commercial buildings were constructed in the floodway zone, which lead to higher flood levels, structural damage, and potential loss of life. Low bridges that cross the stream in this zone could partially block the flow and thereby increase flood levels.

The situation would be quite different in a regulated floodplain. Ideally, the floodway would be zoned for as much open space as possible, such as parks, picnic areas, and golf courses. This would facilitate easier passage of flood waters and restrict buildings to higher ground wherever possible.

Floodplain Extent. It has been estimated that 100-year floodplains make up from 7 to 10 percent of the total land area of the United States. The floodplains with the largest areas are located in the southern portions of the country; those with large populations are located along the north Atlantic coast, the Great Lakes, and California.

Types of Floods. Damaging floods occur in varying locations. Flash floods are associated with quickly developing thunderstorms in mountainous areas. The rapid downslope movement of water, even if relatively shallow, can quickly move vehicles and their occupants to the extent that 50 percent of the flood deaths in the United States are caused by flash floods. If floodwater is flowing at a depth of only 2 feet, a trapped vehicle experiences a lateral force of 1,000 pounds and a buoyant force of 1,500 pounds, more than enough to flip the vehicle over and drown the passengers.

Regional floods are associated with level land that experiences long periods of heavy rain. The resulting damage to homes and properties in the floodplain can be enormous, as evidenced in the August 1993 flood in the lower Missouri and upper Mississippi rivers. About 69 percent of the levees built along the upper Mississippi River to protect the floodplain occupants were overtopped, drowning forty-eight persons, submerging seventy-five towns, destroying 50,000 homes, and even excavating more than 700 coffins from a cemetery in Missouri.

Storm surges powered by tropical storms and hurricanes can wreak havoc along low-lying coasts. The settlers who founded New Orleans in 1718 along the lower Mississippi River experienced their first major flood the same year. Over time, many other floods have occurred in New Orleans and the lower reaches of the Mississippi, aided by the slow sinking of the floodplain as river sediments compact over time.

APPLICATIONS AND PRODUCTS

Structural Flood-Control Objectives. Most structural flood-control measures are physical and expensive. They include construction of large reservoirs, diversion structures, levees and flood walls, channel alterations, modifications to bridges and culverts, and tidal barriers.

Flood-control reservoirs are built so that excess water from storms can be stored and released at a later time, thereby reducing peak discharge. The economic value of protecting property that may be damaged in floods can justify the construction costs of a reservoir. In addition, the stored water can also be used for hydropower, water-supply purposes, and recreation. Diversion structures are designed to reduce peak flows by forcing floods to go to another location via channels or tunnels. Levees and flood walls can physically keep water away from floodplains, thereby providing good potential for damage reduction. Levees are usually made of earth and flood walls of concrete. Both structures are usually set parallel to the stream.

Channel alterations to the stream can lower the height of water so that peak discharges are reduced at one point, but they can be increased downstream. The cost of these measures is substantial, but they can be justified in certain situations by their potential protection for valuable property. Possible long-term negative effects of these structural measures include aggradation or degradation of the downstream channel and sediment deposition in downstream bypass channels or tunnels. Bridge and culvert modifications are useful when the structures have inadequate capacity to handle flood flows. Repairing or raising the structures over the channel can lower damages by allowing more water to flow downstream. Tidal barriers along the coast can prevent high tides from moving upstream and damage developed areas. They are expensive to build and are generally used if large urbanized areas need protection. Note that the potential for tidal flooding is expected to increase as sea level rises because of climate change.

Nonstructural Flood-Control Objectives. These measures are employed in order to lower flooding susceptibility, thereby decreasing possible damage. They include flood warning, flood proofing, and a variety of land-use control procedures. Flood warning can reduce possible loss of life, or even better, eliminate that possibility. Flood-proofing procedures include rearranging the working space in buildings, waterproofing outside walls, and raising the height of buildings that occupy particularly susceptible locations. Land-use controls include many kinds of action within the floodplain that can reduce flood-hazard potential. These controls include proper building codes, purchasing flood insurance, zoning restrictions, or purchase of land and property in the less vulnerable portions of the floodplain.

A backhoe shapes the side of a flood-control levee being built June 1, 2011, on the edge of the Missouri River in Fort Pierre, South Dakota. (AP Photo)

Fascinating Facts About Flood-Control Technology

- The Black Sea was a freshwater lake during the last glacial advance. As the glaciers receded, global sea level rose and Mediterranean seawater flowed through the Bosporus Strait in Turkey into the Black Sea. More than 40,000 square miles of farmland was flooded, displacing an untold number of people.
- Glacial recession at the end of the last Ice Age allowed enormous volumes of meltwater to scour out land in North America to depths as great as 1,600 feet, thereby forming the Great Lakes, which hold 18 percent of all of the freshwater on the Earth's surface.
- Peak stream flow on the Mississippi is generally highest from February to June. However, the greatest flood that was ever recorded on the upper Mississippi River at St. Louis occurred during the summer of 1993, due mostly to a wet winter that was followed by an even wetter summer (double the normal precipitation). About two-thirds of the 1,576 levees were damaged enough to allow water into the floodplain.

- Hurricane Katrina in August 2005 holds the record as the most damaging storm to strike the United States The high winds (category 3) and heavy precipitation of eight to ten inches, in conjunction with a storm surge of twenty-four to twenty-eight feet, resulted in a death toll of 1,836 and property losses of more than $100 billion. Floodwaters with depths approaching twenty feet covered 80 percent of New Orleans.
- Paleoflood hydrology is the study of ancient floods. It is generally used for events within the past 10,000 years, as earlier times experienced very different global climatic changes, namely, the Ice Age.
- Major floods that occur in historical times can be exceeded centuries later: Florence, Italy, was devastated on November 4, 1333, when the Arno River flooded the city to depths of 14 feet. Almost 633 years to the day, on November 3, 1966, another flood occurred, covering the city with depths of 20 feet.

Climate Change. There is growing concern that global climate change could substantially impact flooding in coastal locations. For example, since 1870 sea level has increased about 5.9 inches. The overall sea-ice cover in the Arctic has been decreasing about 3 percent per decade and by 7 percent per decade in the summer. Estimates indicate that average global surface temperatures for the period 1990-2100 will increase between 3.1 and 10.5 degrees Fahrenheit. As a result, sea level could increase between 8.3-18.5 inches by the end of the twenty-first century. In addition, future peak precipitation events are expected to increase in intensity. To make matters worse, some of the climatic models employed suggest that rates of sea-level rise will double along the shorelines of certain portions of the eastern United States and the western North American and Arctic coasts.

Flood Mapping. It would be very useful to be able to generate flood maps in advance of approaching storms. Experimental work has been developed by the USGS and the National Weather Service (NWS) to make forecasting and mapping of potential floods readily available to local officials. The goal is to have NWS provide storm forecasts for a particular area, thereby enabling the USGS to generate maps showing potential flooded areas on the floodplain,

arrival times, and the depth of the flood itself. The combination of new methodologies and techniques, such as light detection and ranging (LIDAR), advanced computer programs, and geographic information systems (GIS) have led to the strong possibility of creating useful flood-forecast maps.

Flood Warning Systems. A useful addition to the regulation of floodplain development occurs when flood-warning systems are employed by local governments. These systems use radar, rainfall, and stream-flow gauging that are connected by satellite transmitters to relay real-time data to computers at some central site, which in turn make the data available to interested parties via the Internet. Automatic warnings are then dispatched to emergency-management officials who can then institute procedures, based on the predicted flood levels, that range from selective road closures to complete evacuation.

IMPACT ON INDUSTRY

U.S. Government Research. Several agencies of the federal government are involved with flooding issues. The U.S. Army Corps of Engineers (Corps) is the oldest agency that was assigned to deal with increasing flooding problems, particularly on the Mississippi. Flood Control Acts of 1928, 1936, and 1944 gave additional authority to the Corps to reduce flooding.

The major source of stream-flow data for the United States has been primarily gathered by the USGS. Its first gauging station was built on the Rio Grande in New Mexico in 1889. The early gauges were mechanical-current devices that measured the velocity of flowing water in conjunction with measurement of the stream's cross-sectional area. Improvements since 1995 have included hydroacoustic meters that have come to constitute about 30 percent of the discharge gauges that now number more than 7,400. The new techniques have lowered the average discharge measurement time from 96 to only 18 minutes. The National Water Information System (NWIS), part of the USGS, provides real-time data for selected surface water, groundwater, and water-quality sites around the country with maps and links to the sites.

Other federal and regional agencies that record specialized stream-flow data include the Tennessee Valley Authority (TVA), Bureau of Reclamation, Natural Resources Conservation Service (NRCS), U.S. Forest Service, and the Agricultural Research Service (ARS).

Government Regulations. The National Flood Insurance Program (NFIP) was created by the United States Congress in 1968, and is now under the control of the Federal Emergency Management Agency (FEMA). FEMA became part of the U.S. Department of Homeland Security (DHS) in March 1, 2003, and employs more than 3,700 workers. The purpose of this agency is to identify and map flood-hazard areas, disseminate this information to those living in floodplains, and provide information about flood insurance. Local governments are required by the NFIP to adopt regulations that would deter proposed housing developments that would not be in accordance with national standards. USGS maps are used by the NFIP to delineate 100- and 500-year floodplains. Banks that lend money for homes and commercial property in the 100-year floodplain require the prospective owners to purchase flood insurance. Developments in the 500-year floodplain do not necessarily have to have flood insurance. The probability of a 100-year and 500-year flood occurring in any 1-year period is 1 percent and 0.2 percent, respectively.

Association of State Floodplain Managers (ASFPM). This organization of 13,000 professionals in the United States was formed in 1976. The major concerns that interest ASFPM include flood-hazard mitigation, floodplain management, the NFIP, and just about anything else that involves flooding issues.

CAREERS AND COURSE WORK

There are a surprising number of varied courses and future jobs that are included in the field of flood-control technology. The most obvious ones include civil engineering for construction of dams and levees, hydraulic engineering to handle channel deepening where appropriate, and surveying. However, there are many other skills that are needed, and it would benefit those interested in this field to study geology, physical geography, hydrology, meteorology (storms), and water-resources management. Other useful courses of study include environmental planning, computer science, economics, and environmental law.

Many governmental organizations, including the Army Corps of Engineers, American Water Resources Association, and FEMA, have job postings on their Web sites that encompass a range of employment opportunities. These agencies employ civil engineers, scientists, natural-resource specialists, administrators, hydrologic technicians, and ecologists, among others.

SOCIAL CONTEXT AND FUTURE PROSPECTS

Floods have occurred on numerous occasions over many millennia. Societies have designed ways to store some of this water in special collecting basins or small reservoirs, and in selected cases by creating diversion canals. In other cases, the floods have been very damaging to settlements and farms. What makes the situation even worse is the population growth on floodplains, the inhabitants of which have the mistaken notion that floods can be contained by structural techniques. Too many damaging floods have occurred in too many countries that renders naive the presumption that one is safe behind the levee.

One step in the right direction has been the growing recognition that floodplain development carries a substantial risk: predicting flood heights is fraught with uncertainty because of a complicated mix of storm tracks, soil-moisture conditions, available upstream reservoir storage, and the elevations of commercial facilities and homes. One can now recognize the benefits of nonstructural measures, such as zoning lands that are in the 100-year floodplain as not suitable for development. To make matters more interesting, and more complicated, how does one factor in shifts that could occur over many decades, such as climate change, river behavior, and sea-level rise that would impact coastal areas as well as the existing homeowners in the floodplain who simply enjoy living close to rivers?

Two things will certainly help out in this situation. One would simply be better explanation of the flooding danger as given by the probability of certain events occurring in any year. These odds could, of course, change as floodplain development continues or decreases. The second element is simply better discussion of future weather patterns, as unpredictable as they are, that could occur on a global scale.

Robert M. Hordon, B.A., M.A., Ph.D.

FURTHER READING

Bedient, Philip B., Wayne C. Huber, and Baxter E. Vieux. *Hydrology and Floodplain Analysis.* 4th ed. Upper Saddle River, N.J.: Prentice Hall, 2008. In addition to chapters on hydrology, floodplain hydraulics, and flood management, the book contains seven pages of hydrology-related Internet links.

Cech, Thomas V. *Principles of Water Resources: History, Development, Management, and Policy.* 3d ed. Hoboken, N.J.: John Wiley & Sons, 2010. A highly readable and very informative book on hydrology, flood events, and federal water agencies with useful diagrams, maps, and photos.

Dunne, Thomas, and Luna B. Leopold. *Water in Environmental Planning.* New York: W. H. Freeman, 1978. A classic text in the water field; contains useful information regarding runoff processes, flood hazards, and human occupancy of flood-prone areas.

Dzurik, Andrew A. *Water Resources Planning.* 3d ed. Lanham, Md.: Rowman and Littlefield, 2003. A readable book on hydrology, federal water agencies and related legislation, and floodplain management.

Mays, Larry W. *Water Resources Engineering.* 2d ed. Hoboken, N.J.: John Wiley & Sons, 2010. A detailed textbook that contains numerous examples and diagrams involving hydrology, surface runoff, and flood control.

Strahler, Alan. *Introducing Physical Geography.* 5th ed. Hoboken, N.J.: John Wiley & Sons, 2010. An excellent textbook with very useful diagrams, maps, and color photos that cover many aspects of hydrology and stream flow.

Viessman, Warren, Jr., and Gary L. Lewis. *Introduction to Hydrology.* 5th ed. Upper Saddle River, N.J.: Prentice Hall, 2003. Includes many chapters on floods and floodplains with good diagrams.

WEB SITES

American Water Resources Association
http://careers.awra.org/jobs

Federal Emergency Management Agency
http://www.fema.gov

U.S. Army Corps of Engineers
http://www.usace.army.mil

U.S. Geological Survey
Water Resources
http://www.usgs.gov/water

See also: Civil Engineering; Climatology; Computer Science; Hydraulic Engineering; Hydrology and Hydrogeology; Meteorology.

FLOW REACTOR SYSTEMS

FIELDS OF STUDY

Chemical reaction engineering; chemical engineering; chemical process modeling; chemistry; physics; mathematics; engineering; mechanical engineering; fluid mechanics; kinetics; heat transfer; control engineering; process engineering; industrial engineering; electrical engineering; safety engineering; thermodynamics; materials science.

SUMMARY

Flow reactor systems provide the heart of chemical reactors that enable the continuous conversion of various reactants, very often in the presence of catalysts, into desired new products. Contemporary chemical reaction engineering has designed a wide variety of flow reactor systems that have been customized for the specific chemical-conversion processes they facilitate. Flow reactors have become essential in process industries as they provide a very effective way of converting raw materials into desired products. The world's demand for gasoline, for example, could hardly be met without the flow reactor enabling fluid catalytic cracking of low-value hydrocarbons into high-value hydrocarbons derived from crude oil.

KEY TERMS AND CONCEPTS

- **Baffle:** Vane or panel to direct the flow inside a reactor.
- **Bubbling Bed Reactor:** Flow reactor where gas is inserted at the bottom and forms small bubbles as it rises to the top, fluidizing the solids inside the reactor.
- **Catalyst:** Substance that aides a chemical reaction without being consumed by it; a catalyst can suffer degeneration during the process it supports.
- **Continuous-Flow Stirred Tank Reactor (CSTR):** Flow reactor designed to achieve a perfect mix of all reactants in its tank; sometimes called a backmix reactor.
- **Fluidized Bed Reactor:** Special type of flow reactor that mixes the catalyst with the reactants in a common fluid phase.
- **Packed Bed Reactor:** Type of flow reactor in which the catalyst remains stationary as the reactants flow over or through it (also called a fixed-bed reactor).
- **Plug Flow Reactor (PFR):** Tubular reactor that fluid moves through at a continuous, steady pace so that the conversion of chemicals are functions of the position within the reactor rather than time. Also called continuous tubular reactor (CTR).
- **Reactant:** Raw material molecules that will react to form a new product.

DEFINITION AND BASIC PRINCIPLES

Flow reactor systems are designed to allow the continuous operation of chemical reactors by having reactions take place while the reactants flow through the reactor. This is possible when the reaction times of the desired processes are so short that they can occur while the reactants move through the reactor. A chemical reactor is a unit that houses the process where raw materials, consisting of feed molecules called reactants, react with each other, often supported by catalysts, to form desired products.

In a flow reactor, reactants are loaded continuously at a steady pace, react while flowing through the reactor, and the ensuing products of the reaction are taken off continuously. This enables a steady-state operation of the reactor and eliminates the downtime associated with a stop-and-go, or batch, process. To the contrary, a batch reactor is loaded and unloaded for each separate reaction run. Generally, the shorter the reaction time of the process is, the more advantageous is a flow reactor compared with its opposite, the batch reactor.

Because their true advantages appear only if flow reactors are customized for the exact reaction they facilitate, there is a great variety of flow reactor systems. Designing contact modes for reactants with catalysts, which is required for most reactors, has led to further design variations. Typically, catalysts either remain stationary as reactants flow along or through them in a packed (or fixed-) bed reactor, or catalysts move with reactants in a fluidized bed reactor.

BACKGROUND AND HISTORY

The history of flow reactor systems is closely related to the development of chemical reaction engineering, a branch of the young discipline of chemical engineering. With the rise of the chemical industry, beginning in the mid-nineteenth century with the industrial revolution, processes were designed that could be run with reactors in either the batch or the flow mode. An early example is Belgian chemist Ernest Solvay's plant that manufactured soda ash from brine and limestone in 1863. Russian-German chemist Wilhelm Ostwald's groundbreaking work on catalysts, for which he won the Nobel Prize in Chemistry in 1909, laid the foundation for catalyst use in chemical reactors for which flow reactors are well suited.

The 1909 invention of the Haber-Bosch process for industrial ammonia production, as well as the invention of coal liquefaction to obtain synthetic fuels by German chemist Friedrich Bergius in 1913 and the Fischer-Tropsch process of 1925, all utilized chemical reactions well suited for flow reactors. In the 1930's, academic work on chemical engineering, including flow reactors, flourished in particular in the United States at the Massachusetts Institute of Technology and in Germany. In Germany, Gerhard Damköhler taught physical chemistry at the University of Göttingen from 1936 to 1944 and discovered the Damköhler numbers, which describe the timescale of chemical reactions in a flow reactor. The start of the world's first commercial fluid catalytic cracker unit in Baton Rouge, Florida, in 1942, marked a milestone in the industrial application of flow reactors. Octave Levenspiel's influential textbook on chemical reactors, *Chemical Reaction Engineering*, first published in 1962 and since updated, has been credited with teaching generations of chemical engineers.

HOW IT WORKS

For all the existing variety in flow reactor systems, there are some basic models. These can be distinguished by how they design the flow of the reactants or how they bring together reactants and catalysts.

Continuous-Flow Stirred Tank Reactor (CSTR). In this flow reactor, a tank reactor is continuously fed with reactants that exist in a single fluid phase, which can be either a gas, liquid, or slurry (solids thickly suspended in liquids). The tank reactor may include a catalyst, and the reactants are mixed with a stirring propeller. Ideally, this complete mix creates the desired product, which is continuously removed from the tank. In practice, perfect mixing can be

hampered by stagnant regions along tank edges or the involuntary creation of a fast bypass flow of little-mixed reactants. There are different techniques to maximize mixing in a CSTR. Baffles are commonly added to the walls of CSTRs to direct flow to the propeller. Mixing can also be done at multiple stages where reactants flow from one mixing zone to the next inside the tank reactor. A pump can create additional circulation of the reactant and product flow.

CSTRs generally provide uniform product quality and their operating temperature can be well controlled. Product yield is often lower than with other reactors. This is often improved by using two or three CSTRs after each other to ensure most reactants form a product.

Plug Flow Reactor (PFR). Typically, in a PFR reactants are mixed in a single fluid phase before they are sent on a journey down a tubular flow reactor. As each reactant of a single plug, or single drop, completely fills the tube's cylinder and flows with the same speed, ideally, there is no upstream or downstream mixing of reactants of different plugs. In an ideal PFR, all reactions occur in each single plug as it flows along the reactor tube. PFR tubes are bent for tubular flow conditions. In practice, there are possible deviations from ideal flow within a PFR. A common form is recirculation of reactants causing dead spots, or the development of laminar (layered) flow, where reactants closer to the reactor wall flow slower than those in the middle of its tube. This requires corrective design resolutions. Another challenge is hot spots in a reactor part that demands special cooling.

PFRs can deal with very high volumes of reactants and quickly deliver very high product yields. Able to operate at high temperatures, PFR heat control has remained challenging. While PFRs have long maintenance intervals, their eventual servicing is more costly than for CSTRs or batch reactors.

Catalytic Multiphase Reactors. Flow reactors can be designed to deal with reactants existing in one or two phases, gas and liquid, and a solid-phase catalyst. There are two basic designs.

In a packed or fixed-bed reactor, with both names used interchangeably, the reactants flow through catalysts that remain stationary. Typically, an area of the reactor tube of a PFR is filled with catalysts in the form of small pellets that are held in place in a stationary bed and can range in size from millimeters to centimeters. The reactants flow through this bed. If both gas and liquid reactants flow downward through the reactor, or gas and liquid move in a countercurrent flow, this is called a trickle bed reactor as the liquid can wet the catalyst for the gas reactants.

The most challenging flow reactors to design are fluidized bed reactors. Here, the catalyst particles begin to move together with the flow of the reactants. The key advantage is that the catalysts can be removed, regenerated, and added without having to stop the process as with a packed bed. Therefore, fluidized bed reactors are a good solution for processes with high catalyst degradation requiring their speedy exchange.

Depending on the speed with which catalysts are entrained and move with the fluid reactants, there are different kinds of fluidized bed reactors. They range from incipiently fluidized bed to bubbling bed and turbulent bed to fast fluidized and finally pneumatic or transport bed.

APPLICATIONS AND PRODUCTS

Flow reactor systems are widely used in process industries, particularly oil and gas, petrochemicals, and chemicals. Their relatively higher cost, compared with batch reactors, is generally warranted wherever their advantage of enabling a continuous process is of material benefit. This is especially true for plants with reactors that truly operate continuously, day and night, sometimes over months or years at a time, without needing a maintenance shutdown. Flow reactors tend to be customized for each process they facilitate. Their complexity tends to increase, from a basic CSTR or PFR to a fluidized bed reactor, for example, wherever the optimization of product quality and yield gained by highly sophisticated flow reactor design and operation significantly exceeds the greater costs of specialization.

Oil and Gas Industry. One of the most important applications of a flow reactor system is to enable fluid catalytic cracking at an oil refinery. Only because long hydrocarbon molecule chains that remain after distillation of crude oil can be broken, or cracked, very efficiently in a fluidized bed reactor, can the world's demand for lighter hydrocarbon products such as gasoline or jet fuel, be satisfied. In addition, oil refineries employ a variety of other flow reactor systems to facilitate many of their other processes. Catalytic reforming, to increase gasoline quality, is done both in packed and fluidized bed reactors.

Fascinating Facts About Flow Reactor Systems

- Some advanced flow reactors at the heart of a refinery's fluid catalytic cracker can run continuously without a single stop for as many as four years.

- Almost all of world's production of 60 million tons of ethylene, 40 million tons of propylene, and 21 million tons of benzene, all base petrochemicals, owes its existence to processes taking place in a flow reactor.

- If imperial Japan had employed the Fischer-Tropsch or the Bergius flow reactor process to turn coal from its possession of Manchuria into gasoline in the late 1930's, it may not have felt so threatened by the U.S. gasoline embargo, which helped precipitate Japan's decision to attack Pearl Harbor in 1941.

- Its lower cost and fewer servicing needs make fixed bed flow reactors a good alternative to more demanding fluidized bed flow reactors, but at the price of less customized processes.

- A fluidized bed reactor used for catalytic cracking must withhold incredibly hot temperatures—up to 1,650 degrees Fahrenheit—without developing fatal hot spots.

Isomerization and polymerization, for higher octane gasoline, and aromatization for petrochemicals, are typically done using packed bed reactors. Hydrocracking, to increase gasoline yield from crude-oil distillates, and desulfurization, essential to clean gasoline and diesel fuels of polluting sulfur oxides, is done both in packed or trickle bed reactors, or in a fluidized bed reactor for desulfurization.

Manufacture of synthetic fuels, such as gasoline or jet fuel from coal gasification in the Fischer-Tropsch process, one of the oldest applications of flow reactors, or the methanol to gasoline process, is done using a bubble bed or a slurry reactor, respectively.

Petrochemical Industry. The petrochemical industry could not operate without the manifold flow reactor systems employed in its processes. The manufacture of such key industrial petrochemicals as ethene (commonly known as ethylene) oxide and dichloride, essential feedstocks for the plastics industry, is done utilizing packed bed reactors. The same is true for the production of maleic and phthalic anhydride, building blocks for resins and plasticizers, respectively, which is done either in packed or fluid bed reactors. Polyethylene, material for plastic bags and other packaging applications, is also made in either tank or tube flow reactors. The same is true for the petrochemical polypropylene, used for plastics, textiles, and even banknotes, and polystyrene and polyvinyl chloride (PVC), essential for the plastics industry. Flow reactors are the blood vessels of the petrochemical industry.

Inorganic Bulk Chemicals. Flow reactors, from less expensive but reliable packed bed reactors to more complex fluid bed reactors, have enabled very economic production of inorganic bulk chemicals used in the chemical industry. The manufacture of ammonia, one of the most common bulk chemicals used especially for fertilizers and a key part of agriculture, saw with the Haber-Bosch process one of the first commercial uses of a flow reactor. A sulfuric acid plant uses a flow reactor for the manufacturing step of oxidation of sulfur dioxide to arrive at the widely used final product employed for lead-acid batteries, in refining, ore processing, or even wastewater treatment. Similarly, the manufacture of nitric acid used for fertilizers, explosives, or chemical synthesis, and hydrogen peroxide, a powerful bleaching agent, and sodium borohydride, which is used to replace chlorine-based bleaching of paper products, all rely on packed bed or fluidized bed flow reactors.

Other Uses. Flow reactors are used in virtually every process industry in a very wide variety of applications. Wood processing relies on customized flow reactors for its key processes, particularly those involving bleaching. Increasingly, organic fine chemicals such as pharmaceuticals, cosmetics, herbicides, and pesticides are produced in flow rather than batch reactors. Biochemical processes including fermentation or biological wastewater treatment have come to employ slurry reactors or packed bed reactors. Metal extraction processes such as roasting of sulfide ores, production of crystalline silicon and titanium dioxide, or uranium processing, use flow reactors. The same holds true for gas cleaning and combustion processes. Even solid waste can be incinerated in a flow reactor.

Research. Apart from industrial applications, flow reactors are used in research, especially on the laboratory or pilot scale. They are employed to analyze the mechanism and kinetics, or study of motion, of chemical reactions. They provide data in process simulation and aid the investigation of process performance before large-scale industrial flow reactors are designed and built for new or improved applications. They are also subject to test new ideas for efficiency and process optimization.

IMPACT ON INDUSTRY

Flow reactor systems are a vital ingredient of the process industries of any industrialized or industrializing nation. Flow reactors operate at the core of industrial plants, manufacturing products worth trillions of dollars each year.

Government Agencies. Historically, both the governments of Germany and the United States supported flow reactor systems research during World Wars I and II. Without government support for plants utilizing the Haber-Bosch process, Germany would have run out of ammunition after the first months of World War I. In World War II, Germany relied on flow reactors for its synthetic fuel production, and the United States used fluid catalytic cracking for gasoline and airplane fuel of sufficient quality for military engines. During the Cold War, Communist-planned economies, with the exception of a few countries such as Romania and East Germany, did not focus as much on the chemical as opposed to the steel industry and did not discover as many advances in flow reactor design as did those in the West and Japan.

By 2011, governments of industrialized and industrializing countries devoted public money to support research in flow reactor systems, particularly through grants aiming at developing more efficient and less polluting processes. The U.S. Department of Energy was as supportive of this research as was, for example, Japan's Ministry of Economy, Trade and Industry (METI), the 2001 successor of the Ministry of International Trade and Industry (MITI), renowned for its close ties to industrial research.

Universities and Research Institutes. Traditionally, Germany and the United States were leaders in flow reactor design and theory. The Massachusetts Institute of Technology established the world's first curriculum in chemical engineering in 1880. American universities have been global leaders in flow reactor design and theory ever since. Flow reactor design research is also carried out very successfully at Germany's University of Erlangen-Nuremberg and Johannes

Gutenberg University Mainz; University of Twente in the Netherlands; and the Universities of Nancy and Lyon in France. Japan, Taiwan, and the People's Republic of China also have universities home to flow reactor engineers of world reputation. An important scientific forum is the International Symposia on Chemical Reaction Engineering (ISCRE), which brings together the research of American, Asian, and European flow reactor scientists.

Industry and Business. Flow reactor systems are vital in any process industry. This includes the oil and gas, petrochemical, and chemical industries. Flow reactors are also important in metallurgy. Increasingly, the pharmaceutical, food, and biochemical industries have made use of flow reactors as well. Flow reactors have also become relevant in the waste-management industry, particularly in biological wastewater treatment and solid-waste incineration plants.

The three key industries utilizing flow reactors to generate much of their value are the chemical, petrochemical, and oil and gas. Because their operations are very capital intensive, large, multinational companies dominate these industries, and the major corporations in the field are active in all three areas.

By 2009, the chemical industry created about $3 trillion in annual revenue. The leading companies by revenue were Germany's BASF, the United States' Dow Chemical, and the United Kingdom's INEOS. If petrochemical production was included, which relies even more heavily on flow reactors, by 2009, the world's leading company was still Germany's BASF, with annual revenues of $62.3 billion. This revenue could not have been generated without flow reactors. America's Dow Chemical and ExxonMobil Chemical, with revenues of $57.5 billion and $55 billion, came in second and third. Among the next seven largest companies, two more, DuPont and Chevron Phillips, came from America, with the others based in the Netherlands (LyondellBasell Industries), the United Kingdom (INEOS), Saudi Arabia (Saudi Basic Industries Corporation), Taiwan (Formosa Plastics Group), and Japan (Sumitomo Chemical).

The multinational oil companies such as ExxonMobil, Shell, BP, or non-Western national oil companies such as the world's largest, Saudi Aramco, have invested significantly in flow reactors, without which they could not really function in the contemporary market. To prove this point: In 2010, at U.S. refineries alone, more than one-third, or exactly 35

percent, or 6.2 million barrels, of the crude oil processed daily into fuel products was subject of fluid catalytic cracking.

CAREERS AND COURSE WORK

Flow reactor systems are designed by chemical reactor engineers. They work in a subfield of chemical engineering and should combine good understanding of chemistry and engineering. A key career decision is whether to work in the chemical industry or in academia. Operation and maintenance of flow reactors also requires very skilled technicians.

Students interested in eventually designing, supervising construction and operation, or operating and servicing a flow reactor should take science courses in high school, particularly chemistry and physics, as well as mathematics and computer courses. The same is true for those who want to pursue research in the field.

There are many opportunities for technicians who benefit from grounding in the sciences. An associate's degree in science or applied science, or in an engineering field, is a very good basis for a flow reactor technician's career path.

A bachelor of science in chemical engineering is the most straightforward basis for an advanced career in flow reactor systems. However, a degree in mechanical, electrical, or computer engineering can also be helpful. At the same time, a bachelor of science or arts in chemistry, with at least some courses in engineering, or an engineering degree with an additional minor in chemistry, is similarly useful. Given the rising importance of computational fluid dynamics and numerical mathematics for the field, a degree in computer science or mathematics can also be advantageous.

For an advanced career, postgraduate study of chemical reaction engineering is suitable preparation. When choosing a university for postgraduate work, a student should look for professors with expertise and renown in chemical reaction engineering who are leading an active research community in the field. A Ph.D. in chemical engineering with an emphasis in computational science and engineering is good preparation for top research in flow reactor systems.

SOCIAL CONTEXT AND FUTURE PROSPECTS

Without flow reactors, there would not be

sufficient fuel for the world's road- and air-transport needs, sufficient fertilizer to enable crops to grow to feed the world's people, nor the many plastics-based applications contemporary humanity utilizes. Flow reactors provide very efficient and economic means to derive desired products from raw materials as varied as crude oil, natural gas, or sulfur, oxygen, and water. The contemporary world depends on the work of flow reactors to such an extent that their absence would decisively change the very nature of society in all industrialized and industrializing nations.

Because flow reactors are so essential for core processes defining contemporary industry, society, and culture, there has been significant global research into improving their operations and customizing them for new chemical reaction processes. On a theoretical level, the rise and development of computational fluid dynamics (CFD) has greatly aided flow-reactor data analysis and simulation for scale-ups from laboratory to industrial size. Contemporary research focuses on design of transitional, or dynamic, and oscillating operation of catalytic flow reactors. Advanced systems engineering of processes has yielded successes such as reactive distillation: integrating the chemical reaction with the subsequent distillation of product in the same reactor. There has been a quest to discover innovative forms for reactors such as monoliths or fiber reactors, to work with microwaves or ultrasound as sources of energy, and utilize supercritical reaction media. Research into microreactors has been especially exciting.

R. C. Lutz, B.A., M.A., Ph.D.

FURTHER READING

Bartholomew, C. H., and R. J. Farrauto. *Fundamentals of Industrial Catalytic Processes*. 2d ed. Hoboken, N.J.: John Wiley & Sons, 2006. Chapter 4 provides a very accessible, clear introduction to flow reactors and distinguishes them from batch reactors. Illustrated with a bibliography.

Fogler, H. Scott. *Essentials of Chemical Reactor Engineering*. Boston: Pearson Education, 2011. Flow reactors are presented next to their alternatives in this useful textbook geared to undergraduate readers. Solid presentation of principles and challenges for flow reactor design and their various possible features and characteristics.

Jakobsen, Hugo A. *Chemical Reactor Modeling: Multiphase Reactive Flows*. Berlin: Springer-Verlag, 2008. Advanced scientific survey. Chapter 6 provides an advanced-level overview of chemical reaction engineering. Presents scientific formulas driving reactor designs.

Mann, Uzi. *Principles of Chemical Reactor Analysis and Design: New Tools for Industrial Chemical Reactor Operations*. 2d ed. Hoboken, N.J.: John Wiley & Sons, 2009. Chapter 1 provides an accessible overview with subsequent chapters presenting flow reactors and their alternatives in more detail, while shedding light on economic considerations.

Nauman, E. Bruce. *Chemical Reactor Design, Optimization, and Scaleup*. 2d ed. Hoboken, N.J.: John Wiley & Sons, 2008. Advanced-level work that also covers many contemporary applications and looks at meso-, micro-, and nanoscales. Useful for readers familiar with basics desiring specialized knowledge in the field.

Salmi, Tapio O., Jyri-Pekka Mikkola, and Johan P. Warna. *Chemical Reaction Engineering and Reactor Technology*. Boca Raton, Fla.: CRC Press, 2011. Textbook that moves from introductory to intermediary material; good introduction and also historical overview of the field. Usefully illustrated and well written.

WEB SITES

American Chemical Society
http://portal.acs.org/portal/acs/corg/content

American Institute of Chemical Engineers
http://www.aiche.org

International Symposia on Chemical Reaction Engineering
http://www.iscre.org

See also: Chemical Engineering; Computer Science; Electrical Engineering; Engineering; Engineering Mathematics; Fluid Dynamics; Fluidized Bed Processing; Mechanical Engineering.

FLUID DYNAMICS

FIELDS OF STUDY

Physics; mathematics; engineering; chemistry; suspension mechanics; hydrodynamics; computational fluid dynamics; microfluidic systems; coating flows; multiphase flows; viscous flows.

SUMMARY

Fluid dynamics is an interdisciplinary field concerned with the behavior of gases, air, and water in motion. An understanding of fluid dynamic principles is essential to the work done in aerodynamics. It informs the design of air and spacecraft. An understanding of fluid dynamic principles is also essential to the field of hydromechanics and the design of oceangoing vessels. Any system with air, gases, or water in motion incorporates the principles of fluid dynamics.

KEY TERMS AND CONCEPTS

- **Aerodynamics:** Study of air in motion.
- **Boundary Layer:** Region between the wall of a flowing fluid and the point where the flow speed is nearly equal to that of the fluid.
- **Continuum:** Continuous flow of fluid.
- **Fluid:** State of matter in which a substance cannot maintain a shape on its own.
- **Hydrodynamics:** Study of water in motion.
- **Ideal Fluids:** Fluids without any internal friction (viscosity).
- **Incompressible Flows:** Those in which density does not change when pressure is applied.
- **Inviscid Fluid:** Fluid without viscosity.
- **Newtonian Fluids:** Fluids that quickly correct for shear strain.
- **Shear Strain:** Stress in a fluid that is parallel to the fluid motion velocity or streamline.
- **Streamline:** Manner in which a fluid flows in a continuum with unbroken continuity.
- **Viscosity:** Internal friction in a fluid.

DEFINITION AND BASIC PRINCIPLES

Fluid dynamics is the study of fluids in motion. Air, gases, and water are all considered to be fluids. When the fluid is air, this branch of science is called aerodynamics. When the fluid is water, it is called hydrodynamics.

The basic principles of fluid dynamics state that fluids are a state of matter in which a substance cannot maintain an independent shape. The fluid will take the shape of its container, forming an observable surface at the highest level of the fluid when it does not completely fill the container. Fluids flow in a continuum, with no breaks or gaps in the flow. They are said to flow in a streamline, with a series of particles following one another in an orderly fashion in parallel with other streamlines. Real fluids have some amount of internal friction, known as viscosity. Viscosity is the phenomenon that causes some fluids to flow more readily than others. It is the reason that molasses flows more slowly than water at room temperature.

Fluids are said to be compressible or incompressible. Water is an incompressible fluid because its density does not change when pressure is applied. Incompressible fluids are subject to the law of continuity, which states that fluid flows in a pipe are constant. This theory explains why the rate of flow increases when the area of the pipe is reduced and vice versa. The viscosity of a fluid is an important consideration when calculating the total resistance on an object.

The point where the fluid flows at the surface of an object is called the boundary layer. The fluid "sticks" to the object, not moving at all at the point of contact. The streamlines further from the surface are moving, but each is impeded by the streamline between it and the wall until the effect of the streamline closest to the wall is no longer a factor. The boundary layer is not obvious to the casual observer, but it is an important consideration in any calculations of fluid dynamics.

Most fluids are Newtonian fluids. Newtonian fluids have a stress-strain relationship that is linear. This means that a fluid will flow around an object in its path and "come together" on the other side without a delay in time. Non-Newtonian fluids do not have a linear stress-strain relationship. When they encounter shear stress their recovery varies with the type of non-Newtonian fluid.

A main consideration in fluid dynamics is the amount of resistance encountered by an object moving through a fluid. Resistance, also known as drag, is made up of several components but all have in common that they occur at the point where the object meets the fluid. The area can be quite large as in the wetted surface of a ship, the portion of a ship that is below the waterline. For an airplane, the equivalent is the body of the plane as it moves through the air. The goal for those who work in the field of fluid dynamics is to understand the effects of fluid flows and minimize their effect on the object in question.

BACKGROUND AND HISTORY

Swiss mathematician Daniel Bernoulli introduced the term "hydrodynamics" with the publication of his book *Hydrodynamica* in 1738. The name referred to water in motion and gave the field of fluid dynamics its first name, but it was not the first time water in action had been noted and studied. Leonardo da Vinci made observations of water flows in a river and was the one who realized that water was an incompressible flow and that for an incompressible flow, V = constant. This law of continuity states that fluid flow in a pipe is constant. In the late 1600's, French physicist Edme Mariotte and Dutch mathematician Christiaan Huygens contributed the velocity-squared law to the science of fluid dynamics. They did not work together but they both reached the conclusion that resistance is not proportional to velocity; it is instead the square of the velocity.

Sir Isaac Newton put forth his three laws in the 1700's. These laws play a fundamental part in many branches of science, including fluid dynamics. In addition to the term hydrodynamics, Bernoulli's contribution to fluid dynamics was the realization that pressure decreases as velocity increases. This understanding is essential to the understanding of lift. Leonhard Euler, the father of fluid dynamics, is considered by many to be the preeminent mathematician of the eighteenth century. He is the one who derived what is today known as the Bernoulli equation from the work of Daniel Bernoulli. Euler also developed equations for inviscid flows. These equations were based on his own work and are still used for compressible and incompressible fluids.

The Navier-Stokes equations result from the work of French engineer Claude-Louis Navier and British physicist George Gabriel Stokes in the mid-nineteenth century. They did not work together, but their equations apply to incompressible flows. The Navier-Stokes equations are still used. At the end of the nineteenth century, Scottish engineer William John Macquorn Rankine changed the understanding of the way fluids

flow with his streamline theory, which states that water flows in a steady current of parallel flows unless disrupted. This theory caused a fundamental shift in the field of ship design because it changed the popular understanding of resistance in oceangoing vessels. Laminar flow is measured today by use of the Reynolds number, developed by British engineer and physicist Osborne Reynolds in 1883. When the number is low, viscous forces dominate. When the number is high, turbulent flows are dominant.

American naval architect David Watson Taylor designed and operated the first experimental model basin in the United States at the start of the twentieth century. His seminal work, *The Speed and Power of Ships*, first published in 1910, is still read. Taylor played a role in the use of bulbous bows on vessels of the navy. He also championed the use of airplanes that would be launched from naval craft underway in the ocean.

The principles of fluid dynamics took to the air in the eighteenth century with the work done by aviators such has the Montgolfier brothers and their hot-air balloons and French physicist Louis-Sébastien Lenormand's parachute. It was not until 1799 when English inventor Sir George Cayley designed the first airplane with an understanding of the roles of lift, drag, and propulsion, that aerodynamics came under scrutiny. Cayley's work was soon followed by the work of American engineer Octave Chanute. In 1875, he designed several biplane gliders, and with the publication of his book *Progress in Flying Machines* in 1894, he became internationally recognized as an aeronautics expert.

The Wright brothers are rightfully called the first aeronautical engineers because of the testing they did in their wind tunnel. By using balances to test a variety of different airfoil shapes, they were able to correctly predict the lift and drag of different wing shapes. This work enabled them to fly successfully at Kitty Hawk, North Carolina, on December 17, 1903.

German physicist Ludwig Prandtl identified the boundary layer in 1904. His work led him to be known as the father of modern aerodynamics. Russian scientist Konstantin Tsiolkovsky and American physicist Robert Goddard followed, and Goddard's first successful liquid propellant rocket launch in 1926 earned him the title of the father of modern rocketry.

All of the principles that applied to hydrodynamics—the study of water in motion—applied to aerodynamics: the study of air in motion. Together these principles comprise the field of fluid dynamics.

Fascinating Facts About Fluid Dynamics

- The pitot tube is a simple device invented by French hydraulic engineer Henri Pitot in the 1700's. It is used to measure air speed in wind tunnels and on aircraft.

- The Bernoulli principle, that pressure decreases as velocity increases, explains why an airplane wing produces lift.

- English engineer William Froude was the first to prove the validity of scale-model tests in the design of full-size vessels. He did this in the 1870's, building and operating a model basin for this purpose.

- The Froude number is a dimensionless number that measures resistance. The greater the Froude number, the greater the resistance.

- Froude performed the seminal work on the rolling of ships.

- Alfred Thayer Mahan's book *The Influence of Sea Power Upon History:1660-1783*, published in 1890, made such a powerful case for the importance of a strong navy that it caused the major powers of that time to invest heavily in new technology for their fleets.

- American naval architect David Watson Taylor was a rear admiral in the U.S. Navy. He was meticulous in his work and developed procedures still in use in model basins. The Taylor Standard Series was a series of trials run with specific models. The results could be used to estimate the resistance of a ship effectively before it was built.

- The bulbous bow is a torpedo-shaped area of the bow of a ship. It is below the waterline. It reduces the resistance of a ship by reducing the impact of the waves on the front bow of a ship underway.

- David Watson Taylor studied the phenomenon of suction between two vessels moving close together in a narrow channel. He was called as an expert witness for the Olympic-Hawke trial in 1911.

- Bioengineers examine fluid flows when designing pacemakers and other medical equipment that will be implanted in the human body.

HOW IT WORKS

When an object moves through a fluid such as gas or water, it encounters resistance. How much resistance depends upon the amount of internal friction in the fluid (the viscosity) as well as the shape of the object. A torpedo, with its streamlined shape, will encounter less resistance than a two-by-four that is neither sanded nor varnished. A ship with a square bow will encounter more resistance than one with a bulbous bow and V shape. All of this is important because with greater resistance comes the need for greater power to cover a given distance. Since power requires a fuel source and a way to carry that fuel, a vessel that can travel with a lighter fuel load will be more efficient. Whether the design under consideration is for a tractor trailer, an automobile, an ocean liner, an airplane, a rocket, or a space shuttle, these basic considerations are of paramount importance in their design.

APPLICATIONS AND PRODUCTS

Fluid dynamics plays a part in the design of everything from automobiles to the space shuttle. Fluid dynamic principles are also used in medical research by bioengineers who want to know how a pacemaker will perform or what effect an implant or shunt will have on blood flow. Fire flows are also being studied to aid in the science of wildfire management. Until now the models have focused on heat transfer but new studies are looking at fire systems and their fluid dynamic properties. Sophisticated models are used to predict fluid flows before model testing is done. This lowers the cost of new designs and allows the people involved to gain a thorough understanding of the trade-off between size and power given a certain design and level of resistance.

IMPACT ON INDUSTRY

Before English engineer William Froude, ships were built based on what had worked and what should work. Once Froude proved that scale-model testing could reliably predict the performance of full-scale vessels, the entire process of vessel design was forever altered. The experience with scale models in a model basin transferred to the testing of scale models of airplanes and automobiles in wind tunnels. Computational fluid dynamic models are used for testing key elements of everything from ships to skyscrapers. The use of fluid dynamic theories to predict performance is ongoing and continues to be vital to engineers.

CAREERS AND COURSE WORK

Fluid dynamics plays a part in a host of careers. Naval architects use fluid dynamic principles to design vessels. Aeronautical engineers use the principles to design aircraft. Astronautical engineers use fluid dynamic principles to design spacecraft. Weapons are constructed with and understanding of fluids in motion. Automotive engineers must understand fluid dynamics to design fuel-efficient cars. Architects must take the motion of air into their design of skyscrapers and other large buildings. Bioengineers use fluid dynamic principles to their advantage in the design of components that will interact with blood flow in the human body. Land-management professionals can use their understanding of fluid flows to develop plans for protecting the areas under their care from catastrophic loss due to fires. Civil engineers take the principles of fluid dynamics into consideration when designing bridges. Fluid dynamics also plays a role in sports: from pitchers who want to improve their curveballs to quarterbacks who are determined to increase the accuracy of their passes.

Students should take substantial course work in more than one of the primary fields of study related to fluid dynamics (physics, mathematics, computer science, and engineering), because the fields that depend upon knowledge of fluid dynamic principles draw from multiple disciplines. In addition, anyone desiring to work in fluid dynamics should possess skills that go beyond the academic, including an aptitude for mechanical details and the ability to envision a problem in more than one dimension. A collaborative mind-set is also an asset, as fluid dynamic applications tend to be created by teams.

SOCIAL CONTEXT AND FUTURE PROSPECTS

The science of fluid dynamics touches upon a number of career fields that range from sports to bioengineering. Anything that moves through liquids such as air, water, or gases is subject to the principles of fluid dynamics. The more thorough the understanding, the more efficient vessel and other designs will be. This will result in the use of fewer resources in the form of power for inefficient designs and help create more efficient aircraft and launch vehicles as well as medical breakthroughs.

Gina Hagler, B.A., M.B.A.

FURTHER READING

Anderson, John D., Jr. *A History of Aerodynamics and Its Impact on Flying Machines.* New York: Cambridge University Press, 1997. Includes several chapters that deal with the theories of fluid dynamics and their application to flight.

Carlisle, Rodney P. *Where the Fleet Begins: A History of the David Taylor Research Center.* Washington, D.C.: Naval Historical Center, 1998. A detailed history of the work done at the David Taylor Research Center.

Çengel, Yunus A., and John M. Cimbala. *Fluid Mechanics: Fundamentals and Applications.* Boston: McGraw-Hill, 2010. An essential text for those seeking familiarity with the principles of fluid dynamics.

Darrigol, Olivier. *Worlds of Flow: A History of Hydrodynamics from the Bernoullis to Prandtl.* New York: Oxford University Press, 2005. A thorough account of the progress in hydrodynamic and fluid dynamic theory.

Eckert, Michael. *The Dawn of Fluid Dynamics: A Discipline Between Science and Technology.* Weinheim, Germany: Wiley-VCH, 2006. An introduction to fluid dynamics and its applications.

Ferreiro, Larrie D. *Ships and Science: The Birth of Naval Architecture in the Scientific Revolution, 1600-1800.* Cambridge, Mass.: MIT Press, 2007. A fully documented account of the transition from art to science in the field of naval architecture.

Johnson, Richard W., ed. *The Handbook of Fluid Dynamics.* Boca Raton, Fla.: CRC Press, 1998. A definitive text on the principles of fluid dynamics.

Mahan, A. T. *The Influence of Sea Power Upon History, 1660-1783.* 1890. Reprint. New York, Barnes & Noble Books, 2004. Mahan wrote the book that changed the way nations viewed the function of their navies.

WEB SITES

American Physical Society
http://www.aps.org

National Agency for Finite Element Methods and Standards
http://www.nafems.org

Society of Naval Architects and Marine Engineers
http://www.sname.org

Von Karman Institute for Fluid Dynamics
https://www.vki.ac.be

See also: Architecture; Civil Engineering; Computer Science; Engineering; Naval Architecture and Marine Engineering; Spacecraft Engineering.

FLUIDIZED BED PROCESSING

FIELDS OF STUDY

Chemistry; engineering; chemical engineering; chemical process modeling; corrosion engineering; control engineering; process engineering; industrial engineering; mechanical engineering; electrical engineering; safety engineering; physics; thermodynamics; mathematics; materials science; food processing; cryogenics.

SUMMARY

Fluidized bed processing is a technique that mixes solid particles with either a liquid or a gaseous stream to form a mix that behaves like a fluid. This mix can be made to flow, keep a uniform temperature, and facilitate interaction among solid and fluid particles, supporting heat and mass transfers. The early 1940's saw its first major application in fluid catalytic cracking units that increase a refinery's yield of high-quality gasoline and aviation fuel. The technique has become widely used in the process industry, including synthesis of polyethylene and polypropylene, key feedstocks of the chemical industry. It is also used in the food and pharmaceutical industries, at power plants, and for waste incineration.

KEY TERMS AND CONCEPTS

- **Aggregative Fluidization:** As a gaseous liquid fluidizes the bed of solids, bubbles form, making the process look like boiling water.
- **Bed:** Area of a processing unit where solid particles are fluidized.
- **Distributor:** Grid, or a plate with holes, through which the fluid is injected into the bed.
- **Entrainment:** Transport of solid particles up out of the bed in the fluid mix.
- **Fluid:** In physics, either a liquid or a gas.
- **Fluidization:** Physical process that converts the state of solid particles to a fluidlike state as a fluid is sent through the solid material.
- **Minimum Fluidization:** Point at which the solid particles of the bed begin to behave like a fluid.
- **Minimum Fluidizing Velocity:** Minimum superficial velocity with which the fluid has to enter the bed to fluidize the solid particles.
- **Particulate Fluidization:** Bed of solids expands continuously and uniformly, typically caused by a liquid fluidization medium; the opposite of aggregative fluidization.
- **Superficial Velocity:** Roughly, the mean velocity with which the fluid hits the solids.

DEFINITION AND BASIC PRINCIPLES

Fluidized bed processing is a technique developed from the fact that if small solid particles encounter a quickly moving fluid of liquid or gaseous nature, the resulting mix including the solids behaves like a fluid itself. This provides numerous advantages in process engineering.

For fluidized bed processing to work properly, the solids have to exist in the form of tiny particles, often called granules, ranging in size from 1 or 2 millimeters to as little as 0.02 millimeter. The smaller the particles are, the easier it is to get them into a fluidized state. The only exceptions are food particles, which can be much larger because of their low density.

To achieve fluidization and to make use of this physical conversion process, a fluid medium is injected into a bed of solid particles. As the fluid rises at its own superficial velocity—roughly meaning its own speed, which is measured by dividing the volumetric flow rate of the fluid by the cross-sectional area of the bed of solids—it encounters the solid particles. Once the rising fluid exerts a drag force greater than the net weight of the solid particle, which is called minimum fluidization, the solid particles begin to behave like the rest of the fluid. Typically, they become entrained in the fluid and rise with it to the top of the unit. Because of the laws of thermodynamics, the solids quickly share the same temperature as the fluid medium. Because of their large shared surface area, solids and fluid medium particles can interact very efficiently.

BACKGROUND AND HISTORY

On December 16, 1921, German chemist Fritz Winkler was the first to use the principle of fluidized bed processing in a coal-gasification experiment based on small-grained lignite. By 1926, the German company BASF that employed Winkler began commercial coal gasification based on Winkler's new technique. Fluidized bed processing has remained an important part of coal gasification and has been applied to gasification of biomass as well.

The second most important discovery was the use of fluidized bed processing for the catalytic cracking of long hydrocarbons at petroleum refineries. This was done to obtain more valuable gasoline and aviation fuel from crude oil distillate. Four American scientists, Donald Campbell, Homer Martin, Eger Murphree, and Charles Tyson filed their patent for the process on December 27, 1940. It was granted on October 19, 1948. Based on their research, a pilot plant started in Baton Rouge in May, 1940, and the commercial plant began operations on May 25, 1942. Since then, a fluid catalytic cracker has been at the heart of every modern refinery.

In each subsequent decade, fluidized bed processing has found new applications in the chemical, petrochemical, pharmaceutical, and food-processing industries. The technique is important in the quest for cleaner power generation—and even the production of carbon nanotubes.

How It Works

Bed Preparation. To begin the process, solid particles are placed or injected onto a bed in a holding unit. This is called a packed or a fixed bed, until fluid is applied. The bed is typically in a reactor, boiler furnace, or another part of a processing unit, such as the catalytic riser of a refinery's fluid catalytic cracker. There are different kinds of beds, the most common being stationary or bubbling beds or more complex circulating beds.

Typically, the solid particles of the bed are distinguished by their size and density. They are commonly classified into four categories, proposed first by English chemical engineer and professor Derek Geldart in 1973, and known as Geldart groups. The larger and denser the particles are, the more energy is needed to fluidize them.

In the case of fluidized bed combustion, there is a difference between solid bed materials (which will not be burned), solid sorbents, and the solid particles that serve as fuel. Solid particles can be inserted into the bed in a continuous process, even if there is already a fluid mix. This is very commonly done, rather than individual batch processing, to save time and energy.

Fluidization. Once the solids are in the bed, fluid is injected into the bed at a high velocity through a distributor, which is commonly either a grid or a porous plate at the bottom of the bed. The fluid can be both in liquid or gas form, as defined in physics. Very often, superheated steam is used or, in the case of fluid catalytic cracking, evaporated hydrocarbon molecule chains.

Because of the velocity with which the fluid flows through the bed to the top of the containing unit, it will exert a drag force on the solids in the bed. Minimum fluidization begins once the solids start to rise with the fluid medium and become entrained in it as they leave the bed for the top. If a gaseous liquid is used as fluid medium, bubbles will form in the bed, giving it the characteristic look of boiling water in a process called aggregative fluidization. If a liquid fluid is used, the bed of solids will expand continuously and uniformly, as solids rise within the liquid fluid to the top. This is called particulate fluidization.

To measure and control fluidization, process engineers measure the so-called superficial velocity with which the fluid charges through the bed. It is obtained by dividing the volumetric flow rate of the fluid by the cross-sectional area of the bed and corresponds roughly to the speed with which the fluid medium is pressed through the bed. It is typically given in meters per second. Each type of solid has its own minimal fluidization velocity that must be reached for the solids to begin to fluidize.

Processing. In the fluid state, processing of solids and fluid occurs as designed for each application. This is aided by the uniform heat of the fluid mix, its transportability through pipes, and the high rate of surface interaction between solid and fluid medium particles. At the end of desired processing, remaining solids and fluid medium are typically separated.

Applications and Products

Fluid Catalytic Cracker (FCC). Traditionally, this is the key industrial application of fluidized bed processing. Both atmospheric and vacuum distillation of crude oil at a refinery leave behind a large percentage of low-value residue instead of desired gasoline, diesel, or aviation fuel. This distillation residue leaves behind long hydrocarbon molecule chains.

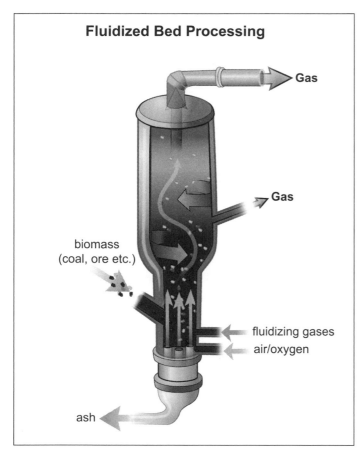

Fluidized Bed Processing

Fluidized bed processing is a technique that mixes solid particles with either a liquid or a gaseous stream to form a mix that behaves like a fluid.

Cracking creates much more of the shorter hydrocarbon chains, which yield desirable fuel such as gasoline. The American invention of the FCC in 1942 quickly spread to refineries all over the world after World War II ended in 1945.

The great technical advantage of the fluid catalytic cracker is that the long hydrocarbon chains, typically light fuel (or gas) oil, come into contact with the solid catalyst particles in a fluidized, rather than a stationary, bed environment. That means that the total surface area between catalyst and oil particles is much larger than possible if the catalyst were just fixed to the floor or sides of the reactor. In the cracker's catalytic riser, tiny catalyst particles are completely surrounded by the oil particles and swim in a fluid stream together for a few crucial fractions of a second.

In this case, it is the fluid medium, the fuel oil preheated to its evaporation point, that undergoes the value-adding conversion. Its long hydrocarbon chains are quickly cracked when encountering the catalyst particles in the fluid mix. The fluid is sent to a reactor. Cyclones at the top of the reactor separate the gaseous hydrocarbon fluid from the spent catalyst and transport it into a distillation unit. The solid catalyst particles are collected at the bottom of the reactor and sent to a regenerator, from where they are fed back into the catalytic riser in a continuous process.

Fluidized Bed Combustion (FBC). This is the oldest application that has gained new attention in the quest for cleaner power generation and waste incineration. Its basic goal is to convert the energy of solid fuels, such as coal, biomass, or waste materials, through gasification, steam generation, or incineration using fluidized bed combustion.

An advanced power plant can use FBC to increase its fuel efficiency and lower harmful emissions. Inside the combustor, there is a bed of inert material that will not be burned itself, typically a form of sand. To the solid fuel particles that are added to the bed is mixed a solid sorbent, which absorbs potentially harmful substances released during combustion. Typically, when coal is used as fuel, limestone or dolomite is used as sorbent to absorb sulfur that would otherwise react and be emitted as sulfur oxide, a strong atmospheric pollutant. The solid particles are fluidized by the injection of a fluid, typically heated air. There are two kinds of arrangements for the bed of solids that is hit by the fluid, either a bubbling fluid bed or a circulating fluid bed.

The two great advantages of the FBC are that it can be fired by a wide variety and mix of solid fuel particles, including coal, solid wastes, biomass, or natural gas; and it generates much less pollution because of the use of sorbent in the fluid to catch sulfur. It is able to burn fuel at lower temperatures where oxygen and nitrogen of the air fluid do not yet react to form the pollutant nitrogen oxide.

A simple FBC operates at atmospheric pressure as the solid fuel particles are burned in the fluidized state and the sorbent particles bind potentially harmful by-products. In a more advanced design, the fluid of generally hot air is pressurized. Now, combustion creates a hot pressurized gas flow that fires a gas turbine, while steam generated during combustion fires a steam turbine for maximum energy yield in what is called a combined-cycle power plant. This accounts for one of the most efficient and least polluting modes of power generation.

Chemical Applications. Fluidized bed processing has become widely used in the chemical industry. It is important in particular for the synthesis of polyethylene and polypropylene, key basic plastics used for packaging, textiles, and plastic components. Fluidized bed reactors are used also for the industrial production of monomers such as vinyl chloride or acrylonitrile, which are both used to make plastics. These reactors are also employed to produce polymers such as synthetic rubber and polystyrene. The advantages of uniform heat transfer, great surface interaction, and transportation as fluid, whether in liquid or gaseous form, have made fluidized bed processing very valuable for contemporary chemical industry processes.

Pharmaceutical and Food Processing. Because of its excellent properties facilitating material and heat transfer, fluidized bed processing has become an important application for the coating or drying of pharmaceuticals. It is also used for batch granulation of pharmaceuticals. In the food-processing industry, the technique is used especially for creating individually quick-frozen (IQF) products. Because of their low density, individual food particles as large as bite-size diced vegetables can be fluidized and frozen quickly to maintain their taste. Complete food packages, as well as individual food particles, are also frozen, blanched, cooked, roasted, or heat sterilized in fluidized bed processing.

Mineral Processing. This older application dating back to the 1950's is used to decompose or purify ores through calcination or the roasting and pre-reduction of ores. Fluidized bed processing is also used in cement manufacture.

IMPACT ON INDUSTRY

The vital global importance of fluidized bed processing for petroleum refining, cleaner and more efficient power generation, as well as often innovative applications in the chemical, food, and pharmaceutical industries have made this a field of considerable research and industry interest. There is both academic and industry inquiry to learn and fully understand all theoretical scientific underpinnings of the technique as well as practical developments to optimize processes and come up with new applications or design solutions.

Government Research. As of 2011, national governments have taken a lead in promoting both basic and applied research into fluidized bed processing in the quest to develop more environmentally friendly and energy-efficient modes of power generation. For example, China promotes research into coal-fired cogeneration plants using fluidized bed processing in its coal-particle combustion units. A particular concern is discovering economically viable alternatives (other than coal washing or flue-gas scrubbing) to reduce the massive sulfur oxide-

---

Fascinating Facts About Fluidized Bed Processing

- The invention and industrial application of fluid catalytic cracking is credited by some historians for contributing significantly to the Allied victory in World War II because of the wide availability of quality automotive and aviation fuel.

- In food processing, solid particles of the fluidized bed can be as large as a potato chip, while the smallest particles used in the chemical industry are as tiny as 20 micrometers.

- It takes less than one second for a catalyst to crack a long hydrocarbon molecule chain in the catalytic riser of a fluid catalyst cracker at the heart of a modern refinery.

- The four American scientists who invented fluid catalytic cracking, Donald Campbell, Homer Martin, Eger Murphree, and Charles Tyson, had to wait almost eight years, from December, 1940, to October, 1948, for their U.S. Patent No. 2,451,804 to be granted after filing it.

- For many years, the chemical engineering department at the University College London used to show its students how fluidization of a bed of sand made a buried light plastic duck rise from the bottom to the top and exchange places with the brass duck that was sinking down.

- Fluidized bed combustion holds some promise for use within zero-emission power generation.

emissions of older-style Chinese power plants, especially since coal washing also causes water pollution.

The United States Department of Energy sponsors many research projects, particularly into next-generation pressurized fluid bed combustion combined-cycle power plants. The goal is to design plants with a net system efficiency of more than 50 percent, extremely low sulfur and nitrogen oxide emissions well below 2010 emission limits, and at a power-generation cost of three-quarters by a conventional coal-fired power plant. The European Union similarly sponsors research in this area, as does Japan and other developed or developing countries.

Universities and Research Institutes. Universities and research institutes are interested in promoting both primary research and innovative process designs for fluidized bed processes for the same reason governments and industry are interested. As the quest to limit human-made carbon and other emissions intensifies, the promise of efficiency and low-emission output gained from fluidized bed processing motivates much research for which government grants tend to be available. In addition, catalyst research affecting FCC performance is a vibrant area, and often finds partners in industrial users. There has been consistent growth internationally in patents awarded for fluidized bed process optimizations resulting from the work of university and other institutional research teams.

Industry and Business. Fluidized bed processing is vital in the petroleum, chemical, petrochemical, pharmaceutical, and other process industries. Its impact is considerable, particularly in the petroleum and power industry.

In refining alone, all over the world in 2007, about four hundred fluid catalytic crackers cracked about 10.6 million barrels of feedstock each day, with half of it occurring in the United States. In 2010, FCCs produced from 35 to 45 percent of the gasoline gained at different U.S. refineries. The exact designs of an FCC, and the catalysts used therein, have been closely held corporate proprietary information.

Major Corporations. Because of the capital-intensive nature of units operating fluidized bed processing in most cases, there is a concentration of major corporations active in the field. All the international oil and gas companies employ engineers and scientists seeking to improve operations at their fluid catalytic crackers. Many FCCs are built under

license by internationally active companies such as Kellogg Brown and Root (KBR) and Universal Oil Products (UOP). These companies hold patents for the unique design of their specialized fluidized bed reactors.

All major engineering companies who provide utility companies with fluidized bed combustion units, such as Germany's Lurgi GmbH or France's Technip, for example, need specialists to design, build, and improve these units. A pharmaceutical or food-processing corporation can gain a significant competitive edge if it develops a new or improved fluidized bed process for its manufacturing process.

CAREERS AND COURSE WORK

Students interested in a career in designing, developing, or supervising the operation of fluidized bed processes should take high school courses in the natural and computer sciences and mathematics. Refineries, power plants, and chemical and pharmaceutical companies employ skilled technicians for their operations of fluidized bed processes, and an associate's degree in science or engineering is a good foundation. Universities and research institutes also hire technicians to assist their researchers in the field.

A bachelor of science or arts degree in physics or chemistry, and perhaps a double major in computer science, is very useful for a career in the field. Students interested in food processing could also major in biology.

An engineering degree is an excellent foundation for an advanced career in the field, especially in chemical, electrical, mechanical, or computer engineering. Students should make sure to take classes in process engineering, chemical process modeling, control and safety engineering, and have a good understanding of thermodynamics.

Any master of engineering or doctoral degree in chemistry, physics, or chemical engineering is good preparation for a high-level career. A Ph.D. in one of the sciences can lead to an advanced research position either with a university, company, or government agency.

For a career in private industry, some sense of economics is also helpful. For those who are interested in applications in the refining or the power-generation industry, a willingness to work outside the United States is a plus as many contemporary Greenfield plants using the latest technology are built abroad.

For this reason, a semester or two of study abroad is advised.

SOCIAL CONTEXT AND FUTURE PROSPECTS

As global demand for high-end petroleum products such as gasoline, diesel, and jet fuel continues, optimization of fluidized bed processes in FCCs will lose none of their importance. The drive to develop more efficient catalysts and processes, as well as to increase desirable product yield, lower energy consumption, and emissions, will provide for an exciting field of research in this technique.

The quest for a zero-emission power-generation plant would bring enormous prestige as well as business benefits for its developer. As of 2011, engineers and scientists in the power industry seek to develop a commercially viable chemical-looping combustion process that would use two fluidized beds. In the first, oxygen for combustion would be gained from a metal oxide bed. Then the metal would be reoxidized in a second bed, ready for return into the first one. This process could greatly revolutionize the power industry.

As the chemical and petrochemical industry has been growing, so have applications for well-designed, specialized fluidized bed processes in their production processes. The same holds true for the pharmaceutical and food-processing industries, where qualitative process advantages have been sought as well as new modes for innovative processes.

R. C. Lutz, B.A., M.A., Ph.D.

FURTHER READING

Basu, Prabir. *Combustion and Gasification in Fluidized Beds.* Boca Raton, Fla.: CRC Press, 2006. Comprehensive account of all aspects of this important application of fluidized bed processing.

Froment, Gilbert, Juray DeWilde, and Kenneth Bischoff. *Chemical Reactor Analysis and Design.* 3d ed. Hoboken, N.J.: John Wiley & Sons, 2011. Includes description of chemical reactors using fluidized bed processing; for undergraduate and graduate students, comprehensive scope.

Occelli, Mario L., ed. *Advances in Fluid Catalytic Cracking: Testing, Characterization, and Environmental Regulations.* Boca Raton, Fla.: CRC Press, 2010. Series of essays concerning technological advances in this key application of fluidized bed processing.

_____. *Fluid Catalytic Cracking VII: Materials, Methods and Process Innovation.* Oxford, England: Elsevier, 2007. Seventeen individual chapters cover key aspects of fluidized bed processing in fluid catalytic cracking at refineries, focus on catalysts used, process optimization, cleaner processes, and future developments.

Smith, P. G. *Application of Fluidization to Food Processing.* Oxford, England: Blackwell Science, 2007. Excellent survey of basic principles of fluid bed processing and the technique's growing application in the food processing industry. Somewhat technical but accessible; glossary.

Tsotsas, Evangelos, and Arun S. Mujumdar, eds. *Modern Drying Technology: Experimental Techniques, Volume 2.* Weinheim, Germany: Wiley-VCH, 2009. Chapter 5.4 focuses on the application of fluidized bed processing in granulation of detergents and batch granulation of pharmaceuticals and methods to measure attrition dust and over-spray in continuous spray processes.

WEB SITES

American Petroleum Institute
http://www.api.org

International Energy Agency
http://www.iea.org

National Petrochemical and Refiners Association
http://www.npra.org

See also: Chemical Engineering; Electrical Engineering; Engineering; Mechanical Engineering.

FOOD PRESERVATION

FIELDS OF STUDY

Chemistry; food science; food technology; nutrition; agriculture; animal and dairy science; agricultural engineering; veterinary science; microbiology; botany; biology; horticulture; industrial engineering; mechanical engineering; electrical engineering; computer science; robotics.

SUMMARY

Food preservation involves the application of scientific methods to prevent agricultural resources, whether eaten raw or further processed, from becoming contaminated or spoiling. Preservation ensures that people have sufficient supplies of food to survive. The availability of preserved foods makes it easier for people to migrate and settle in new areas. Explorers and military forces rely on preserved foods for nourishment. Preservation methods also enable perishable foods to be traded worldwide.

KEY TERMS AND CONCEPTS

- **Aseptic Packaging:** Preservation method in which food is sealed in pathogen-free containers in a sterile environment.
- **Canning:** Preservation method in which food is sealed in airtight containers by using pressure or other scientific techniques.
- **Freeze-Drying:** Preservation method in which food is dried by being quickly frozen and then vacuum processed.
- **Irradiation:** Preservation method in which food is exposed to ionizing radiation to slow decay.
- **Modified Atmosphere Packaging:** Preservation method in which the gas in food containers is altered to improve shelf life; typically, the amount of oxygen is reduced to slow oxidation and the growth of aerobic organisms.
- **Pasteurization:** Preservation methods that use heat, chemicals, or radiation to kill microorganisms and inactivate some enzymes.
- **Shelf Life:** Duration that a food can be stored and still be suitable for consumption.
- **Thermal Processing:** Preservation methods using different types and degrees of heat.
- **Vacuum Packaging:** Preservation method in which air is extracted from sealed containers holding food.

DEFINITION AND BASIC PRINCIPLES

Food preservation is the science of destroying or impeding the growth of harmful microorganisms, slowing oxidation, and controlling chemical reactions that cause foods to decay and become inedible. Food technologists and scientists realize that food will inevitably decay but that preservation strategies can extend its freshness. Food preservation professionals develop procedures and technologies to decrease spoilage, which ruins between one-tenth and one-third of food produced globally. They improve packaging, which keeps out hazardous contaminants, and devise innovative storage and transportation modes to maintain food quality, facilitating the manufacture and distribution of more food varieties worldwide.

Preservation is essential to ensure a consistent and sufficient supply of safe food to nourish global populations. Manufacturing processes to protect foods from contaminants typically involve exposing them to temperature extremes, using chemicals, applying high pressures, placing foods in a vacuum, or passing them through radiation or diverse energy sources. Removing moisture and drying foods is basic to many food preservation methods. Food and beverage packaging makes use of protective films, including edible coatings; of gases injected into or removed from food containers; and of aseptic environments in which to package foods.

Critics of food preservation allege that many of the preservatives added to foods are dangerous and suggest that nonchemical alternatives be developed and employed. They also protest the use of irradiation and other controversial technologies that they believe may be carcinogenic or pose other health risks to consumers. The U.S. Food and Drug Administration (FDA) responds to these criticisms by investigating potential threats involving food preservation and setting standards for those processes, such as establishing permitted irradiation dosages for various foods.

BACKGROUND AND HISTORY

Since ancient times, people have preserved surplus food to eat when fresh foods are not readily available. Early methods included drying or cooling foods and using easily obtainable preservatives such as honey, vinegar, alcohol, fats, sugar, and oils. Hanging foods to dry reduced their moisture content and minimized decay. Foods could be cooled in caves, snow, and streams. Standard preservation methods included salting, pickling, and fermenting. Many of these basic preservation strategies are incorporated into modern methods.

In the late eighteenth century, scientists became aware that microorganisms existed but did not know that microbes could cause food to spoil. By the early nineteenth century, confectioner Nicolas François Appert had devised a technique to preserve food in glass bottles, which served as the foundation for modern industrialized canning. Food preservation techniques improved as people learned how microorganisms made food spoil. In the mid-nineteenth century, French bacteriologist Louis Pasteur developed the pasteurization process, which uses heat to destroy microbes. This process, named after its inventor, became intrinsic to food preservation. In the 1850's, inventor Gail Borden, Jr., used vacuum technology to remove air from containers to create condensed milk, which remained fresh for extended periods while shipping or in storage. In the 1880's, Gustavus F. Swift, founder of a meatpacking company, created refrigerated railroad cars to preserve meat being transported over long distances. In the 1920's, businessman and inventor Clarence Birdseye patented a process to quick freeze foods.

As food preservation methods (some harmful) and patent medicines proliferated with little or no regulation, the U.S. government enacted legislation to ensure the quality of foods and drugs. The Pure Food and Drug Act of 1906 allowed the federal government to inspect meat products and prohibited the "manufacture, sale, or transportation of adulterated or misbranded or poisonous or deleterious foods, drugs, medicines, and liquors." The law's focus was on the accurate labeling of products. Although the FDA can trace its history to 1848, it began acting as a consumer protection agency after the passage of the 1906 act. Many additional laws regarding food and drug safety followed.

HOW IT WORKS

Maintaining the freshness of food as it goes from market to the table is accomplished through chemicals and other substances that act as preservatives, processes that prevent or delay food spoilage, and packaging that guards against the chemical reactions that initiate decay.

Preservatives. Food technologists use many preservatives, both chemical and natural. Natural food preservatives include antimicrobials that attack microorganisms. Enzymes that are effective antimicrobials include glucose oxidase, lactoperoxidase, and lysozymes. Antioxidants, such as vitamins C and E, butylated hydropxytoluene (BHT), and butylated hydroxyyanisole (BHA), block spoilage caused by oxidation. Other food preservation strategies involve adding acids to lower pH levels, which disrupts microbe activity and slows food deterioration.

Thermal Processes. Heat damages or destroys most microbes and harmful enzymes in foods. During pasteurization, heat penetrates food and eliminates most microorganisms. Food technologists adjust the duration and intensity of heat applied to foods depending on the process and the food and microbes or enzymes involved. Many manufacturers use high-temperature short-time (HTST) or ultra-high-temperature (UHT) methods. Foods are transported on conveyor belts into contained areas, where they are heated to the proper temperature and for the appropriate length of time. Automated canning and bottling systems produce thousands of cans or bottles per hour. The bottles or cans are filled with liquid or food and sealed with lids as they move along conveyor belts into heating chambers.

Aseptic food preservation processes differ from canning in that foods and containers are both sterilized separately before packaging occurs. The foods are flash pasteurized then cooled before being packaged and sealed in sterile containers in a sterile environment. This type of packaging, often used for soy milk and drink boxes, allows foods—with no added preservatives—to be kept for long periods without refrigeration. Other thermal food preservation methods include using energy from radio frequencies and microwaves to heat foods during manufacturing.

Nonthermal Processes. Because exposure to heat alters some foods' chemical properties, food technologists have devised alternative methods. Nonthermal techniques control microorganisms with the application of electricity, radiation, pressure, or optics, individually or in combination. Some food manufacturers employ high-pressure processing (HPP), placing foods in chambers that can be pressurized at varying intensities. The ultra-high-pressure processing (UHP) technique pasteurizes foods, typically vegetables and fruits, by spraying them with powerful cold-water jets that remove most microbes.

Irradiation exposes foods to electron beams emitted by gamma rays, X rays, or other radioactive sources producing ions for times ranging from several seconds to an hour. The FDA and U.S. Department of Agriculture regulate irradiation dosages to disinfect specific foods. Other nonthermal food preservation methods include electronic pasteurization or the use of a pulsed electric field (PEF), pulsed magnetic field (PMF), or high intensities of pulsed light (PL) to inactivate microbes. Ultrasound preservation processes inactivate hazardous spores, enzymes, and microorganisms in foods.

Cooling and Freezing Processes. Reducing food temperatures is an effective way to slow biodeterioration because chilling foods alters the kinetic energy of microorganisms. In these processes, food is placed on conveyor belts and moved past chilled metal plates, coils holding refrigerants, compressors, and evaporators. Temperatures of 0 degrees Celsius (freezing point of water) or less kill most microbes or impede their movement. The type of freezing machinery and chemicals used determine how long the freezing process takes; some methods instantly freeze foods and others slowly transform them. Some food manufacturers use cool water to chill vegetables and fruits to preserve their freshness.

Vacuum and Dehydration Processes. Removing air from environments surrounding foods by creating a vacuum in a contained space helps minimize oxidation and other chemical reactions that cause spoilage. Engineers design industrial machinery to remove water efficiently from large quantities of food. Food manufacturers often use mechanical dryers to blow streams of warm, dry air over food. Fans circulate this air, which carries water molecules evaporated from food. Osmotic drying uses chemical solutions containing sodium chloride, sucrose, or other agents.

Freeze-drying involves ice sublimating from frozen food in a vacuum chamber.

Packaging. Food preservation relies on protective containers or wrappings to block contaminants. Films impede oxygen and moisture from contacting foods. Packaging technology includes films containing antimicrobial preservatives and coatings that are safe to eat. The modified atmosphere packaging process involves machinery that places foods in containers that are injected with gases, including nitrogen, oxygen, carbon dioxide, or noble gases such as argon. Variations of modified atmosphere packing include controlled-atmosphere packaging, which uses scrubber technology to attain more precise gas levels and vacuum packaging at higher pressures.

APPLICATIONS AND PRODUCTS

Nutrition. People require consistent access to sufficient nutritious foods to survive. Preservation strategies help provide adequate nourishment to populations worldwide. Aseptic processing and packaging contributes significantly to saving agricultural surpluses, which otherwise would spoil or be discarded. Preservation techniques enable people to have a safe source of food and water when emergency situations interrupt the normal flow of goods. Relief organizations distribute bulk shipments of preserved foods

This is an industrial scale bain marie (water bath) used to heat milk in order to prevent microbial growth. Developed by the French chemist Louis Pasteur, this food preservation process is known as pasteurization. (CCI Archives/Photo Researchers, Inc.)

that do not require refrigeration. French manufacturer Nutriset and U.S. producer Edesia Global Nutrition Solutions produce Plumpy'nut, a high-calorie, vitamin-rich paste preserved in foil pouches, which alleviates malnourishment in areas affected by famine or disasters.

Travel, Transportation, and Trade. Food preservation has aided the movement of people for centuries. Travelers, including nomads and migrant workers, often rely on preserved foods to sustain them on journeys. Transportation of preserved foods takes place over the land and seas and through the air. Trucks move preserved foods and beverages from manufacturers to wholesalers and retailers. Engineers have designed special shipping cartons and vehicles to protect foods. Boxes of food are often wrapped with plastic. Technicians remove air between the plastic and the food, inserting gases to create a modified atmosphere to stabilize the food until delivery. Films placed around egg packages absorb shock. Refrigerated trucks keep foods frozen or chilled, and fans circulate air within cargo areas to keep perishables fresh. Many of those trucks contain microprocessors or are regulated by satellites to assure refrigeration is consistent while traveling.

Many railroad companies use refrigerated cars to move foods. These cars are designed to transport specific preserved foods and can be removed from trains and hauled by trucks to their final destinations. Rail yards near ports often have stacks of thousands of containers waiting to be transported inland or to sea. Cargo ships transport large tanks filled with several million gallons of preserved foods. Aseptic preservation processes enable worldwide shipments of vegetables and fruit products. Many preserved foods are shipped by airplane, with Federal Express carrying much of that cargo worldwide. Innovations in transporting preserved foods have enhanced international trade. Unique foods associated with specific geographical areas can be shipped and sold in distant markets within days of harvests, expanding sales of native produce.

Exploration. The availability of preserved foods often determines whether an expedition to remote areas is successful. Historically, explorers traveling for extended periods across land masses and bodies of water relied on salted, cured, canned, and other preserved food to provide them energy and nutrients. Modern explorers also carry preserved foods.

Fascinating Facts About Food Preservation

- Philip E. Nelson received the 2007 World Food Prize in recognition of his technological contributions to preserving food. He has received numerous patents for aseptic process and packaging inventions.
- In 1938, British biochemist Jack Drummond unsealed canned mutton that Captain William Parry and his crew had taken on their 1824 search for the Northwest Passage. The canned mutton had been left behind when the crew abandoned their icebound ship. Drummond found that the salvaged food was still safe to eat.
- In 1955, France began circulating a twelve-franc postal stamp featuring Nicolas François Appert and his food-preserving bottling process.
- Aseptic packaging ranges in size from small pudding cups and juice boxes commonly found in grocery stores to square tanks, standing as high as a six-story building and having an equivalent width, which can hold more than 1 million gallons.
- In August, 2010, Al and Susie Howell opened a can of Cougar Gold cheese that the Washington State University Creamery had produced in 1987 using technology that the school had created for packaging foods in tin containers. Although the can was twenty-three years old, the cheese retained its flavor.

The National Aeronautics and Space Administration (NASA) asked Pillsbury Company food scientists to help them develop preserved foods that would not produce crumbs. Mercury, Gemini, and early Apollo program crews consumed foods contained in tubes. NASA arranged for Oregon Freeze Dry to use its freeze-drying technology to preserve more elaborate meals for astronauts to eat during later Apollo and space shuttle missions.

NASA encourages commercial businesses to adapt its food preservation and packaging technology, granting permission to those businesses that want to use these innovations for preserving and packaging their food products. In the 1980's, Sky-Lab Foods used NASA technology to manufacture freeze-dried meals in bags. Oregon Freeze Dry produces freeze-dried military rations, fruits, vegetables, meats, and pet treats as well as a line of products for backpackers under the brand name Mountain House. Reflective metals used in space films were adapted for use in packages to protect and insulate foods.

In 2010, NASA sponsored a project examining the feasibility of growing vegetables in space to feed astronauts on possible missions to Mars or colonists on a Moon base. Orbitec Technologies assisted NASA scientists in determining ways to create gardens on spacecraft. This program supplemented ongoing plant experiments on the International Space Station and attempts by Martek Biosciences in the 1990's to cultivate algae to create oil with polyunsaturated fatty acids for lengthy missions.

Military. Since ancient times, armies have preserved foods to serve as military rations. These foods boost morale by providing soldiers familiar meals similar to those they consumed at home. Quartermasters distribute preserved food provisions to wounded soldiers in military hospitals and to prisoners of war. During the late twentieth century, military forces began supplying troops with Meals, Ready-to-Eat (MRE). These preserved foods and drinks, resembling those eaten in space, are packaged in pouches. MRE manufacturers process foods, such as beef stew, barbecued chicken, and cultural favorites such as curry and vegemite, with special techniques to preserve them for consumption under the adverse conditions associated with military deployments. Soldiers often combine MREs to concoct mixtures that have become part of military tradition.

The U.S. Department of Defense secures MREs from several manufacturers, including International Meals Supply, Wornick, Sopakco, and Ameriqual. International Meals Supply makes MREs for military units worldwide and produces MREStars for civilians. The other MRE manufacturers also make versions for nonmilitary customers. Applications of MRE manufacturing technology include producing energy foods for athletes and survival foods for use in emergencies. Governments, hospitals, and safety and relief agencies acquire MRE for use in crises.

IMPACT ON INDUSTRY

In 2008, the global food preservation industry's value exceeded $500 billion. The food preservation market consists of manufacturers that produce technology to perform food preservation processes in factories, businesses, and homes and that develop and distribute synthetic and natural preservatives. Researchers throughout North America, Europe, Asia, Africa, Australia, and New Zealand contribute food preservation advances in academic, government, and industrial settings.

Government and University Research. The USDA provides the majority of government funding for food preservation work in the United States. Researchers at academic laboratories and agricultural experiment stations conduct studies relevant to developing new preservation procedures and identifying natural antimicrobials to counter microorganisms in food. Many researchers participate in interdisciplinary projects, using knowledge and insights from colleagues with expertise in related fields to envision, refine, and design technology that can maintain the quality of food for longer periods and to provide effective packaging to protect foods. Modeling techniques project how microbes will react to preservation methods and how various processes and intensities might affect spoilage rates. Researchers address problems associated with microorganisms becoming resistant to food preservation technologies. They assess if new food preservation techniques are compatible with food safety requirements and how they affect the environment.

American universities with food science departments making significant contributions to food preservation research include Purdue University and Washington State University. In 2000, the USDA's Cooperative State Research, Education, and Extension Service provided funds to establish the National Center for Home Food Preservation based at the University of Georgia. The center's primary mission is to educate consumers and teachers regarding scientific aspects of food preservation and new techniques introduced by researchers. Notable European food preservation work occurs at Wageningen University's Agrotechnological Research Institute in the Netherlands. In 2010, the journal *Applied and Environmental Microbiology* reported on research conducted by German scientists at the University of Applied Sciences into the application of acids to kill the norovirus.

Industry and Business Sectors. According to the Business Communications Company, U.S. industries focusing on food preservation generated more than $250 billion in 2008. Worldwide, food preservation contributed to almost three-fourths of the food manufacturing industry's profits. A 2003 report issued by the company attributed growth in the food preservation industry to successful incorporation of research to minimize spoilage and increase the shelf life of fresh foods.

Manufacturers produce equipment that incorporates food preservation in the process of converting agricultural materials into consumer products. Examples of the successful industrial use of food preservation technology include Fresherized Foods, which outfitted its plants with high-pressure processing machines in 1997 to manufacture guacamole for a variety of commercial foods. Japanese food manufacturers had begun implementing high-pressure processing technology in the 1980's. By the early twenty-first century, about fifty-five businesses worldwide were equipped with this technology.

Businesses manufacture and sell home food preservation devices, such as FoodSaver products, which use a vacuum process to seal foods in bags. Other basic food preservation equipment used in homes and retail businesses includes pressure cookers, canning jars, and seals.

The food preservation industry's packaging sector thrives because 98 percent of food is sold in some form of plastic, polymer, or metal wrapping or container. The U.S. food packaging industry generates more than $110 billion yearly. A 2009 report by the research firm Frost & Sullivan suggested that more food manufacturers are likely to incorporate nanotechnology approaches to food preservation, producing packaging films containing natural antimicrobial agents such as enzymes and cultures with antibiotic properties. A 2009 report released by Food Technology Intelligence noted that methods involving fungal amylases, low pH levels, and zero-oxygen packaging extend shelf live, increasing profits for manufacturers that implement these strategies.

CAREERS AND COURSE WORK

People seeking careers associated with food preservation can pursue several educational options depending on their professional interests and goals. Most industrial and government food preservation career paths require employees to have a bachelor of science degree. Food preservation professionals can find positions with both industrial and academic employers wanting to improve techniques to protect foods. Undergraduate classes focusing on food

science, nutrition, agricultural engineering, or animal and dairy science provide students with a fundamental comprehension of how sciences are applied to cultivating, harvesting, and preserving agricultural resources.

An interdisciplinary approach expands students' employability. Study of related scientific and technological fields, especially microbiology, chemistry, and engineering, can qualify students for entry-level positions at industries using food preservation techniques or prepare them for further study. Food manufacturing industries recruit people with engineering degrees and work experience who can design machinery to preserve food or can incorporate robotics and automation into packaging processes. Computer expertise is needed to write programs monitoring preservation processes.

Graduate degrees, usually a master's degree and sometimes a doctorate, in specialized fields are usually needed for research positions involving food preservation work at industrial, government, or academic laboratories. Veterinary science degrees prepare professionals focusing on food preservation issues associated with livestock products. Faculty positions enable qualified personnel to teach students or perform research at university experiment stations. Government agencies, especially the USDA and FDA, hire employees with food preservation knowledge to work in diverse research, education, and administrative roles.

SOCIAL CONTEXT AND FUTURE PROSPECTS

International humanitarian and agricultural groups emphasize the need for continued food preservation research to mitigate hunger and malnutrition. In 2010, the United Nations stated that 925 million people worldwide suffer from chronic hunger. Aid workers teach impoverished people basic food preservation techniques so that they can stockpile foods. The recession in the early twenty-first century made bulk buying of foods attractive to many people, who then became interested in preserving these foods to store them for longer periods. The number of people growing gardens or purchasing produce from farmers markets increased, making home food preservation more popular. These people were motivated not only by economic considerations but also by a desire to control the quality of their food.

Manufacturers expect food technologists to improve preservation techniques so that their foods remain fresher longer than those of their competitors. Packaging research and creative technological advances, such as aseptic methods and antimicrobial films, represent a growing component of the food preservation industry. Industries also have asked scientists to use more natural preservatives and avoid synthetic chemicals because many consumers refuse to purchase foods containing any additives that might be detrimental to their health. Food preservation researchers strive to advance and patent unique methods incorporating bacteriophages (bacteria-eating viruses), enzymes, and other innovative scientific concepts to combat microorganisms while producing affordable, nutritious foods with long shelf lives.

Elizabeth D. Schafer, Ph.D.

FURTHER READING

Belasco, Warren, and Roger Horowitz, eds. *Food Chains: From Farmyard to Shopping Cart.* Philadelphia: University of Pennsylvania Press, 2009. A collection of case studies describing how various agricultural products—including hogs, poultry, and seafood—are turned into food products.

Branen, A. Larry, et al., eds. *Food Additives.* 2d ed. New York: Marcel Dekker, 2002. Contains chapters on regulations governing additives, hypersensitivity, and children.

Kurlansky, Mark. *Salt: A World History.* New York: Walker, 2002. Explores salt's role as a food preservative and its place in culture, economics, and the military.

Lelieveld, Huub L. M., Servé Notermans, and Sjoerd W. H. De Haan, eds. *Food Preservation by Pulsed Electric Fields: From Research to Application.* Cambridge, England: Woodhead, 2007. Explains how pulsed electric fields are used to preserve foods.

Nelson, Philip E., ed. *Principles of Aseptic Processing and Packaging.* 3d ed. Lafayette, Ind.: Purdue University Press, 2010. A technical discussion of aseptic manufacturing methods and applications and government regulations establishing guidelines.

Rahman, Mohammad Shafiur, ed. *Handbook of Food Preservation.* 2d ed. Boca Raton, Fla.: CRC Press/Taylor & Francis Group, 2007. Comprehensive guide examines traditional techniques and modern developments in technology and processes to preserve foods, suggesting possibilities for future methods.

Shephard, Sue. *Pickled, Potted, and Canned: How the Art and Science of Food Preserving Changed the World.* New York: Simon & Schuster, 2000. Illustrated history discusses methods used by people at various places and times to preserve foods, providing contemporary quotations and food preservatives references.

WEB SITES

American Council for Food Safety and Quality
http://agfoodsafety.org

American Frozen Foods Institute
http://www.affi.com

International Association for Food Protection
http://www.foodprotection.org

National Aeronautics and Space Administration
Space Food
http://spaceflight.nasa.gov/living/spacefood/index.html

National Center for Home Food Preservation
http://www.uga.edu/nchfp

See also: Agricultural Science; Food Science; Nutrition and Dietetics; Pasteurization and Irradiation.

FOOD SCIENCE

FIELDS OF STUDY

Organic chemistry; inorganic chemistry; biochemistry; cell biology; molecular biology; genetics; tissue engineering; microbiology; nutrition; bacteriology; agricultural science; genetics; physiology; medicine; pharmaceutics; horticulture; phytochemistry; engineering; mathematics; calculus; statistics; physics.

SUMMARY

Food science is a field concerned with studying the biological, chemical, and physical properties of food. Food scientists make use of the tools of science, technology, and engineering to develop effective ways of producing and preserving a safe, healthy food supply for communities and nations. Issues addressed by food science include the safe cultivation and harvest of food plants; the healthy and humane breeding and slaughter of livestock; the optimal preservation of food as it is processed, stored, packaged, and distributed; and the manufacture of new food products with maximal nutritional value. Food science is also the discipline responsible for identifying the inherent nutritional properties of various foods and the ways in which foods interact with human biological systems once they have been consumed.

KEY TERMS AND CONCEPTS

- **Aeration:** Process of introducing air into a mixture by chemical, biological, or mechanical means.
- **Commercial Sterility:** Level of sterility in which only one out of every 10,000 products of a certain kind can be expected to contain bacteria or other microorganisms.
- **Cross-Contamination:** Transfer of a harmful microorganism to a food item through contact with food or another object.
- **Denaturization:** Change in molecular structure that occurs when proteins experience an external stress such as heat or contact with an acid, salt, or organic solvent.
- **Emulsifier:** Substance that enables the smooth combining of two liquids, such as milk and oil, which otherwise could not form a homogenous mixture.
- **Pasteurization:** Process of heating a food to a particular temperature and keeping it there for a specified time before cooling it, so as to destroy pathogenic microorganisms.
- **Phytochemical:** Chemical substance derived from a plant.
- **Refrigeration:** To preserve a food product at a temperature lower than 40 degrees Fahrenheit.
- **Shelf Life:** Time it takes for a given food to become chemically degraded.
- **Standard Plate Count (SPC):** Test for detecting the level of microorganisms in food.
- **Sterilization:** Process of heating a food or other material to the point where most or all microorganisms have been killed.
- **Triglyceride:** Fat molecule (or lipid) composed of three fatty acids linked chemically to one molecule of glycerol.
- **Xantham Gum:** Microorganism produced by fermenting corn sugar; used as a thickening and emulsification agent.
- **Zymurgy:** Science and practice of fermentation, such as in beer or wine making.

DEFINITION AND BASIC PRINCIPLES

Food science is a multidisciplinary field in which principles from biology, chemistry, and engineering are applied to the study of the chemical, physical, and microbiological properties of food. Food scientists investigate the elements contained in foods, factors involved in the physical and chemical deterioration of food, and how different foods interact with human physiology. The specific application of knowledge from food science to practical issues such as preserving, processing, manufacturing, packaging, and distributing food, as well as questions of food safety and quality, is sometimes called food technology.

The fact that foods are chemical systems is an important fundamental principle of food science. The same basic elements that form the cells and tissues of the human body are the ones that make up the majority of the foods people eat. These include carbon, oxygen, hydrogen, nitrogen, sulfur, and calcium. The chemical composition of foods and the ways in which the atoms and molecules within a food are arranged and bonded to each other determine properties such as a food's flavor, texture, color, and nutritional value. Food scientists are also concerned with the processes of digestion—how foods are broken down and absorbed by the body. Digestion occurs through the action of enzymes (biological substances that facilitate chemical reactions) found in saliva, pancreatic juices, and the lining of the small intestine. Nutrition—the study of how the chemical composition of different foods contributes to human health, growth, and disease—is another important subfield of food science. The six basic nutritional components of the human diet are carbohydrates, proteins, fats or lipids, vitamins, minerals, and water. Each of these components serves different vital functions within the human body.

BACKGROUND AND HISTORY

Human beings have been working to identify the properties of different foods and experimenting with ways of processing and preserving foods since time immemorial. In ancient China, Rome, Greece, Egypt, and India, fermentation was commonly used to produce alcoholic beverages, and archaeological finds suggest that in the Middle East, as long ago as 12,000 B.C.E., fruits, vegetables, and meats were deliberately laid out in the hot sun in order to preserve them. Many ancient civilizations also had a rudimentary understanding of the nutritional and medicinal properties of specific foods. The Roman writer Pliny the Elder, for instance, asserted that consuming cabbage was good for vision, could relieve headaches and stomach problems, and could even prevent hangovers.

Food science as a formal scientific discipline has a much more recent history. Beginning in about the nineteenth century, researchers working in fields such as biology and chemistry often took an interest in questions of food science. For example, the French bacteriologist Louis Pasteur introduced the technique of pasteurization in 1864, and in 1847, the German chemist Justus von Liebig published his seminal work, *Researches on the Chemistry of Food*. Liebig was among the first scientists to clearly outline the principle that foods are metabolized, or broken down by the body to produce energy. He also understood the nutritional importance of chemicals such as nitrogen and sulfur.

It was not until the twentieth century, however, that food science and the application of technology to the processing and packaging of food became a recognized field of study in its own right. One of the major achievements of twentieth-century food scientists was the invention of vacuum packaging, which helps preserve perishable foods longer by removing air from the containers in which they are kept, thus preventing the action of bacteria. Other breakthroughs were the development of quick-freezing (pioneered by the frozen vegetable manufacturer Clarence Birdseye), the use of thermal processing to improve the safety of canned foods (which had been around since the late eighteenth century), and the use of radiation to kill microorganisms in foods. Dedicated food science departments were created at universities around the United States and the world, and professional organizations such as the Institute of Food Technologists and the European Federation of Food Science and Technology gathered practicing food scientists together to share their knowledge.

By the twenty-first century, food science had progressed to a point where it was enabling researchers to tinker with the production of various foods in ways that had never before been thought possible. For example, in the making of cheese, rennet (curdled milk) has largely been replaced by a synthetically engineered enzyme called chymosin, which is more chemically consistent and pure. Also, advances in food science allow growers to use genetic engineering rather than breeding to create plants with more desirable attributes—such as tomatoes whose DNA is altered so that they take longer to ripen and reach supermarket shelves just at the point when they are ready to eat.

HOW IT WORKS

There are three chemical compounds that form the building blocks of food: carbohydrates, fats (or lipids), and proteins. Carbohydrate molecules, which are found in fruits, vegetables, starches, and dairy products, consist of atoms of carbon, hydrogen, and oxygen, chemically bonded in a ratio of 1:2:1. Monosaccharides and disaccharides such as glucose, fructose, and sucrose have just one or two molecules of this kind and are known as simple sugars. Polysaccharides like starch, glycogen, and cellulose (an important component of dietary fiber) have several carbohydrate molecules and are known as complex

Kevin Keener, an associate professor of food science at Purdue University, removes eggs from a rapid egg cooling system Wednesday, August 25, 2010, at the university in West Lafayette, Indiana. (AP Photo)

carbohydrates. Food science is able to reveal the chemistry behind the behaviors of carbohydrates in different foods. For example, because sugar molecules tend to crystallize (form solid geometric structures) at low temperatures, ice cream with too much milk in it can be gritty because of the crystallization of lactose, a sugar found in milk whey.

Protein molecules, which are found in meats, nuts, eggs, beans, and dairy, are composed of long chains of amino acids. An amino acid is an organic compound that contains at least one carboxyl group (-COOH) and one amino group (-NH$_2$). The amino acids in a protein are linked to each other by peptide bonds and folded into particular shapes. By identifying how a protein denatures (how the shape of its folded amino acids changes) when heat or chemicals are applied to it, food science can demonstrate how different methods of cooking affect foods and why. For example, when heat is applied to meat, the tidy folds created by the chains of amino acids it contains collapse, causing the protein molecules to shrink and release water. This is why a steak seared on a pan is smaller after cooking.

Lipids, such as animal fats and vegetable oils, are large molecules that are not soluble in water but are soluble in organic solvents such as acid. There are many different kinds of naturally occurring lipids, but most are largely composed of various fatty acids and glycerol. Glycerol is a type of alcohol, and a fatty acid is a compound that contains carbon and

hydrogen atoms linked together in a long line and ending in a carboxyl group. Again, food science is concerned with how the chemical structure of lipids contributes to their physical characteristics. For example, the way in which the fatty acids in a lipid molecule are arranged determines whether a given fat will be solid at room temperature (such as lard) or liquid (such as vegetable oil).

Contemporary food scientists use a variety of sophisticated instruments to analyze the chemical composition and physical properties of foods. For example, a tool known as a spectrophotometer is used to detect how much light is absorbed by the atoms and molecules in a given sample of food and how much passes through it. Another food analysis technique is chromatography, which passes a sample of a food substance—either in liquid or gaseous form—through a medium that allows different components of the sample to travel at various rates. Spectrophotometry, liquid chromatography, and gas chromatography are all methods of analysis that enable food scientists to determine exactly what percentage of a specific food is made up of components such as fatty acids, amino acids, cholesterol, and carbohydrates. They also allow researchers to test for the presence of particular vitamins and minerals.

Digestion and Nutrition. The chemical composition of foods affects how they are broken down, or digested, by the body. The primary chemical reaction involved in digestion is called aerobic respiration, which involves tearing apart the bonds between carbon, oxygen, and hydrogen, the elements found in all three major types of food compounds. When aerobic respiration occurs, the energy contained within these bonds is released, and the individual atoms can be rearranged into different forms.

By analyzing the chemical characteristics of foods and how these properties relate to their digestion, food science can reveal why certain foods affect people the way they do. For example, because aerobic respiration is able to act directly on molecules of simple sugars—commonly found in soft drinks or colas—they are very quickly converted into energy. This results in an intense burst of energy that rapidly fades away—the so-called sugar high that a sweet drink can provide. In contrast, complex carbohydrates, commonly found in whole grains such as oats and brown rice, must first be broken down into simpler form by enzymes before aerobic respiration can take place. These foods take longer to digest and provide longer lasting energy, which is why a bowl of oatmeal in the morning can make a person feel full for hours. The structure of proteins is still more complex than that of whole grains, so they provide an even longer lasting source of energy. Lipid molecules, the hardest of all to digest, provide the body with its most long-lasting source of energy.

Although lipids take the longest time to digest, once they have been broken down, they provide the most energy per weight of food. For this reason, excess energy is stored within the body as fat deposits under the skin and in the abdomen. These enable the vital activities of cells to continue even when there is a temporary shortage of food. However, excess fat deposits are also stored inside organs and blood vessels. Here, they can block the normal flow of blood and lead to serious health problems such as heart failure. The question of how people can maximize the positive physiological effects of foods and avoid the harmful effects of an unhealthy diet is one of the primary concerns of food scientists. Among other pieces of knowledge, research has revealed that fiber, a type of carbohydrate found in fruits, vegetables, grains, and legumes, is essential for the proper functioning of the bowels and adheres to fat molecules traveling through the digestive system so they can be more easily disposed of as waste. Food science has also shown that certain kinds of unsaturated fatty acids, including omega-3 fatty acids, are extremely helpful in promoting heart health and preventing the buildup of cholesterol in the blood vessels.

Although it may not satiate a person's hunger, water is one of the most important nutrients required by the body. Water is the major component of every cell in the body and the environment within which every chemical reaction in the body takes place. It serves as a medium of transport for nutrients and waste, and it helps maintain a steady body temperature. Other chemical components found in food and important for human nutrition include vitamins and minerals. These nutrients serve a variety of essential functions. For example, vitamin E is an antioxidant, a substance that inhibits oxidizing reactions that can damage cells, and vitamin C helps the body process amino acids and fats. Calcium and magnesium are both minerals that are important in the formation of strong, healthy bones.

APPLICATIONS AND PRODUCTS

Preservation Techniques. One of the most important applications of food science is the development of preservation techniques that lengthen the time that a food can remain safe to consume and palatable. Among the many factors that cause foods to deteriorate are being exposed to microorganisms such as bacteria, molds, and yeast; experiencing changes in moisture content; being exposed to oxygen or light; undergoing the action of natural enzymes over time; being contaminated by industrial chemicals; and being attacked by insects or animals such as ants or rodents. Any of these factors may cause physical or biochemical reactions in foods that result in changes in texture, color, and taste, or that make them unsafe to eat.

Heating food is an effective method of preservation because high temperatures destroy both microorganisms and enzymes. Too much heat, however, can cause detrimental changes in the flavor, texture, and nutritional content of foods. Two commonly used methods of mild heat treatment are pasteurization and blanching. Pasteurization, which is most often applied to milk but also used to preserve fruit juices, beer, and eggs, involves heating the food to a temperature of 161 degrees Fahrenheit for just a few seconds. Blanching, most often used for vegetables destined to be frozen, dried, or canned, involves briefly dipping the food in water of about 212 degrees Fahrenheit. The most common severe heat treatment is canning. First, the food is placed inside a cylindrical steel or aluminum container and the air drawn out of it using a vacuum. Then, the lid is sealed in place and heat of about 240 to 250 degrees Fahrenheit is applied to the can. The process of canning ensures that the food in question reaches the point of commercial sterility and that it does not contain any live bacterium of the species *Clostridium botulinum*. This potentially deadly pathogen, if ingested, causes a kind of poisoning known as botulism.

Refrigeration and freezing are the two major types of cold preservation. Refrigeration, which takes place at temperatures ranging from 40 to 45 degrees Fahrenheit, does not destroy microorganisms or enzymes but somewhat inhibits the reactions they cause that result in spoiling. In and of itself, refrigeration is not a long-term method of preservation for most foods. Freezing, which takes place at temperatures ranging from 0 to 32 degrees Fahrenheit, is a more effective inhibitor of biochemical reactions than refrigeration and can preserve foods for longer periods of time. However, extremely cold temperatures can cause undesirable chemical changes, such as crystallization, in foods. If the water molecules in a food crystallize, they will rupture cell walls and cause the food to be softer and more liquid when it is eventually thawed.

Besides thermal processing and the use of cold temperatures, food scientists have developed a host of other techniques to combat the deterioration of food, such as dehydration, radiation, fermentation, and the use of natural and artificial preservative agents such as sugars, salts, acids, and inorganic chemicals like sodium benzoate and sulfur dioxide. Preservation is also aided by careful control over the characteristics of the atmosphere in which a food is stored.

Genetically Modified Foods. Genetically modified (GM) foods, also known as genetically engineered foods, are those whose existing genetic structure has been changed by the introduction of a new gene from a different organism. The technology that enables scientists to do this is called gene splicing, and the new genetic information is known as recombinant DNA. The purpose of genetic modification is usually to achieve some specific trait that will increase the food's usefulness for either producers or consumers. For example, DNA from bacteria has been incorporated into many strains of plants to enable them to resist attacks from insects and other pests. Some plants, such as soybeans, have been genetically modified so that they no longer produce particular substances, such as certain proteins, that can cause allergic reactions in people. According to some estimates, more than half of the foods for sale on supermarket shelves in the United States contain some ingredient from a genetically modified, or transgenic, plant.

Food products from genetically modified animals have not been available in the United States; however, in 2009, the Food and Drug Administration determined its requirements and regulations for genetically engineered animals, opening the door for food products from such animals. Food science research in this area has begun but largely remains in the experimental stages. Potential applications of this technology to animals raised for food include pigs with a genetic modification that causes them to produce omega-3 fatty acids, or salmon or chickens that grow much faster than usual.

Manufactured Foods. In vitro meat, also known as cultured meat, is animal muscle tissue that is grown outside of a living organism. It represents a relatively new and very specialized application of food science that makes use of tissue engineering techniques borrowed from cell biology and biotechnology. The production of in vitro meat involves harvesting either muscle cells or stem cells (cells that are pluripotent, or able to give rise to any number of different cell types) from a live animal, such as a chicken, cow, or pig. Alternatively, cells from a slaughtered animal can be used. Then, the cells are cultured within a medium that provides them with a large quantity of the nutrients required for growth. Typically, this includes amino acids, vitamins, minerals, and glucose. In this environment, the cells multiply rapidly. To encourage the cultured cells to fuse and form the three-dimensional structures that make up muscle fibers and tissues, they are placed on a scaffold, usually made of collagen. They may also be stretched or electronically stimulated to help them form the correct structures. At the end of this process, a substance similar to ground meat is produced.

Beverages and Snacks. Food science principles lie behind almost every beverage and snack food found on supermarket shelves. Carbonation, for example, enables the production of nonalcoholic soft drinks such as colas and sparkling water. Carbonation is simply the introduction of carbon dioxide gas into a liquid. It takes place at high pressures and low temperatures, because both of these factors increase the solubility of carbon dioxide. Alcoholic beverages such as beer and wine are produced through a process called fermentation. A substance—barley in the case of most beers, grapes in the case of wines—is chemically broken down by the action of microorganisms such as bacteria or yeast. It takes place under anaerobic conditions (in the absence of oxygen), because this is what causes the microorganisms to react with the carbohydrates in the substance to be fermented. In the process, sugars present in the original substance are converted into ethyl alcohol, the flavor and texture of the beverage is markedly changed, and its shelf life is prolonged.

Candy and other sweet confections are also the products of food science principles. To control the taste and texture of their products, candy manufacturers use a number of clever techniques designed to manipulate the behavior of sugar. For instance, many different candies begin with the same two ingredients: sugar and water. By changing the ratio of sugar to water, changing the boiling temperature of the sugar-water solution, controlling the time it takes for the boiling mixture to cool down, and adding various interfering agents such as butter, gelatin, cocoa, or pectin to the mixture, candy manufacturers can create a host of different textures from the same basic foundation: creamy and smooth, hard and brittle, moist and chewy, or smooth and transparent.

IMPACT ON INDUSTRY

The application of food science technologies can have a tremendous impact on the amount of money a country spends on the production and distribution of food. For example, although it does not tell the whole story, one of the reasons consumers in the United States spend the lowest percentage of their per capita income on food when compared with consumers in other countries is that farms in the United States are able to take advantage of state-of-the-art technologies for growing, harvesting, distributing, and processing foods, such as advanced chemical fertilizers and preservatives that keep fruits and vegetables fresh over long journeys. These techniques help lower the costs of food production.

Government and University Research. In the United States, several large government agencies are involved in funding or directly conducting food science and technology research. Among them are the Food Safety and Inspection Service, a subsidiary of the U.S. Department of Agriculture, and the Food and Drug Administration. Government research scientists are largely concerned with conducting scientific studies related to specific issues of food safety and quality, such as how to prevent the spread of bacteria, such as salmonella or *Escherichia coli*, that can cause serious disease outbreaks within the population. Academic researchers may tackle a slightly different set of questions, some of which may be more basic and less problem-driven, such as investigations into the chemistry of different species of fruits or nuts.

Some academic research is funded by industry partners within agriculture. For example, the grain-farming sector may provide support for university laboratories conducting research into certain crop species that possess a natural genetic resistance to soil-borne viruses, or the dairy industry for research into the health benefits of consuming milk and yogurt. Strict regulations over conflict of interest issues ensure that the integrity of such scientific research is maintained. The outcomes of studies such as these can have broad economic impacts on the agriculture industry.

Industry and Business. Food science treats foods as complex chemical systems whose components have specific properties and give rise to specific nutritional effects. The packaged foods industry has capitalized on the discoveries of food science to create products that are marketed as having particular health benefits. For example, several butter substitutes on the

Fascinating Facts About Food Science

- Food scientists have gone far beyond freeze-dried ice cream when it comes to creating high-tech food for astronauts. Salt and pepper are turned into liquids so that they do not float around in zero gravity, and perishable fruits and meat can be thermostabilized (processed with heat) to allow them to stay fresh for long space journeys.

- Nutrition can come not just from the food that people eat but also from the vessels in which they cook it. Cooking foods in a cast-iron pot rather than a nonstick or glass pan, for instance, increases the amount of iron—an essential mineral found in hemoglobin, or blood molecules—in the finished dish.

- Food science reveals that plants have an incredibly complex chemical composition. Raw spinach, for instance, contains calcium, iron, magnesium, phosphorus, potassium, sodium, zinc, copper, manganese, and selenium—and that is just the minerals. It also contains at least eighteen vitamins and nearly twenty amino acids.

- Freezer burn is caused by the escape of water molecules from food stored at very cold temperatures. When moisture is lost from a food in this way, it becomes dried out and covered in frost.

- Pickled vegetables can be more nutritious than fresh ones. The bacteria that are involved in fermentation break down molecules in the vegetables and, as a by-product, create vitamins.

- Food and beverage manufacturers in the United States are constantly working to combine existing ingredients into new formulations. In total, they come up with nearly 19,000 new products for supermarket shoppers every year.

market are made with plant-derived sterols, which are substances that inhibit the body's ability to absorb cholesterol and thus improve heart health, reduce the risk of stroke, and combat inflammation in the body. Sports drinks containing stimulants, amino acids, vitamins, and other active ingredients are marketed as providing athletes and other consumers with a quick, safe, and effective burst of energy. Thousands of such functional food and beverage items fill grocery store shelves, allowing food retailers to capitalize on the unique properties of these products. Customers may not be able to effectively differentiate among Granny Smith apples from three separate orchards, but they are likely to have an opinion as to whether they want to buy regular flour or flour that has been fortified with additional beneficial nutrients. The restaurant industry is another major business sector that makes use of food science principles. For example, many restaurants around the world are using a type of cooking known as molecular gastronomy. Restaurants use dozens of high-tech gadgets to transform the chemical and physical structure of foods, creating dishes such as chocolate in the form of a jelly and tomato and cheese in the form of foam.

CAREERS AND COURSE WORK

Food science research and related industries such as agriculture, food processing and manufacturing, and food safety and inspection offer a wide variety of career opportunities. For example, food chemists analyze, create, and modify the biochemical compounds and chemical processes involved in the synthesis or development of food products. They may work for food manufacturing companies, biotechnology corporations, government institutions, environmental consulting companies, and agricultural universities. Food bacteriologists, who often work in similar settings, carry out research into the microbiological and molecular basis of food-borne pathogens. Their work supports the safe production, packaging, and distribution of food plants and animals. Food safety inspectors are employed in settings such as meat-processing plants, fisheries, large farms, restaurants, and food production facilities. They may also work for federal agencies. Food safety inspectors make use of their knowledge of biology and chemistry to ensure that food products such as dairy, grains, fishes, fruits, vegetables, meats, and poultry properly conform to national or industry-based standards of sanitation.

Taking high school courses in biology, chemistry, physics, and mathematics provides a good early foundation for an eventual career in food science. At the undergraduate level, courses in organic and inorganic chemistry, nutrition, cell and molecular biology, biochemistry, genetics, bacteriology, microbiology, and agricultural science are especially relevant. For the student who is especially certain of his or her ambitions, a number of colleges and universities offer specialized undergraduate degrees in food science or food technology. To acquire a position as a technician, an associate's degree or certification in an area such as chemical or biochemical technology may be all that is required. Positions conducting independent research at an academic or government institution or food science research in private laboratories require the completion of a bachelor's degree in science and, usually, a master's or doctoral degree in biochemistry, bacteriology, chemical technology, or a related field.

SOCIAL CONTEXT AND FUTURE PROSPECTS

The potential impact of the field of food science is both far reaching and profoundly positive. For example, the advent of in vitro meat has both environmental and ethical implications. If it became widely used, cultured meat could result in huge energy and water savings over traditional methods of bringing meat to market, reduce the amount of pollution produced by farming, and end the raising of animals in factory farms. Similarly, the gene splicing technology used to create genetically modified foods has already been applied to the pressing problem of global poverty. To increase the nutritional content of rice for consumption in developing countries, where rice often makes up a huge portion of the diet of the poor, researchers have created strains of genetically modified rice. The new rice contains large amounts of beta-carotene, a precursor for the synthesis of vitamin A, which contributes to the iron content of blood and helps maintain the structure and functioning of eyes. The product, called Golden Rice, has proven to be beneficial in places where malnutrition often leads to problems such as anemia and blindness.

Emerging food technologies such as these, however, are still surrounded by a cloud of skepticism and debate. Many consumers and advocacy groups argue that the health safety risks and environmental impact of products such as genetically modified foods have not been properly assessed and that these products could have devastating unintended consequences. For example, some people worry that plants containing genes from foreign species may turn out to be allergenic or even toxic to humans. Others are concerned that genetically modified organisms may interact with the environment in unpredictable ways. One study of monarch butterfly larvae found that feeding them on leaves dusted with bioengineered corn pollen caused their growth to be stunted and, in some cases, resulted in their deaths. Nevertheless, many food scientists point out that attempts to modify the genetic information in both plants and animals is hardly new; the process of selective breeding, they say, pursues essentially the same goal and has been going on for centuries.

M. Lee, B.A., M.A.

FURTHER READING

Brown, Amy. *Understanding Food: Principles and Preparation.* 4th ed. Pacific Grove, Calif.: Brooks/Cole, 2010. An introductory textbook largely organized around specific food types, such as meat, cereals, and fats. Also discusses career options within the field.

Heldman, Dennis R., and Daryl B. Lund, eds. *Handbook of Food Engineering.* 2d ed. Boca Raton, Fla.: CRC Press/Taylor & Francis, 2007. Includes chapters on topics such as cleaning and sanitation, heating and cooling processes, food dehydration, and thermal processing of canned foods. Contains figures, relevant equations, and a comprehensive index.

Hui, Yiu H., et al., eds. *Food Biochemistry and Food Processing.* Ames, Iowa: Blackwell Publishing, 2006. Covers the biochemistry of food and the biotechnology involved in food processing. Includes numerous figures, diagrams, and tables illustrating essential concepts.

Wansink, Brian. *Marketing Nutrition: Soy, Functional Foods, Biotechnology, and Obesity.* 2006. Reprint. Urbana: University of Illinois Press, 2007. An analysis of food science, nutrition, and labeling from a business and marketing perspective. Contains a list of suggested further readings.

WEB SITES

European Federation of Food Science and Technology
http://www.effost.com

Institute of Food Technologists
http://www.ift.org

U.S. Department of Agriculture
Food Safety and Inspection Service
http://www.fsis.usda.gov

U.S. Food and Drug Administration
Food
http://www.fda.gov/Food/default.htm

See also: Agricultural Science; Animal Breeding and Husbandry; Cell and Tissue Engineering; Egg Production; Fisheries Science; Food Preservation; Genetically Modified Food Production; Genetically Modified Organisms; Nutrition and Dietetics; Pasteurization and Irradiation; Plant Breeding and Propagation.

FORENSIC SCIENCE

FIELDS OF STUDY

Chemistry; biology; biochemistry; mathematics; microbiology; physics.

SUMMARY

Forensic science is commonly defined as the application of science to legal matters. Although forensic science incorporates numerous disciplines, ranging from accounting to psychology, in the traditional sense, forensic science refers to the scientific analysis of evidence collected at crime scenes, which is also known as "criminalistics." Pattern evidence, such as fingerprints, bullets, and tool marks, is often compared visually, and chemical evidence (such as illicit drugs) and biological evidence (such as DNA, blood, and bodily fluids) are analyzed and compared using scientific instruments.

KEY TERMS AND CONCEPTS

- **Class Evidence:** Evidence that can be identified as belonging to a group containing many members, all with similar characteristics or features.
- **Criminalistics:** Refers to the analysis of pattern, chemical, and biological evidence; often used interchangeably with forensic science.
- **DNA (Deoxyribonucleic Acid):** Nucleic acid that contains the genetic code and is present in nearly every cell in the body.
- **Illicit Drug:** Substance or drug that is prohibited by federal or state laws because of its undesirable effects or high risk of abuse.
- **Impression Evidence:** Evidence, such as fingerprints, tire tracks, and footprints, formed when an object leaves behind a characteristic marking on a surface.
- **Individualizing Evidence:** Evidence that can be identified as belonging to a group containing only itself as a member.
- **Latent Print:** Fingerprint residue that is not easily visible to the naked eye and must be treated, either physically or chemically, to be observed.
- **Toxicology:** Study of drugs and poisons and their effect on the body.
- **Trace Evidence:** General term for microscopic pieces of evidence that are transferred by contact between people and objects. Examples include hairs and fibers, as well as fragments of paint and glass.

DEFINITION AND BASIC PRINCIPLES

Forensic science is the application of scientific principles to the analysis of numerous types of evidence, most commonly evidence collected at a crime scene. Crime scene investigators, usually police officers, collect evidence at the crime scene and submit it to a crime laboratory for analysis by forensic scientists.

Crime laboratories contain different sections, each of which specializes in a particular type of analysis, such as controlled substances, DNA, firearms and tool marks, latent prints, questioned documents, toxicology, and trace evidence. The type of analysis conducted depends on the type of evidence as well as the circumstances of the crime. A single piece of evidence may be analyzed in more than one section. For example, a firearm may be analyzed in the latent prints and DNA sections, as well as the firearms and tool marks section.

Following analysis, forensic scientists may be summoned to present their findings in a court of law. Forensic scientists present their analysis and interpretation of the evidence before a judge and jury, who are charged with determining the guilt or innocence of the defendant. The unbiased, accurate analysis presented by the forensic scientist is an integral part of the criminal proceedings.

BACKGROUND AND HISTORY

Forensic science aims to determine identifying or individualizing characteristics to link people, places, and objects. In the late 1880's, French criminologist Alphonse Bertillon developed a method of identifying humans based on eleven physical measurements including height, head width, and foot length. However, limitations in this method soon became apparent. In 1880, Scottish scientist Henry Faulds published an article in *Nature* that discussed the use of fingerprints as a means of identification. In 1892, Sir Francis Galton published *Fingerprints*, proposing a system of classifying fingerprint patterns. That same year, an Argentine police officer, Juan Vucetich, used fingerprint evidence that resulted in the arrest and conviction of a murder suspect. From 1896 to 1925, Sir Edward Henry, a police official in British India, developed the Henry Classification System for fingerprints, which was based on the pattern on each finger and the two thumbs.

In the late nineteenth and early twentieth centuries, advances were being made in other areas that have become integral to forensic science. Spanish-born French scientist Mathieu Joseph Bonaventure Orfila, often considered as the pioneer of forensic toxicology, is credited with developing and improving methods for the detection of arsenic (a common poison in the nineteenth century) in the body. French scientist Edmond Locard developed the hypothesis that "every contact leaves a trace," which implies that whenever two objects make contact, there is an exchange between them. This hypothesis became known as Locard's exchange principle and is the foundation of modern trace evidence analysis. In the 1920's, the comparison microscope, which analyzes side-by-side specimens, was developed by American chemist Philip Gravelle and popularized by forensic scientist Calvin Goddard. This microscope enabled significant advances in many areas of forensic science, particularly firearms, tool marks, and trace evidence.

A major scientific breakthrough in the 1980's revolutionized the field of forensic DNA analysis. British geneticist Sir Alec Jeffreys developed DNA profiling, enabling individuals to be identified from samples of blood and other body fluids left at a crime scene. The development of the polymerase chain reaction by American biochemist Kary Mullis in 1983 allowed DNA profiling to be conducted on degraded and very small samples of DNA, making it possible for forensic scientists to test a wider range of evidence.

As the field of forensic science evolves, newly developed technologies and instrumentation allow evidence to be analyzed and compared in an increasingly rapid, objective, and reliable manner. Forensic science is a truly dynamic field, constantly seeking further improvements and advancements in its analytical methodologies.

HOW IT WORKS

Forensic science incorporates numerous subdisciplines, but the most common types of analysis

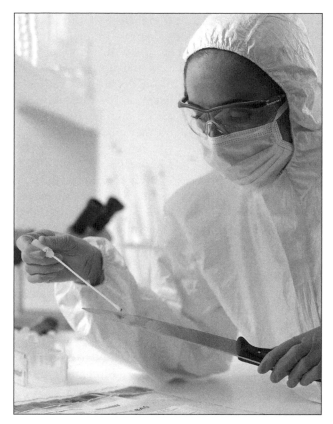

Forensic scientist taking a swab from a knife for DNA (deoxyribonucleic acid) tests. (Tek Image/Photo Researchers, Inc.)

conducted by crime laboratories are the analysis of illicit drugs, biological evidence, latent prints, firearms, footprints, tire marks, tool marks, and trace evidence. Latent prints, footprints, tire marks, and tool marks are considered pattern evidence. The patterns of an unknown sample (usually from the crime scene) and a known sample are visually compared to find similarities between the two. Samples can also be analyzed, either chemically or biologically, with scientific instruments. Some of the more common methods of testing are infrared spectroscopy, ultraviolet/visible microspectrophotometry, gas chromatography-mass spectrometry, and electrophoresis.

Infrared Spectroscopy. In infrared spectroscopy, the chemical structure of a sample is determined based on how the sample interacts with infrared radiation. Chemical bonds can absorb infrared radiation of a specific energy, which causes the bond to vibrate. Additionally, each bond can vibrate in different ways.

Therefore, when infrared radiation is introduced, chemical bonds within the sample absorb different energies, and the results are shown in the form of an infrared spectrum. The spectrum is essentially a graph of radiation transmitted versus wave number, which is related to the energy of the radiation. Additionally, transmission can be mathematically converted to absorbance such that the spectrum can be displayed as absorbance versus wave number. The infrared spectrum of a sample displays numerous absorptions, each corresponding to a particular type of chemical bond and a particular type of vibration. The infrared spectrum of a sample is unique to that sample, and therefore, this technique can be used to definitively identify compounds.

Ultraviolet/Visible Microspectrophotometry. Infrared spectroscopy and ultraviolet/visible microspectrophotometry are both based on the principle of the interaction of radiation with a sample. However, ultraviolet/visible microspectrophotometry is typically used to compare the dye or pigment composition of samples. The technique is used to determine the color of a sample and identify subtle differences in color that cannot be seen with the naked eye.

A microspectrophotometer consists of a microscope with a spectrometer attached, which allows the analysis of microscopic pieces of evidence. The sample is viewed under the microscope, and ultraviolet and/or visible radiation is introduced. Depending on the chemical structure of the sample, wavelengths of light will be absorbed, reflected, or transmitted. The transmitted light is collected in the spectrophotometer, and the intensity of each wavelength is measured. Results are displayed in the form of a spectrum that is a graph of transmittance (or absorbance) versus wavelength. Subtle differences in color between two samples are observed as differences in wavelengths of light transmitted or absorbed in the corresponding spectra. Such differences are caused by differences in chemical composition between the two samples, and therefore, comparison of the resulting spectra can be used to determine if the two samples are similar in color.

Gas Chromatography-Mass Spectrometry. In any chromatography technique, sample mixtures are separated based on differences in interaction between a mobile phase and a stationary phase. In gas chromatography (GC), the mobile phase is a gas, and the stationary phase is a liquid coated on the inner walls of a very thin column. Liquid samples are typically introduced into the system and carried, in the mobile phase, through the stationary phase. Sample components that have a stronger attraction for the stationary phase spend longer in that phase, and components with less attraction spend less time in that phase and move more quickly through the system. The time it takes for sample components to travel through the system and reach the detector is known as the retention time.

In gas chromatography-mass spectrometry (GC-MS), the detector is the mass spectrometer, which contains three major components: the ion source, the mass analyzer, and the detector. On emerging from the GC column, sample components enter the ion source, where each component is first ionized. The resulting ion is known as the molecular ion. This ion is unstable because of its high energy, so it breaks down, or fragments, into smaller ions. Molecular ions and fragment ions then enter the mass analyzer, where the ions are separated according to their mass-to-charge ratio. The separated ions enter the detector, where the number of ions of each mass-to-charge ratio is counted. Results are displayed in the form of a mass spectrum, which is a graph of intensity versus the mass-to-charge ratio. Because molecules break down, or fragment, in a predictable manner, the mass spectrum can be used to determine the structure of the original sample component. Furthermore, because the fragmentation pattern is unique to a molecule, the mass spectrum can be used to definitively identify the component.

On analyzing a sample by GC-MS, two pieces of information are obtained. First, from gas chromatography, a chromatogram is obtained, which is a graph of detector response versus retention time. Each separated component in the sample mixture is shown as a peak on the chromatogram. Components that take longer to reach the detector have greater attraction for the stationary phase and have longer retention times. Additionally, for each separated component, the mass spectrum is also obtained, which can be used to definitively identify the component.

Electrophoresis. Although electrophoresis is also used to separate sample mixtures, the technique is not considered a chromatographic technique because no mobile phase is involved. Instead, sample mixtures are separated based on differences in migration under the influence of an applied electrical potential. Therefore, electrophoresis is used for the analysis of samples that have an electric charge.

Although there are different types of electrophoresis, capillary electrophoresis is most commonly used for DNA profiling purposes. In this technique, a capillary column is filled with a polymer, and the ends of the column are immersed in reservoirs containing a buffer solution. The reservoirs also contain electrodes to allow the application of the electric potential. The sample is introduced to one end of the column, and the sample components move through the column under the influence of the applied potential. Separation occurs based on differences in the migration rate of the components through the column, which depends on size and charge. Separated components pass through a detector at the other end of the column, producing an electropherogram. The electropherogram shows the migration time of the separated components. Smaller components move more quickly, reaching the detector before larger components and have shorter migration times.

APPLICATIONS AND PRODUCTS

The major role of the forensic scientist is to analyze submitted evidence for the purposes of characterization and identification. For example, a blue fiber collected from the scene may be submitted to the trace evidence section, where forensic scientists characterize the fiber (for example, by its dimensions, color, cross-sectional shape) and then identify the type of fiber (for example, nylon, polyester, acrylic). Furthermore, when a known sample is available (such as fibers from the suspect's clothing), forensic scientists compare it with the unknown sample (collected from the crime scene) to determine if the two most likely originated from a common source. This process of characterization, identification, and comparison requires multiple stages of analysis, ranging from visual examination to instrumental analysis.

Infrared Spectroscopy. The technique of infrared spectroscopy is commonly used in the controlled substance and the trace evidence sections of the crime laboratory. This technique can identify illicit drugs present in unknown samples, the type of fiber found at a crime scene or on a person, the polymer present in a paint chip, or the organic compounds present in explosive residues. The evidence is prepared for analysis in several ways, depending on the type of sample.

Solid samples of illicit drugs can be mixed with potassium bromide and pressed into a pellet, which is then placed in the spectrometer. Infrared radiation is passed through the sample, which will absorb at characteristic energies depending on its chemical structure. The transmitted radiation is collected and the infrared spectrum is generated. Because potassium bromide does not absorb infrared radiation, the subsequent infrared spectrum shows only contributions from any drug present in the sample.

For opaque samples, such as fibers or paint chips, attenuated total reflectance-infrared (ATR-IR) spectroscopy is more commonly used. The sample is positioned over a crystal, and pressure is applied to ensure good contact between the sample and crystal. Infrared radiation is passed through the crystal, and because of the close contact, the radiation penetrates a small depth into the sample. Certain energies are absorbed depending on the chemical bonds within the sample, resulting in the characteristic spectrum of the sample.

The infrared spectrum of the questioned sample can be compared to a database containing infrared spectra for known standards (drugs, fibers, paints, and so on) to identify the unknown sample. However, care must be taken when comparing a spectrum to spectra in a database. Although the spectrum of a given compound is unique, it can vary slightly depending on the instrument used to analyze the sample and standard. Rather than relying on a database search, it is often preferable to analyze the unknown sample and known standards on the same day, using the same instrument, to allow for a direct comparison of spectra.

Although samples can be rapidly analyzed using infrared spectroscopy, the technique works best for relatively pure samples. If impurities are present in the sample and they also absorb infrared radiation, the resulting spectrum contains contributions from both the sample and the impurities. This can complicate interpretation of the spectrum and subsequent identification of the sample.

Microspectrophotometry. The comparison and analysis of colored samples is often undertaken using microspectrophotometry. This technique is used in the trace evidence and questioned documents sections to compare the dye or pigment composition of fibers, paints, and inks.

Methods for sample preparation vary depending on the type of sample to be analyzed. Fibers are flattened and mounted on a microscope slide with a drop of immersion oil. Paint samples require more involved preparation, particularly for transmission spectra. The paint chip must be cut into a section so thin that light can be transmitted through it. Spectra of inks can be obtained directly if the paper is sufficiently thin to allow transmission. Otherwise, the ink must be removed from the document. This can be done by removing a small sample of the paper

---

Fascinating Facts About Forensic Science

- Forensic entomology involves the study of insects that invade a body after death to determine the time that has elapsed since the person's death.

- The saliva on a discarded cigarette contains enough DNA to identify the person who smoked it.

- The first use of fingerprint evidence to solve a criminal case was recorded in Argentina in 1892. Police official Juan Vucetich used a bloody fingerprint found at the crime scene to prove that two boys were murdered by their own mother.

- "Forensic" comes from the Latin *forensic*, which means "of the forum." The "forum" relates to the law courts in ancient Rome. In modern times, forensic science is defined as science relating to the law.

- In 1981, a German publishing company purchased what were thought to be Adolf Hitler's diaries. However, forensic document examiners proved that the diaries were fake, based on the presence of a paper-whitening agent that was not used in paper manufacturing until at least 1954.

- Marie Lafarge was the first person to be found guilty of murder based on toxicology evidence. Although she poisoned her husband with arsenic, initial testing did not find any arsenic in his body. However, when French scientist Mathieu Joseph Bonaventure Orfila repeated the tests, he found arsenic in the man's body and proved that the initial testing was inaccurate.

- In 1995, O. J. Simpson was cleared of murdering his wife, Nicole Brown, and her friend Ronald Goldman, despite DNA evidence identifying blood at the crime scene as belonging to the former football player. Furthermore, DNA from Simpson, Brown, and Goldman was found in a leather glove found at the scene.

---

containing the ink and immersing the paper in a solvent to extract the ink. The resulting ink solution is placed on a microscope slide, and the solvent is allowed to evaporate, leaving a residue of ink for analysis. However, this is a destructive procedure because the document is damaged in removing the sample. Alternatively, a piece of clear tape can be placed on an area of the document that contains the ink. When the tape is lifted off, particles of ink adhere to the tape. These particles can be removed from the tape and transferred to a microscope slide for analysis. The document is minimally damaged using this procedure.

Although microspectrophotometry offers a rapid means to investigate the dye or pigment composition of certain samples, no extensive spectral databases are readily available. Therefore, the technique is more useful when known samples are available and the color of the unknown and known samples can be compared directly, based on spectral interpretation.

Gas Chromatography-Mass Spectrometry. As with infrared spectroscopy, gas chromatography-mass spectrometry is commonly used in the controlled substances and trace evidence sections, as well as in the toxicology section, for the determination of drugs and poisons in body fluids.

GC-MS is advantageous over infrared spectroscopy in that samples containing impurities can still be identified because of the separation abilities of gas chromatography. For example, gas chromatography analysis of a drug mixture containing methamphetamine and caffeine separates the two components. In the resulting chromatogram, two peaks are observed: one for methamphetamine and one for caffeine. The mass spectrum of each peak is also obtained, which can be used to definitively identify each component.

In most cases, samples must be in liquid form for GC-MS analysis. This is achieved by adding a suitable solvent to the sample and analyzing the resulting solution. For body fluid or tissue samples, a solid phase extraction or liquid-liquid extraction is necessary to isolate any drugs and poisons from additional components present in the fluids or tissues.

Solid samples can be analyzed using pyrolysis GC-MS. In this case, a pyrolysis unit is attached to the gas chromatography inlet. Solid samples (for example, paint chips or fiber fragments) are placed in a small quartz tube and introduced into the pyrolysis unit, which rapidly heats the sample to a very high temperature. The sample is broken down and vaporized in the pyrolysis unit, and then carried in the flow of carrier gas onto the gas chromatography column, where the sample components are separated.

Before analyzing the sample, it is important to demonstrate that the GC-MS system is free from contamination. This is usually done by injecting a volume of the solvent used to prepare the sample. If the solvent and instrument are not contaminated, the resulting chromatogram should show no peaks. For pyrolysis GC, the empty quartz tube is analyzed to demonstrate that there is no contamination in the tube or instrument.

Because the mass spectrum rather than the retention time is unique to a sample component, the spectrum of an unknown sample is compared to a suitable database of spectra. However, there may be slight differences between the database spectrum and the spectrum obtained for the unknown sample, depending on the instrument used to collect the spectra. It is often preferable to prepare and analyze a known standard in the same way as the unknown sample and then compare the corresponding mass spectra.

Electrophoresis. DNA profiling makes the most use of electrophoresis. Typically, blood, semen, saliva, or another body fluid from the crime scene is used to generate a DNA profile, which is compared with profiles generated from known samples. If known samples are not available, the generated DNA profile can be compared to a database of profiles. The Federal Bureau of Investigation (FBI) maintains a database of DNA profiles submitted by crime laboratories across the United States. This database, the combined DNA index system (CODIS), contains profiles from crime scenes, convicted criminals, and missing persons.

Modern DNA profiling is based on the characterization of short tandem repeats (STRs) that are regions (loci) on the chromosome that repeat at least twice within the DNA. For profiling, the number of repeats at each location on the chromosome is determined. To do this, the DNA is first amplified via the polymerase chain reaction (PCR), in which the double-stranded DNA is split into two single strands and a mixture of enzymes and primers are used to replicate specific STR regions of the DNA. In the United States, STRs at thirteen loci are typically considered. The reaction is repeated many times, generating exact copies of the STRs. Because of this

amplification procedure, profiles can be obtained from very small samples of DNA.

The STRs are analyzed using electrophoresis, most commonly capillary electrophoresis, which allows rapid and automated analysis. The STR mixture is separated based on differences in migration rate through the capillary column, which is related to the size of the STR. The resulting electropherogram displays a series of peaks that correspond to the STRs at each loci. Additionally, for each STR, there are two variants, one inherited from the mother and one from the father; therefore, the electropherogram actually shows a pair of peaks at each loci. A match in the number of STRs for both variants at all loci is considered strong evidence that the unknown and known samples originate from the same person. Because DNA is unique to an individual, this is one type of evidence that is considered individualizing rather than class evidence.

IMPACT ON INDUSTRY

Local, State, and Federal Laboratories. The majority of forensic science laboratories in the United States are funded by the local government, the state, or the federal government. These laboratories offer a variety of services to their customers, who are typically police departments and other law enforcement agencies. The actual services offered vary depending on the size of the laboratory and the geographical area that it covers. Typical services include analysis of illicit drugs, body fluids (often for DNA), fingerprints, firearms, tool marks, and trace evidence. Within any state, there may be only one laboratory that offers all services, with a number of smaller laboratories throughout the state offering two or three services.

The federal government operates numerous specialized forensic science laboratories. The U.S. Department of Justice oversees the FBI, the Drug Enforcement Administration (DEA), and the Bureau of Alcohol, Tobacco, Firearms, and Explosives (ATF), all of which have forensic science laboratories. These laboratories offer analytical services to local and state law enforcement agencies and conduct research to develop new analytical tools and technologies to advance forensic science.

Within the U.S. Department of the Treasury, the Treasury Inspector General for Tax Administration operates a forensic science laboratory that principally analyzes suspected counterfeit documents, using fingerprint and handwriting analyses, along with digital image enhancement procedures. The U.S. Postal Inspection Service operates a forensic science laboratory with the role of analyzing evidence from postal-related crimes. This laboratory mainly conducts fingerprint, document, and chemical analyses, along with digital image enhancement.

The U.S. Fish and Wildlife Forensics Laboratory is operated through the Department of the Interior and is the only laboratory worldwide that focuses solely on crimes against wildlife. Laboratory expertise in genetic and chemical analysis, as well as firearms, trace evidence, latent prints, and pathology is used to analyze wildlife evidence to identify species and determine cause of death.

Private Laboratories. A number of private forensic science laboratories operate throughout the United States. The majority of these laboratories offer expertise in one or two areas (for example, analysis of illicit drugs and trace evidence) rather than a full range of services. The vast majority of these laboratories focus on DNA analysis, particularly paternity testing. Because these laboratories are privately funded, their services are offered to the general public and are not limited to law enforcement agencies.

Federal Funding. The National Institute of Justice (NIJ) is the largest funding agency for forensic science within the United States. The agency funds research that improves methods for the collection, analysis, and interpretation of forensic evidence. Funds for research, development, and evaluation go to projects that improve the analytical tools and technologies available to forensic scientists, from developing new methods for the comparison of evidence to developing new analytical instrumentation for forensic analyses. Forensic laboratory enhancement funds are for projects that improve sample throughput in laboratories, enabling evidence to be analyzed in a timely manner. The agency also awards research fellowships to individuals to conduct specific research that will improve or enhance existing practices in forensic science. In 2009, the institute awarded a total of $284 million.

CAREERS AND COURSE WORK

At a minimum, a forensic scientist must have a bachelor's degree in a natural science, such as biology or chemistry, or in a related discipline, such as microbiology or biochemistry. It is also possible to obtain a bachelor's degree in forensic science; however, care should be taken to ensure that the degree meets minimum credit hour requirements in either biology or chemistry. Successful completion of the bachelor's degree ensures that students have a strong background in the appropriate science for their future field of work.

The popularity of forensic science has made the field highly competitive, and many forensic scientists have a master's degree in forensic science. Although as of 2010, there were more than forty such degree programs in the United States, only thirteen were accredited by the American Academy of Forensic Sciences. Accreditation ensures that rigorous standards have been met in terms of the content and quality of the degree program. In most cases, the degree obtained is a master of science in forensic science with a concentration in a specific discipline, for example, forensic biology, forensic chemistry, or forensic toxicology.

Forensic scientists are not limited to careers in local and state crime laboratories. Many federal agencies, such as the FBI, the DEA, and the ATF, employ forensic scientists. In addition, forensic scientists can be employed by private forensic laboratories (such as paternity testing or sport testing laboratories) or as independent consultants.

SOCIAL CONTEXT AND FUTURE PROSPECTS

Although great advances have been made in forensic science, many more have yet to be achieved. In 2009, the National Research Council published *Strengthening Forensic Science in the United States: A Path Forward*, a report on forensic science in the United States. The report highlighted several deficiencies in the field and recommended improving education, training, and certification for forensic scientists as well as developing standardized procedures and protocols for evidence analysis and reporting. Additionally, the report recommended research into the reliability and validity of many of the procedures used for evidence analysis. The report concluded that more research is necessary, not only to improve existing practices but also to develop new technologies that can be implemented in forensic science laboratories. It called for the development of a national institute of forensic science that would have many objectives, including the development of standards for certification for forensic scientists and accreditation of forensic laboratories, along with improving education and research in the field.

Ruth Waddell Smith, Ph.D.

FURTHER READING

Bertino, Anthony J., and Patricia N. Bertino. *Forensic Science: Fundamentals and Investigations.* Mason, Ohio: South-Western Cengage Learning, 2009. Examines the tests and techniques used for the scientific analysis of various evidence types, including hairs and fibers, DNA, handwriting, and soil.

Brettell, Thomas A., John M. Butler, and José R. Almirall. "Forensic Science." *Analytical Chemistry* 79, no. 12 (2007): 4365-4384. A review of forensic science applications used in common disciplines.

Embar-Seddon, Ayn, and Allan D. Pass, eds. *Forensic Science.* 3 vols. Pasadena, Calif.: Salem Press, 2008. Extensive coverage of forensics, including historical events, famous cases, and types of investigations, evidence, and equipment.

Houck, Max M., and Jay A. Siegel. *Fundamentals of Forensic Science.* 2d ed. Burlington, Mass.: Academic Press, 2010. An introduction to forensic science and common techniques used for the analysis of physical, biological, and chemical evidence.

James, Stuart H., and Jon J. Nordby, eds. *Forensic Science: An Introduction to the Scientific and Investigative Techniques.* 3d ed. Boca Raton, Fla.: CRC Press, 2009. Discusses mass spectrometry techniques in relation to forensic applications, including forensic toxicology, controlled substance identification, and DNA analysis.

Kobilinsky, Lawrence, Thomas F. Liotti, and Jamel Oeser-Sweat. *DNA: Forensic and Legal Applications.* Hoboken, N.J.: Wiley-Interscience, 2005. Presents an overview of DNA analysis, including the historical perspective, scientific principles, and laboratory procedures.

Rudin, Norah, and Keith Inman. *An Introduction to Forensic DNA Analysis.* 2d ed. Boca Raton, Fla.: CRC Press, 2002. Contains an overview of DNA analysis, beginning with its history and examining the principles on which it is based.

Saferstein, Richard. *Criminalistics: An Introduction to Forensic Science.* 10th ed. Upper Saddle River, N.J.: Prentice Hall, 2011. Provides an introduction to forensic science, detailing the techniques to analyze physical, biological, and chemical evidence.

WEB SITES

American Academy of Forensic Sciences
http://www.aafs.org

American Forensic Association
http://www.americanforensics.org

Association of Forensic DNA Analysts and Administrators
http://www.afdaa.org

National Institute of Justice
Forensic Sciences
http://www.ojp.usdoj.gov/nij/topics/forensics/welcome.htm

See also: Criminology; DNA Analysis; Toxicology; Vehicular Accident Reconstruction.

FORESTRY

FIELDS OF STUDY

Botany; ecology; accounting; geographical information systems; civil engineering; statistics; agriculture; policy; applied mathematics; law.

SUMMARY

Forestry is the management of forests for human resource development, extraction, utilization, regeneration, and conservation. It relies heavily on a range of sciences, mathematics, engineering, and administrative procedures. Typical forest products are paper, joinery timbers, composite boards, cardboard, firewood, drinking water, and carbon sequestration. The guiding principle is that both the trees and the supporting resources (such as soil nutrients and hydrology) should not be depleted but be used in such a way that they can persist in perpetuity.

KEY TERMS AND CONCEPTS

- **Afforestation:** Establishment of forest on land previously vacant for a considerable period.
- **Allometric:** Pertaining to the relationship, usually mathematical, between the size of one component of a tree to the size of another of its components.
- **Clearfell:** Felling prescription, or state, of a forest, in which all vegetation types are either felled or pushed over; also known as clearcut.
- **Coppice:** Practice of cutting a trunk near ground level to allow the tree to regenerate with multiple stems.
- **Diameter At Breast Height (DBH):** Diameter of a tree measured at a person's breast height, a universal measurement used in allometrics.
- **Mensuration:** Science of measuring tree and forest stand attributes, especially wood volume, through other parameters, such as diameter at breast height, height, and stand density.
- **Selective Logging:** Felling prescription in which either individual trees or groups of trees are felled, and the demographics of the remainder form a viable population.
- **Senescent Tree:** Living tree of advanced years, beyond its prime mass, with significant crown decay or hollowing of the trunk; senescent trees remain valuable for biodiversity, catchment water, and carbon storage.
- **Silviculture:** Planning and execution of forest yield maintenance through such practices as regeneration, cultivation, thinning, and various harvesting methods.

DEFINITION AND BASIC PRINCIPLES

Forestry is the professional management of forests. It consists of several interlinked processes relying heavily on a range of basic and interdisciplinary sciences (such as soil science, forest ecology, entomology, remote sensing, geographic information systems, and statistics), on engineering (road design and harvesting machinery), on labor (stand inventory, tree felling, haulage, log grading, firefighting, and prescribed burning), on budgeting and marketing, and on product technologies (mill technology and wood processing chemistry). Forestry professionals are guided by government policies, laws, cultural mores, and advances in science.

The purpose of the forestry profession is to provide the long-term sustainability of forest-product yields and financial returns, in contrast to one-time plundering and gradual forest degradation. Sustainability of a variety of forest attributes and products constitutes good planning in forest management. Timber, in some form, is the most common and marketable of forestry outputs. Timber yield and quality is therefore of primary concern to most foresters, with a consequent drive for maximization of timber yields (especially of the more valuable products) and for forecasting those yields.

BACKGROUND AND HISTORY

Forests have been used unsustainably for millennia. Forests have been converted to pasture for livestock, used to smelt ores, or to manufacture timber, tar, and other products. Trees have been turned into charcoal and removed from mountain tops to mine the ores below. Forestry originated in attempts to prevent the overuse of forests. For example, in the eleventh century, William the Conqueror established forest laws governing the usage of vegetation and wild game.

Until the eighteenth century, forest regeneration was accomplished through regrowth, replanting, or coppicing. In the early 1700's, several French scientists and naturalists, Jacques Roger, Henri-Louis Duhamel du Monceau, and René Antoine Ferchault de Réaumur, as well as British naturalist John Turberville Needham, made scientific discoveries in the areas of tree breeding and cultivation, wood strength, and yields. Later that century, the German agriculturalist Georg Ludwig Hartig focused his research on sustained yield and founded the first German school of forestry.

During the nineteenth century, the first forest preservation programs were established in the United States. In 1887, the American Forestry Association created a movement to preserve the forests of the United States. In 1891, the Yellowstone Park Timberland Reserve was created by President William Henry Harrison. Seven years later, Cornell University became the first American college to offer college-level education in forestry. The Organic Administration Act of 1897 established the first national forests; however, they were to be working forests, designed to improve water flows and to provide timber.

The U.S. Forest Service, part of the U.S. Department of Agriculture, was established in 1905 to make sure that the forests continued to provide timber and water. Several national forests, most in the eastern United States, were created by the Weeks Law of 1911, which was designed to reforest areas that had been logged or cleared for farming. In the 1970's, after the start of the environmental movement, forestry began to expand its horizons and encompass environmental concerns. Ecological forestry, although still lacking a standard definition, is generally used to refer to forestry that seeks to make use of forests in a sustainable manner, considering the whole ecosystem and the social and economic environments, as well as incorporating scientific research.

How It Works

Stock and Yield. Forecasting timber yields requires the measurement of existing stock (making an inventory of the forest) and estimating its growth. Growth is estimated either by repeated measurements over many years or by sampling different sites of similar productivity and different times since harvest. A common measure of productivity is the site index, or the mean stand height at the age of fifty years. Yield tables are created, which cross-reference site index and time since harvest, to give harvestable timber volume.

Estimating the standing timber volume requires on-ground measurement of properties such as height and diameter at breast height. Such basic measurements are combined in allometric formulas to provide trunk taper, tree volume, or merchantable wood volume. A technology used at some locations is ground-based lidar (light distance and ranging), a scanning laser providing a map of the trunk surface and, therefore, the volume of several trees per scan.

Statistics are taken and estimates made at several stages.

Stratification. For efficiency, rather than measuring the whole forest estate, inventory is taken of only portions, or stratums, often selected by their site index. This method relies on matching various environmental attributes such as slope and aspect, precipitation, fire history, soil nutrients, and soil depth. Good stratification provides more reliable interpolation between measured strata and thereby improves overall stock and yield estimates.

Stratification is facilitated by on-ground sampling, remote sensing, and geographic information systems (GIS). Relevant remote-sensing platforms include aerial photography, airborne lidar, and satellites. After stratification, the annual yields of the entire forest can be tallied from the yield tables.

Sustainability. Two main time frames are considered in sustainability: harvesting cycles and the long term.

The primary considerations for each harvesting cycle include maintenance or regeneration of the ecosystem type (for native forests). For fauna, this most often refers to the landscape level, as local demographics often depend on time since harvest. Another consideration at this level is the maintenance of near-optimum catchment-level water flows and water-table levels.

Long-term considerations include measures to ensure that there is no downward trend in yields of primary forest products (notably timber), the financial returns from the forest estate, and the levels of vital nutrients such as calcium and phosphorus. Such depletions can occur if successive harvests are too frequent—that is, if they exceed the replacement time needed by nutrient- and site-dependent ecological processes. Another long-term consideration is the

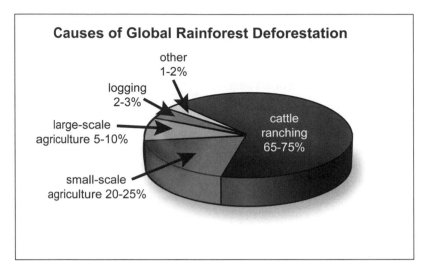

Causes of Global Rainforest Deforestation

other
1-2%

logging
2-3%

large-scale
agriculture 5-10%

small-scale
agriculture 20-25%

cattle
ranching
65-75%

Sustainable logging and ranching are two approaches to reducing rainforest deforestation worldwide.

help in forecasting responses to management efforts, such as spacing of plantings, pruning, commercial thinning, fertilizer addition, provenance selection, genetic improvement, clearfelling, prescribed burns, as well of less controllable affects such as drought and wildfire. Linear programming is one method of optimizing finances, timber yields, sustainability, and carbon sequestration over multiple harvesting areas (coupes) within the forest estate. This allows development of harvesting plans and their approval several years before implementation, which in turn allows employment stability through prior assignment of logging teams and mill deliveries. Forest managers generally prefer error margins in yields to be less than 10 percent. Uncertainties are higher for previously unlogged native forests and mature stands because of buttressing and senescence but lower for more intensively managed forests and plantations.

Product Sourcing. Exotic species form much of the global plantation stock. For example natural rubber comes from the Amazonian tree *Hevea brasiliensis,* which is cultivated extensively in Southeast Asian countries. *Eucalyptus globulus* from Australia is cultivated widely in southern Europe and South America. *Pinus radiata* from California is grown widely in Australia. This translocation reduces browsing by herbivores that evolved alongside the tree species, and some exotic plantation species can out-compete indigenous species, producing higher growth rates than they do in their place of origin. Such plantations form a major source of products consumed at a high rate, such as paper. In some regions, native forests are used unsustainably for such products. The more obvious environmental effects can be observed and unsustainable use can be proven through the same fundamentals of forest sciences used in sustainable yield calculations.

total carbon stock of the wider forest estate; the forest management activity must not constitute a net carbon emission. This consideration has come to the forefront with the rise of concerns over global climate change; this requirement is readily met for plantations on old-field farmlands but is more difficult for native forests newly opened to logging and harvest cycles.

Harvesting Plans. Detailed harvest planning is part of legitimate forest management, entailing, for example, a local inventory, the silvicultural prescription, mapping drag lines and buffer zones, yield expectations, and plan approval from government departments. It ensures that on-ground operations meet all prerequisites, such as application of appropriate engineering expertise in forest areas newly opened to industry and local environmental regulations. The plan also forms part of the site-history record for future forestry operations.

Production Mechanisms. Silviculturalists grow forests by various techniques, such as collecting seed from local stock before harvest, coppicing, genetically engineering saplings, and prescribed burns sometimes followed by aerial seeding. Follow-up treatment to promote growth may be unnecessary or include herbivore and weed control and stand thinning. Growth models, combined with mapped environmental variability, produce catchment-level forecasts of wood volume and carbon sequestration. Models

APPLICATIONS AND PRODUCTS

Internal Products. Some of forestry's outputs are consumed within the industry. These include computer software developed by foresters to aid in

Fascinating Facts About Forestry

- Although the tallest recorded individual trees are no longer standing, scientists determined from taper formulas calibrated on shorter, living specimens that the tallest flowering plant, *Eucalyptus regnans*, a hardwood tree in southeast Australia, can reach 120 meters.

- Since medieval times, coppicing of broadleaf species was practiced in England, often with stands of varying ages, producing timber for infrastructure, weaving, charcoal production, smelting, tools, and tanneries. It is continued in some places for conservation of wildlife that had adapted to the coppices and for craftspeople.

- Nearly every year, cyclists climb Mount Ventoux, a 6,273-foot bald mountain in the Provence region, during the Tour de France. The mountain was once covered by a forest, but it was stripped of trees beginning in the twelfth century to provide wood for shipbuilders in Toulon.

- Cork is produced by peeling the bark from the branches of cork oak trees in the Mediterranean basin at about ten-year intervals, when the tree is between thirty and two hundred years old. It is traditionally used for flooring, bottle stoppers, and buoyancy devices. As the trees are not felled, cork oak forests are an unusual wildlife habitat.

- A variety of remote-sensing platforms are used to study forests. Hyperspectral and lidar combined in a light aircraft provide the most detailed information, such as forest health, species distribution, canopy structure, and individual tree height.

- Urban forests in the northern United States may serve as seed sources for southern species during their northern migration, necessary for their conservation under climate change, but the urban trees may also enable invasion by exotic species.

forestry operations and budgeting, harvesting machinery such as felling mechanisms, virtual reality harvesting simulators, stock valuation and carbon assessments, mill residues to fuel kiln drying of milled timber, and felled trees for construction of bridges on haulage roads.

External Sales. Volume quantities of wood products delivered from the forest can be converted directly to monetary values in terms of dollars per metric ton or cubic meter, depending on the type of product and its destination. Other outputs are not so easily converted to monetary equivalents. These include remediation of land through afforestation, unmonitored firewood collection, maintenance of biodiversity, water quality, salinity remediation, and reduced air pollution. Such calculations are, however, becoming more frequent in the scientific literature and even in government reports.

Mill Products. Two main mill products from logged timber are pulpwood and sawlogs, sometimes from the same forest tree or stand (integrated harvesting) or from entirely separate forests. The proportion of the original tree volume that becomes cut lumber is usually no more than 55 percent and often lower. Consequently, optimization of sawmilling is financially crucial. Taper formulas, which model tree shape, indicate not only total wood volumes but also volumes of different mill products. Computer-automated milling to optimize the cutting pattern within individual logs can reduce wastage. Final product yields are fed back to aid future revenue forecasting. Woodchip and pulp mills are coarser processors, with throughput up to 400 metric tons of woodchips per hour, equivalent to one fully laden log truck every two and one-half minutes. These rates are driven by the global market developed for paper products.

Product Variety. In terms of volume, the material products of forestry are dominated by lumber and paper, but other products such as firewood, charcoal, cork, sandalwood oil, cinnamon, and palm oil can be regionally significant. Some outputs such as firewood, although measurable, can be from forestry harvests or simply collected from forests without forest management, but both come under the heading of forest products in national accounts. Broader-scale outputs of nontraditional forestry include salinity remediation, regeneration of semiarid landscape functionality, amelioration of urban air pollution, and attempts to address climate change. Reducing the emissions from deforestation and forest degradation is an area of increasing investment and requires adaptation of traditional forest sciences to different forest components (such as nonmerchantable tree components, woody debris, and soil) and to a greater range of time scales.

IMPACT ON INDUSTRY

Application of Forestry Principles. Sustainability regulations in forestry vary between private and public land. The distinction between government and industry may be unclear, with forest management on public land operated by corporation-like government agencies. Sustainability principles are applied differently for various forest attributes, such as wood yield and wildlife habitat, and the legality of operations also varies. Therefore, the proportion of annual yields retrieved through application of sustainability principles is difficult to ascertain. Nevertheless, some experts have tried to estimate the amount of illegally sourced lumber. Policies are being implemented to curtail this illegal harvesting, with a view to attenuating environmental damage and establishing civil law; however, one side effect will be a rise in global timber prices.

Global Trade. Trade in forestry is generally perceived in terms of forest products. Annual global trade in forest products in 2009 was about 140 million cubic meters of sawn timber, about 130 million cubic meters of roundwood (premilled timber), 130 million metric tons of paper, 80 million cubic meters of wood panels, and 45 million metric tons of woodpulp. The country with the largest output of roundwood is Russia, and the major importers are Finland, China, Japan, and Korea. Other major exporters of roundwood are the United States, Germany, New Zealand, and Canada. Canada is the lead exporter of both sawn timber and pulpwood, although the United States produces more pulpwood. Africa is the largest producer of firewood. Japan is the leading importer of woodchips, importing hardwood chips principally from Australia, South Africa, and Chile. Within Australia, Tasmania is the leading state-exporter of hardwood chips, most of which are sourced from native forests, although increasingly they are being sourced from plantations.

In 2006, the companies with the largest net sales of fully processed paper products were U.S.-based International Paper ($21.2 billion), Finland's Stora Enso ($16.2 billion), and U.S.-based Procter & Gamble ($12.0 billion). In early 2010, the Nippon Paper Group reported net sales of $11.2 billion and a net profit of $334 million. Paper consumption per capita is highest in the United States, although China and India have the highest growth rates of consumption.

Forestry Research. Scientific research related to forestry is undertaken in colleges and universities, in dedicated research institutes, private companies, and consortiums of research bodies. Funding sources are primarily government and forest agencies, large corporations, educational institutions, and industries. Research areas include primary wood production, manufacturing wood products, and the environment. Most funds from the private sector are devoted to research on manufacturing. Most scientific research data and advances in forestry reach the public arena, although some data, conducted by the forest agencies, forest product companies, or even government departments remain confidential. Demand for comprehensive, high-quality data on forests is intense, as it can advance new research and investment frontiers, especially climate change mitigation.

CAREERS AND COURSE WORK

Trained foresters are in demand for forestry management of both native and plantation forests. The two main drivers are production of wood products and conservation of the environment. The traditional route for those interested in pursuing careers as foresters has been to gain a bachelor's degree at a special forestry school or university department, often with a focus on wood production. Often these studies are combined with environmental sciences or geography, as part of a more general environmental science degree. Nevertheless, dedicated forestry departments are still part of many universities, especially those in major capital cities or those where the forest products industry has regional presence and a substantial stake in the workforce. Some universities link with others to provide a more comprehensive training.

Students who wish to pursue forestry in college should take science classes and at least one mathematics class while in high school. A forestry degree can be combined with biology, botany, accounting, economics, geography, or resource management. A forestry degree will often include at least one class in forest ecology. For more advanced studies in forestry, statistics is necessary at the undergraduate level.

SOCIAL CONTEXT AND FUTURE PROSPECTS

Sustainability of key environmental attributes remains problematic, even with rigorous protocols and planning because advances in forest science must be comprehensively proven and written into policy before integration with demands for productivity. For example, evidence and modeling suggest that in many forests, heat stress from higher summer temperatures and droughts will outweigh the carbon dioxide fertilization and longer growing season accompanying climate change, requiring recalculation of harvesting quotas.

A question remains as to where and how forestry should be implemented. Many plantations, such as rubber and palm oil plantations in Southeast Asia, required clearing of the original forest and have since been reforested. Afforestation can have environmental benefits, such as providing a reprieve for native forests, land rehabilitation, and climate change mitigation. In 2010, although progress had been made toward recycling paper and curtailing illegal logging, the paper industry and some forestry professionals participated in an initiative designed to get people to print more e-mails.

Application of forestry know-how is increasingly sought in the newer, nontraditional industries such as mine-site and rangeland rehabilitation and in urban forestry. Forestry principles are also used in climate change mitigation efforts such as carbon budgeting of practices and calculating emission offsets.

Christopher Dean, B.Sc., Ph.D.

FURTHER READING

Bekessy, Sarah A., et al. "Modelling Human Impacts on the Tasmanian Wedge-Tailed Eagle (*Aquila audax fleayi*)." *Biological Conservation* 142 (2009): 2438-2448. Discusses a case in which forestry and other societal influences affected an endangered species. Notes the steps for conservation and graphs the decline of the population.

Cox, Thomas R. *The Lumberman's Frontier: Three Centuries of Land Use, Society, and Change in America's Forests*. Corvallis: Oregon State University Press, 2010. Examines the history of the lumber industry in the United States and the changing attitudes over time.

Hays, Samuel P. *Wars in the Woods: The Rise of Ecological Forestry in America*. Pittsburgh: University of Pittsburgh Press, 2007. Examines forestry practices and the conflict between ecological forestry and the more traditional approach, which he calls commodity forestry.

Hicke, Jeffrey A., et al. "Spatial Patterns of Forest Characteristics in the Western United States Derived from Inventories." *Ecological Applications* 17, no. 8 (2007): 2387-2402. Shows how continental-scale forest inventory can be combined with remote sensing and modeling to provide a history of management effects on carbon stocks.

Maser, Chris, Andrew W. Claridge, and James M. Trappe. *Trees, Truffles, and Beasts: How Forests Function*. New Brunswick, N.J.: Rutgers University Press, 2008. Describes examples of ecosystem processes and players in forest productivity in the disparate forests of the northern United States and southern Australia.

Pretzsch, Hans. *Forest Dynamics, Growth, and Yield: From Measurement to Model*. New York: Springer, 2009. Presents a modern and detailed review of methods for forest growth and yield calculations, including comparisons with ecological modeling. Provides for carbon accounting. Builds from basics to the professional level.

WEB SITES

International Union of Forest Research Organizations
http://www.iufro.org

National Association of State Foresters
http://www.stateforesters.org

Oregon Forest Resources Institute
http://www.oregonforests.org

Society of American Foresters
http://www.safnet.org

U.S. Forest Service
http://www.fs.fed.us

See also: Agroforestry; Land-Use Management; Silviculture; Soil Science.

FOSSIL FUEL POWER PLANTS

FIELDS OF STUDY

Electrical engineering; mechanical engineering; pressure vessel engineering; industrial electronics; pipe fitting; pressure welding; power plant operations; industrial maintenance; millwrighting; geology; mining; deep borehole technologies; power distribution technology; fluid mechanics; hydraulics; environmental chemistry; combustion chemistry; heat transfer technology; mathematical modeling; computer simulation; controls and switching technology; transportation; management; project coordination; human resources.

SUMMARY

Fossil fuels are the organic residues of geological processes and include the various grades of coal, natural gas, petroleum, and crude oil. By definition, all fossil fuels are nonrenewable resources. Fossil fuel power plants all function in fundamentally the same manner and rely on a principle that has not changed significantly since the earliest applications of steam power. In short, the fuel is combusted to generate heat, producing pressurized steam that in turn drives the turbines of large electric generators.

Coal was the first major fossil fuel to be exploited as an energy source for the steam-powered plants that drove the Industrial Revolution of the eighteenth and nineteenth centuries. When steam technology was applied to the large-scale generation of electricity, coal became the fuel of choice because of its ready availability. Coal remains the major fuel source for fossil fuel power plants, although natural gas and, to a much lesser extent, petroleum-based fuels have been considered.

Significant costs, both economic and environmental, have been identified in the use of fossil fuels for the production of electricity. Economically, the prices of fossil fuels have been driven upward by the market economy, and environmentally, the combustion of large quantities of carbon-based fossil fuels releases a great deal of carbon dioxide and other gases into the atmosphere.

KEY TERMS AND CONCEPTS

- **Carbon Dioxide:** Molecule formed by the combination of one atom of carbon with two atoms of oxygen.
- **Combustion:** Chemical process of oxidation of a fuel material, generally a carbon-based fuel, by molecular oxygen that results in the release of thermal energy in the form of flames.
- **Cooling Tower:** Structure within a power plant that serves to reduce the temperature of the exhaust gas stream from the combustion stage, allowing the recovery of quantities of fly ash, sulfur, and other pollutants.
- **Energy Density:** Amount of usable energy that can be extracted from the unit mass or volume of a material.
- **Fluidization:** Treatment that causes a nonfluid material to behave in a fluidlike manner; typically achieved by passing a strong current of air uniformly through a mass of solid particles so that the particles are suspended in the moving air stream.
- **Fluidized Bed Combustion:** Combustion process carried out by injecting the fuel into a bed of solid particulate matter fluidized with a stream of air.
- **Fossil:** Geologic remnant of a once-living organism; includes mineralized body structures and the carbonaceous materials, such as coal, crude oil, and natural gas, resulting from geologic processes.
- **Fossil Fuel:** Carbonaceous material resulting from geologic processes that is used as a fuel for combustion.
- **Generator:** Device that produces electric current by the interaction of electrical conductors with the moving magnetic field produced by spinning magnets within the device.
- **Heat Exchange:** Transfer of thermal energy from a material at one temperature to a material at a lower temperature.
- **Pulverization:** Process of reducing the physical structure of a material to extremely fine particles.
- **Turbine:** Device that converts linear motion of a pressurized fluid into rotational motion that may then be used to power other devices such as generators.
- **Vaporization:** Conversion of a material from a liquid state to a gas state.

DEFINITION AND BASIC PRINCIPLES

Fossil fuel power plants are generating stations that rely on the combustion of fossil fuels to produce electricity. Only three fossil fuels—coal, petroleum, and natural gas—are used for this purpose. The term "power plant" does not refer only to facilities that generate electricity but rather to any facility whose function is to produce usable power, whether electrical, mechanical, hydraulic, pneumatic, or another type. In common usage, however, power plant generally refers to those facilities that are used to generate electricity.

All fossil fuels are the remnants of organisms that existed, in most cases, many millions of years ago. Time and geologic processes involving heat and pressure chemically and physically altered the form of these organisms, turning them into mineralogical fossils (such as mineralized bones found in sedimentary rock formations) and the carbonaceous forms of coal, crude oil, and natural gas. When these carbonaceous materials are refined, they can be used as combustion fuels in fossil fuel power plants.

The combustion process is carried out in a variety of ways, from standard internal combustion engines using natural gas, gasoline, or diesel oil, to fluidized bed combusters using pulverized coal powder. Internal combustion engines are used to drive a generator directly, while other combustion methods are used to heat water and produce pressurized steam through heat exchange. The steam is then used to drive a turbine that in turn drives electric generators. The spent steam is generally recycled through the system. The exhaust steam from combustion is passed through a treatment process to reduce or eliminate contaminants formed from materials that were in the fuel.

Ideally, the combustion process would produce only carbon dioxide—from the combustion of coal—or carbon dioxide and water—from the combustion of hydrocarbon fuels such as natural gas and refined petroleum fuels. In practice, however, fossil fuels contain a percentage of materials other than carbon and hydrogen, such as sulfur, metals (including iron, mercury, and lead), and nonmetals (including phosphorus, silicon, and arsenic). In addition, air used to supply oxygen for combustion also

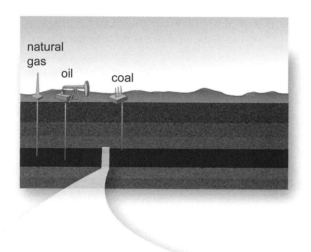

Fossil Fuels

Deposits of plant and animal remains from the first forests on Earth comprise the majority of the existing fossil fuels.

natural gas

oil

coal

Carboniferous Swamp
300 million years ago

contains about 78 percent nitrogen and about 1 percent of other gases. At the temperatures of combustion, these impurities can react with oxygen to produce a variety of pollutant by-products such as sulfur dioxide, nitrogen oxides, and fly ash.

The combustion of fossil fuels results in a very large quantity of carbon dioxide being released into the atmosphere, where it can act as a greenhouse gas. A greenhouse gas has the effect of trapping heat that normally would be radiated out of the atmosphere and into space. Many experts believe that the carbon dioxide released by the burning of fossil fuels has been a primary factor in global warming, which refers to an increase in the mean annual temperature of the planet.

BACKGROUND AND HISTORY

Coal has been used as a fuel for combustion for thousands of years. It was reportedly used by native North Americans when the first European settlers arrived, and it was undoubtedly used by other peoples throughout the world because of the ease with which it could be extracted from the ground in certain areas. It became the fuel of choice beginning in the eighteenth century and had almost completely replaced wood as the dominant fuel of industry by the early 1900's because of its more favorable energy density. Coal's increasing popularity as a fuel also drove the growth of the coal-mining industry, in turn increasing its availability.

With the development of the large-scale generation of electricity and its many applications, coal-fired power plants were used to drive electric generators where suitable water power, such as at voluminous waterfalls, was not available. The convenience and versatility of a common electric grid resulted in the growth of the electric generation industry. Small and localized generation systems ranging from low-output gasoline-powered home generators to large diesel-powered industrial generating stations can provide emergency and local service if the common grid is not available. Large generating stations using fossil fuels have been built and continue to be built in areas where coal or other fossil fuels are readily available.

HOW IT WORKS

Fossil Fuels. Coal and petroleum are the remnants of plants and animals that lived millions of years ago.

Over the years, geologic processes compressed and chemically altered the plants and animals in such a way that coal consists almost entirely of pure carbon, crude petroleum consists almost entirely of a vast assortment of hydrocarbons, and natural gas consists almost entirely of methane, ethane, and propane, which are simple hydrocarbon gases. Coal can be found at varioius depths in the Earth's crust, in veins ranging from only a few centimeters to hundreds of meters thick. It is mined out as a solid, rocky, and relatively lightweight material and used in forms ranging from crude lump coal to a fine powder that is fluid-like in its behavior.

Crude oil and natural gas are found only at depths of hundreds and thousands of meters. As liquids or fluids, these materials have migrated downward through porous rock over a long period of time, until further progress is prevented by an impervious rock layer. There, they collect, often in large pools of oil and gas that can be recovered only after being found through careful exploration and deep borehole drilling. Natural gas requires no further processing before being used as a combustion fuel, unless it is classed as sour gas, meaning it contains an unacceptably large proportion of foul-smelling hydrogen sulfide and other poisonous gases that must be removed. Petroleum, or crude oil, must be heavily refined before it can be used as a fuel. The crude oil is subjected to thermal cracking, which breaks down and separates the various hydrocarbon components into usable portions from light petroleum ethers such as pentanes and hexanes to heavy tars such as asphalt. The most well-known fractions refined from petroleum are gasoline, kerosene, diesel fuel, and waxes, and various grades of lubricating oils and greases.

Combustion. Combustion is a chemical reaction between a material and oxygen. The reaction is an oxidation-reduction process in which one material becomes chemically oxidized and the other becomes chemically reduced. In the context of fossil fuels, the material that becomes oxidized is coal, fuel liquids, or natural gas. Combustion of these carbonaceous materials converts each atom of carbon in the fuel molecules to a molecule of carbon dioxide, according to the general equation:

$$C + O_2 \rightarrow CO_2$$

This conversion (greatly simplified here) releases an amount of energy that can then be transferred

Fascinating Facts About Fossil Fuel Power Plants

- Some 2 billion metric tons of carbon dioxide have been added to the atmosphere from the burning of fossil fuels in the 1800's and 1900's.

- About 70 percent of the world's power plants use fossil fuels. In 1984, fossil fuels accounted for 82 percent of all commercial energy production, and 91 percent of the U.S. energy supply. According to the U.S. Department of Energy, fossil fuels (coal, natural gas, and oil) account for more than 85 percent of the United States' energy consumption, including nearly two-thirds of electricity and nearly all fuels used for transportation.

- Coal was used as a heating fuel by North American native peoples, and presumably other peoples around the world dating back hundreds of years ago.

- Fossil fuel power plants come in all sizes, from little 750-watt gasoline-powered portable generators to large coal-fired or natural gas-fired facilities producing hundreds of megawatts of electricity.

- Carbon dioxide capture and storage technology would take the carbon dioxide produced from fossil fuel combustion and store it a kilometer or more underground. There, slow geologic processes might one day convert it back to free oxygen and carbonaceous deposits that future generations could use as fuel.

- Some fossil fuels are more than 650 million years old.

- In ancient Greece, Archimedes invented a device that directed steam outward in opposing directions, causing the device to rotate. It would be known later as an external combustion steam engine.

to and captured by a moderator—typically water—through the use of heat exchangers (devices that facilitate the transfer of heat from one material to another material). Combustion of hydrocarbons also produces water as an output of the reaction, in which two atoms of hydrogen combine with one atom of oxygen to produce one molecule of water. Other reactions corresponding to the combustion of impurities in the fuel stream also take place, and their products are ejected in the exhaust flow from the combustion process.

Turbines and Generators. Steam under pressure, produced by heating water via the combustion of fuels, is directed into a mechanical device called a turbine. Turbines can basically be described as high-tech versions of the ancient water wheel. The pressure of the flowing gas (steam) pushes against a series of vanes attached to an armature (electric component) in the structure of the turbine, driving them to spin the armature with force. This converts the linear fluid motion of the steam into the rotary mechanical motion of the turbine. Turbines are coupled to an electric generator so that their rotation results in the generation of electricity.

A generator is another rotary device that, in its most basic concept, consists of a magnet spinning inside a cage of conducting wires. The magnetic field of the magnet also spins at the same rate that the magnet spins. The movement of the magnetic field through the conductors in the surrounding cage produces an electromotive force (EMF) in the conductors, which is measured in volts. If the generator is connected into a circuit, this EMF causes current to flow in the circuit. Strictly speaking, generators produce direct current (DC) electricity, while alternators produce alternating current (AC) electricity. AC electricity is the standard form of electricity used in national power grids around the world.

Both generators and alternators are available in various output capacities and are driven by many types of rotary engines, from small internal combustion engines to large industrial steam turbines.

APPLICATIONS AND PRODUCTS

Fossil fuel power plants, in the context of electric generating stations, produce only one product: electricity. Any and all other materials that come from them are considered ancillary or waste by-products. The ultimate goal of operating a fossil fuel power plant is to maximize the output of electricity from each unit of fuel consumed, while also minimizing any and all undesirable outputs. To that end, the efficiencies of control design, the data feedback process, economics, fuel processing, and a host of other aspects of the electric power generation industrial complex are examined each year. Not the least of these considerations is the placement and construction of new facilities and the maintenance of older facilities.

At one time, the competitive cost and the availability of natural gas and crude oil nearly spelled the demise of the coal-fired power plant. However, the prices of natural gas and crude oil were driven upward, both artificially and naturally, making these choices less attractive. Nuclear power plants were initially welcomed by the public, but their popularity declined. Because of these circumstances, the continuing demand for electricity caused fossil fuel power plants—especially coal-fired plants—to regain their position of prominence in electric power generation. The operation of fossil fuel plants spawned related industries: the development of coal-mining methods and machinery, oil and gas exploration and recovery, fossil fuel transportation and preprocessing, specialized construction and trades, environmental assessment and maintenance operations, financial and administrative companies, plant operations and control technology, industrial maintenance, and grid supply and service.

Concern about greenhouse gases resulted in the birth of an industry aimed at capturing and storing the carbon dioxide produced by the power plants. Different approaches are being developed, but possibly the most promising technology is the sequestering of carbon dioxide in deep underground water formations. This has engendered a whole new area of research and development in regard to compression and recovery technology.

The operation of fuel fossil power plants generates chemicals, some of which have been recovered during preprocessing and exhaust gas scrubbing procedures. Sulfur recovered from preprocessing and from entrapment of sulfur dioxide in the exhaust gas stream is used to produce sulfuric acid, an important industrial chemical, as well as numerous other sulfur-containing compounds. Similarly, nitrogen oxides recovered from the combustion process provide nitric acid and other nitrogen-containing compounds. Interestingly, the entire plastics industry grew out of research to find uses for compounds recovered from coal tar, a by-product of the coal processing industry in the nineteenth century. In modern times, however, essentially all plastics are derived from petroleum.

IMPACT ON INDUSTRY

Several countries rely almost exclusively on electricity produced by fossil fuel power plants, especially coal-fired plants. According to the International Energy Agency, in 2009, South Africa and Poland both obtained almost 95 percent of their electric supply from coal-fired power plants, while the United States obtained almost exactly half of its electric supply from coal-fired power plants. Although several major hydroelectric projects are under way in China, coal-fired power plants are also being built at a great rate to help establish an adequate electrical infrastructure. Fossil-fuel power plants have a reputation for high reliability because of their well-established technology, and they are less costly than nuclear and other alternative electricity-generating methods. Additionally, they can be constructed in much less time than would be needed for other types of generating stations of the same capacity.

Modern industry depends on a constant, adequate supply of electric energy for its very existence. The application of electric motors and the ubiquitous nature of electric arc welding in fabrication and manufacturing processes have defined modern industry and created power demands that are hard to match through any other means of power delivery. Stricter regulations regarding the handling of waste materials produced during the combustion of fossil fuels have also had a significant impact on the industry as a whole. Private and governmental power-generating organizations have been required both to reduce the production of waste materials and to assume responsibility for the handling of wastes that were produced in past years. In addition, international agreements for the reduction of greenhouse gas emissions have placed more pressure on the industry.

CAREERS AND COURSE WORK

Students undertaking any program that will lead them into a career related to fossil fuel power plants, whether directly or indirectly, will be required to exit high school with an understanding of physics, chemistry, mathematics, and business and technology. Biology will also be required if the career path chosen is directed toward environmental studies. College and university level course work will depend greatly on the area of specialization chosen, as the range of options at this level is immense. At a minimum, students will continue studies in mathematics, physical sciences, industrial technologies (chemical, electrical, and mechanical), and business as undergraduates or as trade students. More advanced studies will take the form of specialist courses in a chosen field. In addition, as technologies and regulations change, those working in the field of fossil fuel power plants can expect to be required to upgrade their working knowledge on an almost continual basis to keep abreast of changes.

SOCIAL CONTEXT AND FUTURE PROSPECTS

The world's reliance on the ready availability of coal and the industrial convenience of fossil fuel power plants, especially in developing countries, essentially guarantees that those facilities will be part of the landscape for many years to come, despite intensive efforts to develop alternative power sources. New technologies being developed, such as carbon dioxide capture and storage (CCS), combined geothermal-fossil fuel cogeneration, and biomass cogeneration, will require generations of power plant workers who understand both the older and the new technologies as fossil fuel power plants are managed toward a zero emissions platform (ZEP). Emission-abatement plans also include elimination of the heavy metals component of fossil fuels, as the combustion of those materials has the effect of greatly concentrating any heavy metals, including radioactive trace elements such as uranium and thorium, in the ash residues. It is therefore not inconceivable that, in the future, work in a fossil fuel power plant will also require training in working with radioactive materials.

Richard M. J. Renneboog, M.Sc.

FURTHER READING

Borowitz, Sidney. *Farewell Fossil Fuels: Reviewing America's Energy Policy.* New York: Plenum Trade, 1999. Examines the role of fossil fuels in energy production in the United States and other Western nations, and the technologies that can be and are being developed to replace fossil fuels.

Breeze, Paul. *Power Generation Technologies.* Boston: Elsevier, 2005. A comprehensive treatment of the generation of electric energy by different power plant technologies and other methods, with consideration of their economic and environmental impacts

Droege, Peter, ed. *Urban Energy Transition: From Fossil Fuels to Renewable Power.* New York: Elsevier, 2008. Looks at principles, policies, and practices that affect energy use, the technology involved, ways to alter the urban environment, and programs being employed in London and elsewhere.

Evans, Robert L. *Fueling Our Future: An Introduction to Sustainable Energy.* New York: Cambridge University Press, 2007. Looks at energy supply and demand and how to reduce reliance on fossil fuels.

Leyzerovich, Alexander S. *Steam Turbines for Modern Fossil-fuel Power Plants.* Lilburn, Ga.: Fairmont Press, 2008. Closely examines the engineering and operating properties of modern steam turbines in fossil fuel power plants.

WEB SITES

Tennessee Valley Authority
Fossil-fuel Generation
http://www.tva.gov/power/fossil.htm

U.S. Department of Energy
Fossil Energy
http://fossil.energy.gov/index.html

World Coal Institute
Coal and Electricity
http://www.worldcoal.org/coal/uses-of-coal/coal-electricity

Zero Emissions Platform
European Technology Platform for Zero Emission Fossil Fuel Power Plants
http://www.zeroemissionsplatform.eu

See also: Biofuels and Synthetic Fuels; Coal Gasification; Hazardous-Waste Disposal; Hydroelectric Power Plants; Industrial Pollution Control; Petroleum Extraction and Processing; Steam Energy Technology; Wind Power Technologies.

FRACTAL GEOMETRY

FIELDS OF STUDY

Calculus; computer science; geometry; measure theory; metric topology; probability; set theory.

SUMMARY

Fractal geometry is a branch of mathematics that is used to study irregular or fragmented shapes, such as clouds, trees, mountains, and coastlines. Applications of fractal geometry are widespread, from the sciences and medicine to social sciences and the arts, and especially in human anatomy, ecology, physics, geology, economics, and computer graphics. Related areas include chaotic systems and turbulent systems.

KEY TERMS AND CONCEPTS

- **Affine Transformations:** Transformations on the Euclidean plane include translation, scaling, rotation, and reflection. Affine transformations can be used to convert one geometric shape into another.
- **Brownian Motion:** Result of thermal molecular motion in a liquid environment.
- **Chaos Theory:** Behavior of nonlinear dynamic systems, the processes of which are governed by an underlying order and the outcomes of which demonstrate sensitivity to initial conditions.
- **Complex System:** Refers to the properties of a non-linear system, in which the behavior of the system as a whole is different from those of its parts.
- **Diffusion-Limited Aggregation (DLA):** Clustering of particles in random walk due to Brownian motion. DLA is prevalent in diffusion systems. Lichtenberg figures (or electric trees) are a common example of DLA.
- **Dimension:** Space a set occupies near its points.
- **Dynamical System:** Set of possible states with a rule that defines the present in terms of past conditions.
- **Feedback Loop:** Process of a transformational system where the input of a cycle gives a different output, which becomes the input of a new transformation. Feedback processes are fundamental tools for studying natural systems.
- **Fractal:** With reference to a mapped set, the set can be considered fractal if it is detailed at smaller scales, has an irregular structure, demonstrates self-similarity, has a Hausdorff-Besicovitch dimension greater than its topological dimension, and can be defined by a simple statement that is often reiterated. Fractals can be geometric (nonrandom) or random.
- **Fractal Dimension:** Measure of a geometric object that can have fractional values. It refers to the measure of how fast the length, area, or volume of an object increases with a decrease in scale. Fractal dimension can be calculated by box counting or by evaluating the information dimension of an object.
- **Generator:** Collection of scaled copies of an initiator.
- **Hausdorff-Besicovitch Dimension:** Mathematical statement used to obtain a dimension that is not a whole number, commonly written as $d = log (N)/ log (r)$.
- **Initial Condition:** Starting point of a dynamic system.
- **Initiator:** Starting shape of a reiterated self-similar object.
- **Julia Set:** Boundary of the set of points of a function z that escape into infinity under repeated iteration by $f(z)$.
- **Lacuna:** Gap or space within a structure.
- **Lacunarity:** In reference to fractal objects, lacunarity describes the weave and texture of the fractal object, with particular reference to the open and filled spaces of that object.
- **Lindenmayer (L-) System:** Named for Hungarian biologist Aristid Lindenmayer, it is a system used to model the development of plants. It is defined as a parallel rewriting system, one that takes a simple form and then reiteratively replaces its parts by a set of rules. It can generate a fractal with a dimension between one and two.
- **Mandelbrot Set:** Set of all complex numbers c such that iteration of the function $f(x) = x^2 + c$, starting at $x = 0$, does not go into infinity.
- **Noise:** Signal with fractal properties.
- **Rule:** Principle that describes how scaled copies of a generator replace copies of an initiator.
- **Self-Similarity:** Structure that can be broken down into smaller pieces, all of which are individual replicas of the main object under repeated magnification. There is a relationship between

the scaling factor and the number of pieces to which the structure can be reduced.

- **Symmetry:** Universal scientific and philosophical category used in reference to the structure of matter. It is defined by the invariance of particular features under different types of transformations.

DEFINITION AND BASIC PRINCIPLES

Fractal geometry is a workable geometric middle ground between the excessive geometric order of Euclid and the geometric chaos of general mathematics. According to French American mathematician Benoit Mandelbrot in *Proceedings of the Royal Society of London* (May 8,1989), it is based on a form of symmetry previously underused, namely, "invariance under contraction or dilation":

> Fractal geometry is conveniently viewed as a [language that has proven] its value by its uses. Its uses in art and pure mathematics, being without "practical" application, can be said to be poetic. Its uses in various areas of the study of materials and [of] other areas of [engineering] are examples of practical prose. Its uses in [physical] theory, especially in conjunction with the basic equations of mathematical physics, combine poetry and high prose. Several of the problems that fractal [geometry] tackles involve old mysteries, some of them already known to primitive man, others mentioned in the Bible, and others familiar to every landscape artist.

BACKGROUND AND HISTORY

Fractal geometry was first defined in 1975 by Benoit Mandelbrot, a creative and prolific twentieth-century polymath who pioneered a synthesis of mathematical paradigms for interpreting rational surface roughness. *Les Objets Fractals*, translated into English in 1977 under the title *Fractals: Form, Chance, and Dimension*, explored Mandelbrot's innovative approach to understanding the properties of mathematical and natural forms. The manuscript was revised and republished in 1983 under the title *The Fractal Geometry of Nature*. This book illustrated principles that reordered understanding of dimension, symmetry, and scalar transformations. It is considered one of the classic works of twentieth-century mathematics. As a result of rational inquiry into shapes and nonlinear physical phenomena, fractal analysis can be applied with exceptional effectiveness in both pure and

Benoit Mandelbrot, French mathematician, pioneer of fractal geometry, and author of the book The Fractal Geometry of Nature *(1982). Mandelbrot, who worked at IBM for over thirty years, used a computer to plot images of fractals called "Julia sets" in 1979. (Hank Morgan / Photo Researchers, Inc.)*

applied mathematics. Advances in computation during the late twentieth century made it possible to evaluate unique objects constructed after millions of iterations of a function. Repetitive calculations that are not feasible in human time were performed with novel speed and duration via the computer, producing extraordinary mappings of functions, the visual beauty of which astounded its practitioners. The results of fractal analysis transformed and aligned knowledge in many domains, including physics, religion, the mechanical and biological sciences, economics, statistics, music, and many fields of art.

Mandelbrot's investigations regarding the nature of roughness or nonlinearity in the structural composition and generation of mathematical and physical objects are products of the twentieth century. However, fractal analysis is synthetic and embraces systems of thought originating in the preceding centuries. In retrospect, the principles described by fractal theory provide a platform for integrating a variety of human expressions regarding infinity. Throughout history, mathematicians intuitively understood the implicit beauty and symmetry of mathematical systems. Glimpses of the infinite were magnified and demonstrated in fractal forms with particular appeal to the human sense of vision. Moreover, the language of fractals can be grasped and used in a variety of fields requiring varying levels of mathematical and computational ability. Finally, fractal symmetry provides an accessible framework for understanding universal structures and dynamics.

The antecedents to fractal geometry can be seen in the following areas.

Projective Geometry. Projective geometry refers to a class of geometric properties describing invariance under projection. Scholarship suggests that this phenomenon was understood by early geometers—perhaps by Euclid himself. Mathematicians of the seventeenth century are credited for realizing these properties in an attempt to make geometry more practical. French mathematicians Gérard Desargues, Blaise Pascal, and Philippe de la Hire provided mathematical foundations for new applications in the nineteenth century. Subsequent formulations of the works of Desargues are fundamental to novel descriptions of space and time. French mathematicians Jean-Victor Poncelet and Michel Chasles continued the work of Desargues. Similar innovations were made in the fields of analytic geometry, non-Euclidean geometry, and Riemannian geometry.

Geometers have relied on the definitions and assumptions described in Euclid's *Elements of Geometry* (1893), a textbook that established time-honored principles for studying geometric properties and their mathematical relationships. These fundamental truths formed the basis of rational descriptions of shapes and measurements to the present day. During the nineteenth century, scholars found the ideal shapes of Euclidean geometry insufficient to analyze the natural phenomena described by science. In particular, Euclid's Postulate 5 defied proof by the brightest minds. The investigations of German mathematician Carl Friedrich Gauss, Hungarian

mathematician János Bolyai, and Russian mathematician Nikolai Ivanovich Lobachevsky set the groundwork for the descriptions of the hyperbolic plane, a concept that would have stunning implications in descriptions of the mechanics of space and time, a field pioneered by Albert Einstein.

Cartography. In his 2002 essay "A Maverick's Apprenticeship," Mandelbrot mentioned his familiarity with maps as a result of an uncle's tutoring. It is of note that well-established cartographic publishing houses and government agencies supported the works of many nineteenth- and twentieth-century mathematicians who made magnificent contributions to the field of cartography. Expanding centers of global trade, competition for world hegemony, and the remarkable advances of science and technology created new demands for accurate systems of measurement. Novel uses of geometric projections stimulated inventive minds engaged in the proliferation of maps representing expanding enterprises in topographical land surveillance, marine navigation, meteorology, oceanography, climatology, geology, mineralogy, biogeography, and demography. Many two- and three-dimensional projections were developed to represent different sets of data visually. Accurate scalar measurements of coastlines and elevated land features were absolutely essential to mariners and land developers, who relied on visual data to guide their enterprises. Mathematics was an indispensable tool of the trade.

Symmetry. The adoption of the concept of symmetry was a very gradual process. It was an accepted condition of mathematical equations in the nineteenth century, but its laws and functions in relation to physical phenomena were poorly understood. In the field of physics, French mineralogist René Just Haüy's analysis of the geometric forms of crystals and subsequent studies of the symmetry of crystalline structure and properties were eventually applied to the study of other natural and dynamic systems. Mandelbrot's introduction of fractal theory helped to refine core concepts of symmetry, particularly those of proportion and scale. The concept of symmetry is considered a universal property that unifies all natural and aesthetic phenomena.

Artists were among the first groups to embrace the fractal as an exquisite audiovisual medium for creative expression. The processes of fractal symmetry have stimulated profound intellectual and technical affinities among mathematicians, scientists, theologians, software designers, composers, architects, graphic artists, and writers who understand the universal principles of harmony that fractals represent.

Monsters. German mathematician Karl Weierstrass's proof of a non-differential function replicated earlier work by Bernhardt Riemann and opened the door for further analysis of non-differentiable curves. Functional analysis and formal logic were essential practices of the mathematician's craft. Graphic representations of complex functions were not in practice and, in the case of some functions, their representations simply could not be mapped in human time. Subsequently, many leading mathematicians rebuffed the anomalous solutions of nondifferential functions, banishing them to the periphery of mathematics. As Mandelbrot explained in his introduction to *The Fractal Geometry of Nature*, many symbolic solutions of non-differential functions were anomalous to the tradition of differential curves and thus dismissed as "pathological" and belonging to a "gallery of monsters" and relegated to the periphery of mathematics.

Mandelbrot also pointed out that these unique mathematical concepts are now the fundamental tools defining natural phenomena in the world around us. He referenced particular mathematicians who made singular contributions to the theory of fractals; their theories and the objects named after them are standards of the canon of fractal geometry.

In 1883, German mathematician Georg Cantor, one of the founders of point-set topology, introduced the Cantor set (also known as the triadic Cantor dust), a self-similar disconnected function. In 1890, Italian mathematician Giuseppe Peano defined the function of a space-filling curve, and in 1891, German mathematician David Hilbert provided variations of the Peano curve and included graphic representations of the models described. In 1915, Polish mathematician Waclaw Sierpinski introduced the Sierpinski gasket (also called a triangle or sieve).

In 1919, German mathematician Felix Hausdorff developed the concept of fractional dimension, a measure theory originated by Greek mathematician Constantin Carathéodory in 1914. Russian mathematician Abram Samoilovitch Besicovitch developed the idea of fractional dimension between 1929 and 1934. Taken together, the concepts defined by Hausdorff and Besicovitch were used by Mandelbrot to define what he termed the "fractal dimension" of a surface.

In 1938, French mathematician Paul Lévy introduced another geometric object, known as Lévys Dragon, which demonstrated a triangular set that could be configured (or tiled) to fill a curved space. Other contributors to fractal theory mentioned by Mandelbrot are Bohemian mathematician Bernard Bolzano, French mathematician Henri-Leon Lebesgue, American mathematician William Fogg Osgood, and Russian mathematician Pavel Samuilovich Urysohn. Finally, the works of French mathematicians Gaston Maurice Julia, Henrí Poincaré, and Pierre Joseph Louis Fatou are memorialized in Mandelbrot's graphic representations of the Julia set and the Mandelbrot set described below.

Unusual Data Sets and Nonlinear Phenomena. During the late 1950's and 1960's Mandelbrot was an employee at IBM, and computer technologies were in their infancy. One of the first problems Mandelbrot was asked to resolve was that of "noise" during data transmission across telephone wires. Occasional errors were problematic. Mandelbrot graphed the sequence of erratic transmissions and noticed a regular, self-similar series that reminded him of the phenomena described by Georg Cantor many years before. Mandelbrot realized that the binary nature of electronic data transmission permitted intermittent switching of signals that would prove disastrous to the system IBM was interested in marketing. The recognition of these signal patterns confirmed his deep-seated hunches about the nature of the "monsters" that lurked in the works of several turn-of-the-century mathematicians. These transmission data sets illustrated the powerful concepts of reiteration and self-similarity.

However, not all natural phenomena are self-identical, nor are they fixed or static. At the time, Mandelbrot was aware that nonlinear mechanics were an important component of dynamic systems, particularly in the study of turbulence and galaxy formation. Edward Lorenz, a meteorologist at the Massachusetts Institute of Technology, was pioneering new investigations in chaos theory relative to the unreliable prediction of weather patterns. Better working concepts were needed for explaining the mechanics of difficult nonlinear phenomena such as turbulence and clustering. Mandelbrot recognized the potential of the Hausdorff-Besicovitch dimension to describe

---

Fascinating Facts About Fractal Geometry

- In February, 2011, researchers at the department of gerontology at the National Institute for Longevity Science in Aichi, Japan, and the Thayer School of Engineering in New Hampshire reported the use of a portable device that can model the gait of patients with Parkinson's disease, a progressive disorder of the central nervous system. A sensor attached to the device measures the patient's movements in three dimensions and then uses fractal analysis to assign a measure to the movement. This provides a parameter for understanding the progression of the disease. The higher the fractal measure, the more complex are the movements of the patient. A fractal dimension of 1.3 is given to walking patterns of people not afflicted with the disease, while Parkinson's patients have a value of 1.48 or higher.

- Researchers studying the power relationships of the metabolic rate and body mass of different animals were surprised to find a proportional increase with a fractal surface value of 2.25. This value reflects the remarkable efficiency of convoluted, space-filling physical systems with fractal properties on a three-dimensional plane that allows for the compact organization of transport services. Like the coastline of

Great Britain, physiognomic surfaces have an infinite measure.

- In the laboratories of professor Eshel Ben-Jacob of Tel-Aviv University, in collaboration with professor Herbert Levine of University of California, San Diego's Center for Theoretical Biological Physics, bacterial forms grown in petri dishes display remarkable fractal organization, demonstrating principles of intelligent cooperation in response to a variety of environmental stressors.

- In July, 2010, physicists at Rice University in Houston, Texas; the Max Planck Institute for Chemical Physics of Solids and the Max Planck Institute for the Physics of Complex Systems, both in Dresden, Germany; and the Vienna University of Technology have reported that after seven years of research on high-temperature superconductors they have established quantum-critical scaling properties at work during the transition from one quantum phase to another. In experiments with a heavy-fermion metal containing ytterbium, rhodium, and silicon, researchers identified thermodynamic scaling properties as a result of a fermi-volume collapse.

---

irregular phenomena and so chose the coastline of Great Britain as a teaching rubric for working through the concept of fractional dimension. In subsequent studies of scalar proportions, he recognized that the shape of the coastline was similar to the form of the Koch curve, which was developed by Swedish mathematician Helge von Koch in 1904. Manipulating the parameters of the curve, Mandelbrot was able to duplicate mathematically, with exceptional accuracy, the shapes of various landforms. These and other mathematical data sets are the basis of subsequent innovations in image data compression, a technology on which the magnificent computer-generated landscapes of movies such as *Star Wars: Episode VI—Return of the Jedi* and *Star Trek II: The Wrath of Khan* are based.

Self-Similar Systems: The Julia Set and the Mandelbrot Set. In 1918, Gaston Maurice Julia and Pierre Joseph Louis Fatou published independent works describing the process of iterating rational functions. Mandelbrot's uncle gave him original reprints of the articles, suggesting they were worth pursuing, but he

did not follow up on them until much later, when employed at IBM.

Access to computers gave him an exceptional opportunity to test the reiterative models described by Julia and Fatou. Instead of simply solving equations, he ran them as a reiterative feedback loop, using the results of one calculation to serve as the inputs of the next run. After millions of iterations, he mapped the results on a grid and was astounded by the results, now known as a Julia set. Working with the graphics of different functions, Mandelbrot devised his own complex quadratic polynomial, such that $f(x) = x^2 + c$, starting at $x = 0$, does not go into infinity, and then ran it. The result was a point set now known as the Mandelbrot set. The Mandelbrot set is unique in its graphic mapping of recursive self-similarity applicable at all magnitudes examined. Its manifestation had a profound and immediate effect on scholars and artists worldwide, who found in the magnificent representations of nonlinear functions a new and incontrovertible paradigm for understanding the universal shapes and forms of the natural world.

Describing Chaos. The dynamics of complex systems are notoriously difficult to characterize, much less to control. Nevertheless, Mandelbrot's application of the concepts of the random walk, Brownian motion, diffusion-limited aggregation, galaxy clusters, fractal attractors, percolation, fractal nets and lattices, L-systems, box counting, and multi-fractal surface dimensions created new conceptual models for rationalizing chaotic systems. Novel cluster patterns are used in the research design of global feedback systems to regulate chaotic systems. In May, 2001, researchers at the Fritz Haber Institute of the Max Planck Society in Germany studying catalytic reactions announced the controlled design of repeating chemical clusters with fractal-like patterns. These patterns and formations anticipated new methods of controlling chemical systems turbulence.

Fractals and the Natural Sciences. The first conference on Fractals in Natural Sciences was held in Budapest, Hungary, from August 30 to September 2, 1993. The major subjects covered included topics related to DNA sequencing and biorhythms; complex bacterial colony formation; correlations of galaxy distribution; fractal tectonics, erosion, and river networks; the geometry of large diffusion-limited aggregates; the use of wavelets to characterize fractals beyond fractal dimension; fixed-scale transformations; diffusion properties of dynamical systems; transfers across fractal electrodes; crack branching; the fractal structure of electrodeposits; the vibrations of fractals; experiments with surface and kinetic roughness; self-organized criticality; granular segregation and pattern formation; and nonlinear ocean waves. The proceeds of the conference were published by the World Scientific Publishing Company in 1994 and provide an excellent starting point for understanding the immediate reception of fractal theory in the biological sciences, chemistry, earth science, and physics.

How It Works

The Feedback Loop. The iterated feedback loop is a fundamental concept of fractal geometry. A feedback loop is a transformational system where the input of a cycle (or mathematical function) gives a different output, which becomes the input of a new transformation. Feedback processes are fundamental tools for studying natural systems. Examples of iterated mathematical systems can be found in history

over thousands of years. However, the capability of the feedback loop to generate complex systems was not fully understood until electronic and computational technologies were developed. The iterated feedback system is a standard process for generating geometric fractals. The function of an iterated object can be linear or nonlinear and can include affine, projective, and Möbius transformations. It is an essential component of the description of chaotic systems.

Fractal Dimension. Describing the dimensions of a given space was a topic that reordered the landscape of mathematics and physics in the nineteenth and twentieth centuries. Euclidean space is three-dimensional, a definition that refers to the range of motion of objects in a physical space. D = 1 refers to points on a straight line, D = 2 refers to points located on a plane, and D = 3 refers to points located within a cube.

Smoothness is an essential feature of Euclidean geometry. Idealized figures such as the cube, the sphere, and the triangle are abstractions that facilitate human understanding of fundamental spatial relationships. These concepts had enormous use in the applied arts and sciences and are the basic tools of technology. However, natural forms such as a flower, a mountain, or a riverbed defied classic mathematical descriptions. Mandelbrot's breakthrough was the realization that Euclidean parameters could not be used to define natural forms because they are irregular. Roughness is an essential characteristic of the natural world, and it requires a different set of mathematical tools for analysis.

Mandelbrot used the fraction as a description of the non-integer dimensions that pertain to natural surfaces. The dimensions of natural objects are fractional composites of the straight line, the plane, and the cube. That fractal dimension is summarized in the equation $d = log\ (N)/log\ (r)$, where r refers to a scaling factor that indicates the roughness of an object. When these values are coordinated on a log-log plot, the steepness of the slope of the line indicates the fractal dimension or roughness of the object.

Technologists and statisticians working with graphic variables describing biological and physical phenomena immediately recognized similarities in the results of time-series representations and the graphic plots of fractal analytics. Heartbeats, physical motions, geological processes, and the biological progression of species in a particular ecosystem could be defined using fractal nomenclature. Institutions all over the world use fractal analysis as a standard tool for interpreting physical systems. The BENOIT fractal analysis software, patented by TruSoft International, provides methods for measuring data sets that are chaotic and not amenable to traditional analytics. These include methods for plotting self-similar and self-affine fractal characteristics. Self-similar methods apply calculations of box dimension, perimeter-area dimension, information dimension, mass dimension, and the ruler method. Self-affine analytics require tools that calculate the scalar properties of the data aligned on the horizontal and vertical axes at constant sampling intervals. These include measurements of fractal Brownian motion, fractional Gaussian noise, power-spectral analysis, variogram analysis, and wavelet analysis.

Iterated Function Systems. Iterated function systems make it possible to re-create natural objects using mathematical descriptions collected into data sets with accompanying rules for computation. This process is also known as fractal image compression, and it is the technology that makes computer-generated landscapes and visual effects in film possible. Very simply, a table of numbers (or matrix formulation) is created describing the affine transformations desired. These follow the conventional order of scalings, reflections, rotations, and translations.

APPLICATIONS AND PRODUCTS

In his introduction to *The Fractal Geometry of Nature*, Mandelbrot wrote: "More generally, I claim that many patterns of Nature are so irregular and fragmented, that, compared with *Euclid*—a term used in this work to denote all of standard geometry—Nature exhibits not simply a higher degree but an altogether different level of complexity. The number of distinct scales of length of natural patterns is for all practical purposes infinite."

Biological Diversity. Understanding biodiversity and the power laws that govern its rich complexity is one of the great challenges of science in the twenty-first century. Scaling relationships in nature provide a powerful paradigm for understanding how biological systems evolve and their relationship to the physical landscape. Environmental factors have a measurable influence on diffusion distributions of species.

The Earth Sciences. Computational molecular modeling is slowly transforming the metrics of geochemistry, making it possible to analyze the geo-metric structures and properties of materials at microscopic levels. Materials such as clays, composed of fine-grain silicates, are considered complex systems, the dynamics of which are still poorly understood. Water, too, is a substance the surface of which is difficult to characterize. Surface chemistries at the edge of a molecule are often different from that of a flat surface, a reality that challenges accurate manipulations of surface dynamics.

Chemical and Biochemical Processes. In physical chemistry and biochemistry, the fractal dimension of surface porosities is a remarkable statistic for the evaluation of processes affecting adsorption rates, chemical clustering, dispersion and uptake algorithms, spectroscopy, photochemistry, estimates of agglomeration dynamics and electrolyte deposition, the behavior of electrolytic dendrites, and the study of disordered systems and catalytic rates. Controlling for nanoscale surface roughness is an important part of developing effective thermal conductors for use in microelectronics. Similar systems analyses of the properties of granular surfaces are of particular importance to the understanding of the chemistry of interstellar dust particles and how ice crystals adhere to them, essential processes in the evolution of stars and galaxies.

Self-Organized Nanostructures. One of the remarkable properties of nanoparticles (NPs) is their capacity for self-organization. In the early twenty-first century, researchers noted that nanoparticles interact to form rings, linear chains, and hyper-branched (dentritic) structures. Of particular interest is the step-growth organization of NPs into polymer-like aggregates with varying types of isomers. These properties suggest the potential use of nanostructures in a variety of products. In 2008, researchers at the University of Wisconsin-Madison reported the creation of nanotrees, spiraling branched objects. Scientists are developing nanoforests with an array of structures. These have great promise for revolutionizing the production of high-performance integrated circuits, biosensors, solar cells, light-emitting diodes (LEDs), and lasers.

Epigenetics. Evidence of self-organization at the nanoparticle level has provided new insights aligning the structural self-organization of life-forms from the symmetries of atomic structure through the allometric scalings of organisms to the forces arranging the patterns of stars. Nevertheless, continuing research into the dynamics of the epigenome reinforce the enormous complexity of the pathways of genetic inheritance, suggesting that genetic expression is influenced by a recursive and scaling array of multidimensional patterns affecting phenotype expression. Fractional dimension analytics are a valuable tool of bioinformatics and are used to measure the complexity of self-similar biological organisms and their relationship to other organisms in their environment. Time is an essential dimension of epigenetic models.

Software. TruSoft International's BENOIT fractal analysis system enables the user to measure the fractal dimension and the Hurst exponent of data sets using eleven different methods.

IMPACT ON INDUSTRY

Properties of self-similarity and proportion are ubiquitous in the world around us and are deeply imprinted in society's cultures and practices. Understanding those principles and deliberately applying them to the design and production of human systems and product life cycles offer a compelling model for the development of thriving world cultures in balance with particular geographies and climates. Self-similarity, reiteration of motifs, and interesting transformations of properties are novel landscape tools for developing the natural aesthetics. Multidimensional fractal models of urban landforms and the structures and activities that occupy them both in the past and the present can be used to make critical macro-level evaluations of how human structures and production systems harmonize with the wider environment over time. Similar studies of fractal dimensionality can be used to evaluate agroforestry systems design and biochemical efficiencies. These virtual types of macro-level evaluations are essential for modeling a thriving world eco-economy based on sustainable design in harmony with particular locales and cultures.

The beauty of a fractal is its conceptual simplicity, one that even a child can understand. In the article "Science Starts Early" published in the February 25, 2011, issue of *Science*, author Frank C. Keil commented on the ability of infants and young children to make sophisticated judgments about causal relationships and patterns. The fractal is a magnificent tool for encouraging a child's thoughtful and informed understanding of the effectiveness of simple processes to create a range of effects. Persistence is

an essential part of the effort to impress upon future generations the importance of understanding and respecting the complex dynamics of life systems. Industrial leaders are an important part of that process.

CAREERS AND COURSE WORK

Fractal geometry is a subject applicable to a broad range of academic subjects and careers. Dynamic geometry software programs such as Geometer's Sketchpad provide interactive tools for learning fractal structures at the high school level. At the university level, introductory fractals and chaos theory courses have been designed for and taught to undergraduate liberal arts majors. Chaos games, iterated function systems, fractional dimension, cellular automata, and artificial life are basic topics addressed. Familiarity with fractal software programs is essential for careers in which the product is the visual image, including photography, fashion design, graphic design, urban landscape design, the technologies of film and sound, medical imaging, scientific illustration, laser optics, and land topography mapping technologies.

Upper-level undergraduate and graduate courses in fractal mathematics require training in writing mathematical proofs, advanced calculus, metric topology, and measure theory. Fractal mathematics is an essential feature of Internet engineering design, economics, mechanical and electrical engineering, the physical and life sciences, computer-game design, electronics, telecommunications, hydrology, geography, demography, and other statistical disciplines.

SOCIAL CONTEXT AND FUTURE PROSPECTS

In 1993, Benoit Mandelbrot was awarded the Wolf Foundation Prize for Physics. On April 25, 2003, he shared the Japan Prize for Science and Technology of Complexity with James A. Yorke, professor of mathematics and physics at the University of Maryland. Both men were cited for the creation of the science and technology of universal concepts in complex systems, namely fractals and chaos. These structures underlie complex phenomena in a wide range of fields, demonstrating the importance of understanding the behavior of systems as a whole as opposed to the reduction of phenomena into discrete elements for observation. Both the Wolf Foundation Prize and the Japan Prize were given in recognition of unique contributions to the progress of science and technology and the promotion of peace and prosperity to humankind.

Fractal theory continues to integrate human knowledge, providing a workable interface that links very different cognitive domains to examine complex physical phenomena in unusual ways. Computer technologies and interactive software programs search for ways to engage the evolutionary capacity of the human eye to assist in the processing of multiple and continuous streams of data. Fractal mappings and time series analyses are important new components of human logic. These are built on new realms of thinking about and modeling geometric shapes, setting the stage for the creation and manipulation of workable multidimensional geometries and design technologies.

Victoria M. Breting-García, B.A., M.A.

FURTHER READING

Brown, James H., et al. "The Fractal Nature of Nature: Power Laws, Ecological Complexity and Biodiversity." *Philosophical Transactions of the Royal Society B: Biological Sciences* 357, no. 1421 (May 29, 2002): 619-626. This essay provides a good working example of how fractal metrics are applied in primary field research.

Carter, Paul. "Dark With Excess of Bright: Mapping the Coastlines of Knowledge" in *Mappings.* Edited by Denis Cosgrove. London: Reaktion Books, 1999. This essay explores human interactions with coastlines and how that engagement stimulated a formal analysis of the processes governing the natural world. It is an excellent prequel to Mandelbrot's essay "How Long Is the Coast of Great Britain? Statistical Self-Similarity and Fractional Dimension."

Edgar, Gerald A., ed. *Classics on Fractals.* Boulder, Colo.: Westview Press, 2004. This highly regarded textbook is designed as an introduction to the theory and practice of fractal mathematics. It is a unique collection of primary mathematical documents describing non-linear mathematical concepts as they were understood in the nineteenth and early twentieth centuries.

Falconer, Kenneth. *Fractal Geometry: Mathematical Foundations and Applications.* 2d ed. Chichester, England: John Wiley & Sons, 2003. Introduces researchers and graduate-level students to the current mathematical models of fractal geometry. Includes substantial sets of notes and references to other foundational texts and essays.

Frame, Michael, and Benoit B. Mandelbrot, eds. *Fractals, Graphics, and Mathematical Education.* Washington, D.C.: Mathematical Association of America, 2002. An excellent description of how teachers at different levels of mathematical education are incorporating fractal theory into their mathematics curriculum.

Mandelbrot, Benoit B. *The Fractal Geometry of Nature.* San Francisco: W. H. Freeman, 1983. This is considered the foundational text in the development of fractal geometry.

_____. "How Long Is the Coast of Britain? Statistical Self-Similarity and Fractional Dimension." *Science* 156, no. 3775 (May 5, 1967): 636-638. This is one of several classic introductory essays written by Mandelbrot on metrics and fractional dimensions.

Peitgen, Heinz-Otto, Hartmut Jürgens, and Dietmar Saupe. *Chaos and Fractals: New Frontiers of Science.* 2d ed. New York: Springer-Verlag, 2004. This well-organized textbook explains fractal geometry and how it is used to describe chaotic or erratic systems.

Siegmund-Schultze, Reinhard. *Mathematicians Fleeing From Nazi Germany: Individual Fates and Global Impact.* Princeton, N.J.: Princeton University Press, 2009. Describes the magnitude of the intellectual and cultural transformations that occurred in academia in the wake of World War II.

WEB SITES

American Mathematical Society
http://www.ams.org

Fractal Foundation
http://fractalfoundation.org

Mathematical Association of America
http://www.maa.org

See also: Calculus; Chaotic Systems; Computer Graphics; Computer Science; Geometry.

FUEL CELL TECHNOLOGIES

FIELDS OF STUDY

Physics; chemistry; electrochemistry; thermodynamics; heat and mass transfer; fluid mechanics; combustion; materials science; chemical engineering; mechanical engineering; electrical engineering; systems engineering; advanced energy conversion.

SUMMARY

The devices known as fuel cells convert the chemical energy stored in fuel materials directly into electrical energy, bypassing the thermal-energy stage. Among the many technologies used to convert chemical energy to electrical energy, fuel cells are favored for their high efficiency and low emissions. Because of their high efficiency, fuel cells have found applications in spacecraft and show great potential as sources of energy in generating stations.

KEY TERMS AND CONCEPTS

- **Anode:** Electrode through which electric current flows into a polarized electrical device.
- **Carnot Efficiency:** Highest efficiency at which a heat engine can operate between two temperatures: that at which energy enters the cycle and that at which energy exits the cycle.
- **Cathode:** Electrode through which electric current flows out of a polarized electrical device.
- **Cogeneration:** Using a heat engine to generate both electricity and useful heat simultaneously.
- **Electrocatalysis:** Using a material to enhance electrode kinetics and minimize overpotential.
- **Electrode:** Electrical conductor used to make contact with a nonmetallic part of a circuit.
- **Electrolyte:** Substance containing free ions that make the substance electrically conductive.
- **Electron:** Subatomic particle carrying a negative electric charge.
- **In Situ:** Latin for "in position"; here, it refers to being in the reaction mixture.
- **Proton:** Subatomic particle carrying a positive electric charge.

DEFINITION AND BASIC PRINCIPLES

Fuel cells provide a clean and versatile means to convert chemical energy to electricity. The reaction between a fuel and an oxidizer is what generates electricity. The reactants flow into the cell, and the products of that reaction flow out of it, leaving the electrolyte behind. As long as the necessary reactant and oxidant flows are maintained, they can operate continuously. Fuel cells differ from electrochemical cell batteries in that they use reactant from an external source that must be replenished. This is known as a thermodynamically open system. Batteries store electrical energy chemically and are considered a thermodynamically closed system. In general, fuel cells consist of three components: the anode, where oxidation of the fuel occurs; the electrolyte, which allows ions but not electrons to pass through; and the cathode, which consumes electrons from the anode.

A fuel cell does not produce heat as a primary energy conversion mode and is not considered a heat engine. Consequently, fuel cell efficiencies are not limited by the Carnot efficiency. They convert chemical energy to electrical energy essentially in an isothermal manner.

Fuel cells can be distinguished by: reactant type (hydrogen, methane, carbon monoxide, methanol for a fuel and oxygen, air, or chlorine for an oxidizer); electrolyte type (liquid or solid); and working temperature (low temperature, below 120 degrees Celsius, intermediate temperature, 120 degrees to 300 degrees Celsius, or high temperature, more than 600 degrees Celsius).

BACKGROUND AND HISTORY

The first fuel cell was developed by the Welsh physicist and judge Sir William Robert Grove in 1839, but fuel cells did not receive serious attention until the early 1960's, when they were used to produce water and electricity for the Gemini and Apollo space programs. These were the first practical fuel cell applications developed by Pratt & Whitney. In 1989, Canadian geophysicist Geoffrey Ballard's Ballard Power Systems and Perry Oceanographics developed a submarine powered by a polymer electrolyte membrane or proton exchange membrane fuel cell (PEMFC). In

1993, Ballard developed a fuel-cell-powered bus and later a PEMFC-powered passenger car. Also in the late twentieth century, United Technologies (UTC) manufactured a large stationary fuel cell system for the cogeneration power plant, while continuously developing the fuel cells for the U.S. space program. UTC is also developing fuel cells for automobiles. Siemens Westinghouse has successfully operated a 100-kilowatt (kW) cogeneration solid oxide fuel cell (SOFC) system, and 1-megawatt (MW) systems are being developed.

How It Works

Polymer Electrolyte Membranes or Proton Exchange Membrane Fuel Cells (PEMFCs). PEMFCs use a proton conductive polymer membrane as an electrolyte. At the anode, the hydrogen separates into protons and electrons, and only the protons pass through the proton exchange membrane. The excess of electrons on the anode creates a voltage difference that can work across an exterior load. At the cathode, electrons and protons are consumed and water is formed.

For PEMFC, the water management is critical to the fuel cell performance: Excess water at the positive electrode leads to flooding of the membrane; dehydration of the membrane leads to the increase of ohmic resistance. In addition, the catalyst of the membrane is sensitive to carbon monoxide poisoning. In practice, pure hydrogen gas is not economical to mass produce. Thus, hydrogen gas is typically produced by steam reforming of hydrocarbons, which contains carbon monoxide.

Direct Methanol Fuel Cells (DMFCs). Like PEMFCs, DMFCs also use a proton exchange membrane. The main advantage of DMFCs is the use of liquid methanol, which is more convenient and less dangerous than gaseous hydrogen. As of 2011, the efficiency is low for DMFCs, so they are used where the energy and power density are more important than efficiency, such as in portable electronic devices.

At the anode, methanol oxidation on a catalyst layer forms carbon dioxide. Protons pass through the proton exchange membrane to the cathode. Water is produced by the reaction between protons and oxygen at the cathode and is consumed at the anode. Electrons are transported through an external circuit from anode to cathode, providing power to connected devices.

Solid Oxide Fuel Cells (SOFCs). Unlike PEMFCs, SOFCs can use hydrocarbon fuels directly and do not require fuel preprocessing to generate hydrogen prior to utilization. Rather, hydrogen and carbon monoxide are generated in situ, either by partial oxidation or, more typically, by steam reforming of the hydrocarbon fuel in the anode chamber of the fuel cell. SOFCs are all-solid electrochemical devices. There is no liquid electrolyte with its attendant material corrosion and electrolyte management problems. The high operating temperature (typically 500-1,000 degrees Celsius) allows internal reforming, promotes rapid kinetics with nonprecious materials, and yields high-quality byproduct heat for cogeneration. The total efficiency of a cogeneration system can be 80 percent—far beyond the conventional power-production system.

The function of the fuel cell with oxides is based on the activity of oxide ions passing from the cathode region to the anode region, where they combine with hydrogen or hydrocarbons; the freed electrons flow through the external circuit. The ideal performance of an SOFC depends on the electrochemical reaction that occurs with different fuels and oxygen.

Molten Carbonate Fuel Cells (MCFCs). MCFCs use an electrolyte composed of a molten carbonate salt mixture suspended in a porous, chemically inert ceramic matrix. Like SOFCs, MCFCs do not require an external reformer to convert fuels to hydrogen. Because of the high operating temperatures, these fuels are converted to hydrogen within the fuel cell itself by an internal re-forming process.

MCFCs are also able to use carbon oxides as fuel. They are not poisoned by carbon monoxide or carbon dioxide, thus MCFCs are advanced to use gases from coal so that they can be integrated with coal gasification.

Applications and Products

Hydrogen Fuel Cell Vehicles. In recent years, both the automobile and energy industries have had great interest in the fuel cell powered vehicle as an alternative to internal combustion engine vehicles, which are driven by petroleum-based liquid fuels. Many automobile manufacturers, such as General Motors, Renault, Hyundai, Toyota, and Honda, have been developing prototype hydrogen fuel cell vehicles. Energy industries have also been installing prototype hydrogen filling stations in large cities, including Los Angeles; Washington, D.C.; and Tokyo.

The first hydrogen fuel cell passenger vehicle for a private individual was leased by Honda in 2005. However, public buses provide better demonstrations of hydrogen fuel cell vehicles compared with passenger vehicles, since public buses are operated and maintained by professionals and they have more volume for the hydrogen fuel storage than passenger vehicles. A number of bus manufacturers such as Toyota, Man, and Daimler have developed hydrogen fuel cell buses and they have been in service in Palm Springs, California; Nagoya, Japan; Vancouver; and Stockholm.

Despite many advantages of hydrogen fuel cell vehicles, this technology still faces substantial challenges such as high costs of novel metal catalyst, safety of hydrogen fuel, effective storage of hydrogen onboard, and infrastructure needed for public refueling stations.

Stationary Power Plants and Hybrid Power Systems. Siemens Westinghouse and UTC have produced a number of power plant units in the range of about 100 kW by using SOFCs, MCFCs, and phosphoric acid fuel cells (PAFCs). Approximately half of the power plants were MCFC-based plants. They showed that these fuel cell systems have exceeded the research-and-discovery level and already produced an economic benefit. These systems generate power with less fossil fuel and lower emissions of greenhouse gases and other harmful products. Just a small number PEMFC-based power plants were built as the cost of fuel cell materials was prohibitive. In many cases, the fuel-cell-based stationary power plants are used for heat supply in addition to power production, enabling so-called combined heat and power systems. Such systems increase the total efficiency of the power plants and offer an economic benefit.

More recently, many efforts to develop hybrid power plants combining fuel cells and gas turbines were made. While the high-temperature fuel cells, such as SOFCs and MCFCs, produce electrical power, the gas turbines produce additional electrical power from the heat produced by the fuel cells' operation. At the same time, the gas turbines compress the air fed into the fuel cells. The expected overall efficiency for the direct conversion of chemical energy to electrical energy is up to 80 percent.

Small Power Generation for the Portable Electronic Devices. At the end of the twentieth century, the demand for electricity continued to increase in many applications, including portable electronics. Batteries have seen significant advances, but their power density is still far inferior to combustion devices. Typically, hydrocarbon fuels have 50 to 100 times more energy storage density than commercially available batteries. Even with low conversion efficiencies, fuel-driven generators will still have superior energy density. There is considerable interest in miniaturizing thermochemical systems for electrical power generation for remote sensors, micro-robots, unmanned vehicles (UMVs), unmanned aerial vehicles (UAVs), even portable electronic devices such as laptop computers and cell phones.

Fuel cells combine hydrogen and oxygen (from air) to create electricity. The only by-products produced are heat and water, making this technology a clean source of power compared to the pollutant gases produced by burning fossil fuels. (Friedrich Saurer/Photo Researchers, Inc.)

Much work on such systems has been developed by the military. The Defense Advanced Research Projects Agency (DARPA) has initiated and developed many types of portable power concepts using the fuel cells. Industries such as Samsung, Sony, NEC, Toshiba, and Fujitsu have developed fuel cells based portable power generation. Most (about 90 percent) devices were based on PEMFCs or DMFCs, which require lower operating temperatures than SOFC. However, development of SOFC-based portable power generation under the DARPA Microsystems Technology Office showed the feasibility of employing high-temperature fuel cells with appropriate thermal management.

The Military. In addition to the portable power generation for the foot soldiers, the military market has been interested in developing medium-size power plants (a few hundred watts) for recharging various types of storage batteries and high stationary power plants (more than a few kW) for the auxiliary power units.

Military programs in particular have been interested in the direct use of logistic fuel (for example, Jet Propellant 8) for the fuel cells, because of the complexities and difficulties of the re-forming processes. While the new and improved re-forming processes of logistic fuel were being developed to feed hydrogen into the fuel cells, direct jet-fuel SOFCs were also demonstrated by developing new anode materials that had a high resistance to coking and sulfur poisoning.

IMPACT ON INDUSTRY

Government and University Research. One of the biggest sources of funding for fuel cell research in the United States is the Department of Energy (DOE). DOE has developed many programs for the fuel cells and hydrogen. For example, DOE formed Solid State Energy Conversion Alliance (SECA) in 1999 and formulated a program with funding of $1 billion for 10 years. Other government agencies, such as the Department of Defense (DOD), DARPA, Air Force Office of Scientific Research (AFOSR), Office of Naval Research (ONR), and Army Research Laboratory (ARL) have also funded a number of the fuel cell projects taking place in academic and corporate settings to bring about the transfer of the energy technologies to those fighting wars.

Professional societies have also noticed the importance of the energy security and advanced energy

---

Fascinating Facts About Fuel Cell Technologies

- In 2003, U.S. president George Bush launched the Hydrogen Fuel Initiative (HFI), which was later implemented by legislation through the 2005 Energy Policy Act and the 2006 Advanced Energy Initiative. President Bush stated that "the first car driven by a child born today could be powered by hydrogen and pollution free."
- The Department of Energy is the largest funder of fuel cell science and technology in the United States.
- As of 2011, 191 states have signed Kyoto Protocol, which is a legally binding international agreement to reduce greenhouse-gas emissions by 5.2 percent of 1990 levels by the year 2012.
- In 2008, Boeing announced that it has, for the first time in aviation history, flown a manned airplane powered by hydrogen fuel cells. The Fuel Cell Demonstrator Airplane used a proton exchange membrane fuel cell and lithium-ion battery hybrid system to power an electric motor, which was coupled to a conventional propeller.
- In 2002, typical fuel cell systems cost $1,000 per kilowatt. But, by 2009, the fuel cell system costs had been reduced with volume production (estimated at 500,000 units per year) to $61 per kilowatt.
- Top international universities built their own hydrogen fuel cell racing vehicle to compete against one another on a mobile track in a race called Formula Zero Championship. More advanced races are planned for 2011 Street Edition: The race class will scale up to hydrogen racers, which will compete globally on street circuits in city centers. In the 2015 Circuit Edition, full-size hydrogen fuel cell racing cars built by car manufacturers will compete on racing circuits around the world.

---

technologies. In 2003, the American Institute of Aeronautics and Astronautics (AIAA) and American Society of Mechanical Engineers (ASME) brought in new international conferences: AIAA International Energy Conversion Engineering Conference (IECEC) and ASME International Fuel Cell Science, Engineering, and Technology Conference. The conferences' goals are to expand international cooperation, understanding, promotion of efforts, and disciplines in the area of energy conversion technology, advanced energy and power systems and devices, and the policies, programs, and environmental impacts associated with the development and utilization of energy technologies.

As of 2011, the Korean fuel cell market is in a nascent stage and is expected to witness rapid growth as a result of government-supported policies. Korea has nine fuel cell units installed in various regions. The major driver behind the future development of the hydrogen and fuel cell industry in Korea is the country's need to achieve energy security.

Industry and Business. Almost every car manufacturer has developed a fuel cell vehicle powered by PEMFCs or SOFCs. They hope that the fuel cell vehicles double the efficiency of internal combustion engine vehicles. Some (Honda, General Motors, Toyota) are using their own developed fuel cells, but most companies buy the fuel cell systems from the fuel cell manufacturers such as UTC and De Nora. Many electronic companies such as Motorola, NEC, Toshiba, Samsung, and Matsushita are rushing to develop their own small fuel cells that will provide power up to ten times longer on a single charge than conventional batteries for small portable electronic devices.

The U.S. fuel cell market is growing rapidly. By the end of 2009, 620 fuel cell power units were installed. Government-supported promotion of clean energy is responsible for this, and the tax credits permitted under Energy Policy Act of 2005 continue to drive the U.S. fuel cell market. Fuel cell manufacturers are also working on development of small fuel cell power systems intended to be used in homes and office buildings. For example, a 200 kW PAFC system was installed to power a remote police station in New York City's Central Park.

CAREERS AND COURSE WORK

Courses in chemistry, physics, electrochemistry, materials science, chemical engineering, and mechanical engineering make up the foundational requirements for students interested in pursuing careers in fuel cell research. Earning a bachelor of science degree in any of these fields would be appropriate preparation for graduate work in a similar area. In most circumstances, either a master's or doctorate degree is necessary for the most advanced career opportunities in both academia and industry.

Careers in the fuel cells field can take several different shapes. Fuel cell industries are the biggest employers of fuel cell engineers, who focus on developing and manufacturing new fuel cell units as well as maintaining or repairing fuel cell units. Other industries in which fuel cell engineers often find work include aviation, automotive, electronics, telecommunications, and education.

Many fuel cell engineers prefer employment within the national laboratories and government agencies such as the Pacific Northwest National Laboratory, the National Renewable Energy Laboratory, the Argonne National Laboratory, the National Aeronautics and Space Administration (NASA), DOE, and DARPA. Others find work in academia. Such professionals divide their time between teaching university classes on fuel cells and conducting their own research.

SOCIAL CONTEXT AND FUTURE PROSPECTS

In the future, it is not likely that sustainable transportation will involve use of conventional petroleum. Transportation energy technologies should be developed with both the goal of providing an alternative to the petroleum-based internal combustion engine vehicles. People evaluate vehicles not only on the basis of fuel economy but also performance. Vehicles using an alternative energy source should be designed with these parameters.

One of the most promising energy sources for the future will be hydrogen. The hydrogen fuel cell vehicles face cost and technical challenges, especially the fuel cell stack and onboard hydrogen storage.

For the fuel cell power plants, the economic and lifetime related issues hinder the acceptance of fuel cell technologies. Such problems were not associated with fuel cells but with auxiliary fuel cell units such as thermal management, reactant storage, and water management. Therefore, the auxiliary units of fuel cell systems should be further developed to address these issues.

The fundamental problems of fuel cells related to electrocatalysis also need to be addressed for improvement in performance, as highly selective catalysts will provide better electrochemical reactions.

Lastly, once the new fuel cell technologies are successfully developed and meet the safety requirements, the infrastructure to distribute and to recycle fuel cells will also be necessary.

Jeongmin Ahn, B.S., M.S., Ph.D.

FURTHER READING

Bagotsky, Vladimir S. *Fuel Cells: Problems and Solutions.* Hoboken, N.J.: John Wiley & Sons, 2009. Provides extensive explanations of the various types of fuel cells operation.

Hoogers, Gregor, ed. *Fuel Cell Technology Handbook.* Boca Raton, Fla.: CRC Press, 2003. Recognizes the part played by the change in Gibb's potential.

Kotas, T.J. *The Exergy Method of Thermal Plant Analysis.* Malabar, Fla.: Krieger Publications, 1995. Proves that the fuel chemical exergy and the lower calorific value of the fuel, with different units, are numerically equal.

Larminie, James, and Andrew Dicks. *Fuel Cell Systems Explained.* 2d ed. Hoboken, N.J.: John Wiley & Sons, 2000. This text provides construction details of the various types of fuel cells.

O'Hayre, Ryan P., et al. *Fuel Cell Fundamentals.* 2d ed. Hoboken, N.J.: John Wiley & Sons, 2008. Includes extensive discussions on thermodynamics, transport science, and chemical kinetics in the early chapters with a supporting appendix on quantum-mechanical issues. Also addresses modeling and characterization of fuel cells and fuel cell systems and their environmental impact.

Reddy, Thomas B., ed. *Linden's Handbook of Batteries.* 4th ed. New York: McGraw-Hill, 2011. Includes detailed technical descriptions of chemistry, electrical characteristics, construction details, applications, and pros and cons charts.

WEB SITES

Battery Council International
http://www.batterycouncil.org

Fuel Cell and Hydrogen Energy Association
http://www.fchea.org

Fuel Cell Europe
http://www.fuelcelleurope.org

See also: Chemical Engineering; Electrical Engineering; Electric Automobile Technology; Electronics and Electronic Engineering; Hybrid Vehicle Technologies; Mechanical Engineering.